X

BRITISH STANDARD

incorporating
Amendment
No 1: 2011

Requirements for Electrical Installations

IET Wiring Regulations
Seventeenth Edition

CONFINED

Not For Loan

© The Institution of Engineering and Technology and BSI
NO COPYING IN ANY FORM WITHOUT WRITTEN PERMISSION

Published by: the Institution of Engineering and Technology, LONDON, UK in agreement with BSI

The Institution of Engineering and Technology is registered as a Charity in England & Wales (no. 211014) and Scotland (no. SC038698).

The Institution of Engineering and Technology is the new institution formed by the joining together of the IEE (the Institution of Electrical Engineers) and the IIE (the Institution of Incorporated Engineers). The new institution is the inheritor of the IEE brand and all its products and services, such as this one, which we hope you will find useful.

British Standards Institution is the independent national body responsible for preparing British Standards. It presents the UK view on standards in Europe and at the international level. It is incorporated by Royal Charter.

This British Standard was published under the joint authority of the IET and of the Standards Policy and Strategy Committee on 1st July 2011.

While the publisher and contributors believe that the information and guidance given in this work is correct, all parties must rely upon their own skill and judgement when making use of it. Neither the publisher nor any contributor assume any liability to anyone for any loss or damage caused by any error or omission in the work, whether such error or omission is the result of negligence or any other cause. Any and all such liability is disclaimed.

The UK participation in its preparation was entrusted to Joint Technical Committee JPEL/64 Electrical Installations. A list of organizations represented on this committee can be obtained on request to its secretary.

This publication does not purport to include all the necessary provisions of a contract. Users are responsible for its correct application. Compliance with a British Standard cannot confer immunity from legal obligations.

It is the constant aim of the IET and BSI to improve the quality of our products and services. We should be grateful if anyone finding an inaccuracy or ambiguity while using this British Standard would inform the Secretary, G D Cronshaw {gcronshaw@theiet.org}, The IET, Six Hills Way, STEVENAGE, SG1 2AY, UK.

IET Standards & Compliance staff
as at June 2011

JPEL/64	Joint IET/BSI Technical Committee	G D Cronshaw CEng FIET
	Electrical Installations	Eur Ing D Locke BEng (Hons) CEng MIET
		I M Reeve BTech(Hons) CEng MIET
JPEL/64/A	Verification	R J Townsend MIET
JPEL/64/B	Thermal Effects	M Coles BEng(Hons) MIET
JPEL/64/C	Shock Protection	P Bicheno BSc(Hons) MIET
JPEL/64/D	External Influences	M Coles BEng(Hons) MIET

ISBN-13: 978-1-84919-269-9 (Paperback)

Printed in the United Kingdom by Polestar Wheatons

CONTENTS

Foreword

This British Standard is published under the direction of the British Electrotechnical Committee (BEC) and the Institution of Engineering and Technology (IET).

Following a full review, this Standard replaced the 16th Edition of the IEE Wiring Regulations BS 7671:2001 as amended. Copyright is held jointly by the IET and BSI.

Technical authority for this Standard is vested in the Joint IET/BSI Technical Committee JPEL/64. This Joint Technical Committee, which is responsible for the work previously undertaken by the IEE Wiring Regulations Committee and the BSI Technical Committee PEL/64, meets the constitutional and operational requirements of both parent bodies. JPEL/64 has the responsibility for the content of this British Standard under the joint authority of the IET and the BSI Standards Board.

All references in this text to the Wiring Regulations or the Regulation(s), where not otherwise specifically identified, shall be taken to refer to BS 7671:2008 Requirements for Electrical Installations as amended by Amendment No 1, 2011.

Regulations changed owing to the issue of Amendment 1 are indicated by a side bar in the margin.

Introduction to BS 7671:2008

BS 7671:2008 Requirements for Electrical Installations was issued on 1st January 2008 and came into effect on 1st July 2008. Installations designed after 30th June 2008 are to comply with BS 7671:2008.

The Regulations apply to the design, erection and verification of electrical installations, also additions and alterations to existing installations. Existing installations that have been installed in accordance with earlier editions of the Regulations may not comply with this edition in every respect. This does not necessarily mean that they are unsafe for continued use or require upgrading.

BS 7671:2008 includes changes necessary to maintain technical alignment with CENELEC harmonization documents. A summary of the main changes is given below.

NOTE 1: This is not an exhaustive list.

NOTE 2: Particular attention is drawn to Section 701. This section now allows socket-outlets (other than SELV and shaver supply units to BS EN 61558-2-5) to be installed in locations containing a bath or shower 3m horizontally beyond the boundary of zone 1.

Regulation 131.6 adds requirements to protect against voltage disturbances and implement measures against electromagnetic disturbances. In doing so, the design shall take into consideration the anticipated electromagnetic emissions, generated by the installation or the installed equipment, which shall be suitable for the current-using equipment used with, or connected to, the installation.

Regulation 132.13 requires that documentation for the electrical installation, including that required by Chapter 51, Part 6 and Part 7, is provided for every electrical installation.

Chapter 35 Safety services, recognizes the need for safety services as they are frequently regulated by statutory authorities whose requirements have to be observed, e.g. emergency escape lighting, fire alarm systems, installations for fire pumps, fire rescue service lifts, smoke and heat extraction equipment.

Chapter 36 Continuity of service, requires that an assessment be made for each circuit of any need for continuity of service considered necessary during the intended life of the installation.

Chapter 41 Protection against electric shock, now refers to basic protection, which is protection under normal conditions (previously referred to as protection against direct contact), and fault protection, which is protection under fault conditions (previously referred to as protection against indirect contact). Chapter 41 now includes those requirements previously given in Section 471 of BS 7671:2001.

> Chapter 41 now requires that for the protective measure of automatic disconnection of supply for an a.c. system, additional protection by means of an RCD with a rated residual operating current ($I_{\Delta n}$) not exceeding 30 mA and an operating time not exceeding 40 ms at a residual current of 5 $I_{\Delta n}$ be provided for socket-outlets with a rated current not exceeding 20 A that are for use by ordinary persons and are intended for general use, and for mobile equipment with a current rating not exceeding 32 A for use outdoors. This additional protection is now to be provided in the event of failure of the provision for basic protection and/or the provision for fault protection or carelessness by users of the installation. Note that certain exceptions are permitted – refer to Regulation 411.3.3.

Chapter 41 includes Tables 41.2, 41.3 and 41.4 for earth fault loop impedances (replacing Tables 41B1, 41B2 and 41D). These new tables are based on a nominal voltage of 230 V (not 240 V), hence the values are slightly reduced. It has been clarified that where an RCBO is referred to in these Tables, the overcurrent characteristic of the device is being considered.

Chapter 41 includes a new Table 41.5 giving maximum values of earth fault loop impedance for RCDs to BS EN 61008-1 and BS EN 61009-1.

FELV is recognized as a protective measure and the new requirements are detailed in Regulation 411.7.

Chapter 41 includes the UK reduced low voltage system. Requirements are given in Regulation 411.8.

Chapter 42   Protection against thermal effects, includes requirements in Section 422 Precautions where particular risks of fire exist (These requirements were previously stated in Section 482 of BS 7671:2001).

Chapter 43   Protection against overcurrent, includes those requirements previously given in Section 473 of BS 7671:2001. Information on the overcurrent protection of conductors in parallel is given in Appendix 10.

Chapter 44   Protection against voltage disturbances, includes a new Section 442, Protection of low voltage installations against temporary overvoltages due to earth faults in the high voltage system and due to faults in the low voltage system. This new section provides for the safety of the low voltage system under fault conditions including faults in the high voltage system, loss of the supply neutral in the low voltage system and short-circuit between a line conductor and neutral in the low voltage installation.

Section 443   Protection against overvoltages of atmospheric origin or due to switching, retains the existing text from BS 7671 and adds regulations enabling designers to use a risk assessment approach when designing installations which may be susceptible to overvoltages of atmospheric origin.

Chapter 52   Selection and erection of wiring systems, now includes busbar trunking systems and powertrack systems.

It is now required to protect cables concealed in a wall or partition (at a depth of less than 50 mm) by a 30 mA RCD where the installation is not intended to be under the supervision of a skilled or instructed person, if the normal methods of protection including use of cables with an earthed metallic covering, mechanical protection (including use of cables with an earthed metallic covering, or mechanical protection) cannot be employed. This applies to a cable in a partition where the construction includes metallic parts other than fixings <u>irrespective of the depth of the cable.</u>

Table 52.2 Cable surrounded by thermal insulation, gives slightly reduced derating factors, to take account of the availability of material with improved thermal insulation.

Chapter 53   Protection, isolation, switching, control and monitoring. Simplification means that requirements previously in Chapter 46, Sections 476 and 537 of BS 7671:2001 are now in this single chapter. Chapter 53 also includes a new Section 532 Devices for protection against the risk of fire, and a new Section 538 Monitoring devices.

Chapter 54   Earthing arrangements and protective conductors. The requirement that a metallic pipe of a water utility supply shall not be used as an earth electrode is retained in Regulation 542.2.4 which also states that other metallic water supply pipework shall not be used as an earth electrode unless precautions are taken against its removal and it has been considered for such a use. An example of other metallic water supply pipework could be a privately owned water supply network.

A note to Regulation 543.4.1 states that in Great Britain, Regulation 8(4) of the Electricity Safety, Quality and Continuity Regulations 2002 prohibits the use of PEN conductors in consumers' installations. Regulation 543.7 has earthing requirements for the installation of equipment having high protective conductor currents, previously in Section 607 of BS 7671:2001.

Chapter 55   Other equipment, includes new additional requirements in Regulation 551.7 to ensure the safe connection of low voltage generating sets including small-scale embedded generators (SSEGs).

Section 559   Luminaires and lighting installations, is a new series of regulations giving requirements for fixed lighting installations, outdoor lighting installations, extra-low voltage lighting installations, lighting for display stands and highway power supplies and street furniture (previously in Section 611 of BS 7671:2001).

Chapter 56   Safety services, has been expanded in line with IEC standardization.

Part 6   Inspection and testing, was Part 7 of BS 7671:2001. Changes have been made to the requirements for insulation resistance; when testing SELV and PELV circuits at 250 V, the minimum insulation resistance is raised to 0.5 MΩ; for systems up to and including 500 V, including FELV, the minimum insulation resistance is raised to 1.0 MΩ.

Part 7 Special installations or locations, was Part 6 of BS 7671:2001. The structure of Part 7 includes the following changes.

Section 607 in BS 7671:2001 relating to high protective conductor currents has been incorporated into Chapter 54.

Section 608 in BS 7671:2001 relating to caravans, motor caravans and caravan parks has been incorporated into Section 708: Electrical installations in caravan/camping parks and similar locations and Section 721: Electrical installations in caravans and motor caravans.

Section 611 in BS 7671:2001 relating to highway power supplies is now incorporated into Section 559.

The following major changes are incorporated in Part 7:

Section 701 Locations containing a bath tub or shower basin.

Zone 3 is no longer defined.

Each circuit in the special location must have 30 mA RCD protection.

Supplementary bonding is no longer required providing the installation has main bonding in accordance with Chapter 41.

This section now allows socket-outlets (other than SELV and shaver supply units to BS EN 61558-2-5) to be installed in locations containing a bath or shower 3m horizontally beyond the boundary of zone 1.

Section 702 Swimming pools and other basins. This special location now includes basins of fountains. Zones A, B and C in BS 7671:2001 are replaced by zones 0, 1 and 2.

Section 703 Rooms and cabins containing sauna heaters. Zones A, B, C and D in BS 7671:2001 are replaced by zones 1, 2 and 3 (with changed dimensions).

Section 704 Construction and demolition site installations. The reduced disconnection times (0.2 s) and the 25 V equation no longer appear.

Section 705 Agricultural and horticultural premises. The reduced disconnection times (0.2 s) and the 25 V equation no longer appear. Additional requirements applicable to life support systems are included.

Section 706 Conducting locations with restricted movement, was Section 606 in BS 7671:2001.

Section 708 Electrical installations in caravan/camping parks and similar locations, now includes the requirement that each socket-outlet must be provided individually with overcurrent and RCD protection.

The following new sections are now included in Part 7:

Section 709 Marinas and similar locations

Section 711 Exhibitions, shows and stands

Section 712 Solar photovoltaic (pv) power supply systems

Section 717 Mobile or transportable units

Section 721 Electrical installations in caravans and motor caravans – previously in Section 608 of BS 7671:2001

Section 740 Temporary electrical installations for structures, amusement devices and booths at fairgrounds, amusement parks and circuses

Section 753 Floor and ceiling heating systems.

Appropriate changes have been made to **Appendices 1 to 7**, in particular the methods and tables used in **Appendix 4**.

The following new appendices are now included:

Appendix 8 Current-carrying capacity and voltage drop for busbar trunking and powertrack systems

Appendix 9 Definitions – multiple source, d.c. and other systems

Appendix 10 Protection of conductors in parallel against overcurrent

Appendix 11 Effect of harmonic currents on balanced three-phase systems

Appendix 12 Voltage drop in consumers' installations

Appendix 13 Methods for measuring the insulation resistance/impedance of floors and walls to Earth or to the protective conductor system

Appendix 14 Measurement of earth fault loop impedance: consideration of the increase of the resistance of conductors with increase of temperature

Appendix 15 Ring and radial final circuit arrangements, Regulation 433.1

Introduction to the first amendment

The first amendment to BS 7671:2008 Requirements for Electrical Installations was issued on 1st July 2011 and is intended to come into effect on 1st Jan 2012. Installations designed after 31st Dec 2011 are to comply with BS 7671:2008 incorporating Amendment No 1, 2011.

A summary of the main changes is given below.

NOTE: This is not an exhaustive list.

Numbering system The 17th edition introduced a new IEC decimal point numbering system to make it easier to embody future changes and additions resulting from ongoing international standards work within IEC and CENELEC. In order to accommodate future IEC changes it has been decided to have a 100 numbering system for UK only regulations. This means for example, the former Regulation 422.3.14 is now Regulation 422.3.100 etc. This only applies to UK only regulations.

Part 2 Definitions have been expanded and modified. The system diagrams for TN-S, TN-C-S, TT have been moved to Part 3, while TN–C and IT have been deleted.

Part 3 Assessment of general characteristics, now includes the system diagrams for TN-S, TN-C-S and TT. In addition, diagrams have been included for single-phase 2-wire, single-phase 3-wire, two-phase 3-wire, three-phase 3-wire, three-phase 4-wire a.c. and 2- and 3-wire d.c.

Chapter 41 Table 41.5 has been simplified.

Chapters 42 and 43 Contain minor changes.

Chapter 44 Protection against voltage disturbances and electromagnetic disturbances, now includes modifications to Section 443. A number of important notes have been added to Section 443. The section makes reference to Section 534. The user is also made aware that work is currently in progress at IEC level on Section 443.

A new Section 444 has been added, which deals with measures against electromagnetic disturbances. Electromagnetic interference (EMI) may disturb or damage information technology equipment/systems as well as equipment with electronic components or circuits. Currents due to lightning, switching operations, short-circuits and other electromagnetic phenomena may cause overvoltages and electromagnetic interference. Section 444 provides basic requirements and recommendations to enable the avoidance and reduction of electromagnetic disturbances.

Chapter 51 Common rules, includes a number of changes. The requirements concerning the connection of cables to switchgear in relation to temperature rating have been modified. Orange as a means of identification of an electrical conduit has been deleted. The regulation concerning the warning notice for voltage has been modified.

Chapter 52 Regulation 522.6.100 includes an additional indent.

Regulation 522.6.101 now permits an exception for a cable forming part of a SELV or PELV circuit.

Chapter 53 Protection, isolation, switching, control and monitoring, now includes a new Section 534 which deals with the installation of surge protective devices (SPD). The requirements of Section 534 are for the selection and erection of SPDs for electrical installations of buildings in order to limit transient overvoltages of atmospheric origin transmitted via the supply distribution system and against switching overvoltages. The requirements are also intended to protect against transient overvoltages caused by direct lightning strikes or lightning strikes in the vicinity of buildings protected by a lightning protection system. The requirements do not take into account surge protective components, which may be incorporated in the appliances connected to the installation. (See also Appendix 16.)

Chapter 56 Safety services, incorporates editorial changes, updated standards, notes added, and requirements for switchgear and controlgear for use on a.c. and d.c. added.

Part 6 Inspection and testing. Mainly editorial changes to Chapter 61 and notes added. Changes have been made to Chapter 63 to take account of the change from periodic inspection report to condition report and other minor changes.

Part 7 Special installations or locations. There are minor changes throughout Part 7, mainly in connection with PME supplies.

Section 717 Mobile or transportable units, includes a new Regulation 717.313 giving requirements for supplies.

The following new sections are now included in Part 7:

Section 710 Medical locations

Section 710 applies to electrical installations in medical locations, so as to ensure safety of patients and medical staff. These requirements, in the main, refer to hospitals, private clinics, medical and dental practices, healthcare centres and dedicated medical rooms in the workplace. This section also applies to electrical installations in locations designed for medical research. The requirements of this section do not apply to medical electrical equipment.

Section 729 Operating and maintenance gangways

Section 729 applies to restricted areas for authorised persons. These are areas such as switchrooms with switchgear and controlgear assemblies with a need for operating and maintenance gangways. The scope of Section 729 applies to basic protection and other aspects in restricted access areas with switchgear and controlgear assemblies, and gives requirements for operating and maintenance gangways such as the width of gangways and access areas for operational access, emergency access, emergency evacuation and for transport of equipment.

The following main changes have been made within the appendices:

Appendix 3 Time/current characteristics of overcurrent protective devices and RCDs, includes minor changes.

Appendix 4 Current-carrying capacity and voltage drop for cables, now includes rating factors for triple harmonic currents in four-core and five-core cables with four cores carrying current, and the voltage drop requirements from Appendix 12. In addition, changes to the tables on rating factors and correction factors and notes with tables.

Appendix 5 Classification of external influences, changes to the tables on utilisation category.

Appendix 6 Model forms for certification and reporting, now includes a new electrical installation condition report that replaces the periodic inspection report. The new condition report includes a new inspection schedule. Also the form has been redesigned. Within the observations section the four codes, 1 (requires urgent attention), 2 (requires improvement), 3 (requires further investigation), 4 (does not comply with BS 7671:2008 amended to …) have been replaced by three codes: C1 - Danger present, C2 - Potentially dangerous, C3 - Improvement recommended.

Appendix 11 Effect of harmonic currents on balanced three-phase systems, has been deleted and the content moved to Appendix 4.

Appendix 12 Voltage drop in consumers' installations, has been deleted and the content moved to Appendix 4.

Appendix 16 Devices for protection against overvoltage, has been added.

Editions

The following editions have been published:

FIRST EDITION	Entitled 'Rules and Regulations for the Prevention of Fire Risks Arising from Electric Lighting'. Issued in 1882.
SECOND EDITION	Issued in 1888.
THIRD EDITION	Entitled 'General Rules recommended for Wiring for the Supply of Electrical Energy'. Issued in 1897.
FOURTH EDITION	Issued in 1903.
FIFTH EDITION	Entitled 'Wiring Rules'. Issued in 1907.
SIXTH EDITION	Issued in 1911.
SEVENTH EDITION	Issued in 1916.
EIGHTH EDITION	Entitled 'Regulations for the Electrical Equipment of Buildings'. Issued in 1924.
NINTH EDITION	Issued in 1927.
TENTH EDITION	Issued in 1934.
ELEVENTH EDITION	Issued in 1939.
	Revised, issued in 1943.
	Reprinted with minor Amendments, 1945.
	Supplement issued, 1946.
	Revised Section 8 issued, 1948.
TWELFTH EDITION	Issued in 1950.
	Supplement issued, 1954.
THIRTEENTH EDITION	Issued in 1955.
	Reprinted 1958, 1961, 1962 and 1964.
FOURTEENTH EDITION	Issued in 1966.
	Reprinted incorporating Amendments, 1968.
	Reprinted incorporating Amendments, 1969.
	Supplement on use in metric terms issued, 1969.
	Amendments issued, 1970.
	Reprinted in metric units incorporating Amendments, 1970.
	Reprinted 1972.
	Reprinted 1973.
	Amendments issued, 1974.
	Reprinted incorporating Amendments, 1974.
	Amendments issued, 1976.
	Reprinted incorporating Amendments, 1976.
FIFTEENTH EDITION	Entitled 'Regulations for Electrical Installations'.
	Issued in 1981. (Red Cover)
	Amendments issued, 1 January 1983.
	Reprinted incorporating Amendments, 1983. (Green Cover)
	Amendments issued, 1 May 1984.
	Reprinted incorporating Amendments, 1984. (Yellow Cover)
	Amendments issued, 1 January 1985.
	Amendments issued, 1 January 1986.
	Reprinted incorporating Amendments, 1986. (Blue Cover)
	Amendments issued, 12 June 1987.
	Reprinted incorporating Amendments, 1987. (Brown Cover)
	Reprinted with minor corrections, 1988. (Brown Cover)
SIXTEENTH EDITION	Issued in 1991. (Red Cover)
	Reprinted with minor corrections, 1992. (Red Cover)
	Reprinted as BS 7671:1992. (Red Cover)
	Amendment No 1 issued, December 1994.
	Reprinted incorporating Amendment No 1, 1994. (Green Cover)
	Amendment No 2 issued, December 1997.
	Reprinted incorporating Amendment No 2, 1997. (Yellow Cover)
	Amendment No 3 issued, April 2000.
	BS 7671:2001 issued, June 2001. (Blue Cover)
	Amendment No 1 issued, February 2002.
	Amendment No 2 issued, March 2004.
	Reprinted incorporating Amendments 1 and 2, 2004. (Brown Cover)
SEVENTEENTH EDITION	BS 7671:2008 issued, January 2008. (Red Cover)
	Reprinted incorporating Amendment No 1, 2011. (Green Cover)

Joint IET/BSI Technical Committee JPEL/64
CONSTITUTION
as at June 2011

P D Galbraith IEng MIET (Chairman)

for IET
G Digilio IEng FIEE ACIBSE MSLL
P E Donnachie BSc CEng FIET
R F B Lewington MIET
Eur Ing G Stokes BSc(Hons) CEng FIEE FCIBSE
G G Willard DipEE CEng MIEE JP
W H Wright MA CEng FIEE

and
K Siriwardhana BScEng(Hons) MBA MSc CEng MIEE MCIBSE MIHEEM (ACE)
Eur Ing K J Hawken CEng FIET MIAgrE MCGI MemASABE (Agricultural Engineers Association)
S A MacConnacher BSc CEng MIEE MInstR (Association of Manufacturers of Domestic Appliances)
S Archer MIDGTE (Association of Manufacturers of Power Generating Systems)
R Jefferis (ASTA British Electrotechnical Approvals Board)
C K Reed IEng MIIE (British Cables Association)
J M R Hagger BTech(Hons) AMIMMM (British Cables Association)
Eur Ing M H Mullins BA CEng FIEE (BEAMA Installation)
P Sayer IEng MIET GCGI (BEAMA Installation Ltd)
S R Willmore MIEE (BEAMA Installation Ltd)
D Stead (British Standards Institution)
H R Lovegrove IEng FIET (City & Guilds of London Institute)
B Sensecall (Central Heating Installations RHE/24)
Dr A C M Sung BSc(Hons) MSc PhD CEng FIET FCIBSE SrMIEEE MEI MHKIE(Ir) (CIBSE and EI)
C Flynn IEng MIET LCGI (Competent person scheme operators)
M W Coates BEng (co-opted)
D Irwin BEng(Hons) (DC Power Users Forum)
B O'Connell BSc BA(Hons) MEng(Hons) CEng MIEE (Department of Health)
Eur Ing D J Williams BTech CEng MIET (Department for Transport)
C Flynn IEng MIET LCGI (Electrical Contractors' Association)
R Plum MIET (Electrical Contractors' Association)
R Cairney IEng MIET (Electrical Contractors' Association of Scotland t/a SELECT)
D Spillett (Energy Networks Association)
Eur Ing J T Bradley BSc CEng FIET FCIBSE (Electrical Safety Council)
M C Clark BSc(Eng) MSc CEng FIEE FIMechE FCIBSE (Electrical Safety Council)
K Morriss (The GAMBICA Association Ltd)
K J Morton BSc CEng FIET (Health and Safety Executive)
I P Andrews (Heating, Ventilation and Air Conditioning Manufacturers' Association)
Eur Ing P Harris BEng(Hons) FIHEEM MIEE MCIBSE (Institute of Healthcare Engineering and Estate Management)
G C T Pritchard BTech(Hons) CEng FILP TechIOSH (ASLEC & ILP)
L C Barling (The Lighting Association)
I Craig (Lighting Industry Federation Ltd)
F Bertie MIET (NAPIT)
A Wells IEng MIET (NICEIC - Part of the Ascertiva Group Ltd)
J Eade BEng(Hons) MIEE AMIMechE (Professional Lighting and Sound Association)
G Brindle BSc(Hons) CEng MIET CMILT (Railway Industry Association)
J Reed ARIBA (Royal Institute of British Architects)
I Trueman CEng MSOE MBES MIET (Safety Assessment Federation - SAFed)
C J Tanswell CEng MIEE MCIBSE (Society of Electrical and Mechanical Engineers serving Local Government)
R Bates BSc(Hons) (UNITE)
Eur Ing J Pettifer BSc CEng MIET FCQI CQP (United Kingdom Accreditation Service)

Secretary
G D Cronshaw CEng FIET

Preface

BS 7671 Requirements for Electrical Installations takes account of the technical substance of agreements reached in CENELEC. In particular, the technical intent of the following CENELEC Harmonization Documents is included:

CENELEC Harmonization Document reference			BS 7671 reference
HD 60364-1	2008	Fundamental principles, assessment of general ..., definitions	Parts 1, 2, 3
HD 60364-4-41	2007	Protection against electric shock	Chapter 41
HD 60364-4-42	2001	Protection against thermal effects	Chapter 42
HD 60364-4-42	2001	Protection against fire	Chapter 42
HD 60364-4-43	2008	Protection against overcurrent	Chapter 43
HD 60364-4-43	2008	Measures against overcurrent	Chapter 43
IEC 60364-4-44	2008	Introduction to voltage & electro disturbances	Section 440
HD 60364-4-442	1997	Protection of low voltage installations against temporary overvoltages ...	Section 442
HD 60364-4-443	2006	Protection against overvoltages	Section 443
FprHD 60364-4-444	200X	Measures against electromagnetic disturbances	Section 444
IEC 60364-4-44	2008	Introduction to voltage & electro disturbances	Section 445
HD 60364-5-51	2006	Selection and erection - Common rules	Chapter 51
HD 60364-5-534	2008	Devices for protection against overvoltage	Section 534
HD 60364-5-54	2007	Earthing arrangements	Chapter 54
HD 384.5.551	1997	Low voltage generating sets	Section 551
HD 60364-5-559	2005	Outdoor lighting installations	Section 559
HD 60364-6	2007	Initial verification	Part 6
HD 60364-7-701	2007	Locations containing a bath or shower	Section 701
FprHD 60364-7-702	2009	Swimming pools and other basins	Section 702
HD 60364-7-703	2005	Sauna heaters	Section 703
HD 60364-7-704	2007	Construction and demolition site installations	Section 704
HD 60364-7-705	2007	Agricultural and horticultural premises	Section 705
HD 60364-7-706	2007	Locations with restricted movement	Section 706
HD 60364-7-708	2009	Caravan parks, camping parks and similar locations	Section 708
HD 60364-7-709	2009	Marinas and similar locations	Section 709
FprHD 60364-7-710	2010	Medical locations	Section 710
HD 384.7.711 SI	2003	Exhibitions, shows and stands	Section 711
HD 60364-7-712	2005	Solar photovoltaic (PV) power supply systems	Section 712
FprHD 60364-7-717	2009	Mobile or transportable units	Section 717
HD 60364-7-721	2009	Electrical installations in caravans and motor caravans	Section 721
HD 60364-7-729	2009	Operating and maintenance gangways	Section 729
HD 60364-7-740	2006	Temporary electrical installations for structures, amusement devices ...	Section 740
HD 60364-5-51	2006	External influences	Appx 5
HD 308 S2	2001	Identification of cores, cords ...	Appx 7

The dates in brackets refer to the year of issue of amendments to the Harmonization Document (HD).

Where the above HDs contain UK special national conditions, those conditions have been incorporated within BS 7671. If BS 7671 is be applied in other countries the above HDs should be consulted to confirm the status of a particular regulation.

BS 7671 will continue to be amended from time to time to take account of the publication of new or amended CENELEC standards. The opportunity has been taken to revise regulations that experience has shown require clarification or to allow for new technology and methods.

Reference is made throughout BS 7671 to publications of the British Standards Institution, both specifications and codes of practice. Appendix 1 lists these publications and gives their full titles whereas throughout BS 7671 they are referred to only by their numbers.

Where reference is made in BS 7671 to a British Standard which takes account of a CENELEC Harmonization Document or European Norm, it is understood that the reference also relates to any European national standard similarly derived from the CENELEC standard, although account needs to be taken of any national exemptions.

Note by the Health and Safety Executive

The Health and Safety Executive (HSE) welcomes the publication of BS 7671:2008, Requirements for Electrical Installations, IET (formerly IEE) Wiring Regulations 17th Edition, and its updating with the first amendment. BS 7671 and the IEE Wiring Regulations have been extensively referred to in HSE guidance over the years. Installations which conform to the standards laid down in BS 7671:2008 are regarded by HSE as likely to achieve conformity with the relevant parts of the Electricity at Work Regulations 1989. Existing installations may have been designed and installed to conform to the standards set by earlier editions of BS 7671 or the IEE Wiring Regulations. This does not mean that they will fail to achieve conformity with the relevant parts of the Electricity at Work Regulations 1989.

Notes on the plan of the 17th Edition

The edition is based on the plan agreed internationally for the arrangement of safety rules for electrical installations.

The regulation numbering follows the pattern and corresponding references of HD 60364. The numbering does not, therefore, necessarily follow sequentially. The numbering system used in Part 7 is explained in Section 700.

In the numbering system used, the first digit signifies a Part, the second digit a Chapter, the third digit a Section and the subsequent digits the Regulation number. For example, the Section number 413 is made up as follows:

> PART 4 – PROTECTION FOR SAFETY
> Chapter 41 (first chapter of Part 4) – PROTECTION AGAINST ELECTRIC SHOCK.
> Section 413 (third section of Chapter 41) – PROTECTIVE MEASURE: ELECTRICAL SEPARATION

Part 1 sets out the scope, object and fundamental principles.

Part 2 defines the sense in which certain terms are used throughout the Regulations, and provides a list of symbols used.

The subjects of the subsequent parts are as indicated below:

Part	Subject
Part 3	Identification of the characteristics of the installation that will need to be taken into account in choosing and applying the requirements of the subsequent Parts. These characteristics may vary from one part of an installation to another and should be assessed for each location to be served by the installation.
Part 4	Description of the measures that are available for the protection of persons, livestock and property, and against the hazards that may arise from the use of electricity.
Part 5	Precautions to be taken in the selection and erection of the equipment of the installation.
Part 6	Inspection and testing.
Part 7	Special installations or locations – particular requirements.

The sequence of the plan should be followed in considering the application of any particular requirement of the Regulations. The general index provides a ready reference to particular regulations by subject, but in applying any one regulation the requirements of related regulations should be borne in mind. Cross-references are provided, and the index is arranged to facilitate this.

In many cases, a group of associated regulations is covered by a side heading which is identified by a two-part number, e.g. 544.2. Throughout the Regulations where reference is made to such a two-part number, that reference is to be taken to include all the individual regulation numbers which are covered by that side heading and include that two-part number.

Numbering system The 17th edition introduced a new IEC decimal point numbering system to make it easier to embody future changes and additions resulting from ongoing international standards work within IEC and CENELEC. In order to accommodate future IEC changes it has been decided to have a 100 numbering system for UK only regulations. This means for example, the former Regulation 422.3.14 is now Regulation 422.3.100 etc. This only applies to UK only regulations.

PART 1

SCOPE, OBJECT AND FUNDAMENTAL PRINCIPLES

CONTENTS

PART 1

SCOPE, OBJECT AND FUNDAMENTAL PRINCIPLES

CHAPTER 11

SCOPE

110.1   GENERAL

110.1.1   The Regulations apply to the design, erection and verification of electrical installations such as those of:

 (i) residential premises

 (ii) commercial premises

 (iii) public premises

 (iv) industrial premises

 (v) agricultural and horticultural premises

 (vi) prefabricated buildings

 (vii) caravans, caravan parks and similar sites

 (viii) construction sites, exhibitions, shows, fairgrounds and other installations for temporary purposes including professional stage and broadcast applications

 (ix) marinas

 (x) external lighting and similar installations

 (xi) mobile or transportable units

 (xii) photovoltaic systems

 (xiii) low voltage generating sets

 (xiv) highway equipment and street furniture

 (xv) medical locations

 (xvi) operating and maintenance gangways.

NOTE: "Premises" covers the land and all facilities including buildings belonging to it.

110.1.2   The Regulations include requirements for:

 (i) circuits supplied at nominal voltages up to and including 1000 V a.c. or 1500 V d.c. For a.c., the preferred frequencies which are taken into account in this Standard are 50 Hz, 60 Hz and 400 Hz. The use of other frequencies for special purposes is not excluded

 (ii) circuits, other than the internal wiring of equipment, operating at voltages exceeding 1000 V and derived from an installation having a voltage not exceeding 1000 V a.c., e.g. discharge lighting, electrostatic precipitators

 (iii) wiring systems and cables not specifically covered by the standards for appliances

 (iv) all consumer installations external to buildings

 (v) fixed wiring for information and communication technology, signalling, control and the like (excluding internal wiring of equipment)

 (vi) additions and alterations to installations and also parts of the existing installation affected by an addition or alteration.

110.1.3   The Regulations are intended to be applied to electrical installations generally but, in certain cases, they may need to be supplemented by the requirements or recommendations of other British or Harmonized Standards or by the requirements of the person ordering the work.

Such cases include the following:

 (i) Electric signs and high voltage luminous discharge tube installations - BS 559 and BS EN 50107

 (ii) Emergency lighting - BS 5266

 (iii) Electrical apparatus for explosive gas atmospheres - BS EN 60079

 (iv) Electrical apparatus for use in the presence of combustible dust - BS EN 50281 and BS EN 61241

 (v) Fire detection and fire alarm systems in buildings - BS 5839

 (vi) Telecommunications systems - BS 6701

 (vii) Electric surface heating systems - BS EN 60335-2-96

 (viii) Electrical installations for open-cast mines and quarries - BS 6907

(ix) Code of practice for temporary electrical systems for entertainment and related purposes – BS 7909

(x) Life safety and firefighting applications – BS 8519.

110.2 EXCLUSIONS FROM SCOPE

The Regulations do not apply to the following installations:

(i) Systems for the distribution of electricity to the public

(ii) Railway traction equipment, rolling stock and signalling equipment

(iii) Equipment of motor vehicles, except those to which the requirements of the Regulations concerning caravans or mobile units are applicable

(iv) Equipment on board ships covered by BS 8450

(v) Equipment of mobile and fixed offshore installations

(vi) Equipment of aircraft

(vii) Those aspects of mines and quarries specifically covered by Statutory Regulations

(viii) Radio interference suppression equipment, except so far as it affects safety of the electrical installation

(ix) Lightning protection systems for buildings and structures covered by BS EN 62305

(x) Those aspects of lift installations covered by relevant parts of BS 5655 and BS EN 81-1

(xi) Electrical equipment of machines covered by BS EN 60204

(xii) Electric fences covered by BS EN 60335-2-76.

111 *Not used*

112 *Not used*

113 EQUIPMENT

113.1 The Regulations apply to items of electrical equipment only so far as selection and application of the equipment in the installation are concerned. The Regulations do not deal with requirements for the construction of assemblies of electrical equipment, which are required to comply with appropriate standards.

114 RELATIONSHIP WITH STATUTORY REGULATIONS

114.1 The Regulations are non-statutory. They may, however, be used in a court of law in evidence to claim compliance with a statutory requirement. The relevant statutory provisions are listed in Appendix 2 and include Acts of Parliament and Regulations made thereunder. In some cases statutory Regulations may be accompanied by Codes of Practice approved under Section 16 of the Health and Safety at Work etc. Act 1974. The legal status of these Codes is explained in Section 17 of the 1974 Act.

For a supply given in accordance with the Electricity Safety, Quality and Continuity Regulations 2002, it shall be deemed that the connection with Earth of the neutral of the supply is permanent. Outside England, Scotland and Wales, confirmation shall be sought from the distributor that the supply conforms to requirements corresponding to those of the Electricity Safety, Quality and Continuity Regulations 2002, in this respect. Where the ESQCR do not apply, equipment for isolation and switching shall be selected accordingly as specified in Chapter 53.

115 INSTALLATIONS IN PREMISES SUBJECT TO LICENSING

115.1 For installations in premises over which a licensing or other authority exercises a statutory control, the requirements of that authority shall be ascertained and complied with in the design and execution of the installation.

CHAPTER 12

OBJECT AND EFFECTS

120   **GENERAL**

120.1   This Standard contains the rules for the design and erection of electrical installations so as to provide for safety and proper functioning for the intended use.

120.2   Chapter 13 states the fundamental principles. It does not include detailed technical requirements, which may be subject to modification because of technical developments.

120.3   This Standard sets out technical requirements intended to ensure that electrical installations conform to the fundamental principles of Chapter 13, as follows:

     Part 3   Assessment of general characteristics

     Part 4   Protection for safety

     Part 5   Selection and erection of equipment

     Part 6   Inspection and testing

     Part 7   Special installations or locations.

Any intended departure from these Parts requires special consideration by the designer of the installation and shall be noted on the Electrical Installation Certificate specified in Part 6. The resulting degree of safety of the installation shall be not less than that obtained by compliance with the Regulations.

120.4   ***Moved by BS 7671:2008 Amendment No 1 to 133.5***

CHAPTER 13

FUNDAMENTAL PRINCIPLES

131   **PROTECTION FOR SAFETY**

131.1   **General**

The requirements of this chapter are intended to provide for the safety of persons, livestock and property against dangers and damage which may arise in the reasonable use of electrical installations. The requirements to provide for the safety of livestock are applicable in locations intended for them.

In electrical installations, risk of injury may result from:

  (i)   shock currents

 (ii)   excessive temperatures likely to cause burns, fires and other injurious effects

(iii)   ignition of a potentially explosive atmosphere

(iv)   undervoltages, overvoltages and electromagnetic disturbances likely to cause or result in injury or damage

 (v)   mechanical movement of electrically actuated equipment, in so far as such injury is intended to be prevented by electrical emergency switching or by electrical switching for mechanical maintenance of non-electrical parts of such equipment

(vi)   power supply interruptions and/or interruption of safety services

(vii)   arcing or burning, likely to cause blinding effects, excessive pressure and/or toxic gases.

131.2 Protection against electric shock

131.2.1 Basic protection (protection against direct contact)

NOTE: For low voltage installations, systems and equipment, 'basic protection' generally corresponds to protection against 'direct contact'.

Persons and livestock shall be protected against dangers that may arise from contact with live parts of the installation.

This protection can be achieved by one of the following methods:

(i) Preventing a current from passing through the body of any person or any livestock

(ii) Limiting the current which can pass through a body to a non-hazardous value.

131.2.2 Fault protection (protection against indirect contact)

NOTE: For low voltage installations, systems and equipment, 'fault protection' generally corresponds to protection against 'indirect contact', mainly with regard to failure of basic insulation.

Persons and livestock shall be protected against dangers that may arise from contact with exposed-conductive-parts during a fault.

This protection can be achieved by one of the following methods:

(i) Preventing a current resulting from a fault from passing through the body of any person or any livestock

(ii) Limiting the magnitude of a current resulting from a fault, which can pass through a body, to a non-hazardous value

(iii) Limiting the duration of a current resulting from a fault, which can pass through a body, to a non-hazardous time period.

In connection with fault protection, the application of the method of protective equipotential bonding is one of the important principles for safety.

131.3 Protection against thermal effects

131.3.1 The electrical installation shall be so arranged that the risk of ignition of flammable materials due to high temperature or electric arc is minimized. In addition, during normal operation of the electrical equipment, there shall be minimal risk of burns to persons or livestock.

131.3.2 Persons, livestock, fixed equipment and fixed materials adjacent to electrical equipment shall be protected against harmful effects of heat or thermal radiation emitted by electrical equipment, and in particular the following:

(i) Combustion, ignition, or degradation of materials

(ii) Risk of burns

(iii) Impairment of the safe function of installed equipment.

Electrical equipment shall not present a fire hazard to adjacent materials.

131.4 Protection against overcurrent

Persons and livestock shall be protected against injury, and property shall be protected against damage, due to excessive temperatures or electromechanical stresses caused by any overcurrents likely to arise in live conductors.

NOTE: Protection can be achieved by limiting the overcurrent to a safe value and/or duration.

131.5 Protection against fault current

Conductors other than live conductors, and any other parts intended to carry a fault current, shall be capable of carrying that current without attaining an excessive temperature. Electrical equipment, including conductors, shall be provided with mechanical protection against electromechanical stresses of fault currents as necessary to prevent injury or damage to persons, livestock or property.

131.6 Protection against voltage disturbances and measures against electromagnetic disturbances

131.6.1 Persons and livestock shall be protected against injury, and property shall be protected against any harmful effects, as a consequence of a fault between live parts of circuits supplied at different voltages, in accordance with Section 442.

131.6.2 Persons and livestock shall be protected against injury, and property shall be protected against damage, as a consequence of overvoltages such as those originating from atmospheric events or from switching, in accordance with Section 443.

NOTE: For protection against lightning strikes, refer to the BS EN 62305 series.

131.6.3 Persons and livestock shall be protected against injury, and property shall be protected against damage, as a consequence of undervoltage and any subsequent voltage recovery, in accordance with Section 445.

131.6.4 The installation shall have an adequate level of immunity against electromagnetic disturbances so as to function correctly in the specified environment, in accordance with Section 444. The installation design shall take into consideration the anticipated electromagnetic emissions, generated by the installation or the installed equipment, which shall be suitable for the current-using equipment used with, or connected to, the installation.

131.7 Protection against power supply interruption

Where danger or damage is expected to arise due to an interruption of supply, suitable provisions shall be made in the installation or installed equipment.

131.8 Moved by BS 7671:2008 Amendment No 1 to 132.16

132 DESIGN

132.1 General

The electrical installation shall be designed to provide for:

  (i) the protection of persons, livestock and property in accordance with Section 131

  (ii) the proper functioning of the electrical installation for the intended use.

The information required as a basis for design is stated in Regulations 132.2 to 5. The requirements with which the design shall comply are stated in Regulations 132.6 to 16.

132.2 Characteristics of available supply or supplies

Information on the characteristics of the available supply or supplies shall be determined by calculation, measurement, enquiry or inspection.

The following characteristics shall be included in the documentation referred to in Regulation 132.13 to show conformity with the Regulations:

  (i)   Nature of current: a.c. and/or d.c.
  (ii) Purpose and number of conductors:
        – for a.c.   line conductor(s)
                     neutral conductor
                     protective conductor
                     PEN conductor
        – for d.c.   conductors equivalent to those listed above (outer/middle/earthed live conductors, protective conductor, PEN conductor)
  (iii) Values and tolerances:
        – nominal voltage and voltage tolerances
        – nominal frequency and frequency tolerances
        – maximum current allowable
        – prospective short-circuit current
        – external earth fault loop impedance
  (iv) Protective measures inherent in the supply, e.g. earthed neutral or mid-wire
  (v)   Particular requirements of the distributor.

NOTE: If the distributor changes the characteristics of the power supply this may affect the safety of the installation.

132.3 Nature of demand

The number and type of circuits required for lighting, heating, power, control, signalling, communication and information technology, etc shall be determined from knowledge of:
- (i) location of points of power demand
- (ii) loads to be expected on the various circuits
- (iii) daily and yearly variation of demand
- (iv) any special conditions, such as harmonics
- (v) requirements for control, signalling, communication and information technology, etc.
- (vi) anticipated future demand if specified.

132.4 Electrical supply systems for safety services or standby electrical supply systems

Where a supply for safety services or standby electrical supply systems is specified the following shall be determined:
- (i) Characteristics of the supply
- (ii) Circuits to be supplied by the safety source.

132.5 Environmental conditions

132.5.1 The design of the electrical installation shall take into account the environmental conditions to which it will be subjected.

132.5.2 Equipment in surroundings susceptible to risk of fire or explosion shall be so constructed or protected, and such other special precautions shall be taken, as to prevent danger.

132.6 Cross-sectional area of conductors

The cross-sectional area of conductors shall be determined for both normal operating conditions and, where appropriate, for fault conditions according to:
- (i) the admissible maximum temperature
- (ii) the admissible voltage drop
- (iii) the electromechanical stresses likely to occur due to short-circuit and earth fault currents
- (iv) other mechanical stresses to which the conductors are likely to be exposed
- (v) the maximum impedance for correct operation of short-circuit and earth fault protection
- (vi) the method of installation
- (vii) harmonics
- (viii) thermal insulation.

132.7 Type of wiring and method of installation

The choice of the type of wiring system and the method of installation shall include consideration of the following:
- (i) The nature of the location
- (ii) The nature of the structure supporting the wiring
- (iii) Accessibility of wiring to persons and livestock
- (iv) Voltage
- (v) The electromechanical stresses likely to occur due to short-circuit and earth fault currents
- (vi) Electromagnetic interference
- (vii) Other external influences (e.g. mechanical, thermal and those associated with fire) to which the wiring is likely to be exposed during the erection of the electrical installation or in service.

132.8 Protective equipment

The characteristics of protective equipment shall be determined with respect to their function, including protection against the effects of:
- (i) overcurrent (overload and/or short-circuit)
- (ii) earth fault current
- (iii) overvoltage
- (iv) undervoltage and no-voltage.

The protective devices shall operate at values of current, voltage and time which are suitably related to the characteristics of the circuits and to the possibilities of danger.

132.9　Emergency control

An interrupting device shall be installed in such a way that it can be easily recognized and effectively and rapidly operated where, in the case of danger, there is a necessity for immediate interruption of the supply.

132.10　Disconnecting devices

Disconnecting devices shall be provided so as to permit switching and/or isolation of the electrical installation, circuits or individual items of equipment as required for operation, inspection, testing, fault detection, maintenance and repair.

132.11　Prevention of mutual detrimental influence

The electrical installation shall be arranged in such a way that no mutual detrimental influence will occur between electrical installations and non-electrical installations.

Electromagnetic interference shall be taken into account.

132.12　Accessibility of electrical equipment

Electrical equipment shall be arranged so as to afford as may be necessary:
  (i)　sufficient space for the initial installation and later replacement of individual items of electrical equipment
  (ii)　accessibility for operation, inspection, testing, fault detection, maintenance and repair.

132.13　Documentation for the electrical installation

Every electrical installation shall be provided with appropriate documentation, including that required by Regulation 514.9, Part 6 and where applicable Part 7.

132.14　Protective devices and switches

132.14.1　A single-pole fuse, switch or circuit-breaker shall be inserted in the line conductor only.

132.14.2　No switch or circuit-breaker, except where linked, or fuse, shall be inserted in an earthed neutral conductor. Any linked switch or linked circuit-breaker inserted in an earthed neutral conductor shall be arranged to break all the related line conductors.

132.15　Isolation and switching

132.15.1　Effective means, suitably placed for ready operation, shall be provided so that all voltage may be cut off from every installation, from every circuit thereof and from all equipment, as may be necessary to prevent or remove danger.

132.15.2　Every fixed electric motor shall be provided with an efficient means of switching off, readily accessible, easily operated and so placed as to prevent danger.

132.16　Additions and alterations to an installation

No addition or alteration, temporary or permanent, shall be made to an existing installation, unless it has been ascertained that the rating and the condition of any existing equipment, including that of the distributor, will be adequate for the altered circumstances. Furthermore, the earthing and bonding arrangements, if necessary for the protective measure applied for the safety of the addition or alteration, shall be adequate.

133　SELECTION OF ELECTRICAL EQUIPMENT

133.1　General

133.1.1　Every item of equipment shall comply with the appropriate British or Harmonized Standard. In the absence of such a standard, reference shall be made to the appropriate International (IEC) standard or the appropriate standard of another country.

133.1.2　Where there are no applicable standards, the item of equipment concerned shall be selected by special agreement between the person specifying the installation and the installer.

133.1.3　Where equipment to be used is not in accordance with Regulation 133.1.1 or is used outside the scope of its standard, the designer or other person responsible for specifying the installation shall confirm that the equipment provides at least the same degree of safety as that afforded by compliance with the Regulations.

133.2 Characteristics

Every item of electrical equipment selected shall have suitable characteristics appropriate to the values and conditions on which the design of the electrical installation (see Section 132) is based and shall, in particular, fulfil the requirements of Regulations 133.2.1 to 4.

133.2.1 Voltage

Electrical equipment shall be suitable with respect to the maximum steady-state voltage (rms value for a.c.) likely to be applied, as well as overvoltages likely to occur.

NOTE: For certain equipment, it may also be necessary to take account of the lowest voltage likely to occur.

133.2.2 Current

Electrical equipment shall be selected with respect to the maximum steady current (rms value for a.c.) which it has to carry in normal service and with respect to the current likely to be carried in abnormal conditions and the period (e.g. operating time of protective devices, if any) during which it may be expected to flow.

133.2.3 Frequency

Equipment shall be suitable for the frequencies likely to occur in the circuit.

133.2.4 Power

Electrical equipment which is selected on the basis of its power characteristics shall be suitable for the duty demanded of the equipment, taking into account the load factor and the normal service conditions.

133.3 Conditions of installation

Electrical equipment shall be selected so as to withstand safely the stresses, the environmental conditions (see Regulation 132.5) and the characteristics of its location. An item of equipment which does not by design have the properties corresponding to its location may be used where adequate further protection is provided as part of the completed electrical installation.

133.4 Prevention of harmful effects

All electrical equipment shall be selected so that it will not cause harmful effects on other equipment or impair the supply during normal service, including switching operations.

NOTE: Examples of characteristics which are likely to have harmful effects are given in Chapter 33.

133.5 New materials and inventions

Where the use of a new material or invention leads to departures from the Regulations, the resulting degree of safety of the installation shall be not less than that obtained by compliance with the Regulations. Such use is to be noted on the Electrical Installation Certificate specified in Part 6.

134 ERECTION AND INITIAL VERIFICATION OF ELECTRICAL INSTALLATIONS

134.1 Erection

134.1.1 Good workmanship by competent persons or persons under their supervision and proper materials shall be used in the erection of the electrical installation. Electrical equipment shall be installed in accordance with the instructions provided by the manufacturer of the equipment.

134.1.2 The characteristics of the electrical equipment, as determined in accordance with Section 133, shall not be impaired by the process of erection.

134.1.3 Conductors shall be identified in accordance with Section 514. Where identification of terminals is necessary, they shall be identified in accordance with Section 514.

134.1.4 Every electrical joint and connection shall be of proper construction as regards conductance, insulation, mechanical strength and protection.

134.1.5 Electrical equipment shall be installed in such a manner that the design temperatures are not exceeded.

134.1.6 Electrical equipment likely to cause high temperatures or electric arcs shall be placed or guarded so as to minimize the risk of ignition of flammable materials.

Where the temperature of an exposed part of electrical equipment is likely to cause injury to persons or livestock that part shall be so located or guarded as to prevent accidental contact therewith.

134.1.7 Where necessary for safety purposes, suitable warning signs and/or notices shall be provided.

134.2 Initial verification

134.2.1 During erection and on completion of an installation or an addition or alteration to an installation, and before it is put into service, appropriate inspection and testing shall be carried out by competent persons to verify that the requirements of this Standard have been met.

Appropriate certification shall be issued in accordance with Sections 631 and 632.

134.2.2 The designer of the installation shall make a recommendation for the interval to the first periodic inspection and test as detailed in Part 6.

NOTE: The requirements of Chapter 34 (maintainability) should be taken into consideration.

135 PERIODIC INSPECTION AND TESTING

135.1 It is recommended that every electrical installation is subjected to periodic inspection and testing, in accordance with Chapter 62.

PART 2

DEFINITIONS

For the purposes of the Regulations, the following definitions shall apply. As far as practicable the definitions align with the International Electrotechnical Vocabulary and BS 4727 - 'Glossary of electrotechnical, power, telecommunication, electronics, lighting and colour terms'.

NOTE: Where a section number is listed, e.g. {444}, the definition only applies within that section.

8/20 Current impulse, {534}. A current impulse with a virtual front time of 8 µs and a time to half-value of 20 µs where:

   (i)   the front time is defined as $1.25(t_{90} - t_{10})$, where t_{90} and t_{10} are the 90% and 10% points on the leading edge of the waveform

   (ii)   the time to half-value is defined as the time between the virtual origin and the 50% point on the tail. The virtual origin is the point where a straight line drawn through the 90% and 10% points on the leading edge of the waveform intersects the $I = 0$ line.

Accessory. A device, other than current-using equipment, associated with such equipment or with the wiring of an installation.

Agricultural and horticultural premises. Rooms, locations or areas where:
   – livestock are kept, or
   – feed, fertilizers, vegetable and animal products are produced, stored, prepared or processed, or
   – plants are grown, such as greenhouses.

Ambient temperature. The temperature of the air or other medium where the equipment is to be used.

Amusement device. Ride, stand, textile or membrane building, side stall, side show, tent, booth or grandstand intended for the entertainment of the public.

Appliance. An item of current-using equipment other than a luminaire or an independent motor.

Arm's reach. A zone of accessibility to touch, extending from any point on a surface where persons usually stand or move about to the limits which a person can reach with a hand in any direction without assistance. (See Figure 417.)

Arrangements for livestock keeping. Buildings and rooms (housing for animals), cages, runs or other containers used for continuous accommodation of livestock.

Back-up protection. Protection which is intended to operate when a system fault is not cleared, or abnormal condition not detected, in the required time because of failure or inability of other protection to operate or failure of the appropriate circuit-breaker(s) to trip.

Barrier. A part providing a defined degree of protection against contact with live parts from any usual direction of access.

Basic insulation. Insulation applied to live parts to provide basic protection and which does not necessarily include insulation used exclusively for functional purposes.

Basic protection. Protection against electric shock under fault-free conditions.

NOTE: For low voltage installations, systems and equipment, basic protection generally corresponds to protection against direct contact, that is "contact of persons or livestock with live parts".

Basin of fountain. A basin not intended to be occupied by persons and which cannot be accessed (reached by persons) without the use of ladders or similar means. For basins of fountains which may be occupied by persons, the requirements for swimming pools apply.

Bonding conductor. A protective conductor providing equipotential bonding.

Bonding network (BN), {444}. A set of interconnected conductive parts that provide a path for currents at frequencies from direct current (d.c.) to radio frequency (RF) intended to divert, block or impede the passage of electromagnetic energy.

Bonding ring conductor (BRC), {444}. A bus earthing conductor in the form of a closed ring.

NOTE: Normally the bonding ring conductor, as part of the bonding network, has multiple connections to the common bonding network (CBN) that improves its performance.

Booth. Non-stationary unit, intended to accommodate equipment generally for pleasure or demonstration purposes.

Building void, accessible. A space within the structure or the components of a building accessible only at certain points. Such voids include the space within partitions, suspended floors, ceilings and certain types of window frame, door frame and architrave.

Building void, non-accessible. A space within the structure or the components of a building which has no ready means of access.

Bunched. Cables are said to be bunched when two or more are contained within a single conduit, duct, ducting, or trunking or, if not enclosed, are not separated from each other by a specified distance.

Busbar trunking system. A type-tested assembly, in the form of an enclosed conductor system comprising solid conductors separated by insulating material. The assembly may consist of units such as:
  – busbar trunking units, with or without tap-off facilities
  – tap-off units where applicable
  – phase-transposition, expansion, building-movement, flexible, end-feeder and adaptor units.

NOTE: Other system components may include tap-off units.

Bypass equipotential bonding conductor, {444}. Bonding conductor connected in parallel with the screens of cables.

Cable bracket. Deleted by BS 7671:2008 Amendment No 1

Cable channel. An enclosure situated above or in the ground, ventilated or closed, and having dimensions which do not permit the access of persons but allow access to the conduits and/or cables throughout their length during and after installation. A cable channel may or may not form part of the building construction.

Cable cleat. A component of a support system, which consists of elements spaced at intervals along the length of the cable or conduit and which mechanically retains the cable or conduit.

Cable coupler. A means of enabling the connection or disconnection, at will, of two flexible cables. It consists of a connector and a plug.

Cable ducting. An enclosure of metal or insulating material, other than conduit or cable trunking, intended for the protection of cables which are drawn in after erection of the ducting.

Cable ladder. A cable support consisting of a series of transverse supporting elements rigidly fixed to main longitudinal supporting members.

Cable tray. A cable support consisting of a continuous base with raised edges and no covering. A cable tray may or may not be perforated.

Cable trunking. A closed enclosure normally of rectangular cross-section, of which one side is removable or hinged, used for the protection of cables and for the accommodation of other electrical equipment.

Cable tunnel. A corridor containing supporting structures for cables and joints and/or other elements of wiring systems and whose dimensions allow persons to pass freely throughout the entire length.

Caravan. A trailer leisure accommodation vehicle, used for touring, designed to meet the requirements for the construction and use of road vehicles (see also definitions of Motor caravan and Leisure accommodation vehicle).

Caravan park / camping park. Area of land that contains two or more caravan pitches and/or tents.

Caravan pitch. Plot of ground intended to be occupied by a leisure accommodation vehicle.

Caravan pitch electrical supply equipment. Equipment that provides means of connecting and disconnecting supply cables from leisure accommodation vehicles or tents with a mains electrical supply.

Cartridge fuse link. A device comprising a fuse element or two or more fuse elements connected in parallel enclosed in a cartridge usually filled with arc-extinguishing medium and connected to terminations (see fuse link).

Central power supply system. A system supplying the required emergency power to essential safety equipment.

Central power supply system (low power output). Central power supply system with a limitation of the power output of the system at 500 W for 3 h or 1 500 W for 1 h.

NOTE: A low power supply system normally comprises a maintenance-free battery and a charging and testing unit.

Circuit. An assembly of electrical equipment supplied from the same origin and protected against overcurrent by the same protective device(s).

Circuit-breaker. A device capable of making, carrying and breaking normal load currents and also making and automatically breaking, under predetermined conditions, abnormal currents such as short-circuit currents. It is usually required to operate infrequently although some types are suitable for frequent operation.

Circuit-breaker, linked. A circuit-breaker the contacts of which are so arranged as to make or break all poles simultaneously or in a definite sequence.

Circuit protective conductor (cpc). A protective conductor connecting exposed-conductive-parts of equipment to the main earthing terminal.

Class I equipment. Equipment in which protection against electric shock does not rely on basic insulation only, but which includes means for the connection of exposed-conductive-parts to a protective conductor in the fixed wiring of the installation (see BS EN 61140).

Class II equipment. Equipment in which protection against electric shock does not rely on basic insulation only, but in which additional safety precautions such as supplementary insulation are provided, there being no provision for the connection of exposed metalwork of the equipment to a protective conductor, and no reliance upon precautions to be taken in the fixed wiring of the installation (see BS EN 61140) .

Class III equipment. Equipment in which protection against electric shock relies on supply at SELV and in which voltages higher than those of SELV are not generated (see BS EN 61140).

Cold tail. The interface between the fixed installation and a heating unit.

Common equipotential bonding system, common bonding network (CBN), {444}. Equipotential bonding system providing both protective equipotential bonding and functional equipotential bonding.

Competent person. A person who possesses sufficient technical knowledge, relevant practical skills and experience for the nature of the electrical work undertaken and is able at all times to prevent danger and, where appropriate, injury to him/herself and others.

Complementary floor heating. Direct heating system integrated into the floor construction, for example, in the border zones close to outer walls, which complements the heat dissipation of a thermal storage floor heating system.

Conducting location with restricted movement. A location comprised mainly of metallic or conductive surrounding parts, within which it is likely that a person will come into contact through a substantial portion of their body with the conductive surrounding parts and where the possibility of preventing this contact is limited.

Conduit. A part of a closed wiring system for cables in electrical installations, allowing them to be drawn in and/or replaced, but not inserted laterally.

Connector. The part of a cable coupler or of an appliance coupler which is provided with female contacts and is intended to be attached to the end of the flexible cable remote from the supply.

Consumer unit (may also be known as a consumer control unit or electricity control unit). A particular type of distribution board comprising a type-tested co-ordinated assembly for the control and distribution of electrical energy, principally in domestic premises, incorporating manual means of double-pole isolation on the incoming circuit(s) and an assembly of one or more fuses, circuit-breakers, residual current operated devices or signalling and other devices proven during the type-test of the assembly as suitable for such use.

Continuous operating voltage (U_c), {534}. Maximum rms voltage which may be continuously applied to an SPD's mode of protection. This is equal to the rated voltage.

Controlgear *(see Switchgear).*

Conventional impulse withstand voltage. The peak value of an impulse test voltage at which insulation does not show any disruptive discharge when subjected to a specified number of applications of impulses of this value, under specified conditions.

Current-carrying capacity of a conductor. The maximum current which can be carried by a conductor under specified conditions without its steady-state temperature exceeding a specified value.

Current-using equipment. Equipment which converts electrical energy into another form of energy, such as light, heat or motive power.

d.c. system - see Appendix 9.

Danger. Risk of injury to persons (and livestock where expected to be present) from:

(i) fire, electric shock, burns, arcing and explosion arising from the use of electrical energy, and

(ii) mechanical movement of electrically controlled equipment, in so far as such danger is intended to be prevented by electrical emergency switching or by electrical switching for mechanical maintenance of non-electrical parts of such equipment.

Design current (of a circuit). The magnitude of the current (rms value for a.c.) to be carried by the circuit in normal service.

Device for connecting a luminaire (DCL). System comprising an outlet and a connector providing a fixed luminaire with electrical connection to and disconnection from a fixed installation but not providing mechanical support for a luminaire.

Direct contact *Deleted by BS 7671:2008 (see Basic protection).*

Direct heating system. Heating system which generates heat from electrical energy and dissipates it to the room to be heated with a response time being as low as possible.

Disconnector. A mechanical switching device which, in the open position, complies with the requirements specified for the isolating function.

NOTE 1: A disconnector is otherwise known as an isolator.

NOTE 2: A disconnector is capable of opening and closing a circuit when either a negligible current is broken or made, or when no significant change in the voltage across the terminals of each pole of the disconnector occurs. It is also capable of carrying currents under normal circuit conditions and carrying for a specified time current under abnormal conditions such as those of short-circuit.

Discrimination. Ability of a protective device to operate in preference to another protective device in series.

Distribution board. An assembly containing switching or protective devices (e.g. fuses, circuit-breakers, residual current operated devices) associated with one or more outgoing circuits fed from one or more incoming circuits, together with terminals for the neutral and circuit protective conductors. It may also include signalling and other control devices. Means of isolation may be included in the board or may be provided separately.

Distribution circuit. A circuit supplying a distribution board or switchgear.

A distribution circuit may also connect the origin of an installation to an outlying building or separate installation, when it is sometimes called a sub-main.

Distributor. A person who distributes electricity to consumers using electrical lines and equipment that he/she owns or operates.

Double insulation. Insulation comprising both basic insulation and supplementary insulation.

Duct, Ducting *(see Cable ducting).*

Earth. The conductive mass of the Earth, whose electric potential at any point is conventionally taken as zero.

Earth electrode. Conductive part, which may be embedded in the soil or in a specific conductive medium, e.g. concrete or coke, in electrical contact with the Earth.

Earth electrode network, {444}. Part of an earthing arrangement comprising only the earth electrodes and their interconnections.

Earth electrode resistance. The resistance of an earth electrode to Earth.

Earth fault current. A current resulting from a fault of negligible impedance between a line conductor and an exposed-conductive-part or a protective conductor.

Earth fault loop impedance. The impedance of the earth fault current loop starting and ending at the point of earth fault. This impedance is denoted by the symbol Z_s.

The earth fault loop comprises the following, starting at the point of fault:
- the circuit protective conductor, and
- the consumer's earthing terminal and earthing conductor, and
- for TN systems, the metallic return path, and
- for TT and IT systems, the Earth return path, and
- the path through the earthed neutral point of the transformer, and

- the transformer winding, and
- the line conductor from the transformer to the point of fault.

Earth leakage current *(see Protective conductor current).*

Earthed concentric wiring. A wiring system in which one or more insulated conductors are completely surrounded throughout their length by a conductor, for example a metallic sheath, which acts as a PEN conductor.

Earthing. Connection of the exposed-conductive-parts of an installation to the main earthing terminal of that installation.

Earthing conductor. A protective conductor connecting the main earthing terminal of an installation to an earth electrode or to other means of earthing.

Electric shock. A dangerous physiological effect resulting from the passing of an electric current through a human body or livestock.

Electrical circuit for safety services. Electrical circuit intended to be used as part of an electrical supply system for safety services.

Electrical equipment (abbr: *Equipment*). Any item for such purposes as generation, conversion, transmission, distribution or utilisation of electrical energy, such as machines, transformers, apparatus, measuring instruments, protective devices, wiring systems, accessories, appliances and luminaires.

Electrical installation (abbr: *Installation*). An assembly of associated electrical equipment having co-ordinated characteristics to fulfil specific purposes.

Electrical source for safety services. Electrical source intended to be used as part of an electrical supply system for safety services.

Electrical supply system for safety services. A supply system intended to maintain the operation of essential parts of an electrical installation and equipment:

   (i)   for the health and safety of persons and livestock, and

   (ii)   to avoid damage to the environment and to other equipment.

NOTE: The supply system includes the source and the circuit(s) up to the terminals of the electrical equipment.

Electrically independent earth electrodes. Earth electrodes located at such a distance from one another that the maximum current likely to flow through one of them does not significantly affect the potential of the other(s).

Electrode boiler (or electrode water heater). Equipment for the electrical heating of water or electrolyte by the passage of an electric current between electrodes immersed in the water or electrolyte.

Electronic convertor (static convertor). A convertor having no moving parts and notably using semiconductor rectifiers.

Emergency stopping. Emergency switching intended to stop an operation.

Emergency switching. An operation intended to remove, as quickly as possible, danger, which may have occurred unexpectedly.

Enclosure. A part providing protection of equipment against certain external influences and in any direction providing basic protection.

Equipment *(see Electrical equipment).*

Equipotential bonding. Electrical connection maintaining various exposed-conductive-parts and extraneous-conductive-parts at substantially the same potential. (See also *Protective equipotential bonding.*)

Escape route. Path to follow for access to a safe area in the event of an emergency.

Exhibition. Event intended for the purpose of displaying and/or selling products etc., which can take place in any suitable location, either a room, building or temporary structure.

Exposed-conductive-part. Conductive part of equipment which can be touched and which is not normally live, but which can become live under fault conditions.

External influence. Any influence external to an electrical installation which affects the design and safe operation of that installation.

Extra-low voltage *(see Voltage, nominal).*

Extraneous-conductive-part. A conductive part liable to introduce a potential, generally Earth potential, and not forming part of the electrical installation.

Fairground. Area where one or more stands, amusement devices or booths are erected for leisure use.

Fault. A circuit condition in which current flows through an abnormal or unintended path. This may result from an insulation failure or a bridging of insulation. Conventionally the impedance between live conductors or between live conductors and exposed- or extraneous-conductive-parts at the fault position is considered negligible.

Fault current. A current resulting from a fault.

Fault protection. Protection against electric shock under single fault conditions.

NOTE: For low voltage installations, systems and equipment, fault protection generally corresponds to protection against indirect contact, mainly with regard to failure of basic insulation. Indirect contact is "contact of persons or livestock with exposed-conductive-parts which have become live under fault conditions".

Final circuit. A circuit connected directly to current-using equipment, or to a socket-outlet or socket-outlets or other outlet points for the connection of such equipment.

Fixed equipment. Equipment designed to be fastened to a support or otherwise secured in a specific location.

Flexible cable. A cable whose structure and materials make it suitable to be flexed while in service.

Flexible cord (see Flexible cable). Deleted by BS 7671:2008 Amendment No 1.

Flexible sheet heating element. Heating element consisting of sheets of electrical insulation laminated with electrical resistance material, or a base material on which electrically insulated heating wires are fixed.

Flexible wiring system. A wiring system designed to provide mechanical flexibility in use without degradation of the electrical components.

Follow-current interrupting rating, {534}. The level of prospective short-circuit current that an SPD is able to interrupt without back-up protection.

Functional bonding conductor, {444}. Conductor provided for functional equipotential bonding.

Functional earth. Earthing of a point or points in a system or in an installation or in equipment, for purposes other than electrical safety, such as for proper functioning of electrical equipment.

Functional extra-low voltage (FELV). An extra-low voltage system in which not all of the protective measures required for SELV or PELV have been applied.

Functional switching. An operation intended to switch 'on' or 'off' or vary the supply of electrical energy to all or part of an installation for normal operating purposes.

Fuse. A device which, by the melting of one or more of its specially designed and proportioned components, opens the circuit in which it is inserted by breaking the current when this exceeds a given value for a sufficient time. The fuse comprises all the parts that form the complete device.

Fuse carrier. The movable part of a fuse designed to carry a fuse link.

Fuse element. A part of a fuse designed to melt when the fuse operates.

Fuse link. A part of a fuse, including the fuse element(s), which requires replacement by a new or renewable fuse link after the fuse has operated and before the fuse is put back into service.

Fused connection unit. A device associated with the fixed wiring of an installation by which appliances may be connected, and having provision for a replaceable cartridge fuse link.

Gas installation pipe. Any pipe, not being a service pipe (other than any part of a service pipe comprised in a primary meter installation) or a pipe comprised in a gas appliance, for conveying gas for a particular consumer and including any associated valve or other gas fitting.

Harmonized Standard. A standard which has been drawn up by common agreement between national standards bodies notified to the European Commission by all member states and published under national procedures.

Hazardous-live-part. A live part which can give, under certain conditions of external influence, an electric shock.

Heating cable. Cable with or without a shield or a metallic sheath, intended to give off heat for heating purposes.

Heating-free area. Unheated floor or ceiling area which is completely covered when placing pieces of furniture or kept free for built-in furniture.

Heating unit. Heating cable or flexible sheet heating element with rigidly fixed cold tails or terminal fittings which are connected to the terminals of the electrical installation.

High-density livestock rearing. Breeding and rearing of livestock for which the use of automatic systems for life support is necessary.

NOTE: Examples of automatic life support systems are those for ventilation, feeding and air-conditioning.

High voltage *(see Voltage, nominal).*

Highway. A highway means any way (other than a waterway) over which there is public passage and includes the highway verge and any bridge over which, or tunnel through which, the highway passes.

Highway distribution board. A fixed structure or underground chamber, located on a highway, used as a distribution point, for connecting more than one highway distribution circuit to a common origin. Street furniture which supplies more than one circuit is defined as a highway distribution board. The connection of a single temporary load to an item of street furniture shall not in itself make that item of street furniture into a highway distribution board.

Highway distribution circuit. A Band II circuit connecting the origin of the installation to a remote highway distribution board or items of street furniture. It may also connect a highway distribution board to street furniture.

Highway power supply. An electrical installation comprising an assembly of associated highway distribution circuits, highway distribution boards and street furniture, supplied from a common origin.

Houseboat. Floating decked structure which is designed or adapted for use as a place of permanent residence often kept in one place on inland water.

Impulse current (I$_{imp}$), {534}. A parameter used for the classification test for SPDs; it is defined by three elements, a current peak value, a charge Q and a specific energy W/R.

Impulse withstand voltage, {534}. The highest peak value of impulse voltage of prescribed form and polarity which does not cause breakdown of insulation under specified conditions.

Indirect contact. *Deleted by BS 7671:2008 (see Fault protection).*

Inspection. Examination of an electrical installation using all the senses as appropriate.

Installation. *(see Electrical installation).*

Instructed person. A person adequately advised or supervised by skilled persons to enable him/her to avoid dangers which electricity may create.

Insulation. Suitable non-conductive material enclosing, surrounding or supporting a conductor.

Insulation co-ordination, {534}. The selection of the electric strength of equipment in relation to the voltages which can appear on the system for which the equipment is intended, taking into account the service environment and the characteristics of the available protective devices.

Isolation. A function intended to cut off for reasons of safety the supply from all, or a discrete section, of the installation by separating the installation or section from every source of electrical energy.

Isolator. A mechanical switching device which, in the open position, complies with the requirements specified for the isolating function. An isolator is otherwise known as a disconnector.

Ladder *(see Cable ladder).*

Leakage current. Electric current in an unwanted conductive path under normal operating conditions.

Leisure accommodation vehicle. Unit of living accommodation for temporary or seasonal occupation which may meet requirements for construction and use of road vehicles.

Lightning protection zone (LPZ), {534}. Zone where the lightning electromagnetic environment is defined.

Line conductor. A conductor of an a.c. system for the transmission of electrical energy other than a neutral conductor, a protective conductor or a PEN conductor. The term also means the equivalent conductor of a d.c. system unless otherwise specified in the Regulations.

Live conductor. *(see Live part).*

Live part. A conductor or conductive part intended to be energized in normal use, including a neutral conductor but, by convention, not a PEN conductor.

Low voltage *(see Voltage, nominal).*

Luminaire. Equipment which distributes, filters or transforms the light transmitted from one or more lamps and which includes all the parts necessary for supporting, fixing and protecting the lamps, but not the lamps themselves, and where necessary, circuit auxiliaries together with the means for connecting them to the supply.

NOTE: Lamps includes devices such as light emitting diodes.

Luminaire supporting coupler (LSC). A means, comprising an LSC outlet and an LSC connector, providing mechanical support for a luminaire and the electrical connection to and disconnection from a fixed wiring installation.

LV switchgear and controlgear assembly. A combination of one or more low voltage switching devices together with associated control, measuring, signalling, protective, regulating equipment, etc., completely assembled under the responsibility of the manufacturer with all the internal electrical and mechanical interconnection and structural parts. The components of the assembly may be electromechanical or electronic. The assembly may be either type-tested or partially type-tested (see BS EN 60439-1).

Main earthing terminal. The terminal or bar provided for the connection of protective conductors, including protective bonding conductors, and conductors for functional earthing, if any, to the means of earthing.

Maintenance. Combination of all technical and administrative actions, including supervision actions, intended to retain an item in, or restore it to, a state in which it can perform a required function.

Marina. Facility for mooring and servicing of pleasure craft with fixed wharves, jetties, piers or pontoon arrangements capable of berthing one or more pleasure craft.

Mechanical maintenance. The replacement, refurbishment or cleaning of lamps and non-electrical parts of equipment, plant and machinery.

Medical location, {710}. Location intended for purposes of diagnosis, treatment including cosmetic treatment, monitoring and care of patients.

- **Applied part.** Part of medical electrical equipment that in normal use necessarily comes into physical contact with the patient for ME equipment or an ME system to perform its function.

- **Group 0.** Medical location where no applied parts are intended to be used and where discontinuity (failure) of the supply cannot cause danger to life.

- **Group 1.** Medical location where discontinuity of the electrical supply does not represent a threat to the safety of the patient and applied parts are intended to be used:
  - externally
  - invasively to any part of the body except where group 2 applies.

- **Group 2.** Medical location where applied parts are intended to be used, and where discontinuity (failure) of the supply can cause danger to life, in applications such as:
  - intracardiac procedures
  - vital treatments and surgical operations.

  NOTE: An intracardiac procedure is a procedure whereby an electrical conductor is placed within the heart of a patient or is likely to come into contact with the heart, such conductor being accessible outside the patient's body. In this context, an electrical conductor includes insulated wires such as cardiac pacing electrodes or intracardiac ECG electrodes, or insulated tubes filled with conducting fluids.

- **Medical electrical equipment (ME equipment).** Electrical equipment having an applied part or transferring energy to or from the patient or detecting such energy transfer to or from the patient and which is

  (a) provided with not more than one connection to a particular supply mains, and

  (b) intended by the manufacturer to be used

  - in the diagnosis, treatment or monitoring of a patient, or
  - for compensation or alleviation of disease, injury or disability.

  NOTE: ME equipment includes those accessories as defined by the manufacturer that are necessary to enable the normal use of the ME equipment.

- **Medical electrical system (ME system).** Combination, as specified by the manufacturer, of items of equipment, at least one of which is medical electrical equipment to be interconnected by functional connection or by use of a multiple socket-outlet.

  NOTE: The system includes those accessories which are needed for operating the system and are specified by the manufacturer.

- **Medical IT system.** IT electrical system fulfilling specific requirements for medical applications.

  NOTE: These supplies are also known as isolated power supply systems.

- **Patient.** Living being (person or animal) undergoing a medical, surgical or dental procedure.

  NOTE: A person under treatment for cosmetic purposes may be considered a patient.

- **Patient environment.** Any volume in which intentional or unintentional contact can occur between a patient and parts of the medical electrical equipment or medical electrical system or between a patient and other persons touching parts of the medical electrical equipment or medical electrical system.

  NOTE 1: For illustration see Figure 710.1.

  NOTE 2: This applies when the patient's position is predetermined; if not, all possible patient positions should be considered.

Meshed bonding network (MESH-BN), {444}. Bonding network in which all associated equipment frames, racks and cabinets and usually the d.c. power return conductor are bonded together as well as at multiple points to the CBN and may have the form of a mesh.

NOTE: A MESH-BN improves the performance of a common bonding network.

Minimum illuminance. Illuminance for emergency lighting at the end of the rated operating time.

Minor works. Additions and alterations to an installation that do not extend to the provision of a new circuit.

NOTE: Examples include the addition of socket-outlets or lighting points to an existing circuit, the relocation of a light switch etc.

Mobile and offshore installations. Installations used for the exploration or development of liquid or gaseous hydrocarbon resources.

Mobile equipment (portable equipment (deprecated)). Electrical equipment which is moved while in operation or which can easily be moved from one place to another while connected to the supply.

Mobile home. A transportable leisure accommodation vehicle which includes means for mobility but does not meet the requirements for construction and use of road vehicles.

Monitoring. Observation of the operation of a system or part of a system to verify correct functioning or detect incorrect functioning by measuring system variables and comparing the measured values with specified values.

Motor caravan. Self-propelled leisure accommodation vehicle, used for touring, that meets the requirements for the construction and use of road vehicles.

NOTE: It is either adapted from a series production vehicle, or designed and built on an existing chassis, with or without the driving cab, the accommodation being either fixed or dismountable.

Neutral conductor. A conductor connected to the neutral point of a system and contributing to the transmission of electrical energy. The term also means the equivalent conductor of an IT or d.c. system unless otherwise specified in the Regulations and also identifies either the mid-wire of a three-wire d.c. circuit or the earthed conductor of a two-wire earthed d.c. circuit.

Nominal discharge current (I_{nspd}), {534}. A parameter used for the classification test for Class I SPDs and for preconditioning of an SPD for Class I and Class II tests; it is defined by the crest value of current through an SPD, having a current waveform of 8/20.

Nominal voltage *(see Voltage, nominal).*

Obstacle. A part preventing unintentional contact with live parts but not preventing deliberate contact.

Open-circuit voltage under standard test conditions $U_{oc\ STC}$. Voltage under standard test conditions across an unloaded (open) generator or on the d.c. side of the convertor.

Operating and maintenance gangway, {729}. Gangway providing access to facilitate operations such as switching, controlling, setting, observation and maintenance of electrical equipment.

Ordinary person. A person who is neither a skilled person nor an instructed person.

Origin of an installation. The position at which electrical energy is delivered to an electrical installation.

Origin of the temporary electrical installation. Point on the permanent installation or other source of supply from which electrical energy is delivered to the temporary electrical installation.

Overcurrent. A current exceeding the rated value. For conductors the rated value is the current-carrying capacity.

Overcurrent detection. A method of establishing that the value of current in a circuit exceeds a predetermined value for a specified length of time.

Overload current. An overcurrent occurring in a circuit which is electrically sound.

PEL. A conductor combining the functions of both a protective earthing conductor and a line conductor.

PELV (protective extra-low voltage). An extra-low voltage system which is not electrically separated from Earth, but which otherwise satisfies all the requirements for SELV.

PEM. A conductor combining the functions of both a protective earthing conductor and a midpoint conductor.

PEN conductor. A conductor combining the functions of both protective conductor and neutral conductor.

Phase conductor *(see Line conductor).*

Pleasure craft. Any boat, vessel, yacht, motor launch, houseboat or other floating craft used exclusively for sport or leisure.

Plug. Accessory having pins designed to engage with the contacts of a socket-outlet, and incorporating means for the electrical connection and mechanical retention of a flexible cable.

Point (in wiring). A termination of the fixed wiring intended for the connection of current-using equipment.

Portable equipment *(see Mobile equipment).*

Powertrack. A system component, which is generally a linear assembly of spaced and supported busbars, providing electrical connection of accessories.

Powertrack system (PT system). An assembly of system components including a powertrack by which accessories may be connected to an electrical supply at one or more points (predetermined or otherwise) along the powertrack.

NOTE: The maximum current rating of a powertrack system is 63A.

Prefabricated wiring system. Wiring system consisting of wiring sections incorporating the means of interconnection designed to allow sections to be connected together to form a given system, and incorporating installation couplers conforming to BS EN 61535.

Prospective fault current (I_{pf}). The value of overcurrent at a given point in a circuit resulting from a fault of negligible impedance between live conductors having a difference of potential under normal operating conditions, or between a live conductor and an exposed-conductive-part.

Protective bonding conductor. Protective conductor provided for protective equipotential bonding

Protective conductor (PE). A conductor used for some measures of protection against electric shock and intended for connecting together any of the following parts:

   (i)   Exposed-conductive-parts

   (ii)   Extraneous-conductive-parts

   (iii)   The main earthing terminal

   (iv)   Earth electrode(s)

   (v)   The earthed point of the source, or an artificial neutral.

Protective conductor current. Electric current appearing in a protective conductor, such as leakage current or electric current resulting from an insulation fault.

Protective earthing. Earthing of a point or points in a system or in an installation or in equipment for the purposes of safety.

Protective equipotential bonding. Equipotential bonding for the purposes of safety.

Protective multiple earthing (PME). An earthing arrangement, found in TN-C-S systems, in which the supply neutral conductor is used to connect the earthing conductor of an installation with Earth, in accordance with the Electricity Safety, Quality and Continuity Regulations 2002 (see also Figure 3.9).

Protective separation. Separation of one electric circuit from another by means of:
  (i) double insulation, or
  (ii) basic insulation and electrically protective screening (shielding), or
  (iii) reinforced insulation.

PV, {712}. Solar photovoltaic.

- **PV a.c. module**. Integrated module/convertor assembly where the electrical interface terminals are a.c. only. No access is provided to the d.c. side.

- **PV array.** Mechanically and electrically integrated assembly of PV modules, and other necessary components, to form a d.c. power supply unit.

- **PV array cable.** Output cable of a PV array.

- **PV array junction box.** Enclosure where PV strings of any PV array are electrically connected and where devices can be located.

- **PV cell.** Basic PV device which can generate electricity when exposed to light such as solar radiation.

- **PV convertor.** Device which converts d.c. voltage and d.c. current into a.c. voltage and a.c. current.

- **PV d.c. main cable.** Cable connecting the PV generator junction box to the d.c. terminals of the PV convertor.

- **PV generator.** Assembly of PV arrays.

- **PV generator junction box.** Enclosure where PV arrays are electrically connected and where devices can be located.

- **PV installation.** Erected equipment of a PV power supply system.

- **PV module.** Smallest completely environmentally protected assembly of interconnected PV cells.

- **PV string.** Circuit in which PV modules are connected in series, in order for a PV array to generate the required output voltage.

- **PV string cable.** Cable connecting PV modules to form a PV string.

- **PV supply cable.** Cable connecting the a.c. terminals of the PV convertor to a distribution circuit of the electrical installation.

Rated current. Value of current used for specification purposes, established for a specified set of operating conditions of a component, device, equipment or system.

Rated impulse withstand voltage level (U_W), {534}. The level of impulse withstand voltage assigned by the manufacturer to the equipment, or to part of it, characterizing the specified withstand capability of its insulation against overvoltages.
NOTE: For the purposes of BS 7671, only withstand voltage between live conductors and earth is considered.

Reduced low voltage system. A system in which the nominal line-to-line voltage does not exceed 110 volts and the nominal line to Earth voltage does not exceed 63.5 volts.

Reinforced insulation. Single insulation applied to live parts, which provides a degree of protection against electric shock equivalent to double insulation under the conditions specified in the relevant standard. The term 'single insulation' does not imply that the insulation must be one homogeneous piece. It may comprise two or more layers which cannot be tested singly as supplementary or basic insulation.

Reporting. Communicating the results of periodic inspection and testing of an electrical installation to the person ordering the work.

Residences and other locations belonging to agricultural and horticultural premises. Residences and other locations which have a conductive connection to the agricultural and horticultural premises by either protective conductors of the same installation or by extraneous-conductive-parts.
NOTE: Examples of other locations include offices, social rooms, machine-halls, workrooms, garages and shops.

Residential park home. A factory produced relocatable dwelling designed for permanent residence which may be used for leisure purposes.

Residual current. Algebraic sum of the currents in the live conductors of a circuit at a point in the electrical installation.

Residual current device (RCD). A mechanical switching device or association of devices intended to cause the opening of the contacts when the residual current attains a given value under specified conditions.

Residual current operated circuit-breaker with integral overcurrent protection (RCBO). A residual current operated switching device designed to perform the functions of protection against overload and/or short-circuit.

Residual current operated circuit-breaker without integral overcurrent protection (RCCB). A residual current operated switching device not designed to perform the functions of protection against overload and/or short-circuit.

Residual operating current. Residual current which causes the RCD to operate under specified conditions.

Resistance area (for an earth electrode only). The surface area of ground (around an earth electrode) on which a significant voltage gradient may exist.

Response time. The time that elapses between the failure of the normal power supply and the ability of the auxiliary power supply to energize the equipment.

Restrictive conductive location. *(see Conducting location with restricted movement).*

Ring final circuit. A final circuit arranged in the form of a ring and connected to a single point of supply.

Safety service. An electrical system for electrical equipment provided to protect or warn persons in the event of a hazard, or essential to their evacuation from a location.

Sauna. A room or location in which air is heated, in service, to high temperatures where the relative humidity is normally low, rising only for a short period of time when water is poured over the heater.

SELV (separated extra-low voltage). An extra-low voltage system which is electrically separated from Earth and from other systems in such a way that a single fault cannot give rise to the risk of electric shock.

Selectivity *(see Discrimination).*

Shock *(see Electric shock).*

Shock current. A current passing through the body of a person or livestock such as to cause electric shock and having characteristics likely to cause dangerous effects.

Short-circuit current. An overcurrent resulting from a fault of negligible impedance between live conductors having a difference in potential under normal operating conditions.

Short-circuit current under standard test conditions $I_{sc \, STC}$. Short-circuit current of a PV module, PV string, PV array or PV generator under standard test conditions.

Show. Display or presentation in any suitable location, either a room, building or temporary structure.

Simple separation. Separation between circuits or between a circuit and Earth by means of basic insulation.

Simultaneously accessible parts. Conductors or conductive parts which can be touched simultaneously by a person or, in locations specifically intended for them, by livestock.

NOTE: Simultaneously accessible parts may be: live parts, exposed-conductive-parts, extraneous-conductive-parts, protective conductors or earth electrodes.

Skilled person. A person with technical knowledge or sufficient experience to enable him/her to avoid dangers which electricity may create.

Socket-outlet. A device, provided with female contacts, which is intended to be installed with the fixed wiring, and intended to receive a plug. A luminaire track system is not regarded as a socket-outlet system.

Spur. A branch from a ring or radial final circuit.

Stand. Area or temporary structure used for display, marketing or sales.

Standard test conditions (STC). Test conditions specified in BS EN 60904-3 for PV cells and PV modules.

Standby electrical source. Electrical source intended to maintain, for reasons other than safety, the supply to an electrical installation or a part or parts thereof, in case of interruption of the normal supply.

Standby electrical supply system. Supply system intended to maintain, for reasons other than safety, the functioning of an electrical installation or a part or parts thereof, in case of interruption of the normal supply.

Static convertor. A convertor having no moving parts and notably using semiconductor rectifiers.

Stationary equipment. Electrical equipment which is either fixed or which has a mass exceeding 18 kg and is not provided with a carrying handle.

Street furniture. Fixed equipment located on a highway.

NOTE: Street furniture includes street located equipment.

Street located equipment. Fixed equipment, located on a highway, the purpose of which is not directly associated with the use of the highway.

Supplementary insulation. Independent insulation applied in addition to basic insulation for fault protection.

Supplier *(see Distributor).*

Surge current, {534}. A transient wave appearing as an overcurrent caused by a lightning electromagnetic impulse.

Surge protective device (SPD), {534}. A device that is intended to limit transient overvoltages and divert surge currents. It contains at least one non-linear component.

Switch. A mechanical device capable of making, carrying and breaking current under normal circuit conditions, which may include specified operating overload conditions, and also of carrying for a specified time currents under specified abnormal circuit conditions such as those of short-circuit. It may also be capable of making, but not breaking, short-circuit currents.

Switch, linked. A switch the contacts of which are so arranged as to make or break all poles simultaneously or in a definite sequence.

Switch-disconnector. A switch which, in the open position, satisfies the isolating requirements specified for a disconnector.

NOTE: A switch-disconnector is otherwise known as an isolating switch.

Switchboard. An assembly of switchgear with or without instruments, but the term does not apply to groups of local switches in final circuits.

Switchgear. An assembly of main and auxiliary switching equipment for operation, regulation, protection or other control of an electrical installation.

System. An electrical system consisting of a single source or multiple sources running in parallel of electrical energy and an installation. See Part 3. For certain purposes of the Regulations, types of system are identified as follows, depending upon the relationship of the source, and of exposed-conductive-parts of the installation, to Earth:

- **TN system.** A system having one or more points of the source of energy directly earthed, the exposed conductive-parts of the installation being connected to that point by protective conductors.

- **TN-C system.** A system in which neutral and protective functions are combined in a single conductor throughout the system.

- **TN-S system.** A system having separate neutral and protective conductors throughout the system (see Figure 3.8).

- **TN-C-S system.** A system in which neutral and protective functions are combined in a single conductor in part of the system (see Figure 3.9).

- **TT system.** A system having one point of the source of energy directly earthed, the exposed-conductive-parts of the installation being connected to earth electrodes electrically independent of the earth electrodes of the source (see Figure 3.10).

- **IT system.** A system having no direct connection between live parts and Earth, the exposed-conductive-parts of the electrical installation being earthed (see Appendix 9 Figure 9C).

- **Multiple source and d.c. systems** - see Appendix 9.

Temporary electrical installation. Electrical installation erected for a particular purpose and dismantled when no longer required for that purpose.

Temporary overvoltage (U_{TOV}), {534}. A fundamental frequency overvoltage occurring on the network at a given location, of relatively long duration.

NOTE 1: TOVs may be caused by faults inside the LV system ($U_{TOV,LV}$) or inside the HV system ($U_{TOV,HV}$)

NOTE 2: Temporary overvoltages, typically lasting up to several seconds, usually originate from switching operations or faults (for example, sudden load rejection, single-phase faults, etc.) and/or from non-linearity (ferroresonance effects, harmonics, etc.)

Temporary structure. A unit or part of a unit, including mobile portable units, situated indoors or outdoors, designed and intended to be assembled and dismantled.

Temporary supply unit. An enclosure containing equipment for the purpose of taking a temporary electrical supply safely from an item of street furniture.

Testing. Implementation of measures to assess an electrical installation by means of which its effectiveness is proved. This includes ascertaining values by means of appropriate measuring instruments, where measured values are not detectable by inspection.

Thermal storage floor heating system. Heating system in which, due to a limited charging period, a restricted availability of electrical energy is converted into heat and dissipated mainly through the surface of the floor to the room to be heated with an intended time delay.

Triplen harmonics. The odd multiples of the 3rd harmonic of the fundamental frequency (e.g. 3rd, 9th, 15th, 21st)

Trunking *(see Cable trunking).*

Verification. All measures by means of which compliance of the electrical installation with the relevant requirements of BS 7671 are checked, comprising inspection, testing and certification.

Voltage, nominal. Voltage by which an installation (or part of an installation) is designated. The following ranges of nominal voltage (rms values for a.c.) are defined:

- **Extra-low.** Not exceeding 50 V a.c. or 120 V ripple-free d.c., whether between conductors or to Earth.

- **Low.** Exceeding extra-low voltage but not exceeding 1000 V a.c. or 1500 V d.c. between conductors, or 600 V a.c. or 900 V d.c. between conductors and Earth.

- **High.** Normally exceeding low voltage.

NOTE: The actual voltage of the installation may differ from the nominal value by a quantity within normal tolerances, see Appendix 2.

Voltage, reduced *(see Reduced low voltage system).*

Voltage band

Band I
Band I covers:
- installations where protection against electric shock is provided under certain conditions by the value of voltage;
- installations where the voltage is limited for operational reasons (e.g. telecommunications, signalling, bell, control and alarm installations).

Extra-low voltage (ELV) will normally fall within voltage Band I.

Band II
Band II contains the voltages for supplies to household and most commercial and industrial installations. Low voltage (LV) will normally fall within voltage Band II.

NOTE: Band II voltages do not exceed 1000 V a.c. rms or 1500 V d.c.

Voltage protection level (U_p), {534}. A parameter that characterizes the performance of an SPD in limiting the voltage across its terminals, which is selected from a list of preferred values; this value is greater than the highest value of the measured limiting voltages.

Wiring system. An assembly made up of cable or busbars and parts which secure and, if necessary, enclose the cable or busbars.

Fig 2.1 – Illustration of earthing and protective conductor terms (see Chapter 54)

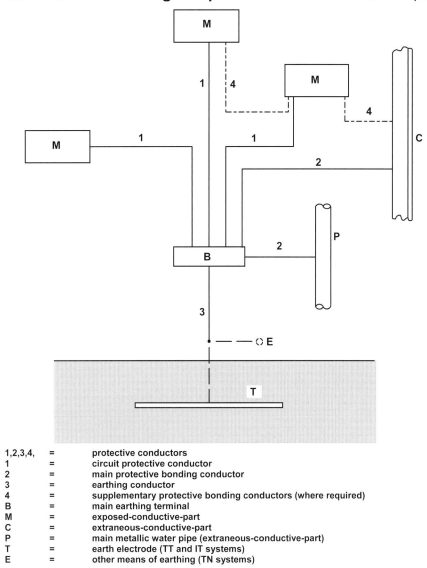

1,2,3,4,	=	protective conductors
1	=	circuit protective conductor
2	=	main protective bonding conductor
3	=	earthing conductor
4	=	supplementary protective bonding conductors (where required)
B	=	main earthing terminal
M	=	exposed-conductive-part
C	=	extraneous-conductive-part
P	=	main metallic water pipe (extraneous-conductive-part)
T	=	earth electrode (TT and IT systems)
E	=	other means of earthing (TN systems)

Figs 2.2 to 2.6   *Moved/deleted by BS 7671:2008 Amendment No 1 (see Regulation 312.2)*

SYMBOLS USED IN THE REGULATIONS
including first reference

C	rating factor - general		Appx 4 sec 3
C_a	rating factor for ambient temperature		Appx 4 sec 3
C_c	rating factor for circuits buried in the ground		Appx 4 sec 3
C_d	rating factor for depth of burial		Appx 4 sec 5.3
C_f	rating factor for semi-enclosed fuse to BS 3036		Appx 4 sec 5.3
C_g	rating factor for grouping		Appx 4 sec 3
C_h	rating factor for higher harmonic currents in line conductors		Appx 4 sec 5.6
C_i	rating factor for conductors embedded in thermal insulation		Appx 4 sec 3
C_s	rating factor for thermal soil resistivity		Appx 4 sec 5.3
C_t	rating factor for operating temperature of conductor		Appx 4 sec 6.1
c	battery capacity	Ah	A721.525
D_e	external cable diameter	mm	Appx 4, Table 4A2
d	conventional length of the supply line	km	443.2.4
d_1	the length of the low voltage overhead supply line to the structure	km	443.2.4
d_2	the length of the low voltage underground unscreened line to the structure	km	443.2.4
d_3	the length of the high voltage overhead supply line to the structure	km	443.2.4
d_c	critical length	km	443.2.4
ΣI_{zk}	the sum of the continuous current-carrying capacities of m conductors in parallel	A	Appx 10 sec 2
F	group rating factor		Appx 4 sec 2.3.3.1
gG	class 'gG' utilisation category of fuses to BS 88-2 - general use		411.4.6
gM	class 'gM' utilisation category of fuses to BS 88-2 motor circuit application		411.4.6
	frequency in cycles per second	Hz	110.12
I	current (general term)	A	
I_a	current causing automatic operation of protective device within the time stated	A	411.4.5
I_b	design current of circuit	A	433.1.1

I_{bh}	design current including the effect of third harmonic currents	A	Appx 10 sec 5.5.2
I_{bk}	design current for conductor k	A	Appx 10 sec 2
I_c	charging current	A	A721.525
I_{cw}	rated short-time withstand current	A	434.5.3
I_d	fault current of first fault (IT system)	A	411.6.2
I_E	part of the earth fault current in the high voltage system that flows through the earthing arrangement of the transformer substation	A	442.1.2
I_f	fault current	A	442.2.1
I_h	the fault current that flows through the earthing arrangement of the exposed-conductive-parts of the equipment of the low voltage installation during a period when there is a high voltage fault and a first fault in the low voltage installation	A	442.1.2
I_{h5}	5th harmonic current	A	Appx 4 sec 5.6
I_{hn}	nth harmonic current	A	Appx 4 sec 5.6
I_{imp}	impulse current, {534}	A	534.2.3.4
I_n	rated current or current setting of protective device	A	Table 41.3
$I_{\Delta n}$	rated residual operating current of the protective device in amperes	A	411.5.3
I_{nk}	rated current of the protective for conductor k	A	Appx 10 sec 2
I_{nspd}	nominal discharge current, {534}	A	534.2.3.4
I_{pf}	prospective fault current	A	Appx 6
I_{sc}	short-circuit current	A	712.433.1
$I_{sc\,STC}$	short-circuit current under standard test conditions	A	712.433.1
I_t	tabulated current-carrying capacity of a cable	A	Appx 4 sec 3
I_z	current-carrying capacity of a cable for continuous service under the particular installation conditions concerned	A	433.1.1
I_{zk}	the continuous current-carrying capacity of conductor k	A	Appx 10 sec 2
I^2t	energy let-through value of device	A^2s	434.5.2
I_2	current causing effective operation of the overload protective device	A	433.1.1

Symbol	Description	Units	Reference
λ	thermal conductivity	$Wm^{-1}K^{-1}$	523.9
k	material factor taken from Table 43.1	$As^{1/2}mm^{-2}$	434.5.2
$k^2 S^2$	energy withstand of cable	A^2s	434.5.2
k_g	reduction factor based on the ratio of strikes between overhead lines and underground screened cables		443.2.4
k_t	typical reduction factor for a transformer		443.2.4
$(mV/A/m)_r$	resistive voltage drop per ampere per metre	$mVA^{-1}m^{-1}$	Appx 4 sec 6
$(mV/A/m)_x$	reactive voltage drop per ampere per metre	$mVA^{-1}m^{-1}$	Appx 4 sec 6
$(mV/A/m)_z$	impedance voltage drop per ampere per metre	$mVA^{-1}m^{-1}$	Appx 4 sec 6
n	number of circuits in a group		Appx 4 sec 2.3.3.1
R	resistance of supplementary bonding conductor	Ω	415.2.2
R_A	the sum of the resistances of the earth electrode and the protective conductor connecting it to the exposed-conductive-parts	Ω	411.5.3
	also defined as "the resistance of the earthing arrangement of the exposed-conductive-parts of the equipment of the low voltage installation"		442.1.2
R_B	resistance of the earthing arrangement of the low voltage system neutral, for low voltage systems in which the earthing arrangements of the transformer substation and of the low voltage system neutral are electrically independent	Ω	442.1.2
R_E	resistance of the earthing arrangement of the transformer substation	Ω	442.1.2
R_f	resistance of the fault	Ω	442.2.1
R_1	resistance of line conductor of a distribution or final circuit	Ω	Appx 6 Schedule of test results
R_2	resistance of circuit protective conductor (cpc) of a distribution or final circuit.	Ω	Appx 6 Schedule of test results
S	size (nominal cross-sectional area of the conductor)	mm^2	543.1.3
$S_1,...S_m$	cross-sectional area of parallel conductors	mm^2	Appx 10 sec 2
S_k	cross-sectional area of conductor k	mm^2	Appx 10 sec 2
t	time (period)	h	434.5.2
t_p	maximum permitted normal operating conductor temperature	$°C$	Appx 4 sec 6.1
U	voltage between lines	V	411.6.4
U_c	continuous operating voltage, {534}	V	534.2.3.2
U_f	power frequency fault voltage that appears in the low voltage system between exposed-conductive-parts and earth for the duration of the fault	V	442.1.2
U_{oc}	open-circuit voltage	V	Table 16A
$U_{oc\,STC}$	open-circuit voltage under standard test conditions	V	712.44.1.1
U_p	voltage protection level, {534}	V	534.2.3.1
U_{TOV}	temporary overvoltage, {534}	V	Part 2 TOV
U_w	rated impulse withstand voltage level, {534}	V	Table 44.3
U_x	voltage at test electrode to Earth (when measuring insulation resistance of floors and walls)	V	Appx 13 sec 2
U_0	nominal a.c. rms line voltage to Earth	V	Table 41.1
$U_{\alpha\,spd}$	nominal a.c. rms line voltage of the low voltage system to Earth, {534}	V	Table 53.3
U_1	power frequency stress voltage between the line conductor and the exposed-conductive-parts of the low voltage equipment of the transformer substation during the fault	V	442.1.2
U_2	power frequency stress voltage between the line conductor and the exposed-conductive-parts of the low voltage equipment of the low voltage installation during the fault	V	442.1.2
Z	the impedance between the low voltage system and an earthing arrangement	Ω	442.1.2
Z_1	the impedance of parallel conductor l	Ω	Appx 10 sec 2
Z_e	that part of the earth fault loop impedance which is external to the installation	Ω	313.1
Z_k	the impedance of conductor k	Ω	Appx 10 sec 2
Z_m	the impedance of parallel conductor m	Ω	Appx 10 sec 2
Z_s	earth fault loop impedance	Ω	411.4.5
Z_x	impedance of floor insulation	Ω	Appx 13 sec 2
Z^1_s	neutral-earth loop impedance (IT systems with distributed neutral only)	Ω	411.6.4
$\emptyset$	phase angle		Appx 4 sec 6.2
$\cos\emptyset$	power factor (sinusoidal)		Appx 4 sec 6.2

PART 3

ASSESSMENT OF GENERAL CHARACTERISTICS

CONTENTS

PART 3

ASSESSMENT OF GENERAL CHARACTERISTICS

CHAPTER 30

301   ASSESSMENT OF GENERAL CHARACTERISTICS

301.1   An assessment shall be made of the following characteristics of the installation in accordance with the chapters indicated:

   (i)   The purpose for which the installation is intended to be used, its general structure and its supplies (Chapter 31)

   (ii)   The external influences to which it is to be exposed (Chapter 32)

   (iii) The compatibility of its equipment (Chapter 33)

   (iv)   Its maintainability (Chapter 34)

   (v)   Recognized safety services (Chapter 35)

   (vi)   Assessment for continuity of service (Chapter 36).

These characteristics shall be taken into account in the choice of methods of protection for safety (see Chapters 41 to 44) and the selection and erection of equipment (Chapters 51 to 56).

CHAPTER 31

PURPOSE, SUPPLIES AND STRUCTURE

311   MAXIMUM DEMAND AND DIVERSITY

311.1   For economic and reliable design of an installation within thermal limits and admissible voltage drop, the maximum demand shall be determined. In determining the maximum demand of an installation or part thereof, diversity may be taken into account.

312   CONDUCTOR ARRANGEMENT AND SYSTEM EARTHING

The following characteristics shall be assessed:

   (i)   Arrangement of current-carrying conductors under normal operating conditions

   (ii)   Type of system earthing.

312.1   General

The following arrangements of current-carrying conductors under normal operating conditions are taken into account in this Standard:

312.1.1   Current-carrying conductors in a.c. circuits

Fig 3.1 – Single-phase 2-wire

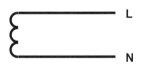

Fig 3.2 – Single-phase 3-wire

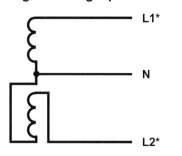

Phase angle 0°
* Numbering of conductors optional

Fig 3.3 – Two-phase 3-wire

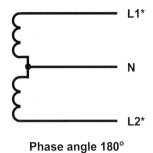 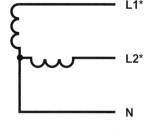 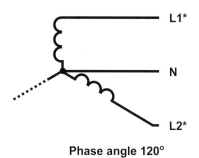

Phase angle 180°   **Phase angle 90°**   **Phase angle 120°**

* Numbering of conductors optional

Fig 3.4 – Three-phase 3-wire

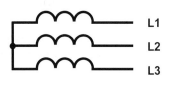 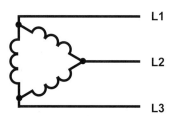

Star connection   **Delta connection**

Fig 3.5 – Three-phase 4-wire

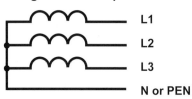

Three-phase, 4-wire with neutral conductor or PEN conductor. By definition, the PEN conductor is not a live conductor but a conductor carrying an operating current.

NOTE 1: In the case of a single-phase 2-wire arrangement which is derived from a three-phase 4-wire arrangement, the two conductors are either two line conductors or a line conductor and a neutral conductor or a line conductor and a PEN conductor.

NOTE 2: In installations with all loads connected between phases, the installation of the neutral conductor may not be necessary.

312.1.2   Current-carrying conductors in d.c. circuits

Fig 3.6 – 2-wire

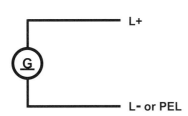

Fig 3.7 – 3-wire

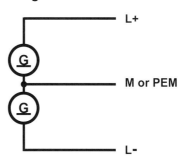

NOTE: PEL and PEM conductors are not live conductors although they carry operating current. Therefore, the designation 2-wire arrangement or 3-wire arrangement applies.

312.2   Types of system earthing

The following types of system earthing are taken into account in this Standard.

NOTE 1: Figures 3.8 to 10 show examples of commonly used three-phase systems. For IT, multiple source, d.c. and other systems see Appendix 9.

NOTE 2: For private systems, the source and/or the distribution system may be considered as part of the installation within the meaning of this standard.

NOTE 3: The codes used have the following meanings:

First letter – Relationship of the power system to Earth:

T = direct connection of one point to Earth;

I = all live parts isolated from Earth, or one point connected to Earth through a high impedance.

Second letter – Relationship of the exposed-conductive-parts of the installation to Earth:

T = direct electrical connection of exposed-conductive-parts to Earth, independently of the earthing of any point of the power system;

N = direct electrical connection of the exposed-conductive-parts to the earthed point of the power system (in a.c. systems, the earthed point of the power system is normally the neutral point or, if a neutral point is not available, a line conductor).

Subsequent letter(s) (if any) – Arrangement of neutral and protective conductors:

S = protective function provided by a conductor separate from the neutral conductor or from the earthed line (or, in a.c. systems, earthed phase) conductor.

C = neutral and protective functions combined in a single conductor (PEN conductor).

312.2.1   TN systems

312.2.1.1   Single-source systems

TN systems have one point directly earthed at the source, the exposed-conductive parts of the installation(s) being connected to that point by protective conductors. Two types of TN system are considered according to the arrangement of neutral and protective conductors, as follows:

Fig 3.8 – TN-S system

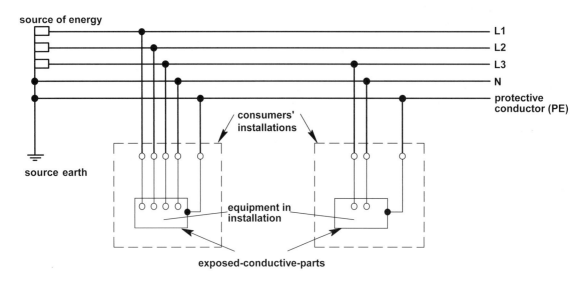

Separate neutral and protective conductors throughout the system.

The protective conductor (PE) is the metallic covering of the cable supplying the installations or a separate conductor.

All exposed-conductive-parts of an installation are connected to this protective conductor via the main earthing terminal of the installation.

Fig 3.9 – TN-C-S (PME) system

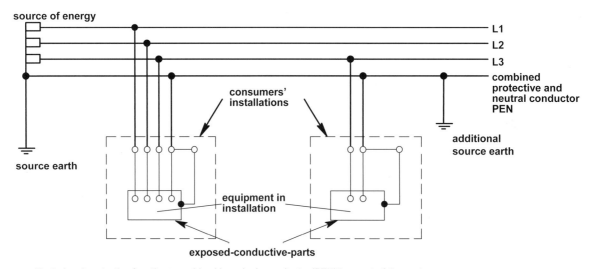

Neutral and protective functions combined in a single conductor (PEN) in a part of the system.

The usual form of a TN-C-S system is as shown, where the supply is TN-C and the arrangement in the installations is TN-S.

This type of distribution is known also as protective multiple earthing (PME).

The supply system PEN conductor is earthed at two or more points and an earth electrode may be necessary at or near a consumer's installation.

All exposed-conductive-parts of an installation are connected to the PEN conductor via the main earthing terminal and the neutral terminal, these terminals being linked together.

312.2.2   TT system

312.2.2.1   Single-source system

A TT system has only one point directly earthed at the source, the exposed-conductive-parts of the installation(s) being connected to earth electrodes electrically independent of the earth electrode of the supply system (the source earth).

Fig 3.10 – TT system

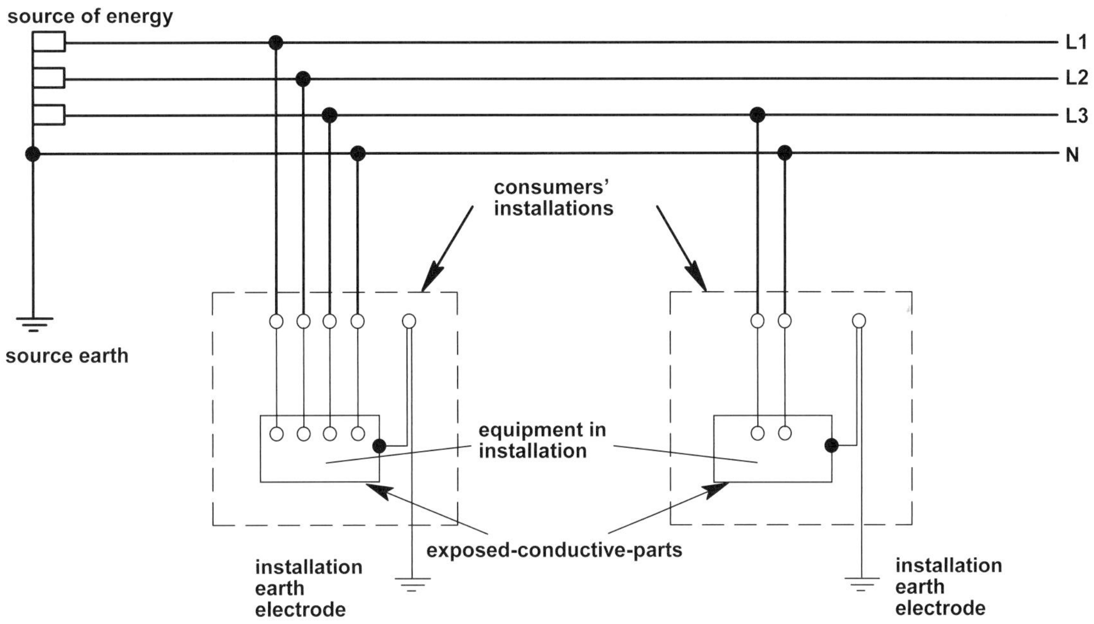

All exposed-conductive-parts of an installation are connected to an earth electrode which is electrically independent of the source earth.

Separate neutral and protective conductors throughout the system.

NOTE: Additional earthing of the PE in the installation may be provided.

312.3   *Deleted by BS 7671:2008 Amendment No 1*

312.4   IT, multiple source, d.c. and other systems

See Appendix 9.

313   SUPPLIES

313.1   General

The following characteristics of the supply or supplies, from whatever source, and the normal range of those characteristics where appropriate, shall be determined by calculation, measurement, enquiry or inspection:

  (i)   The nominal voltage(s) and its characteristics including harmonic distortion

  (ii)   The nature of the current and frequency

  (iii)   The prospective short-circuit current at the origin of the installation

  (iv)   The earth fault loop impedance of that part of the system external to the installation, Z_e

  (v)   The suitability for the requirements of the installation, including the maximum demand

  (vi)   The type and rating of the overcurrent protective device(s) acting at the origin of the installation.

These characteristics shall be ascertained for an external supply and shall be determined for a private source. These requirements are equally applicable to main supplies and to safety services and standby supplies.

NOTE: The above information should be provided by distributors on request (see Appendix 2 sec 2).

313.2 Supplies for safety services and standby systems

Where the provision of safety services is required, for example, by the authorities concerned with fire precautions and other conditions for emergency evacuation of the premises, and/or where the provision of standby supplies is required by the person specifying the installation, the characteristics of the source or sources of supply for safety services and/or standby systems shall be separately assessed. Such supplies shall have adequate capacity, reliability and rating and appropriate changeover time for the operation specified.

NOTE 1: For further requirements for supplies for safety services, see Chapter 35 hereafter and Chapter 56.

NOTE 2: For standby systems, there are no particular requirements in these Regulations.

314 DIVISION OF INSTALLATION

314.1 Every installation shall be divided into circuits, as necessary, to:

  (i)   avoid danger and minimize inconvenience in the event of a fault

  (ii) facilitate safe inspection, testing and maintenance (see also Section 537)

  (iii) take account of hazards that may arise from the failure of a single circuit such as a lighting circuit

  (iv) reduce the possibility of unwanted tripping of RCDs due to excessive protective conductor (PE) currents not due to a fault

  (v)   mitigate the effects of electromagnetic disturbances (see also Chapter 44)

  (vi) prevent the indirect energizing of a circuit intended to be isolated.

314.2 Separate circuits shall be provided for parts of the installation which need to be separately controlled, in such a way that those circuits are not affected by the failure of other circuits, and due account shall be taken of the consequences of the operation of any single protective device.

314.3 The number of final circuits required, and the number of points supplied by any final circuit, shall be such as to facilitate compliance with the requirements of Chapter 43 for overcurrent protection, Section 537 for isolation and switching and Chapter 52 as regards current-carrying capacities of conductors.

314.4 Where an installation comprises more than one final circuit, each final circuit shall be connected to a separate way in a distribution board. The wiring of each final circuit shall be electrically separate from that of every other final circuit, so as to prevent the indirect energizing of a final circuit intended to be isolated.

CHAPTER 32

CLASSIFICATION OF EXTERNAL INFLUENCES

Refer to Chapter 51 and Appendix 5.

CHAPTER 33

COMPATIBILITY

331   COMPATIBILITY OF CHARACTERISTICS

331.1   An assessment shall be made of any characteristics of equipment likely to have harmful effects upon other electrical equipment or other services or likely to impair the supply, for example, for co-ordination with concerned parties e.g. petrol stations, kiosks and shops within shops. Those characteristics include, for example:

   (i)   transient overvoltages

   (ii)   undervoltage

   (iii)   unbalanced loads

   (iv)   rapidly fluctuating loads

   (v)   starting currents

   (vi)   harmonic currents

   (vii)   earth leakage current

 (viii)   excessive PE conductor current not due to a fault

   (ix)   d.c. feedback

   (x)   high-frequency oscillations

   (xi)   necessity for additional connections to Earth

   (xii)   power factor.

For an external source of energy the distributor shall be consulted regarding any equipment of the installation having a characteristic likely to have significant influence on the supply.

332   ELECTROMAGNETIC COMPATIBILITY

332.1   All electrical installations and equipment shall be in accordance with the EMC regulations and with the relevant EMC standard.

332.2   Consideration shall be given by the designer of the electrical installation to measures reducing the effect of induced voltage disturbances and electromagnetic interferences (EMI). Measures are given in Chapter 44.

CHAPTER 34

MAINTAINABILITY

341   GENERAL

341.1   An assessment shall be made of the frequency and quality of maintenance the installation can reasonably be expected to receive during its intended life. The person or body responsible for the operation and/or maintenance of the installation shall be consulted. Those characteristics are to be taken into account in applying the requirements of Parts 4 to 7 so that, having regard to the frequency and quality of maintenance expected:

   (i)   any periodic inspection and testing, maintenance and repairs likely to be necessary during the intended life can be readily and safely carried out, and

   (ii)   the effectiveness of the protective measures for safety during the intended life shall not diminish, and

   (iii)   the reliability of equipment for proper functioning of the installation is appropriate to the intended life.

NOTE: There may be particular statutory requirements relating to maintenance.

CHAPTER 35

SAFETY SERVICES

351   GENERAL

NOTE 1: The need for safety services and their nature are frequently regulated by statutory authorities whose requirements have to be observed.

NOTE 2: Examples of safety services are: emergency escape lighting, fire detection and fire alarm systems, installations for fire pumps, fire rescue service lifts, smoke and heat extraction equipment.

351.1   The following electrical sources for safety services are recognized:

(i)   Storage batteries

(ii) Primary cells

(iii) Generator sets independent of the normal supply

(iv) A separate feeder of the supply network that is effectively independent of the normal feeder (see Regulation 560.6.5).

352   CLASSIFICATION

Refer to Regulation 560.4.

CHAPTER 36

CONTINUITY OF SERVICE

361   GENERAL

361.1   An assessment shall be made for each circuit of any need for continuity of service considered necessary during the intended life of the installation e.g. life-support systems. The following characteristics shall be considered:

(i)   Selection of the system earthing

(ii) Selection of the protective device in order to achieve discrimination

(iii) Number of circuits

(iv) Multiple power supplies

(v)   Use of monitoring devices.

PART 4

PROTECTION FOR SAFETY

CONTENTS

CHAPTER 41

PROTECTION AGAINST ELECTRIC SHOCK

CONTENTS

CHAPTER 41
PROTECTION AGAINST ELECTRIC SHOCK

410 INTRODUCTION

This chapter deals with protection against electric shock as applied to electrical installations. It is based on BS EN 61140, which is a basic safety standard that applies to the protection of persons and livestock. BS EN 61140 is intended to give fundamental principles and requirements that are common to electrical installations and equipment or are necessary for their co-ordination.

The fundamental rule of protection against electric shock, according to BS EN 61140, is that hazardous-live-parts shall not be accessible and accessible conductive parts shall not be hazardous-live, either in use without a fault or in single fault conditions.

According to 4.2 of BS EN 61140, protection under normal conditions is provided by basic protective provisions and protection under single fault conditions is provided by fault protective provisions. Alternatively, protection against electric shock is provided by an enhanced protective provision which provides protection in use without a fault and under single fault conditions.

In BS 7671:2001:

  (i) protection in use without a fault (now designated basic protection) was referred to as protection against direct contact, and

  (ii) protection under fault conditions (now designated fault protection) was referred to as protection against indirect contact.

410.1 Scope

Chapter 41 specifies essential requirements regarding protection against electric shock, including basic protection and fault protection of persons and livestock. It deals also with the application and co-ordination of these requirements in relation to external influences.

Requirements are given for the application of additional protection in certain cases.

410.2 Not used

410.3 General requirements

410.3.1   The following specification of voltage is intended unless stated otherwise:
  – a.c. voltages are rms
  – d.c. voltages are ripple-free.

410.3.2   A protective measure shall consist of:
  (i) an appropriate combination of a provision for basic protection and an independent provision for fault protection, or
  (ii) an enhanced protective provision which provides both basic protection and fault protection.

Additional protection is specified as part of a protective measure under certain conditions of external influence and in certain special locations (see the corresponding Sections of Part 7).

NOTE 1: For special applications, protective measures which do not follow this concept are permitted (see Regulations 410.3.5 and 410.3.6).

NOTE 2: An example of an enhanced protective measure is reinforced insulation.

410.3.3   In each part of an installation one or more protective measures shall be applied, taking account of the conditions of external influence.

The following protective measures generally are permitted:
  (i)   Automatic disconnection of supply (Section 411)
  (ii) Double or reinforced insulation (Section 412)
  (iii) Electrical separation for the supply to one item of current-using equipment (Section 413)
  (iv) Extra-low voltage (SELV and PELV) (Section 414).

The protective measures applied in the installation shall be considered in the selection and erection of equipment.

For particular installations see Regulations 410.3.4 to 9.

NOTE: In electrical installations the most commonly used protective measure is automatic disconnection of supply.

410.3.4 For special installations or locations, the additional protective measures specified in the corresponding section of Part 7 shall be applied.

410.3.5 The protective measures specified in Section 417, i.e. the use of obstacles and placing out of reach, shall be used only in installations where access is restricted to:

(i) skilled persons, or

(ii) instructed persons under the supervision of skilled persons.

410.3.6 The protective measures specified in Section 418, i.e.

(i) non-conducting location

(ii) earth-free local equipotential bonding

(iii) electrical separation for the supply to more than one item of current-using equipment

shall be applied only where the installation is under the supervision of skilled or instructed persons so that unauthorized changes cannot be made.

410.3.7 If certain conditions of a protective measure cannot be met, supplementary provisions shall be applied so that the protective provisions together achieve the same degree of safety.

NOTE: An example of the application of this regulation is given in Regulation 411.7 (FELV).

410.3.8 Different protective measures applied to the same installation or part of an installation or within equipment shall have no influence on each other such that failure of one protective measure could impair the other protective measure or measures.

410.3.9 The provision for fault protection may be omitted for the following equipment:

(i) Metal supports of overhead line insulators which are attached to the building and are placed out of arm's reach

(ii) Steel reinforced concrete poles of overhead lines in which the steel reinforcement is not accessible

(iii) Exposed-conductive-parts which, owing to their reduced dimensions (approximate maximum of 50 mm x 50 mm) or their disposition, cannot be gripped or come into significant contact with a part of the human body and provided that connection with a protective conductor could only be made with difficulty or would be unreliable

   NOTE: This exemption applies, for example, to bolts, rivets, nameplates, cable clips, screws and other fixings.

(iv) Metal enclosures protecting equipment in accordance with Section 412

(v) Unearthed street furniture supplied from an overhead line and inaccessible in normal use.

411   PROTECTIVE MEASURE: AUTOMATIC DISCONNECTION OF SUPPLY

411.1   General

Automatic disconnection of supply is a protective measure in which:

(i) basic protection is provided by basic insulation of live parts or by barriers or enclosures, in accordance with Section 416, and

(ii) fault protection is provided by protective earthing, protective equipotential bonding and automatic disconnection in case of a fault, in accordance with Regulations 411.3 to 6.

Where this protective measure is applied, Class II equipment may also be used.

Where specified, the requirements for additional protection shall be provided by an RCD having the characteristics specified in Regulation 415.1.1.

NOTE: A residual current monitor (RCM) is not a protective device but it may be used to monitor residual currents in an electrical installation. An RCM produces an audible or audible and visual signal when a preselected value of residual current is reached.

411.2   Requirements for basic protection

All electrical equipment shall comply with one of the provisions for basic protection described in Section 416 (basic insulation; barriers or enclosures) or, where appropriate, Section 417 (obstacles; placing out of reach).

411.3 Requirements for fault protection

411.3.1 Protective earthing and protective equipotential bonding

411.3.1.1 Protective earthing

Exposed-conductive-parts shall be connected to a protective conductor under the specific conditions for each type of system earthing as specified in Regulations 411.4 to 6.

Simultaneously accessible exposed-conductive-parts shall be connected to the same earthing system individually, in groups or collectively.

Conductors for protective earthing shall comply with Chapter 54.

A circuit protective conductor shall be run to and terminated at each point in wiring and at each accessory except a lampholder having no exposed-conductive-parts and suspended from such a point.

411.3.1.2 Protective equipotential bonding

In each installation main protective bonding conductors complying with Chapter 54 shall connect to the main earthing terminal extraneous-conductive-parts including the following:
- (i) Water installation pipes
- (ii) Gas installation pipes
- (iii) Other installation pipework and ducting
- (iv) Central heating and air conditioning systems
- (v) Exposed metallic structural parts of the building.

Connection of a lightning protection system to the protective equipotential bonding shall be made in accordance with BS EN 62305.

Where an installation serves more than one building the above requirement shall be applied to each building.

To comply with the requirements of these Regulations it is also necessary to apply equipotential bonding to any metallic sheath of a telecommunication cable. However, the consent of the owner or operator of the cable shall be obtained.

411.3.2 Automatic disconnection in case of a fault

411.3.2.1 Except as provided by Regulations 411.3.2.5 and 411.3.2.6, a protective device shall automatically interrupt the supply to the line conductor of a circuit or equipment in the event of a fault of negligible impedance between the line conductor and an exposed-conductive-part or a protective conductor in the circuit or equipment within the disconnection time required by Regulation 411.3.2.2, 411.3.2.3 or 411.3.2.4.

411.3.2.2 The maximum disconnection time stated in Table 41.1 shall be applied to final circuits not exceeding 32 A.

TABLE 41.1 –
Maximum disconnection times

System	$50 V < U_0 \leq 120 V$ seconds		$120 V < U_0 \leq 230 V$ seconds		$230 V < U_0 \leq 400 V$ seconds		$U_0 > 400 V$ seconds	
	a.c.	d.c.	a.c.	d.c.	a.c.	d.c.	a.c.	d.c.
TN	0.8	NOTE	0.4	5	0.2	0.4	0.1	0.1
TT	0.3	NOTE	0.2	0.4	0.07	0.2	0.04	0.1

Where, in a TT system, disconnection is achieved by an overcurrent protective device and protective equipotential bonding is connected to all the extraneous-conductive-parts within the installation in accordance with Regulation 411.3.1.2, the maximum disconnection times applicable to a TN system may be used.

U_0 is the nominal a.c. rms or d.c. line voltage to Earth.

Where compliance with this regulation is provided by an RCD, the disconnection times in accordance with Table 41.1 relate to prospective residual fault currents significantly higher than the rated residual operating current of the RCD.

NOTE: Disconnection is not required for protection against electric shock but may be required for other reasons, such as protection against thermal effects.

411.3.2.3   In a TN system, a disconnection time not exceeding 5 s is permitted for a distribution circuit and for a circuit not covered by Regulation 411.3.2.2.

411.3.2.4   In a TT system, a disconnection time not exceeding 1 s is permitted for a distribution circuit and for a circuit not covered by Regulation 411.3.2.2.

411.3.2.5   For a system with a nominal voltage U_0 greater than 50 V a.c. or 120 V d.c., automatic disconnection in the time required by Regulation 411.3.2.2, 411.3.2.3 or 411.3.2.4, as appropriate, is not required if, in the event of a fault to a protective conductor or Earth, the output voltage of the source is reduced in not more than that time to 50 V a.c. or 120 V d.c. or less. In such a case consideration shall be given to disconnection as required for reasons other than electric shock.

411.3.2.6   Where automatic disconnection according to Regulation 411.3.2.1 cannot be achieved in the time required by Regulation 411.3.2.2, 411.3.2.3 or 411.3.2.4, as appropriate, supplementary equipotential bonding shall be provided in accordance with Regulation 415.2.

411.3.3   Additional protection

In a.c. systems, additional protection by means of an RCD in accordance with Regulation 415.1 shall be provided for:

  (i)   socket-outlets with a rated current not exceeding 20 A that are for use by ordinary persons and are intended for general use, and

  (ii) mobile equipment with a current rating not exceeding 32 A for use outdoors.

An exception to (i) is permitted for:

  (a)   socket-outlets for use under the supervision of skilled or instructed persons, or

  (b)   a specific labelled or otherwise suitably identified socket-outlet provided for connection of a particular item of equipment.

NOTE 1: See also Regulations 314.1(iv) and 531.2.4 concerning the avoidance of unwanted tripping.

NOTE 2: The requirements of Regulation 411.3.3 do not apply to FELV systems according to Regulation 411.7 or reduced low voltage systems according to Regulation 411.8.

411.4   TN system

411.4.1   In a TN system, the integrity of the earthing of the installation depends on the reliable and effective connection of the PEN or PE conductors to Earth. Where the earthing is provided from a public or other supply system, compliance with the necessary conditions external to the installation is the responsibility of the distributor.

411.4.2   The neutral point or the midpoint of the power supply system shall be earthed. If a neutral point or midpoint is not available or not accessible, a line conductor shall be connected to Earth.

Each exposed-conductive-part of the installation shall be connected by a protective conductor to the main earthing terminal of the installation, which shall be connected to the earthed point of the power supply system.

NOTE: The PE and PEN conductors may additionally be connected to Earth, such as at the point of entry into the building.

411.4.3   In a fixed installation, a single conductor may serve both as a protective conductor and as a neutral conductor (PEN conductor) provided that the requirements of Regulation 543.4 are satisfied.

NOTE: In Great Britain, Regulation 8(4) of the Electricity Safety, Quality and Continuity Regulations 2002 prohibits the use of PEN conductors in consumers' installations.

411.4.4   The following types of protective device may be used for fault protection:

  (i)   An overcurrent protective device

  (ii) An RCD.

NOTE 1: Where an RCD is used for fault protection the circuit should also incorporate an overcurrent protective device in accordance with Chapter 43.

An RCD shall not be used in a TN-C system.

Where an RCD is used in a TN-C-S system, a PEN conductor shall not be used on the load side. The connection of the protective conductor to the PEN conductor shall be made on the source side of the RCD.

NOTE 2: Where discrimination between RCDs is necessary, see Regulation 531.2.9.

411.4.5   The characteristics of the protective devices (see Regulation 411.4.4) and the circuit impedances shall fulfil the following requirement:

$$Z_s \times I_a \leq U_0$$

where:

Z_s is the impedance in ohms (Ω) of the fault loop comprising:
- the source
- the line conductor up to the point of the fault, and
- the protective conductor between the point of the fault and the source.

I_a is the current in amperes (A) causing the automatic operation of the disconnecting device within the time specified in Table 41.1 of Regulation 411.3.2.2 or, as appropriate, Regulation 411.3.2.3.Where an RCD is used this current is the rated residual operating current providing disconnection in the time specified in Table 41.1 or Regulation 411.3.2.3.

U_0 is the nominal a.c. rms or d.c. line voltage to Earth in volts (V).

NOTE: Where compliance with this regulation is provided by an RCD, the disconnection times in accordance with Table 41.1 relate to prospective residual fault currents significantly higher than the rated residual operating current of the RCD.

411.4.6   Where a fuse is used to satisfy the requirements of Regulation 411.3.2.2, maximum values of earth fault loop impedance (Z_s) corresponding to a disconnection time of 0.4 s are stated in Table 41.2 for a nominal voltage (U_0) of 230 V. For types and rated currents of general purpose (gG) and motor circuit application (gM) fuses other than those mentioned in Table 41.2, reference should be made to the appropriate British or Harmonized Standard to determine the value of I_a for compliance with Regulation 411.4.5.

TABLE 41.2 –
Maximum earth fault loop impedance (Z_s) for fuses, for 0.4 s disconnection time with U_0 of 230 V
(see Regulation 411.4.6)

(a) General purpose (gG) and motor circuit application (gM) fuses to BS 88-2 – fuse systems E (bolted) and G (clip in)

Rating (amperes)	6	10	16	20	25	32
Z_s (ohms)	8.21	4.89	2.56	1.77	1.35	1.04

(b) Fuses to BS 88-3 fuse system C

Rating (amperes)	5	16	20	32
Z_s (ohms)	10.45	2.42	2.04	0.96

(c) Fuses to BS 3036

Rating (amperes)	5	15	20	30
Z_s (ohms)	9.58	2.55	1.77	1.09

(d) Fuses to BS 1362

Rating (amperes)	3	13
Z_s (ohms)	16.4	2.42

NOTE: The circuit loop impedances given in the table should not be exceeded when the conductors are at their normal operating temperature. If the conductors are at a different temperature when tested, the reading should be adjusted accordingly. See Appendix 14.

411.4.7   Where a circuit-breaker is used to satisfy the requirements of Regulation 411.3.2.2 or Regulation 411.3.2.3, the maximum value of earth fault loop impedance (Z_s) shall be determined by the formula in Regulation 411.4.5. Alternatively, for a nominal voltage (U_0) of 230 V and a disconnection time of 0.4 s in accordance with Regulation 411.3.2.2 or 5 s in accordance with Regulation 411.3.2.3, the values specified in Table 41.3 for the types and ratings of overcurrent devices listed may be used instead of calculation.

TABLE 41.3 –
Maximum earth fault loop impedance (Z$_S$) for circuit-breakers with U$_0$ of 230 V, for instantaneous operation giving compliance with the 0.4 s disconnection time of Regulation 411.3.2.2 and 5 s disconnection time of Regulation 411.3.2.3
(for RCBOs see also Regulation 411.4.9)

(a) Type B circuit-breakers to BS EN 60898 and the overcurrent characteristics of RCBOs to BS EN 61009-1

Rating (amperes)	3	6	10	16	20	25	32	40	50	63	80	100	125	I$_n$
Z$_S$ (ohms)	15.33	7.67	4.60	2.87	2.30	1.84	1.44	1.15	0.92	0.73	0.57	0.46	0.37	46/I$_n$

(b) Type C circuit-breakers to BS EN 60898 and the overcurrent characteristics of RCBOs to BS EN 61009-1

Rating (amperes)	6	10	16	20	25	32	40	50	63	80	100	125	I$_n$
Z$_S$ (ohms)	3.83	2.30	1.44	1.15	0.92	0.72	0.57	0.46	0.36	0.29	0.23	0.18	23/I$_n$

(c) Type D circuit-breakers to BS EN 60898 and the overcurrent characteristics of RCBOs to BS EN 61009-1

Rating (amperes)	6	10	16	20	25	32	40	50	63	80	100	125	I$_n$
Z$_S$ (ohms)	1.92	1.15	0.72	0.57	0.46	0.36	0.29	0.23	0.18	0.14	0.11	0.09	11.5/I$_n$

NOTE: The circuit loop impedances given in the table should not be exceeded when the conductors are at their normal operating temperature. If the conductors are at a different temperature when tested, the reading should be adjusted accordingly. See Appendix 14.

411.4.8 Where a fuse is used for a distribution circuit or a final circuit in accordance with Regulation 411.3.2.3, maximum values of earth fault loop impedance (Z$_s$) corresponding to a disconnection time of 5 s are stated in Table 41.4 for a nominal voltage (U$_0$) of 230 V. For types and rated currents of general purpose (gG) and motor circuit application (gM) fuses other than those mentioned in Table 41.4, reference should be made to the appropriate British or Harmonized Standard to determine the value of I$_a$ for compliance with Regulation 411.4.5.

TABLE 41.4 –
Maximum earth fault loop impedance (Z$_S$) for fuses, for 5 s disconnection time with U$_0$ of 230 V
(see Regulation 411.4.8)

(a) General purpose (gG) and motor circuit application (gM) fuses to BS 88-2 – fuse systems E (bolted) and G (clip in)

Rating (amperes)	6	10	16	20	25	32	40	50
Z$_S$ (ohms)	12.8	7.19	4.18	2.95	2.30	1.84	1.35	1.04

Rating (amperes)	63	80	100	125	160	200
Z$_S$ (ohms)	0.82	0.57	0.46	0.34	0.28	0.19

(b) Fuses to BS 88-3 fuse system C

Rating (amperes)	5	16	20	32	45	63	80	100
Z$_S$ (ohms)	15.33	4.11	3.38	1.64	1.04	0.72	0.53	0.40

(c) Fuses to BS 3036

Rating (amperes)	5	15	20	30	45	60	100
Z$_S$ (ohms)	17.7	5.35	3.83	2.64	1.59	1.12	0.53

(d) Fuses to BS 1362

Rating (amperes)	3	13
Z$_S$ (ohms)	23.2	3.83

NOTE: The circuit loop impedances given in the table should not be exceeded when the conductors are at their normal operating temperature. If the conductors are at a different temperature when tested, the reading should be adjusted accordingly. See Appendix 14.

411.4.9   Where an RCD is used to satisfy the requirements of Regulation 411.3.2.2, the maximum values of earth fault loop impedance in Table 41.5 may be applied for non-delayed RCDs to BS EN 61008-1 and BS EN 61009-1 for a nominal voltage U_0 of 230 V. In such cases, an overcurrent protective device shall provide protection against overload current and fault current in accordance with Chapter 43.

411.5   TT system

411.5.1   Every exposed-conductive-part which is to be protected by a single protective device shall be connected, via the main earthing terminal, to a common earth electrode. However, if two or more protective devices are in series, the exposed-conductive-parts may be connected to separate earth electrodes corresponding to each protective device.

411.5.2   One or more of the following types of protective device shall be used, the former being preferred:

(i) An RCD

(ii) An overcurrent protective device.

NOTE 1: An appropriate overcurrent protective device may be used for fault protection provided a suitably low value of Z_S is permanently and reliably assured.

NOTE 2: Where an RCD is used for fault protection the circuit should also incorporate an overcurrent protective device in accordance with Chapter 43.

411.5.3   Where an RCD is used for fault protection, the following conditions shall be fulfilled:

(i) The disconnection time shall be that required by Regulation 411.3.2.2 or 411.3.2.4, and

(ii) $R_A \times I_{\Delta n} \leq 50$ V

where:

R_A is the sum of the resistances of the earth electrode and the protective conductor connecting it to the exposed-conductive-parts (in ohms).

$I_{\Delta n}$ is the rated residual operating current of the RCD.

The requirements of this regulation are met if the earth fault loop impedance of the circuit protected by the RCD meets the requirements of Table 41.5.

NOTE 1: Where discrimination between RCDs is necessary refer also to Regulation 531.2.9.

NOTE 2: Where R_A is not known, it may be replaced by Z_S.

TABLE 41.5 –
Maximum earth fault loop impedance (Z_S) for non-delayed RCDs
to BS EN 61008-1 and BS EN 61009-1 for U_0 of 230 V (see Regulation 411.5.3)

Rated residual operating current (mA)	Maximum earth fault loop impedance Z_S (ohms)
30	1667*
100	500*
300	167
500	100

NOTE 1: Figures for Z_S result from the application of Regulation 411.5.3(i) and (ii). Disconnection must be ensured within the times stated in Table 41.1.

NOTE 2: * The resistance of the installation earth electrode should be as low as practicable. A value exceeding 200 ohms may not be stable. Refer to Regulation 542.2.4.

411.5.4   Where an overcurrent protective device is used the following condition shall be fulfilled:

$$Z_S \times I_a \leq U_0$$

where:

Z_S is the impedance in ohms (Ω) of the earth fault loop comprising:
- the source
- the line conductor up to the point of the fault
- the protective conductor from the exposed-conductive-parts
- the earthing conductor
- the earth electrode of the installation, and
- the earth electrode of the source.

I_a is the current in amperes (A) causing the automatic operation of the disconnecting device within the time specified in Table 41.1 of Regulation 411.3.2.2 or, as appropriate, Regulation 411.3.2.4.

U_0 is the nominal a.c. rms or d.c. line voltage to Earth in volts (V).

411.6   IT system

411.6.1   In an IT system, live parts shall be insulated from Earth or connected to Earth through a sufficiently high impedance. This connection may be made either at the neutral point or midpoint of the system, or at an artificial neutral point. The latter may be connected directly to Earth if the resulting impedance to Earth is sufficiently high at the system frequency. Where no neutral point or midpoint exists, a line conductor may be connected to Earth through a high impedance.

Under the above conditions, the fault current is low in the event of a single fault to an exposed-conductive-part or to Earth and automatic disconnection in accordance with Regulation 411.3.2 is not required provided the appropriate condition in Regulation 411.6.2 is fulfilled. Precautions shall be taken to avoid the risk of harmful effects on a person in contact with simultaneously accessible exposed-conductive-parts in the event of two faults existing simultaneously.

It is strongly recommended that IT systems with distributed neutrals should not be employed.

411.6.2   Exposed-conductive-parts shall be earthed individually, in groups, or collectively.

The following condition shall be fulfilled:
  (i)   In a.c. systems $R_A \times I_d \leq 50$ V
  (ii) In d.c. systems $R_A \times I_d \leq 120$ V
where:

R_A is the sum of the resistances of the earth electrode and the protective conductor connecting it to the exposed-conductive-parts (in ohms).

I_d is the fault current in amperes of the first fault of negligible impedance between a line conductor and an exposed-conductive-part. The value of I_d takes account of leakage currents and the total earthing impedance of the electrical installation.

411.6.3   The following monitoring devices and protective devices may be used:
  (i)   Insulation monitoring devices (IMDs)
  (ii)   Residual current monitoring devices (RCMs)
  (iii) Insulation fault location systems
  (iv)   Overcurrent protective devices
  (v)   RCDs.

411.6.3.1   Where an IT system is used for reasons of continuity of supply, an insulation monitoring device shall be provided to indicate the occurrence of a first fault from a live part to an exposed-conductive-part or to Earth. This device shall initiate an audible and/or visual signal which shall continue as long as the fault persists.

If there are both audible and visual signals it is permissible for the audible signal to be cancelled.

411.6.3.2   Except where a protective device is installed to interrupt the supply in the event of the first earth fault, an RCM or an insulation fault location system shall be provided to indicate the occurrence of a first fault from a live part to an exposed-conductive-part or to Earth. This device shall initiate an audible and/or visual signal, which shall continue as long as the fault persists.

If there are both audible and visual signals it is permissible for the audible signal to be cancelled.

411.6.4   After the occurrence of a first fault, conditions for automatic disconnection of supply in the event of a second fault occurring on a different live conductor shall be as follows:
  (i)   Where the exposed-conductive-parts are interconnected by a protective conductor and collectively earthed to the same earthing system, the conditions similar to a TN system apply and the following conditions shall be fulfilled where the neutral conductor is not distributed in a.c. systems and in d.c. systems where the midpoint conductor is not distributed:

$$Z_s \leq \frac{U}{2I_a}$$

or, where the neutral conductor or midpoint conductor respectively is distributed:

$$Z_s^1 \leq \frac{U_0}{2I_a}$$

where:

U is the nominal a.c. rms or d.c. voltage, in volts, between line conductors.

U_0 is the nominal a.c. rms or d.c. voltage, in volts, between a line conductor and the neutral conductor or midpoint conductor, as appropriate.

Z_s is the impedance in ohms of the fault loop comprising the line conductor and the protective conductor of the circuit.

Z_s^1 is the impedance in ohms of the fault loop comprising the neutral conductor and the protective conductor of the circuit.

I_a is the current in amperes causing automatic operation of the protective device within the time specified in Table 41.1 of Regulation 411.3.2.2, or as appropriate, Regulation 411.3.2.3, for a TN system.

The time stated in Table 41.1 for a TN system is applicable to an IT system with a distributed or non-distributed neutral conductor or midpoint conductor.

(ii)   Where the exposed-conductive-parts are earthed in groups or individually, the following condition applies:

$$R_A \times I_a \leq 50 \text{ V}$$

where:

R_A is the sum of the resistances of the earth electrode and the protective conductor connecting it to the exposed-conductive-parts (in ohms).

I_a is the current in amperes causing automatic disconnection of the protective device within the time specified for a TT system in Table 41.1 of Regulation 411.3.2.2, or as appropriate Regulation 411.3.2.4.

411.7   Functional Extra-Low Voltage (FELV)

411.7.1   General

Where, for functional reasons, a nominal voltage not exceeding 50 V a.c. or 120 V d.c. is used but all the requirements of Section 414 relating to SELV or to PELV are not fulfilled, and where SELV or PELV is not necessary, the supplementary provisions described in Regulations 411.7.2 and 411.7.3 shall be applied to ensure basic protection and fault protection.

This combination of provisions is known as FELV.

NOTE: Such conditions may, for example, be encountered where the circuit contains equipment (such as transformers, relays, remote-control switches, contactors) which is insufficiently insulated with respect to circuits at higher voltage.

411.7.2   Requirements for basic protection

Basic protection shall be provided by either:

(i)   basic insulation according to Regulation 416.1 corresponding to the nominal voltage of the primary circuit of the source, or

(ii)   barriers or enclosures in accordance with Regulation 416.2.

411.7.3   Requirements for fault protection

The exposed-conductive-parts of the equipment of the FELV circuit shall be connected to the protective conductor of the primary circuit of the source, provided that the primary circuit is subject to protection by automatic disconnection of supply in accordance with Regulations 411.3 to 6.

411.7.4   Sources

The source of the FELV system shall either be a transformer with at least simple separation between windings or shall comply with Regulation 414.3.

If an extra-low voltage system is supplied from a higher voltage system by equipment which does not provide at least simple separation between that system and the extra-low voltage system, such as an autotransformer, a potentiometer or a semiconductor device, the output circuit is not part of a FELV system and is deemed to be an extension of the input circuit and shall be protected by the protective measure applied to the input circuit.

NOTE: This does not preclude connecting a conductor of the FELV circuit to the protective conductor of the primary circuit.

411.7.5 Plugs, socket-outlets, LSCs, DCLs and cable couplers

Every plug, socket-outlet, luminaire supporting coupler (LSC), device for connecting a luminaire (DCL) and cable coupler in a FELV system shall have a protective conductor contact and shall not be dimensionally compatible with those used for any other system in use in the same premises.

411.8 Reduced low voltage systems

411.8.1 General

411.8.1.1 Where, for functional reasons, the use of extra-low voltage is impracticable and there is no requirement for the use of SELV or PELV, a reduced low voltage system may be used, for which the provisions described in Regulations 411.8.2 to 5 shall be made to ensure basic protection and fault protection.

411.8.1.2 The nominal voltage of the reduced low voltage circuits shall not exceed 110 V a.c. rms between lines (three-phase 63.5 V to earthed neutral, single-phase 55 V to earthed midpoint).

411.8.2 Requirements for basic protection

Basic protection shall be provided by either:

(i) basic insulation according to Regulation 416.1 corresponding to the maximum nominal voltage of the reduced low voltage system given in Regulation 411.8.1.2, or

(ii) barriers or enclosures in accordance with Regulation 416.2.

411.8.3 Requirements for fault protection

Fault protection by automatic disconnection of supply shall be provided by means of an overcurrent protective device in each line conductor or by an RCD, and all exposed-conductive-parts of the reduced low voltage system shall be connected to Earth. The earth fault loop impedance at every point of utilisation, including socket-outlets, shall be such that the disconnection time does not exceed 5 s.

Where a circuit-breaker is used, the maximum value of earth fault loop impedance (Z_s) shall be determined by the formula in Regulation 411.4.5. Alternatively, the values specified in Table 41.6 may be used instead of calculation for the nominal voltages (U_0) and the types and ratings of overcurrent device listed therein.

Where a fuse is used, the maximum values of earth fault loop impedance (Z_s) corresponding to a disconnection time of 5 s are stated in Table 41.6 for nominal voltages (U_0) of 55 V and 63.5 V.

For types and rated currents of fuses other than those mentioned in Table 41.6, reference should be made to the appropriate British or Harmonized Standard to determine the value of I_a for compliance with Regulation 411.4.5, according to the appropriate value of the nominal voltage (U_0).

Where fault protection is provided by an RCD, the product of the rated residual operating current ($I_{\Delta n}$) in amperes and the earth fault loop impedance in ohms shall not exceed 50.

TABLE 41.6 –
Maximum earth fault loop impedance (Z_s) for 5 s disconnection time
and U_0 of 55 V (single-phase) and 63.5 V (three-phase)
(see Regulations 411.8.1.2 and 411.8.3)

	Circuit-breakers to BS EN 60898 and the overcurrent characteristics of RCBOs to BS EN 61009-1 Type						General purpose (gG) fuses to BS 88-2 – fuse systems E and G	
	B		C		D			
U_0 (Volts)	55	63.5	55	63.5	55	63.5	55	63.5
Rating amperes				Z_s ohms				
6	1.83	2.12	0.92	1.07	0.47	0.53	3.06	3.53
10	1.10	1.27	0.55	0.64	0.28	0.32	1.72	1.98
16	0.69	0.79	0.34	0.40	0.18	0.20	1.00	1.15
20	0.55	0.64	0.28	0.32	0.14	0.16	0.71	0.81
25	0.44	0.51	0.22	0.26	0.11	0.13	0.55	0.64
32	0.34	0.40	0.17	0.20	0.09	0.10	0.44	0.51
40	0.28	0.32	0.14	0.16	0.07	0.08	0.32	0.37
50	0.22	0.25	0.11	0.13	0.06	0.06	0.25	0.29
63	0.17	0.20	0.09	0.10	0.04	0.05	0.20	0.23
80	0.14	0.16	0.07	0.08	0.04	0.04	0.14	0.16
100	0.11	0.13	0.05	0.06	0.03	0.03	0.11	0.12
125	0.09	0.10	0.04	0.05	0.02	0.03	0.08	0.09
I_n	$\dfrac{11}{I_n}$	$\dfrac{12.7}{I_n}$	$\dfrac{5.5}{I_n}$	$\dfrac{6.4}{I_n}$	$\dfrac{2.8}{I_n}$	$\dfrac{3.2}{I_n}$		

NOTE: The circuit loop impedances given in the table should not be exceeded when the conductors are at their normal operating temperature. If the conductors are at a different temperature when tested, the reading should be adjusted accordingly. See Appendix 14.

411.8.4   Sources

411.8.4.1   The source of supply to a reduced low voltage circuit shall be one of the following:

(i)   A double-wound isolating transformer complying with BS EN 61558-1 and BS EN 61558-2-23

(ii)   A motor-generator having windings providing isolation equivalent to that provided by the windings of an isolating transformer

(iii)   A source independent of other supplies, e.g. an engine driven generator.

411.8.4.2   The neutral (star) point of the secondary windings of three-phase transformers and generators, or the midpoint of the secondary windings of single-phase transformers and generators, shall be connected to Earth.

411.8.5   Requirements for circuits

Every plug, socket-outlet, luminaire supporting coupler (LSC), device for connecting a luminaire (DCL) and cable coupler of a reduced low voltage system shall have a protective conductor contact and shall not be dimensionally compatible with those used for any other system in use in the same premises.

412   PROTECTIVE MEASURE: DOUBLE OR REINFORCED INSULATION

412.1   General

412.1.1   Double or reinforced insulation is a protective measure in which:

(i)   basic protection is provided by basic insulation and fault protection is provided by supplementary insulation, or

(ii)   basic and fault protection are provided by reinforced insulation between live parts and accessible parts.

NOTE: This protective measure is intended to prevent the appearance of a dangerous voltage on the accessible parts of electrical equipment through a fault in the basic insulation.

412.1.2   The protective measure of double or reinforced insulation is applicable in all situations, unless limitations are given in the corresponding section of Part 7.

412.1.3   Where this protective measure is to be used as the sole protective measure (i.e. where a whole installation or circuit is intended to consist entirely of equipment with double insulation or reinforced insulation), it shall be verified that the installation or circuit concerned will be under effective supervision in normal use so that no change is made that would impair the effectiveness of the protective measure. This protective measure shall not therefore be applied to any circuit that includes a socket-outlet, luminaire supporting coupler (LSC), device for connecting a luminaire (DCL) or cable coupler, or where a user may change items of equipment without authorization.

412.2   Requirements for basic protection and fault protection

412.2.1   Electrical equipment

Where the protective measure of double or reinforced insulation is used for the complete installation or part of the installation, electrical equipment shall comply with one of the following:

    (i)   Regulation 412.2.1.1, or

    (ii)   Regulations 412.2.1.2 and 412.2.2, or

    (iii)   Regulations 412.2.1.3 and 412.2.2.

412.2.1.1   Electrical equipment shall be of the following types, type-tested and marked to the relevant standards:

    (i)   Electrical equipment having double or reinforced insulation (Class II equipment)

    (ii)   Electrical equipment declared in the relevant product standard as equivalent to Class II, such as assemblies of electrical equipment having total insulation (see BS EN 60439-1).

NOTE: This equipment is identified by the symbol ▣   Refer to BS EN 60417: Class II equipment.

412.2.1.2   Electrical equipment having basic insulation only shall have supplementary insulation applied in the process of erecting the electrical installation, providing a degree of safety equivalent to electrical equipment according to Regulation 412.2.1.1 and complying with Regulations 412.2.2.1 to 3.

The symbol ⌀ shall be fixed in a visible position both on the exterior and interior of the enclosure.

412.2.1.3   Electrical equipment having uninsulated live parts shall have reinforced insulation applied in the process of erecting the electrical installation, providing a degree of safety equivalent to electrical equipment according to Regulation 412.2.1.1 and complying with Regulations 412.2.2.2 and 412.2.2.3, such insulation being recognized only where constructional features prevent the application of double insulation.

The symbol ⌀ shall be fixed in a visible position both on the exterior and interior of the enclosure.

412.2.2   Enclosures

412.2.2.1   The electrical equipment being ready for operation, all conductive parts separated from live parts by basic insulation only shall be contained in an insulating enclosure affording at least the degree of protection IPXXB or IP2X.

412.2.2.2   The following requirements apply as specified:

    (i)   The insulating enclosure shall not be traversed by conductive parts likely to transmit a potential, and

    (ii)   the insulating enclosure shall not contain any screws or other fixing means which might need to be removed, or are likely to be removed during installation and maintenance and whose replacement by metallic screws or other fixing means could impair the enclosure's insulation.

Where the insulating enclosure must be traversed by mechanical joints or connections (e.g. for operating handles of built-in equipment), these should be arranged in such a way that protection against shock in case of a fault is not impaired.

412.2.2.3   Where a lid or door in an insulating enclosure can be opened without the use of a tool or key, all conductive parts which are accessible if the lid or door is open shall be behind an insulating barrier (providing a degree of protection not less than IPXXB or IP2X) preventing persons from coming unintentionally into contact with those conductive parts. This insulating barrier shall be removable only by the use of a tool or key.

412.2.2.4   No conductive part enclosed in the insulating enclosure shall be connected to a protective conductor. However, provision may be made for connecting protective conductors which necessarily run through the enclosure in order to serve other items of electrical equipment whose supply circuit also runs through the enclosure. Inside the enclosure, any such conductors and their terminals shall be insulated as though they were live parts, and their terminals shall be marked as protective conductor (PE) terminals.

No exposed-conductive-part or intermediate part shall be connected to any protective conductor unless specific provision for this is made in the specification for the equipment concerned.

412.2.2.5   The enclosure shall not adversely affect the operation of the equipment protected in this way.

412.2.3   Installation

412.2.3.1   The installation of equipment mentioned in Regulation 412.2.1 (fixing, connection of conductors, etc.) shall be effected in such a way as not to impair the protection afforded in compliance with the equipment specification.

412.2.3.2   Except where Regulation 412.1.3 applies, a circuit supplying one or more items of Class II equipment shall have a circuit protective conductor run to and terminated at each point in wiring and at each accessory.

NOTE: This requirement is intended to take account of the possible replacement by the user of Class II equipment by Class I equipment.

412.2.4   Wiring systems

412.2.4.1   Wiring systems installed in accordance with Chapter 52 are considered to meet the requirements of Regulation 412.2 if:

(i)   the rated voltage of the cable(s) is not less than the nominal voltage of the system and at least 300/500 V, and

(ii) adequate mechanical protection of the basic insulation is provided by one or more of the following:

   (a)   The non-metallic sheath of the cable

   (b)   Non-metallic trunking or ducting complying with the BS EN 50085 series of standards, or non-metallic conduit complying with the BS EN 61386 series of standards.

NOTE 1: Cable product standards do not specify impulse withstand capability. However, it is considered that the insulation of the cabling system is at least equivalent to the requirement in BS EN 61140 for reinforced insulation.

NOTE 2: A wiring system should not be identified by the symbol ▣ or by the symbol ⌥.

413   PROTECTIVE MEASURE: ELECTRICAL SEPARATION

413.1   General

413.1.1   Electrical separation is a protective measure in which:

(i)   basic protection is provided by basic insulation of live parts or by barriers or enclosures in accordance with Section 416, and

(ii) fault protection is provided by simple separation of the separated circuit from other circuits and from Earth.

413.1.2   Except as permitted by Regulation 413.1.3, this protective measure shall be limited to the supply of one item of current-using equipment supplied from one unearthed source with simple separation.

NOTE: Where this protective measure is used, it is particularly important to ensure compliance of the basic insulation with the product standard.

413.1.3   Where more than one item of current-using equipment is supplied from an unearthed source with simple separation, the requirements of Regulation 418.3 shall be met.

413.2   Requirements for basic protection

All electrical equipment shall be subject to one of the basic protective provisions in Section 416 or to the protective measures in Section 412.

413.3   Requirements for fault protection

413.3.1   Protection by electrical separation shall be ensured by compliance with Regulations 413.3.2 to 6.

413.3.2   The separated circuit shall be supplied through a source with at least simple separation, and the voltage of the separated circuit shall not exceed 500 V.

413.3.3   Live parts of the separated circuit shall not be connected at any point to another circuit or to Earth or to a protective conductor.

To ensure electrical separation, arrangements shall be such that basic insulation is achieved between circuits in compliance with Regulation 416.1.

413.3.4   Flexible cables shall be visible throughout any part of their length liable to mechanical damage.

413.3.5   For separated circuits the use of separate wiring systems is recommended. If separated circuits and other circuits are in the same wiring system, multi-conductor cables without metallic covering or insulated conductors in insulating conduit, non-metallic ducting or non-metallic trunking shall be used, provided that:

(i)   the rated voltage is not less than the highest nominal voltage, and

(ii)   each circuit is protected against overcurrent.

413.3.6   No exposed-conductive-part of the separated circuit shall be connected either to the protective conductor or exposed-conductive-parts of other circuits, or to Earth.

NOTE: If the exposed-conductive-parts of the separated circuit are liable to come into contact, either intentionally or fortuitously, with the exposed-conductive-parts of other circuits, protection against electric shock no longer depends solely on protection by electrical separation but also on the protective provisions to which the latter exposed-conductive-parts are subject.

414   PROTECTIVE MEASURE: EXTRA-LOW VOLTAGE PROVIDED BY SELV OR PELV

414.1   General

414.1.1   Protection by extra-low voltage is a protective measure which consists of either of two different extra-low voltage systems:

(i)   SELV, or

(ii)   PELV.

Protection by extra-low voltage provided by SELV or PELV requires:

(iii)   limitation of voltage in the SELV or PELV system to the upper limit of voltage Band I, 50 V a.c. or 120 V d.c. (see IEC 60449), and

(iv)   protective separation of the SELV or PELV system from all circuits other than SELV and PELV circuits, and basic insulation between the SELV or PELV system and other SELV or PELV systems, and

(v)   for SELV systems only, basic insulation between the SELV system and Earth.

414.1.2   The use of SELV or PELV according to Section 414 is considered as a protective measure in all situations.

NOTE: In certain locations the requirements of Part 7 limit the value of the extra-low voltage to a value lower than 50 V a.c. or 120 V d.c.

414.2   Requirements for basic protection and fault protection

Basic protection and fault protection is deemed to be provided where:

(i)   the nominal voltage cannot exceed the upper limit of voltage Band I, and

(ii)   the supply is from one of the sources listed in Regulation 414.3, and

(iii)   the conditions of Regulation 414.4 are fulfilled.

NOTE 1:   If the system is supplied from a higher voltage system by equipment which provides at least simple separation between that system and the extra-low voltage system but which does not meet the requirements for SELV and PELV sources in Regulation 414.3, the requirements for FELV may be applicable, see Regulation 411.7.

NOTE 2:   d.c. voltages for ELV circuits generated by a semiconductor convertor (see BS EN 60146-2) require an internal a.c. voltage circuit to supply the rectifier stack. This internal a.c. voltage exceeds the d.c. voltage. The internal a.c. circuit is not to be considered as a higher voltage circuit within the meaning of this regulation. Between internal circuits and external higher voltage circuits, protective separation is required.

NOTE 3:   In d.c. systems with batteries, the battery charging and floating voltages exceed the battery nominal voltage, depending on the type of battery. This does not require any protective provisions in addition to those specified in this regulation. The charging voltage should not exceed a maximum value of 75 V a.c. or 150 V d.c. as appropriate according to the environmental situation as given in Table 1 of PD 6536 (IEC 61201).

414.3   **Sources for SELV and PELV**

The following sources may be used for SELV and PELV systems:

(i)   A safety isolating transformer in accordance with BS EN 61558-2-6

(ii)   A source of current providing a degree of safety equivalent to that of the safety isolating transformer specified in (i) (e.g. motor-generator with windings providing equivalent isolation)

(iii)   An electrochemical source (e.g. a battery) or another source independent of a higher voltage circuit (e.g. a diesel-driven generator)

(iv)   Certain electronic devices complying with appropriate standards, where provisions have been taken to ensure that, even in the case of an internal fault, the voltage at the outgoing terminals cannot exceed the values specified in Regulation 414.1.1. Higher voltages at the outgoing terminals are, however, permitted if it is ensured that, in case of contact with a live part or in the event of a fault between a live part and an exposed-conductive-part, the voltage at the output terminals is immediately reduced to the value specified in Regulation 414.1.1 or less.

> **NOTE 1:**   Examples of such devices include insulation testing equipment and monitoring devices.
>
> **NOTE 2:**   Where higher voltages exist at the outgoing terminals, compliance with this regulation may be assumed if the voltage at the outgoing terminals is within the limits specified in Regulation 414.1.1 when measured with a voltmeter having an internal resistance of at least 3 000 ohms.

A mobile source supplied at low voltage, e.g. a safety isolating transformer or a motor-generator, shall be selected and erected in accordance with the requirements for protection by the use of double or reinforced insulation (see Section 412).

414.4   **Requirements for SELV and PELV circuits**

414.4.1   SELV and PELV circuits shall have:

(i)   basic insulation between live parts and other SELV or PELV circuits, and

(ii)   protective separation from live parts of circuits not being SELV or PELV provided by double or reinforced insulation or by basic insulation and protective screening for the highest voltage present.

SELV circuits shall have basic insulation between live parts and Earth.

The PELV circuits and/or exposed-conductive-parts of equipment supplied by the PELV circuits may be earthed.

> **NOTE 1:** In particular, protective separation is necessary between the live parts of electrical equipment such as relays, contactors and auxiliary switches, and any part of a higher voltage circuit or a FELV circuit.
>
> **NOTE 2:** The earthing of PELV circuits may be achieved by a connection to Earth or to an earthed protective conductor within the source itself.

414.4.2   Protective separation of wiring systems of SELV or PELV circuits from the live parts of other circuits, which have at least basic insulation, shall be achieved by one of the following arrangements:

(i)   SELV and PELV circuit conductors enclosed in a non-metallic sheath or insulating enclosure in addition to basic insulation

(ii)   SELV and PELV circuit conductors separated from conductors of circuits at voltages higher than Band I by an earthed metallic sheath or earthed metallic screen

(iii)   Circuit conductors at voltages higher than Band I may be contained in a multi-conductor cable or other grouping of conductors if the SELV and PELV conductors are insulated for the highest voltage present

(iv)   The wiring systems of other circuits are in compliance with Regulation 412.2.4.1

(v)   Physical separation.

414.4.3   Every socket-outlet and luminaire supporting coupler in a SELV or PELV system shall require the use of a plug which is not dimensionally compatible with those used for any other system in use in the same premises.

Plugs and socket-outlets in a SELV system shall not have a protective conductor contact.

414.4.4   Exposed-conductive-parts of a SELV circuit shall not be connected to Earth, or to protective conductors or exposed-conductive-parts of another circuit.

> **NOTE:** If the exposed-conductive-parts of SELV circuits are liable to come into contact, either fortuitously or intentionally, with the exposed-conductive-parts of other circuits, the requirements of the protective measure SELV have not been met and protection against electric shock no longer depends solely on protection by SELV, but also on the protective provisions to which the latter exposed-conductive-parts are subject.

414.4.5   If the nominal voltage exceeds 25 V a.c. or 60 V d.c., or if the equipment is immersed, basic protection shall be provided for SELV and PELV circuits by:

(i)   insulation in accordance with Regulation 416.1 or

(ii)   barriers or enclosures in accordance with Regulation 416.2.

Basic protection is generally unnecessary in normal dry conditions for:

(iii) SELV circuits where the nominal voltage does not exceed 25 V a.c. or 60 V d.c.

(iv)   PELV circuits where the nominal voltage does not exceed 25 V a.c. or 60 V d.c. and exposed-conductive-parts and/or the live parts are connected by a protective conductor to the main earthing terminal.

In all other cases, basic protection is not required if the nominal voltage of the SELV or PELV system does not exceed 12 V a.c. or 30 V d.c.

415   ADDITIONAL PROTECTION

NOTE: Additional protection in accordance with Section 415 may be specified with the protective measure. In particular, additional protection may be required with the protective measure under certain conditions of external influence and in certain special locations (see the corresponding section of Part 7).

415.1   Additional protection: Residual current devices (RCDs)

415.1.1   The use of RCDs with a rated residual operating current ($I_{\Delta n}$) not exceeding 30 mA and an operating time not exceeding 40 ms at a residual current of 5 $I_{\Delta n}$ is recognized in a.c. systems as additional protection in the event of failure of the provision for basic protection and/or the provision for fault protection or carelessness by users.

415.1.2   The use of RCDs is not recognized as a sole means of protection and does not obviate the need to apply one of the protective measures specified in Sections 411 to 414.

415.2   Additional protection: Supplementary equipotential bonding

NOTE 1   Supplementary equipotential bonding is considered as an addition to fault protection.

NOTE 2   The use of supplementary bonding does not exclude the need to disconnect the supply for other reasons, for example protection against fire, thermal stresses in equipment, etc.

NOTE 3   Supplementary bonding may involve the entire installation, a part of the installation, an item of equipment, or a location.

NOTE 4   Additional requirements may be necessary for special locations (see the corresponding section of Part 7), or for other reasons.

415.2.1   Supplementary equipotential bonding shall include all simultaneously accessible exposed-conductive-parts of fixed equipment and extraneous-conductive-parts including, where practicable, the main metallic reinforcement of constructional reinforced concrete. The equipotential bonding system shall be connected to the protective conductors of all equipment including those of socket-outlets.

415.2.2   Where doubt exists regarding the effectiveness of supplementary equipotential bonding, it shall be confirmed that the resistance R between simultaneously accessible exposed-conductive-parts and extraneous-conductive-parts fulfils the following condition:

$$R \leq 50 \text{ V}/I_a \qquad \text{in a.c. systems}$$

$$R \leq 120 \text{ V}/I_a \qquad \text{in d.c. systems}$$

where I_a is the operating current in amperes of the protective device -
for RCDs, $I_{\Delta n}$.
for overcurrent devices, the current causing automatic operation in 5 s.

416   PROVISIONS FOR BASIC PROTECTION

NOTE: Provisions for basic protection provide protection under normal conditions and are applied where specified as a part of the chosen protective measure.

416.1   Basic insulation of live parts

Live parts shall be completely covered with insulation which can only be removed by destruction.

For equipment, the insulation shall comply with the relevant standard for such electrical equipment.

NOTE: The insulation is intended to prevent contact with live parts.

Paint, varnish, lacquer or similar products are generally not considered to provide adequate insulation for basic protection in normal service.

416.2　Barriers or enclosures

NOTE:　Barriers or enclosures are intended to prevent contact with live parts.

416.2.1　Live parts shall be inside enclosures or behind barriers providing at least the degree of protection IPXXB or IP2X except that, where larger openings occur during the replacement of parts, such as certain lampholders or fuses, or where larger openings are necessary to allow the proper functioning of equipment according to the relevant requirements for the equipment:

(i)　suitable precautions shall be taken to prevent persons or livestock from unintentionally touching live parts, and

(ii)　it shall be ensured, as far as practicable, that persons will be aware that live parts can be touched through the opening and should not be touched intentionally, and

(iii)　the opening shall be as small as is consistent with the requirement for proper functioning and for replacement of a part.

416.2.2　A horizontal top surface of a barrier or enclosure which is readily accessible shall provide a degree of protection of at least IPXXD or IP4X.

416.2.3　A barrier or enclosure shall be firmly secured in place and have sufficient stability and durability to maintain the required degree of protection and appropriate separation from live parts in the known conditions of normal service, taking account of relevant external influences.

416.2.4　Where it is necessary to remove a barrier or open an enclosure or remove parts of enclosures, this shall be possible only:

(i)　by the use of a key or tool, or

(ii)　after disconnection of the supply to live parts against which the barriers or enclosures afford protection, restoration of the supply being possible only after replacement or reclosure of the barrier or enclosure, or

(iii)　where an intermediate barrier providing a degree of protection of at least IPXXB or IP2X prevents contact with live parts, by the use of a key or tool to remove the intermediate barrier.

NOTE:　This regulation does not apply to:
　　　a ceiling rose complying with BS 67
　　　a cord operated switch complying with BS EN 60669-1
　　　a bayonet lampholder complying with BS EN 61184
　　　an Edison screw lampholder complying with BS EN 60238.

416.2.5　If, behind a barrier or in an enclosure, an item of equipment such as a capacitor is installed which may retain a dangerous electrical charge after it has been switched off, a warning label shall be provided. Small capacitors such as those used for arc extinction and for delaying the response of relays, etc shall not be considered dangerous.

NOTE:　Unintentional contact is not considered dangerous if the voltage resulting from static charge falls below 120 V d.c. in less than 5 s after disconnection from the power supply.

417　OBSTACLES AND PLACING OUT OF REACH

417.1　Application

The protective measures of obstacles and placing out of reach provide basic protection only. They are for application in installations, with or without fault protection, that are controlled or supervised by skilled persons.

The conditions of supervision under which the basic protective provisions of Section 417 may be applied as part of the protective measure are given in Regulation 410.3.5.

417.2　Obstacles

NOTE:　Obstacles are intended to prevent unintentional contact with live parts but not intentional contact by deliberate circumvention of the obstacle.

417.2.1　Obstacles shall prevent:

(i)　unintentional bodily approach to live parts, and

(ii)　unintentional contact with live parts during the operation of live equipment in normal service.

417.2.2　An obstacle may be removed without the use of a key or tool but shall be secured so as to prevent unintentional removal.

417.3 Placing out of reach

NOTE: Protection by placing out of reach is intended only to prevent unintentional contact with live parts.

A bare or insulated overhead line for distribution between buildings and structures shall be installed to the standard required by the Electricity Safety, Quality and Continuity Regulations 2002.

417.3.1 Simultaneously accessible parts at different potentials shall not be within arm's reach.

A bare live part other than an overhead line shall not be within arm's reach or within 2.5 m of the following:

   (i)   An exposed-conductive-part

   (ii)   An extraneous-conductive-part

   (iii)   A bare live part of any other circuit.

NOTE: Two parts are deemed to be simultaneously accessible if they are not more than 2.50 m apart (see Figure 417).

417.3.2 If a normally occupied position is restricted in the horizontal direction by an obstacle (e.g. handrail, mesh screen) affording a degree of protection less than IPXXB or IP2X, arm's reach shall extend from that obstacle. In the overhead direction, arm's reach is 2.50 m from the surface, S, not taking into account any intermediate obstacle providing a degree of protection less than IPXXB.

NOTE: The values of arm's reach apply to contact directly with bare hands without assistance (e.g. tools or ladder).

Fig 417 – Arm's reach

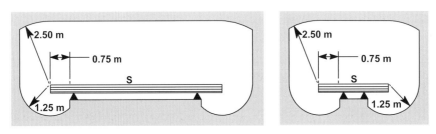

S = surface expected to be occupied by persons

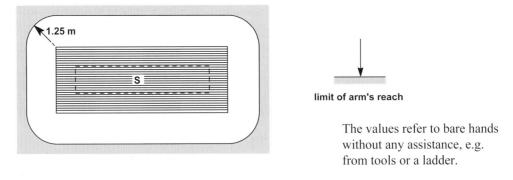

limit of arm's reach

The values refer to bare hands without any assistance, e.g. from tools or a ladder.

417.3.3 In places where bulky or long conductive objects are normally handled, the distances required by Regulations 417.3.1 and 417.3.2 shall be increased, taking account of the relevant dimensions of those objects.

418   PROTECTIVE MEASURES FOR APPLICATION ONLY WHERE THE INSTALLATION IS CONTROLLED OR UNDER THE SUPERVISION OF SKILLED OR INSTRUCTED PERSONS

NOTE: The conditions of supervision under which the fault protective provisions of Section 418 may be applied as part of the protective measure are given in Regulation 410.3.6.

418.1   Non-conducting location

This method of protection is not recognized for general application.

NOTE: This protective measure is intended to prevent simultaneous contact with parts which may be at different potentials through failure of the basic insulation of live parts.

418.1.1   All electrical equipment shall comply with one of the provisions for basic protection described in Section 416.

418.1.2   Exposed-conductive-parts shall be arranged so that under ordinary circumstances persons will not come into simultaneous contact with:

(i)   two exposed-conductive-parts, or

(ii)   an exposed-conductive-part and any extraneous-conductive-part

if these parts are liable to be at different potentials through failure of the basic insulation of a live part.

418.1.3   In a non-conducting location there shall be no protective conductor.

418.1.4   Regulation 418.1.2 is fulfilled if the location has an insulating floor and walls and one or more of the following arrangements applies:

(i)   Relative spacing of exposed-conductive-parts and of extraneous-conductive-parts as well as spacing of exposed-conductive-parts

This spacing is sufficient if the distance between two parts is not less than 2.5 m; this distance may be reduced to 1.25 m outside the zone of arm's reach.

(ii)   Interposition of effective obstacles between exposed-conductive-parts and extraneous-conductive-parts

Such obstacles are sufficiently effective if they extend the distances to be surmounted to the values stated in point (i) above. They shall not be connected to Earth or to exposed-conductive-parts; as far as possible they shall be of insulating material.

(iii)   Insulation or insulating arrangements of extraneous-conductive-parts.

The insulation shall be of sufficient mechanical strength and be able to withstand a test voltage of at least 2 000 V. Leakage current shall not exceed 1 mA in normal conditions of use.

418.1.5   The resistance of insulating floors and walls at every point of measurement under the conditions specified in Part 6 shall be not less than:

(i)   50 kΩ, where the nominal voltage of the installation does not exceed 500 V, or

(ii)   100 kΩ, where the nominal voltage of the installation exceeds 500 V.

NOTE: If at any point the resistance is less than the specified value, the floors and walls are deemed to be extraneous-conductive-parts for the purposes of protection against electric shock.

418.1.6   The arrangements made shall be permanent and it shall not be possible to make them ineffective. The arrangements shall also ensure protection where the use of mobile equipment is envisaged.

NOTE 1: Attention is drawn to the risk that, where electrical installations are not under effective supervision, further conductive parts may be introduced at a later date (e.g. mobile Class I equipment, or extraneous-conductive-parts such as metallic water pipes), which may invalidate compliance with Regulation 418.1.6.

NOTE 2: It is essential to ensure that the insulation of floor and walls cannot be affected by humidity.

418.1.7   Precautions shall be taken to ensure that extraneous-conductive-parts cannot cause a potential to appear external to the location concerned.

418.2 Protection by earth-free local equipotential bonding

This method of protection shall be used only in special circumstances.

NOTE: Earth-free local equipotential bonding is intended to prevent the appearance of a dangerous touch voltage.

418.2.1 All electrical equipment shall comply with one of the provisions for basic protection described in Section 416.

418.2.2 Protective bonding conductors shall interconnect every simultaneously accessible exposed-conductive-part and extraneous-conductive-part.

418.2.3 The local protective bonding conductors shall neither be in electrical contact with Earth directly, nor through exposed-conductive-parts, nor through extraneous-conductive-parts.

NOTE: Where this requirement cannot be fulfilled, protection by automatic disconnection of supply is applicable (see Section 411).

418.2.4 Precautions shall be taken to ensure that persons entering the equipotential location cannot be exposed to a dangerous potential difference, in particular, where a conductive floor insulated from Earth is connected to the earth-free protective bonding conductors.

418.2.5 Where this measure is applied, a warning notice complying with Regulation 514.13.2 shall be fixed in a prominent position adjacent to every point of access to the location concerned.

418.3 Electrical separation for the supply to more than one item of current-using equipment

Where the measure is used to supply two or more items of equipment from a single source, a warning notice complying with Regulation 514.13.2 shall be fixed in a prominent position adjacent to every point of access to the location concerned.

NOTE: Electrical separation of an individual circuit is intended to prevent shock currents through contact with exposed-conductive-parts that may be energized by a fault in the basic insulation of the circuit.

418.3.1 All electrical equipment shall comply with one of the provisions for basic protection described in Section 416.

418.3.2 Protection by electrical separation for the supply to more than one item of equipment shall be ensured by compliance with all the requirements of Section 413 except Regulation 413.1.2, and with the requirements in Regulations 418.3.3 to 8.

418.3.3 Precautions shall be taken to protect the separated circuit from damage and insulation failure.

418.3.4 The exposed-conductive-parts of the separated circuit shall be connected together by insulated, non-earthed protective bonding conductors. Such conductors shall not be connected to the protective conductor or exposed-conductive-parts of any other circuit or to any extraneous-conductive-parts.

NOTE: See Note to Regulation 413.3.6.

418.3.5 Every socket-outlet shall be provided with a protective conductor contact which shall be connected to the equipotential bonding system provided in accordance with Regulation 418.3.4.

418.3.6 Except where supplying equipment with double or reinforced insulation, all flexible cables shall embody a protective conductor for use as a protective bonding conductor in accordance with Regulation 418.3.4.

418.3.7 It shall be ensured that if two faults affecting two exposed-conductive-parts occur and these are fed by conductors of different polarity, a protective device shall disconnect the supply in a disconnection time conforming with Table 41.1.

418.3.8 The product of the nominal voltage of the circuit in volts and length of the wiring system in metres shall not exceed 100 000 Vm, and the length of the wiring system shall not exceed 500 m.

CHAPTER 42

PROTECTION AGAINST THERMAL EFFECTS

CONTENTS

CHAPTER 42

420.1   Scope

This chapter applies to electrical installations and equipment with regard to measures for the protection of persons, livestock and property:

(i)   against the harmful effects of heat or thermal radiation developed by electrical equipment

(ii)   against the ignition, combustion or degradation of materials

(iii)   against flames and smoke where a fire hazard could be propagated from an electrical installation to other nearby fire compartments, and

(iv)   against safety services being cut off by the failure of electrical equipment.

NOTE 1: For protection against thermal effects and fire, statutory requirements may be applicable. Refer to Appendix 2.

NOTE 2: Protection against overcurrent is dealt with in Chapter 43 of these Regulations.

420.2   *Not used*

420.3   *Moved by BS 7671:2008 Amendment No 1 to 421.1.1*

421   PROTECTION AGAINST FIRE CAUSED BY ELECTRICAL EQUIPMENT

421.1   General requirements

421.1.1   Persons, livestock and property shall be protected against harmful effects of heat or fire which may be generated or propagated in electrical installations.

Manufacturers' instructions shall be taken into account in addition to the requirements of BS 7671.

NOTE 1: Harmful effects of heat or fire may be caused by:

- heat accumulation, heat radiation, hot components or equipment
- failure of electrical equipment such as protective devices, switchgear, thermostats, temperature limiters, seals of cable penetrations and wiring systems
- overcurrent
- insulation faults or arcs, sparks and high temperature particles
- harmonic currents
- external influences such as lightning surge.

NOTE 2: Lightning strikes and overvoltages are covered in BS EN 62305 and Section 443 of these Regulations.

421.1.2   Fixed electrical equipment shall be selected and erected such that its temperature in normal operation will not cause a fire. This shall be achieved by the construction of the equipment or by additional protective measures taken during erection.

The heat generated by electrical equipment shall not cause danger or harmful effects to adjacent fixed material or to material which may foreseeably be in proximity to such equipment.

Where fixed equipment may attain surface temperatures which could cause a fire hazard to adjacent materials, one or more of the following installation methods shall be adopted. The equipment shall:

(i)   be mounted on a support which has low thermal conductance or within an enclosure which will withstand, with minimal risk of fire or harmful thermal effect, such temperatures as may be generated, or

(ii) be screened by materials of low thermal conductance which can withstand, with minimal risk of fire or harmful thermal effect, the heat emitted by the electrical equipment, or

(iii) be mounted so as to allow safe dissipation of heat and at a sufficient distance from adjacent material on which such temperatures could have deleterious effects. Any means of support shall be of low thermal conductance.

421.1.3   Where arcs, sparks or particles at high temperature may be emitted by fixed equipment in normal service, the equipment shall meet one or more of the following requirements. It shall be:

(i)   totally enclosed in arc-resistant material

(ii) screened by arc-resistant material from materials upon which the emissions could have harmful effects

(iii) mounted so as to allow safe extinction of the emissions at a sufficient distance from materials upon which the emissions could have harmful effects

(iv) in compliance with its standard.

Arc-resistant material used for this protective measure shall be non-ignitable, of low thermal conductivity and of adequate thickness to provide mechanical stability.

421.1.4   Fixed equipment causing a concentration and focusing of heat shall be at a sufficient distance from any fixed object or building element so that the object or element is not subjected to a dangerous temperature in normal conditions.

421.1.5   Where electrical equipment in a single location contains flammable liquid in significant quantity, adequate precautions shall be taken to prevent the spread of liquid, flame and the products of combustion.

NOTE 1: Examples of such precautions are:

(a)   a retention pit to collect any leakage of liquid and ensure extinction in the event of fire

(b)   installation of the equipment in a chamber of adequate fire-resistance and the provision of sills or other means of preventing burning liquid spreading to other parts of the building, such a chamber being ventilated solely to the external atmosphere.

NOTE 2: The generally accepted lower limit for a significant quantity is 25 litres.

NOTE 3: For quantities less than 25 litres, it is sufficient to take precautions to prevent the escape of liquid.

NOTE 4: Products of combustion of liquid are considered to be smoke and gas.

421.1.6   Materials used for the construction of enclosures of electrical equipment shall comply with the resistance to heat and fire requirements in an appropriate product standard.

Where no product standard exists, the materials of an enclosure shall withstand the highest temperature likely to be produced by the electrical equipment in normal use.

421.7   *Deleted by BS 7671:2008 Amendment No 1*

422   PRECAUTIONS WHERE PARTICULAR RISKS OF FIRE EXIST

422.1   General

The requirements of this regulation shall be applied in addition to those of Section 421 for installations in locations where any of the conditions of external influence described in Regulations 422.2 to 6 exist.

422.1.1   Except for wiring systems meeting the requirements of Regulation 422.3.5, electrical equipment shall be restricted to that necessary to the use of the locations given in Regulation 422.1.

422.1.2   Electrical equipment shall be so selected and erected that its normal temperature rise and foreseeable temperature rise during a fault cannot cause a fire. This shall be achieved by the construction of the equipment or by additional protective measures taken during erection.

Special measures are not necessary where the temperature of surfaces is unlikely to cause combustion of nearby substances.

422.1.3   A temperature cut-out device shall have manual reset only.

422.2   Conditions for evacuation in an emergency

The following regulations refer to conditions:

   BD2: Low density occupation, difficult conditions of evacuation

   BD3: High density occupation, easy conditions of evacuation

   BD4: High density occupation, difficult conditions of evacuation

(Refer to Appendix 5.)

NOTE: Authorities such as those responsible for building construction, public gatherings, fire prevention, hospitals, etc. may specify which BD condition is applicable.

422.2.1   In conditions BD2, BD3 or BD4, wiring systems shall not encroach on escape routes unless the wiring in the wiring system is provided with sheaths or enclosures, provided by the cable management system itself or by other means.

Wiring systems encroaching on escape routes shall not be within arm's reach unless they are provided with protection against mechanical damage likely to occur during an evacuation.

Wiring systems in escape routes shall be as short as practicable.

Wiring systems shall be non-flame propagating. Compliance with this requirement is ensured through one or more of the following, as relevant:

   (i)   Cable shall meet the relevant part of the BS EN 60332-3 series and the requirements of BS EN 61034-2

   (ii)   Non-flame propagating conduit systems shall meet the requirements of BS EN 61386

   (iii)   Non-flame propagating cable trunking systems and cable ducting systems shall meet the requirements of BS EN 50085

   (iv)   Cable tray systems and cable ladder systems classified as non-flame propagating according to BS EN 61537

   (v)   Powertrack systems meeting the requirements of BS EN 61534.

In conditions BD2, BD3 or BD4, wiring systems that are supplying safety circuits shall have a resistance to fire rating of either the time authorized by regulations for building elements or one hour in the absence of such a regulation.

Wiring within escape routes shall have a limited rate of smoke production. Cables meeting a minimum of 60% light transmittance in accordance with BS EN 61034-2 shall be selected.

422.2.2   In conditions BD2, BD3 or BD4, switchgear or controlgear shall be accessible only to authorized persons. If switchgear or controlgear is placed in an escape route, it shall be enclosed in a cabinet or an enclosure constructed of non-combustible or not readily combustible material.

These requirements do not apply to items of switchgear or controlgear installed to facilitate evacuation, such as fire alarm call points.

422.2.3   In escape routes where conditions BD3 or BD4 exist, the use of electrical equipment containing flammable liquids is not permitted.

This requirement does not apply to individual capacitors incorporated in equipment, such as a capacitor installed in a discharge luminaire or a motor starter.

422.3   Locations with risks of fire due to the nature of processed or stored materials

The requirements of this regulation shall be applied in addition to those of Section 421 in locations where BE2 conditions exist.

BE2 conditions exist where there is a risk of fire due to the manufacture, processing or storage of flammable materials including the presence of dust.

(Refer to Appendix 5.)

This regulation does not apply to selection and erection of installations in locations with explosion risks, see BS EN 60079-14 and BS EN 61241-14.

NOTE 1: Examples of locations presenting BE2 conditions include barns (due to the accumulation of dust and fibres), woodworking facilities, paper mills and textile factories (due to the storage and processing of combustible materials).

NOTE 2: Quantities of flammable materials or the surface or volume of the location may be regulated by national authorities.

422.3.1   Except for equipment for which an appropriate product standard specifies requirements, a luminaire shall be kept at an adequate distance from combustible materials. Unless otherwise recommended by the manufacturer, a small spotlight or projector shall be installed at the following minimum distance from combustible materials:

(i)   Rating up to 100 W   0.5 m

(ii)   Over 100 and up to 300 W 0.8 m

(iii) Over 300 and up to 500 W 1.0 m

Lamps and other components of luminaires shall be protected against foreseeable mechanical stresses. Such protective means shall not be fixed to lampholders unless they form an integral part of the luminaire or are fitted in accordance with the manufacturer's instructions.

A luminaire with a lamp that could eject flammable materials in case of failure shall be constructed with a safety protective shield for the lamp in accordance with the manufacturer's instructions.

422.3.2   Measures shall be taken to prevent an enclosure of electrical equipment such as a heater or resistor from exceeding the following temperatures:

(i)   90 °C under normal conditions, and

(ii)   115 °C under fault conditions.

Where materials such as dust or fibres sufficient to cause a fire hazard could accumulate on an enclosure of electrical equipment, adequate measures shall be taken to prevent an enclosure of electrical equipment from exceeding the temperatures stated above.

NOTE: Luminaires marked ▽D in compliance with BS EN 60598-2-24 have limited surface temperature.

422.3.3   Switchgear or controlgear shall be installed outside the location unless:

(i)   it is suitable for the location, or

(ii)   it is installed in an enclosure providing a degree of protection of at least IP4X or, in the presence of dust, IP5X or, in the presence of electrically conductive dust, IPX6, except where Regulation 422.3.11 applies.

422.3.4   A cable shall, as a minimum, satisfy the test under fire conditions specified in BS EN 60332-1-2.

A cable not completely embedded in non-combustible material such as plaster or concrete or otherwise protected from fire shall meet the flame propagation characteristics as specified in BS EN 60332-1-2 .

A conduit system shall satisfy the test under fire conditions specified in BS EN 61386-1.

A cable trunking system or cable ducting system shall satisfy the test under fire conditions specified in BS EN 50085.

A cable tray system or cable ladder shall satisfy the test under fire conditions specified in BS EN 61537.

A powertrack system shall satisfy the test under fire conditions specified in the BS EN 61534 series.

Precautions shall be taken such that the cable or wiring system cannot propagate flame.

Where the risk of flame propagation is high the cable shall meet the flame propagation characteristics specified in the appropriate part of the BS EN 60332-3 series.

NOTE: The risk of flame propagation can be high where cables are bunched or installed in long vertical runs.

422.3.5   A wiring system which passes through the location but is not intended to supply electrical equipment in the location shall:

(i)   meet the requirements of Regulation 422.3.4, and

(ii)   have no connection or joint within the location, unless the connection or joint is installed within an enclosure that does not adversely affect the flame propagation characteristics of the wiring system, and

(iii) be protected against overcurrent in accordance with the requirements of Regulation 422.3.10, and

(iv)   not employ bare live conductors.

422.3.6   *Not used*

422.3.7   A motor which is automatically or remotely controlled or which is not continuously supervised shall be protected against excessive temperature by a protective device with manual reset. A motor with star-delta starting shall be protected against excessive temperature in both the star and delta configurations.

422.3.8   Every luminaire shall:

   (i)   be appropriate for the location, and

   (ii) be provided with an enclosure providing a degree of protection of at least IP4X or, in the presence of dust, IP5X or, in the presence of electrically conductive dust, IPX6, and

   (iii) have a limited surface temperature in accordance with BS EN 60598-2-24, and

   (iv) be of a type that prevents lamp components from falling from the luminaire.

In locations where there may be fire hazards due to dust or fibres, luminaires shall be installed so that dust or fibres cannot accumulate in dangerous amounts.

422.3.9   Wiring systems, other than mineral insulated cables, busbar trunking systems or powertrack systems shall be protected against insulation faults:

   (i)   in a TN or TT system, by an RCD having a rated residual operating current ($I_{\Delta n}$) not exceeding 300 mA according with Regulation 531.2.4 and to relevant product standards.

   Where a resistive fault may cause a fire, e.g. for overhead heating with heating film elements, the rated residual operating current shall not exceed 30 mA.

   (ii) in an IT system, by an insulation monitoring device with audible and visual signals provided in accordance with Regulation 538.1. Disconnection times in the event of a second fault are given in Chapter 41. Alternatively, RCDs with a rated residual operating current as specified in (i) may be used. In the event of a second fault see Chapter 41 for disconnection times.

422.3.10   Circuits supplying or traversing locations where BE2 conditions exist shall be protected against overload and against fault current by protective devices located outside and on the supply side of these locations. Circuits originating inside these locations shall be protected against overcurrent by protective devices located at their origin.

422.3.11   Regardless of the nominal voltage of a circuit supplied at extra-low voltage, live parts shall be either:

   (i)   contained in enclosures affording a degree of protection of at least IPXXB or IP2X, or

   (ii) provided with insulation capable of withstanding a test voltage of 500 V d.c. for 1 minute.

These requirements are in addition to those of Section 414.

422.3.12   A PEN conductor shall not be used. This requirement does not apply to a circuit traversing the location.

422.3.13   Except as permitted by Regulation 537.1.2, every circuit shall be provided with a means of isolation from all live supply conductors by a linked switch or a linked circuit-breaker.

NOTE: Provision may be made for isolation of a group of circuits by a common means, if the service conditions allow this.

422.3.100   Flexible cables shall be of the following construction:

   (i)   Heavy duty type having a voltage rating of not less than 450/750 V, or

   (ii) suitably protected against mechanical damage.

422.3.101   A heating appliance shall be fixed.

422.3.102   A heat storage appliance shall be of a type which prevents the ignition of combustible dusts or fibres by the heat storing core.

422.4   Combustible constructional materials

The requirements of this regulation shall be applied in addition to those of Section 421 in locations where CA2 conditions exist.

CA2 conditions exist where a building is mainly constructed of combustible materials, such as wood.

(Refer to Appendix 5.)

422.4.1   Precautions shall be taken to ensure that electrical equipment cannot cause the ignition of walls, floors or ceilings. In prefabricated hollow walls containing a pre-installed wiring system including accessories, all boxes and enclosures shall have a degree of protection of at least IP3X where the wall is liable to be drilled during erection.

422.4.2   Except for equipment for which an appropriate product standard specifies requirements, a luminaire shall be kept at an adequate distance from combustible materials. Unless otherwise recommended by the manufacturer, a small spotlight or projector shall be installed at the following minimum distance from combustible materials:

(i)   Rating up to 100 W   0.5 m

(ii)   Over 100 and up to 300 W   0.8 m

(iii) Over 300 and up to 500 W   1.0 m

Lamps and other components of luminaires shall be protected against foreseeable mechanical stresses. Such protective means shall not be fixed to lampholders unless they form an integral part of the luminaire or are fitted in accordance with the manufacturer's instructions.

A luminaire with a lamp that could eject flammable materials in case of failure shall be constructed with a safety protective shield for the lamp in accordance with the manufacturer's instructions.

NOTE: Refer to Table 55.2 regarding the marking of luminaires and their installation or mounting on normally flammable surfaces.

422.4.100   Electrical equipment, e.g. installation boxes and distribution boards, installed on or in a combustible wall shall comply with the relevant standard for enclosure temperature rise.

422.4.101   Electrical equipment that does not comply with Regulation 422.4.100 shall be enclosed with a suitable thickness of non-flammable material. The effect of the material on the heat dissipation from electrical equipment shall be taken into account.

422.4.102   Cables shall comply with the requirements of BS EN 60332-1-2.

422.4.103   Conduit and trunking systems shall be in accordance with BS EN 61386-1 and BS EN 50085-1 respectively and shall meet the fire-resistance tests within these standards.

422.5   Fire propagating structures

The requirements of this regulation shall be applied in addition to those of Section 421 in locations where CB2 conditions exist.

CB2 conditions relate to the propagation of fire and exist where a building has a shape and dimensions which facilitate the spread of fire (e.g. chimney effect), such as high-rise buildings or where a building has a forced ventilation system.

NOTE: Fire detectors may be provided which ensure the implementation of measures for preventing propagation of fire, for example, the closing of fireproof shutters in ducts, troughs or trunking.

422.5.1   In structures where the shape and dimensions are such as will facilitate the spread of fire, precautions shall be taken to ensure that the electrical installation cannot propagate a fire (e.g. chimney effect).

422.6   Selection and erection of installations in locations of national, commercial, industrial or public significance

The requirements of Regulation 422.1 shall apply to locations that include buildings or rooms with assets of significant value. Examples include national monuments, museums and other public buildings. Buildings such as railway stations and airports are generally considered to be of public significance. Buildings or facilities such as laboratories, computer centres and certain industrial and storage facilities can be of commercial or industrial significance.

The following measures may be considered:

(i)   Installation of mineral insulated cables according to BS EN 60702

(ii)   Installation of cables with improved fire-resisting characteristics in case of a fire hazard

(iii) Installation of cables in non-combustible solid walls, ceilings and floors

(iv)   Installation of cables in areas with constructional partitions having a fire-resisting capability for a time of 30 minutes or 90 minutes, the latter in locations housing staircases and needed for an emergency escape.

Where these measures are not practicable improved fire protection may be possible by the use of reactive fire protection systems.

423 PROTECTION AGAINST BURNS

423.1 Excepting equipment for which a Harmonized Standard specifies a limiting temperature, an accessible part of fixed electrical equipment within arm's reach shall not attain a temperature in excess of the appropriate limit stated in Table 42.1. Each such part of the fixed installation likely to attain under normal load conditions, even for a short period, a temperature exceeding the appropriate limit in Table 42.1 shall be guarded so as to prevent accidental contact.

TABLE 42.1 –
Temperature limit under normal load conditions for an
accessible part of equipment within arm's reach

Accessible part	Material of accessible surfaces	Maximum temperature (°C)
A hand-held part	Metallic Non-metallic	55 65
A part intended to be touched but not hand-held	Metallic Non-metallic	70 80
A part which need not be touched for normal operation	Metallic Non-metallic	80 90

CHAPTER 43

PROTECTION AGAINST OVERCURRENT

CONTENTS

CHAPTER 43

PROTECTION AGAINST OVERCURRENT

430   INTRODUCTION

430.1   Scope

This chapter provides requirements for the protection of live conductors from the effects of overcurrent.

This chapter describes how live conductors are protected by one or more devices for the automatic disconnection of the supply in the event of overload current (Section 433) and fault current (Section 434), except in cases where the overcurrent is limited in accordance with Section 436 or where the conditions described in Regulation 433.3 (omission of devices for protection against overload) or Regulation 434.3 (omission of devices for protection against fault current) are met. Co-ordination of overload current protection and fault current protection is also covered (Section 435).

NOTE 1: Live conductors protected against overload in accordance with Section 433 are also considered to be protected against faults likely to cause overcurrents of a magnitude similar to overload currents.

NOTE 2: The requirements of this chapter do not take account of external influences.

NOTE 3: Protection of conductors according to these regulations does not necessarily protect the equipment connected to the conductors.

NOTE 4: Flexible cables connecting equipment by plugs and socket-outlets to a fixed installation are outside the scope of this chapter.

NOTE 5: Disconnection does not necessarily mean isolation in this chapter.

430.2   Not used

430.3   General requirement

A protective device shall be provided to break any overcurrent in the circuit conductors before such a current could cause a danger due to thermal or mechanical effects detrimental to insulation, connections, joints, terminations or the surroundings of the conductors.

The protection against overload current and the protection against fault current shall be co-ordinated in accordance with Section 435.

NOTE: An overcurrent may be an overload current or a fault current.

431   PROTECTION ACCORDING TO THE NATURE OF THE CIRCUITS AND THE DISTRIBUTION SYSTEM

431.1   Protection of line conductors

431.1.1   Except where Regulation 431.1.2 applies, detection of overcurrent shall be provided for all line conductors and shall cause the disconnection of the conductor in which the overcurrent is detected, but not necessarily the disconnection of the other line conductors except where the disconnection of one line conductor could cause damage or danger.

If disconnection of a single phase may cause danger, for example in the case of a three-phase motor, appropriate precautions shall be taken.

431.1.2   In a TN or TT system, for a circuit supplied between line conductors and in which the neutral conductor is not distributed, overcurrent detection need not be provided for one of the line conductors, provided that both the following conditions are simultaneously fulfilled:

(i)   There exists, in the same circuit or on the supply side, differential protection intended to detect unbalanced loads and cause disconnection of all the line conductors, and

(ii) the neutral conductor is not distributed from an artificial neutral point of the circuits situated on the load side of the differential protective device mentioned in (i).

431.2 Protection of the neutral conductor

431.2.1 TN or TT system

The neutral conductor shall be protected against short-circuit current.

Where the cross-sectional area of the neutral conductor is at least equivalent to that of the line conductors, and the current in the neutral is not expected to exceed the value in the line conductors, it is not necessary to provide overcurrent detection for the neutral conductor or a disconnecting device for that conductor.

Where the cross-sectional area of the neutral conductor is less than that of the line conductors, it is necessary to provide overcurrent detection for the neutral conductor, appropriate to the cross-sectional area of the conductor. The overcurrent detection shall cause the disconnection of the line conductors, but not necessarily of the neutral conductor.

Except for disconnection complying with Regulation 537.1.2 the requirements for a neutral conductor apply to a PEN conductor.

Where the current in the neutral conductor is expected to exceed that in the line conductors refer to Regulation 431.2.3.

431.2.2 IT system

The neutral conductor shall not be distributed unless one of the following is met:

(i) Overcurrent detection is provided for the neutral conductor of every circuit. The overcurrent detection shall cause the disconnection of all the live conductors of the corresponding circuit, including the neutral conductor

(ii) The particular neutral conductor is effectively protected against short-circuit by a protective device installed on the supply side, for example at the origin of the installation, in accordance with Regulation 434.5

(iii) The particular circuit is protected by an RCD with a rated residual operating current ($I_{\Delta n}$) not exceeding 0.2 times the current-carrying capacity of the corresponding neutral conductor. The RCD shall disconnect all the live conductors of the corresponding circuit, including the neutral conductor. The device shall have sufficient breaking capacity for all poles.

431.2.3 Harmonic currents

Overcurrent detection shall be provided for the neutral conductor in a multiphase circuit where the harmonic content of the line currents is such that the current in the neutral conductor may exceed the current-carrying capacity of that conductor. The overcurrent detection shall cause disconnection of the line conductors but not necessarily the neutral conductor. Where the neutral is disconnected the requirements of Regulation 431.3 are applicable.

431.3 Disconnection and reconnection of the neutral conductor

Where disconnection of the neutral conductor is required, disconnection and reconnection shall be such that the neutral conductor shall not be disconnected before the line conductors and shall be reconnected at the same time as or before the line conductors.

432 NATURE OF PROTECTIVE DEVICES

A protective device shall be of the appropriate type indicated in Regulations 432.1 to 3.

432.1 Protection against both overload current and fault current

Except as permitted by Regulation 434.5.1, a device providing protection against both overload and fault current shall be capable of breaking, and for a circuit-breaker making, any overcurrent up to and including the maximum prospective fault current at the point where the device is installed.

432.2 Protection against overload current only

A device providing protection against overload current is generally an inverse-time-lag protective device whose rated short-circuit breaking capacity may be below the value of the maximum prospective fault current at the point where the device is installed. Such a device shall satisfy the relevant requirements of Section 433.

432.3 Protection against fault current only

A device providing protection against fault current only shall be installed where overload protection is achieved by other means or where Section 433 permits overload protection to be dispensed with. Except as permitted by Regulation 434.5.1, a device shall be capable of breaking, and for a circuit-breaker making, the fault current up to and including the prospective fault current. Such a device shall satisfy the relevant requirements of Section 434.

NOTE: Such a device may be:
    (i)   a circuit-breaker with a short-circuit release, or
    (ii)   a fuse.

432.4 Characteristics of protective devices

The time/current characteristics of an overcurrent protective device shall comply with those specified in BS 88 series, BS 3036, BS EN 60898, BS EN 60947-2 or BS EN 61009-1.

The use of another device is not precluded provided that its time/current characteristics provide a level of protection not less than that given by the devices listed above.

433 PROTECTION AGAINST OVERLOAD CURRENT

433.1 Co-ordination between conductor and overload protective device

Every circuit shall be designed so that a small overload of long duration is unlikely to occur.

433.1.1   The operating characteristics of a device protecting a conductor against overload shall satisfy the following conditions:

  (i)   The rated current or current setting of the protective device (I_n) is not less than the design current (I_b) of the circuit, and

  (ii)   the rated current or current setting of the protective device (I_n) does not exceed the lowest of the current-carrying capacities (I_z) of any of the conductors of the circuit, and

  (iii) the current (I_2) causing effective operation of the protective device does not exceed 1.45 times the lowest of the current-carrying capacities (I_z) of any of the conductors of the circuit.

For adjustable protective devices, the rated current (I_n) is the current setting selected.

The current (I_2) ensuring effective operation of the protective device is given in the product standard or may be provided by the manufacturer.

NOTE 1: Where overload protection is provided by BS 3036 fuses, refer to Regulation 433.1.101.

NOTE 2: Protection in accordance with this regulation may not ensure protection in all cases, for example, where sustained overcurrents less than I_2 occur.

433.1.100   Where the protective device is a general purpose type (gG) fuse to BS 88-2, a fuse to BS 88-3, a circuit-breaker to BS EN 60898, a circuit-breaker to BS EN 60947-2 or a residual current circuit-breaker with integral overcurrent protection (RCBO) to BS EN 61009-1, compliance with conditions (i) and (ii) also results in compliance with condition (iii) of Regulation 433.1.1.

433.1.101   Where the protective device is a semi-enclosed fuse to BS 3036 compliance with condition (iii) of Regulation 433.1.1 is afforded if its rated current (I_n) does not exceed 0.725 times the current-carrying capacity (I_z) of the lowest rated conductor in the circuit protected.

433.1.102   For direct buried cables or cables in buried ducts where the tabulated current-carrying capacity is based on an ambient temperature of 20 °C compliance with condition (iii) of Regulation 433.1.1 is afforded if the rated current or current setting of the protective device (I_n) does not exceed 0.9 times the current-carrying capacity (I_z) of the lowest rated conductor in the circuit protected.

433.1.103   Accessories to BS 1363 may be supplied through a ring final circuit, with or without unfused spurs, protected by a 30 A or 32 A protective device complying with BS 88 series, BS 3036, BS EN 60898, BS EN 60947-2 or BS EN 61009-1 (RCBO). The circuit shall be wired with copper conductors having line and neutral conductors with a minimum cross-sectional area of 2.5 mm² except for two-core mineral insulated cables complying with BS EN 60702-1, for which the minimum cross-sectional area is 1.5 mm². Such circuits are deemed to meet the requirements of Regulation 433.1.1 if the current-carrying capacity (I_z) of the cable is not less than 20 A and if, under the intended conditions of use, the load current in any part of the circuit is unlikely to exceed for long periods the current-carrying capacity (I_z) of the cable.

433.2   Position of devices for protection against overload

433.2.1   Except where Regulation 433.2.2 or 433.3 applies, a device for protection against overload shall be installed at the point where a reduction occurs in the value of the current-carrying capacity of the conductors of the installation.

NOTE: A reduction in current-carrying capacity may be due to a change in cross-sectional area, method of installation, type of cable or conductor, or in environmental conditions.

433.2.2   The device protecting a conductor against overload may be installed along the run of that conductor if the part of the run between the point where a change occurs (in cross-sectional area, method of installation, type of cable or conductor, or in environmental conditions) and the position of the protective device has neither branch circuits nor outlets for connection of current-using equipment and fulfils at least one of the following conditions:

(i)   It is protected against fault current in accordance with the requirements stated in Section 434

(ii)   Its length does not exceed 3 m, it is installed in such a manner as to reduce the risk of fault to a minimum, and it is installed in such a manner as to reduce to a minimum the risk of fire or danger to persons (see also Regulation 434.2.1).

433.3   Omission of devices for protection against overload

This regulation shall not be applied to installations situated in locations presenting a fire risk or risk of explosion or where the requirements for special installations and locations specify different conditions.

433.3.1   General

A device for protection against overload need not be provided:

(i)   for a conductor situated on the load side of the point where a reduction occurs in the value of current-carrying capacity, where the conductor is effectively protected against overload by a protective device installed on the supply side of that point

(ii)   for a conductor which, because of the characteristics of the load or the supply, is not likely to carry overload current, provided that the conductor is protected against fault current in accordance with the requirements of Section 434

(iii)   at the origin of an installation where the distributor provides an overload device and agrees that it affords protection to the part of the installation between the origin and the main distribution point of the installation where further overload protection is provided.

433.3.2   Position or omission of devices for protection against overload in IT systems

433.3.2.1   The provisions in Regulations 433.2.2 and 433.3 for an alternative position or omission of devices for protection against overload are not applicable to IT systems unless each circuit not protected against overload is protected by one of the following means:

(i)   Use of the protective measures described in Regulation 413.2

(ii)   An RCD that will operate immediately on the second fault

(iii)   For permanently supervised systems only, the use of an insulation monitoring device which either:

    (a)   causes the disconnection of the circuit when the first fault occurs, or

    (b)   gives a signal indicating the presence of a fault. The fault shall be corrected in accordance with operational requirements and recognition of the consequences of a second fault.

433.3.2.2   In an IT system without a neutral conductor it is permitted to omit the overload protective device in one of the line conductors if an RCD is installed in each circuit.

433.3.3   Omission of devices for protection against overload for safety reasons

The omission of devices for protection against overload is permitted for circuits supplying current-using equipment where unexpected disconnection of the circuit could cause danger or damage.

Examples of such circuits are:

(i)   the exciter circuit of a rotating machine

(ii)   the supply circuit of a lifting magnet

(iii)   the secondary circuit of a current transformer

(iv)   a circuit supplying a fire extinguishing device

(v) a circuit supplying a safety service, such as a fire alarm or a gas alarm

(vi) a circuit supplying medical equipment used for life support in specific medical locations where an IT system is incorporated.

NOTE: In such situations consideration should be given to the provision of an overload alarm.

433.4 Overload protection of conductors in parallel

Where a single protective device protects two or more conductors in parallel there shall be no branch circuits or devices for isolation or switching in the parallel conductors.

This regulation does not preclude the use of ring final circuits with or without spur connections.

433.4.1 Equal current sharing between parallel conductors

Except for a ring final circuit, where spurs are permitted, where a single device protects conductors in parallel and the conductors are sharing currents equally, the value of I_z to be used in Regulation 433.1.1 is the sum of the current-carrying capacities of the parallel conductors.

It is deemed that current sharing is equal if the requirements of the first indent of Regulation 523.7(i) are satisfied.

433.4.2 Unequal current sharing between parallel conductors

Where the use of a single conductor is impractical and the currents in the parallel conductors are unequal, the design current and requirements for overload protection for each conductor shall be considered individually.

NOTE: Currents in parallel conductors are considered to be unequal if the difference between the currents is more than 10 % of the design current for each conductor. Refer to paragraph 2 of Appendix 10.

434 PROTECTION AGAINST FAULT CURRENT

This section only considers the case of a fault between conductors belonging to the same circuit.

434.1 Determination of prospective fault current

The prospective fault current shall be determined at every relevant point of the installation. This shall be done by calculation, measurement or enquiry.

434.2 Position of devices for protection against fault current

A device providing protection against fault current shall be installed at the point where a reduction in the cross-sectional area or other change causes a reduction in the current-carrying capacity of the conductors, except where Regulation 434.2.1, 434.2.2 or 434.3 applies.

The requirements in Regulations 434.2.1 and 434.2.2 shall not be applied to installations situated in locations presenting a fire risk or risk of explosion or where special requirements for certain locations specify different conditions.

434.2.1 Except where Regulation 434.2.2 or 434.3 applies, a device for protection against fault current may be installed other than as specified in Regulation 434.2, under the following conditions:

In the part of the conductor between the point of reduction of cross-sectional area or other change and the position of the protective device there shall be no branch circuits or socket-outlets and that part of the conductor shall:

(i) not exceed 3 m in length, and

(ii) be installed in such a manner as to reduce the risk of fault to a minimum, and

NOTE: This condition may be obtained, for example, by reinforcing the protection of the wiring against external influences.

(iii) be installed in such a manner as to reduce to a minimum the risk of fire or danger to persons.

434.2.2 The device protecting a conductor may be installed on the supply side of the point where a change occurs (in cross-sectional area, method of installation, type of cable or conductor, or in environmental conditions) provided that it possesses an operating characteristic such that it protects the wiring situated on the load side against fault current, in accordance with Regulation 434.5.2.

434.3 Omission of devices for protection against fault current

A device for protection against fault current need not be provided for:

(i) a conductor connecting a generator, transformer, rectifier or an accumulator battery to the associated control panel where the protective device is placed in the panel

(ii) a circuit where disconnection could cause danger for the operation of the installation concerned, such as those quoted in Regulation 433.3.3

(iii) certain measuring circuits

(iv) the origin of an installation where the distributor installs one or more devices providing protection against fault current and agrees that such a device affords protection to the part of the installation between the origin and the main distribution point of the installation where further protection against fault current is provided,

provided that both of the following conditions are simultaneously fulfilled:

(v) The wiring is carried out in such a way as to reduce the risk of fault to a minimum (see item (ii) of Regulation 434.2.1), and

(vi) the wiring is installed in such a manner as to reduce to a minimum the risk of fire or danger to persons.

434.4   Fault current protection of conductors in parallel

A single protective device may protect conductors in parallel against the effects of fault currents provided that the operating characteristic of the device results in its effective operation should a fault occur at the most onerous position in one of the parallel conductors. Account shall be taken of the sharing of the fault currents between the parallel conductors. A fault can be fed from both ends of a parallel conductor.

If operation of a single protective device may not be effective then one or more of the following measures shall be taken:

(i) The wiring shall be installed in such a manner as to reduce to a minimum the risk of a fault in any parallel conductor, for example, by the provision of protection against mechanical damage. In addition, conductors shall be installed in such a manner as to reduce to a minimum the risk of fire or danger to persons

(ii) For two conductors in parallel, a fault current protective device shall be provided at the supply end of each parallel conductor

(iii) For more than two conductors in parallel, a fault current protective device shall be provided at the supply and load ends of each parallel conductor.

NOTE: Further information is given in paragraph 3 of Appendix 10.

434.5   Characteristics of a fault current protective device

Every fault current protective device shall meet the requirements of this regulation.

434.5.1   Except where the following paragraph applies, the rated short-circuit breaking capacity of each device shall be not less than the maximum prospective fault current at the point at which the device is installed.

A lower breaking capacity is permitted if another protective device or devices having the necessary rated short-circuit breaking capacity is installed on the supply side. In this situation, the characteristics of the devices shall be co-ordinated so that the energy let-through of these devices does not exceed that which can be withstood, without damage, by the device(s) on the load side.

NOTE: Technical data for the selection of protective devices can be requested from the manufacturer.

434.5.2   A fault occurring at any point in a circuit shall be interrupted within a time such that the fault current does not cause the permitted limiting temperature of any conductor or cable to be exceeded.

For a fault of very short duration (less than 0.1 sec), for current limiting devices $k^2 S^2$ shall be greater than the value of let-through energy (I^2t) quoted for the Class of protective device to BS EN 60898-1, BS EN 60898-2 or BS EN 61009-1, or as quoted by the manufacturer.

The time, t, in which a given fault current will raise the live conductors from the highest permissible temperature in normal duty to the limiting temperature, can, as an approximation, be calculated from the formula:

$$t = \frac{k^2 S^2}{I^2}$$

where:

t   is the duration in seconds

S   is the cross-sectional area of conductor in mm^2

I   is the effective fault current, in amperes, expressed for a.c. as the rms value, due account being taken of the current limiting effect of the circuit impedances k is a factor taking account of the resistivity, temperature coefficient and heat capacity of the conductor material, and the appropriate initial and final temperatures. For common materials, the values of k are shown in Table 43.1.

TABLE 43.1 –
Values of k for common materials, for calculation of the effects of fault current
for disconnection times up to 5 seconds

Conductor insulation	Thermoplastic				Thermosetting		Mineral insulated	
	90 °C		70 °C		90 °C	60 °C	Thermoplastic sheath	Bare (unsheathed)
Conductor cross-sectional area	≤ 300 mm²	> 300 mm²	≤ 300 mm²	> 300 mm²				
Initial temperature	90 °C		70 °C		90 °C	60 °C	70 °C	105 °C
Final temperature	160 °C	140 °C	160 °C	140 °C	250 °C	200 °C	160 °C	250 °C
Copper conductor	k = 100	k = 86	k = 115	k = 103	k = 143	k = 141	k = 115	k = 135/115[a]
Aluminium conductor	k = 66	k = 57	k = 76	k = 68	k = 94	k = 93		
Tin soldered joints in copper conductors	k = 100	k = 86	k = 115	k = 103	k = 100	k = 122		

[a] This value shall be used for bare cables exposed to touch.

NOTE 1: The rated current or current setting of the fault current protective device may be greater than the current-carrying capacity of the cable.

NOTE 2: Other values of k can be determined by reference to BS 7454.

434.5.3　For a busbar trunking system complying with BS EN 60439-2 or a powertrack system complying with BS EN 61534, one of the following requirements shall apply:

(i) The rated short-time withstand current (I_{cw}) and the rated peak withstand current of a busbar trunking system or powertrack system shall be not lower than the rms value of the prospective fault current and the prospective fault peak current value, respectively. The maximum time for which the I_{cw} is defined for the busbar trunking system shall be greater than the maximum operating time of the protective device

(ii) The rated conditional short-circuit current of the busbar trunking system or powertrack system associated with a specific protective device shall be not lower than the prospective fault current.

435　CO-ORDINATION OF OVERLOAD CURRENT AND FAULT CURRENT PROTECTION

435.1　Protection afforded by one device

A protective device providing protection against both overload current and fault current shall fulfil the requirements of the relevant regulations in Sections 433 and 434.

Except as required by Regulation 434.4 or 434.5.2, where an overload protective device complying with Regulation 433.1 is to provide fault current protection and has a rated short-circuit breaking capacity not less than the value of the maximum prospective fault current at its point of installation, it may be assumed that the requirements of this section are satisfied as regards fault current protection of the conductors on the load side of that point.

The validity of the assumption shall be checked, where there is doubt, for conductors in parallel and for certain types of circuit-breaker e.g. non-current-limiting types.

435.2　Protection afforded by separate devices

The requirements of Sections 433 and 434 apply, respectively, to the overload current protective device and the fault current protective device.

The characteristics of the devices shall be co-ordinated so that the energy let through by the fault current protective device does not exceed that which can be withstood without damage by the overload protective device (see Regulation 536.1). This requirement does not exclude the type of co-ordination specified in BS EN 60947-4-1. For a circuit incorporating a motor starter, this requirement does not preclude the type of co-ordination described in BS EN 60947-4-1, in respect of which the advice of the manufacturer of the starter shall be sought.

436   LIMITATION OF OVERCURRENT BY THE CHARACTERISTICS OF THE SUPPLY

Conductors are considered to be protected against overload current and fault current where they are supplied from a source incapable of supplying a current exceeding the current-carrying capacity of the conductors (e.g. certain bell transformers, certain welding transformers and certain types of thermoelectric generating set).

CHAPTER 44
PROTECTION AGAINST VOLTAGE DISTURBANCES AND ELECTROMAGNETIC DISTURBANCES

CONTENTS

CHAPTER 44

PROTECTION AGAINST VOLTAGE DISTURBANCES AND ELECTROMAGNETIC DISTURBANCES

440 INTRODUCTION

440.1 Scope

These requirements are intended to provide requirements for the safety of electrical installations in the event of voltage disturbances and electromagnetic disturbances generated for different specified reasons.

The requirements are not intended to apply to systems for distribution of energy to the public, or power generation and transmission for such systems, although such disturbances may be conducted into or between electrical installations via these supply systems. The requirements of this chapter are in addition to those of Chapter 43.

440.2 General

This chapter covers the protection of electrical installations and measures against voltage disturbances and electromagnetic disturbances. The requirements are arranged into four sections as follows:

(i) Section 442   Protection of low voltage installations against temporary overvoltages due to earth faults in the high voltage system and due to faults in the low voltage system

(ii) Section 443   Protection against overvoltages of atmospheric origin or due to switching

(iii) Section 444   Measures against electromagnetic disturbances

(iv) Section 445   Protection against undervoltage.

441   NOT USED

442 PROTECTION OF LOW VOLTAGE INSTALLATIONS AGAINST TEMPORARY OVERVOLTAGES DUE TO EARTH FAULTS IN THE HIGH VOLTAGE SYSTEM AND DUE TO FAULTS IN THE LOW VOLTAGE SYSTEM

442.1 Scope and object

This regulation provides requirements for the safety of the low voltage installation in the event of:

(i) a fault between the high voltage system and Earth in the transformer substation that supplies the low voltage installation

(ii) loss of the supply neutral in the low voltage system

(iii) short-circuit between a line conductor and neutral in the low voltage installation

(iv) accidental earthing of a line conductor of a low voltage IT system.

NOTE: In Great Britain, the requirements for the earthing of transformers that provide a supply from a system for distribution of electricity in accordance with the Electricity Safety, Quality and Continuity Regulations 2002 are addressed in the Distribution Code.

442.1.1 General

Section 442 gives rules for the designer and installer of the substation. It is necessary to have the following information on the high voltage system:

(i) Quality of the system earthing

(ii) Maximum level of earth fault current

(iii) Resistance of the earthing arrangement.

The following regulations consider four situations which generally cause the most severe temporary overvoltages:

(iv) Fault between the high voltage system(s) and Earth (see Regulation 442.2)

(v) Loss of the neutral in a low voltage system (see Regulation 442.3)

(vi) Accidental earthing of a low voltage IT system (see Regulation 442.4)

(vii) Short-circuit in the low voltage installation (see Regulation 442.5).

442.1.2   Symbols

In Section 442 the following symbols are used (see Figure 44.1):

I_E part of the earth fault current in the high voltage system that flows through the earthing arrangement of the transformer substation

R_E resistance of the earthing arrangement of the transformer substation

R_A resistance of the earthing arrangement of the exposed-conductive-parts of the equipment of the low voltage installation

R_B resistance of the earthing arrangement of the low voltage system neutral, for low voltage systems in which the earthing arrangements of the transformer substation and of the low voltage system neutral are electrically independent

U_0 nominal a.c. rms line voltage to Earth

U_f power frequency fault voltage that appears in the low voltage system between exposed-conductive-parts and Earth for the duration of the fault

U_1 power frequency stress voltage between the line conductor and the exposed-conductive-parts of the low voltage equipment of the transformer substation during the fault

U_2 power frequency stress voltage between the line conductor and the exposed-conductive-parts of the low voltage equipment of the low voltage installation during the fault.

NOTE 1: The power frequency stress voltages (U_1 and U_2) are the voltages that appear across the insulation of low voltage equipment and across surge protective devices connected to the low voltage system.

The following additional symbols are used in respect of IT systems in which the exposed-conductive-parts of the equipment of the low voltage installation are connected to an earthing arrangement that is electrically independent of the earthing arrangement of the transformer substation.

I_h the fault current that flows through the earthing arrangement of the exposed-conductive-parts of the equipment of the low voltage installation during a period when there is a high voltage fault and a first fault in the low voltage installation (see Table 44.1)

I_d the fault current, in accordance with Regulation 411.6.2, that flows through the earthing arrangement of the exposed-conductive-parts of the low voltage installation during the first fault in a low voltage system (see Table 44.1)

Z is the impedance (for example, the IMD internal impedance or the artificial neutral impedance) between the low voltage system and an earthing arrangement.

NOTE 2: An earthing arrangement may be considered electrically independent of another earthing arrangement if a rise of potential with respect to Earth in one earthing arrangement does not cause an unacceptable rise of potential with respect to Earth in the other earthing arrangement.

442.2   Overvoltages in low voltage (LV) systems during a high voltage (HV) earth fault

In case of a fault to Earth in the HV side of the substation the following types of overvoltage may affect the LV installation:

(i)   Power frequency fault voltage (U_f)

(ii)   Power frequency stress voltages (U_1 and U_2).

Table 44.1 provides the relevant methods of calculation for the different types of overvoltage.

Fig 44.1 – Representative diagram for connections to Earth in the substation and the LV installation and the overvoltages occurring in case of faults

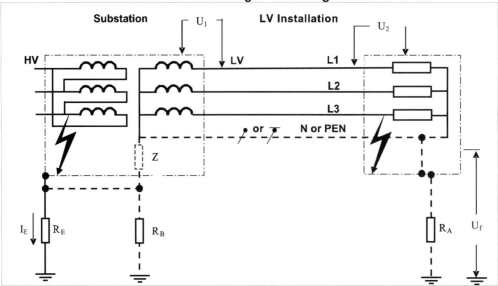

Where high and low voltage earthing systems exist in proximity to each other, two practices are presently used:

- interconnection of all high voltage (R_E) and low voltage (R_B) earthing systems
- separation of high voltage (R_E) from low voltage (R_B) earthing systems.

The general method used is interconnection. The high and low voltage earthing systems shall be interconnected if the low voltage system is totally confined within the area covered by the high voltage earthing system (see BS 7430).

NOTE 1: Details of the different types of system earthing are shown in Part 3 (TN, TT) and Appendix 9 (IT).

NOTE 2: In Great Britain, the requirements for the earthing of transformers that provide a supply from a system for distribution of electricity in accordance with the Electricity Safety, Quality and Continuity Regulations 2002 are addressed in the Distribution Code.

TABLE 44.1 – Power frequency stress voltages and power frequency fault voltage in the low voltage system

Type of system earthing	Type of earth connections	U_1	U_2	U_f
TT	R_E and R_B connected	U_0	$R_E.I_E + U_0$	0
	R_E and R_B separated	$R_E.I_E + U_0$	U_0	0
TN	R_E and R_B connected	U_0	U_0	$R_E.I_E$
	R_E and R_B separated	$R_E.I_E + U_0$	U_0	0
IT	R_E and Z connected R_E and R_A separated	U_0 $U_0.\sqrt{3}$	$R_E.I_E + U_0$ $R_E.I_E + U_0.\sqrt{3}$	0 $R_A.I_h$
	R_E and Z connected R_E and R_A interconnected	U_0 $U_0.\sqrt{3}$	U_0 $U_0.\sqrt{3}$	$R_E.I_E$ $R_E.I_E$
	R_E and Z separated R_E and R_A separated	$R_E.I_E + U_0$ $R_E.I_E + U_0.\sqrt{3}$	U_0 $U_0.\sqrt{3}$	0 $R_A.I_d$

☐ **With existing earth fault in the installation.**

NOTE 1: The requirements for U_1 and U_2 are derived from design criteria for insulation of low voltage equipment with regard to temporary power frequency overvoltage (see also Table 44.2).

NOTE 2: In a system whose neutral is connected to the earthing arrangement of the transformer substation, such temporary power frequency overvoltage is also to be expected across insulation which is not in an earthed enclosure where the equipment is outside a building.

NOTE 3: In TT and TN systems the terms 'connected' and 'separated' refer to the electrical connection between R_E and R_B. For IT systems the terms refer to the electrical connection between R_E and Z and the connection between R_E and R_A.

442.2.1   Magnitude and duration of power frequency fault voltage

The magnitude and duration of the fault voltage U_f, where specified in Table 44.1, which appears in the LV installation between exposed-conductive-parts and Earth shall not exceed the values given for U_f by the curve of Figure 44.2 for the duration of the fault. If the PEN conductor of the low voltage system is connected to Earth at more than one point it is permitted to double the value of U_f given in Figure 44.2.

Normally, the PEN conductor of the low voltage system is connected to Earth at more than one point. In this case, the total resistance is reduced. For these multiple earthed PEN conductors, U_f can be calculated as:

$$U_f = 0.5 \ R_f . I_f$$

Fig 44.2 – Tolerable fault voltage due to an earth fault in the HV system

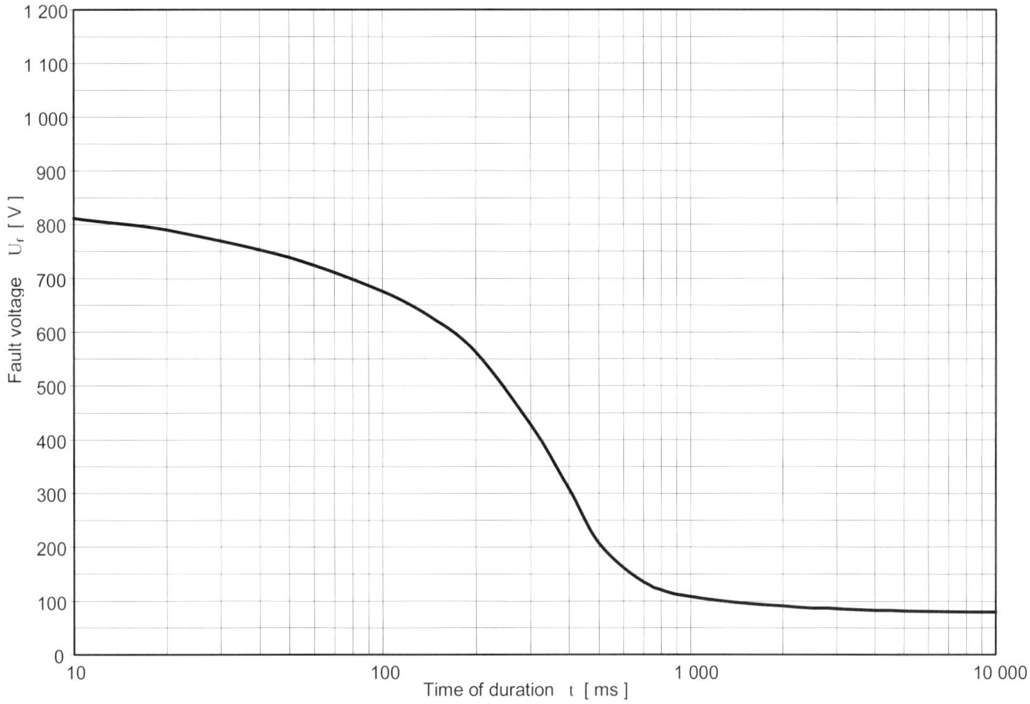

NOTE: The curve shown in Figure 44.2 is taken from IEC 61936-1. On the basis of probabilistic and statistical evidence this curve represents a low level of risk for the simple worst case where the low voltage system neutral conductor is earthed only at the transformer substation earthing arrangements. Guidance is provided in IEC 61936-1 concerning other situations.

442.2.2   Magnitude and duration of power frequency stress voltages

The magnitude and duration of the power frequency stress voltages (U_1 and U_2), where specified in Table 44.1, of the low voltage equipment in the low voltage installation due to an earth fault in the high voltage system, shall not exceed the requirements given in Table 44.2.

TABLE 44.2 – Permissible power frequency stress voltage

Duration of the earth fault in the high voltage system t	Permissible power frequency stress voltage on equipment in low voltage installations U
>5 s	$U_0 + 250$ V
≤5 s	$U_0 + 1200$ V

In systems without a neutral conductor U_0 shall be the line-to-line voltage.

NOTE 1:   The first line of the table relates to high voltage systems having long disconnection times, for example, isolated neutral and resonant earthed high voltage systems. The second line relates to high voltage systems having short disconnection times, for example, low-impedance earthed high voltage systems. Both lines together are relevant design criteria for insulation of low voltage equipment with regard to temporary power frequency overvoltage, BS EN 60664-1.

NOTE 2:   In a system whose neutral is connected to the earthing arrangement of the transformer substation, such temporary power frequency overvoltage is also to be expected across insulation which is not in an earthed enclosure where the equipment is outside a building.

442.2.3 Requirements for calculation of limits

The requirements of Regulations 442.2.1 and 442.2.2 are deemed to be fulfilled for installations receiving a supply at low voltage from a system for distribution of electricity to the public.

To fulfil the above requirements co-ordination between the HV system operator and the LV system installer is necessary. Compliance with the above requirements mainly falls into the responsibility of the substation installer/owner/operator who needs also to fulfil requirements provided by IEC 61936-1. Therefore, the calculation for U_1, U_2 and U_f is normally not necessary for the LV system installer.

Possible measures to fulfil the above requirements are, for example:

    (i)   separation of HV and LV earthing arrangements

    (ii)   change of LV system earthing

    (iii) reduction of earth resistance, R_E.

442.3   Power frequency stress voltage in case of loss of the neutral conductor in a TN or TT system

Consideration shall be given to the fact that, if the neutral conductor in a three-phase TN or TT system is interrupted, basic, double and reinforced insulation as well as components rated for the voltage between line and neutral conductors can be temporarily stressed with the line-to-line voltage. The stress voltage can reach up to $U = \sqrt{3}\ U_0$.

442.4   Power frequency stress voltage in the event of an earth fault in an IT system with distributed neutral

Consideration shall be given to the fact that, if a line conductor of an IT system is earthed accidentally, insulation or components rated for the voltage between line and neutral conductors can be temporarily stressed with the line-to-line voltage. The stress voltage can reach up to $U = \sqrt{3}\ U_0$.

442.5   Power frequency stress voltage in the event of short-circuit between a line conductor and the neutral conductor

Consideration shall be given to the fact that, if a short-circuit occurs in the low voltage installation between a line conductor and the neutral conductor, the voltage between the other line conductors and the neutral conductor can reach the value of $1.45 \times U_0$ for a time up to 5s.

443   PROTECTION AGAINST OVERVOLTAGES OF ATMOSPHERIC ORIGIN OR DUE TO SWITCHING

443.1   Scope and object

443.1.1 This section deals with protection of electrical installations against transient overvoltages of atmospheric origin transmitted by the supply distribution system and against switching overvoltages generated by the equipment within the installation.

NOTE 1: BS EN 62305 provides a comprehensive risk assessment based system for lightning protection. This includes protection for electrical and electronic systems using a number of methods such as surge protective devices.

NOTE 2: This section takes into consideration the technical intent of HD 60364-4-443:2006, which was not fully aligned to BS EN 62305. The IEC are currently reviewing Section 443 and realigning it with IEC 62305.

Protection according to this section can be expected if the relevant equipment product standards require at least the values of withstand voltage of Table 44.3 according to the overvoltage category of equipment in the installation.

NOTE 3: Transient overvoltages transmitted by the supply distribution system are not significantly attenuated downstream in most installations.

Where protection against overvoltages is by the use of surge protective devices (SPDs) they shall be selected and erected in accordance with Section 534.

Examples of equipment with various impulse withstand categories are given in Table 44.4.

NOTE 4: Some electronic equipment may have protection levels lower than Category I of Table 44.3 – see Section 534.

Direct lightning strikes on the low voltage lines of the supply network or on electrical installations are not taken into account (conditions of external influence AQ3). See also BS EN 62305.

This section does not cover overvoltages transmitted by data transmission systems. See BS EN 50174.

NOTE 5: When the need for power SPDs is identified, additional SPDs on other services such as telecom lines are also recommended. See BS EN 62305 and BS EN 61643.

443.2   Arrangements for overvoltage control

443.2.1   Where an installation is supplied by a low voltage system containing no overhead lines, no additional protection against overvoltage of atmospheric origin is necessary if the impulse withstand voltage of equipment is in accordance with Table 44.3.

A suspended cable having insulated conductors with earthed metallic covering is deemed to be an underground cable for the purposes of this section.

443.2.2   Where an installation is supplied by a low voltage network which includes overhead lines or where the installation includes an overhead line and in either case the condition of external influences AQ1 ($\leq$ 25 thunderstorm days per year) exists, no additional protection against overvoltages of atmospheric origin is required if the impulse withstand voltage of equipment is in accordance with Table 44.3.

NOTE: Irrespective of the AQ value, protection against overvoltages may be necessary in applications where a higher reliability or higher risks (e.g. fire) are expected. See BS EN 62305.

443.2.3   Where an installation is supplied by or includes a low voltage overhead line, a measure of protection against overvoltages of atmospheric origin shall be provided if the ceraunic level of the location corresponds to the condition of external influences AQ2 (> 25 thunderstorm days per year). The protection level of the surge protective device shall not exceed the level of overvoltage Category II given in Table 44.3.

443.2.4   As an alternative to the AQ criteria in Regulations 443.2.2 and 443.2.3, the use of surge protection may be based on a risk assessment method.

NOTE 1: As far as Section 443 is concerned, a simple method, based on the critical length d_c of the incoming lines and the level of consequences as described below.

The following are different consequential levels of protection:

    (i)   Consequences related to human life, e.g. safety services, medical equipment in hospitals

    (ii)   Consequences related to public services, e.g. loss of public services, IT centres, museums

    (iii)   Consequences to commercial or industry activity, e.g. hotels, banks, industries, commercial markets, farms

    (iv)   Consequences to groups of individuals, e.g. large residential buildings, churches, offices, schools

    (v)   Consequences to individuals, e.g. small or medium residential buildings, small offices.

For levels of consequences (i) to (iii) protection against overvoltage shall be provided.

NOTE 2: There is no need to perform a risk assessment calculation for levels of consequences (i) to (iii) because this calculation always leads to the result that the protection is required.

For levels of consequences (iv) and (v) requirements for protection depend on the result of a calculation. The calculation shall be carried out using the following formula for the determination of the length d, which is based on a convention and called conventional length.

The configuration of the low voltage distribution line, its earthing, insulation level and the phenomena considered (induced coupling, resistive coupling) lead to different choices for d. The determination proposed below represents, by convention, the worst case.

NOTE 3: This simplified method is based on IEC 61662.

$$d = d_1 + d_2/k_g + d_3/k_t$$

By convention d is limited to 1 km:

where:

    d_1 is the length of the low voltage overhead supply line to the structure, limited to 1 km

    d_2 is the length of the low voltage underground unscreened line to the structure, limited to 1 km
         The length of a screened low voltage underground line is neglected.

    d_3 is the length of the high voltage overhead supply line to the structure, limited to 1 km
         The length of the high voltage underground line is neglected.

$k_g = 4$ is the reduction factor based on the ratio on the influence of strikes between the overhead lines and underground unscreened cables, calculated for a resistivity of soil of 250 Ωm

$k_t = 4$ is the typical reduction factor for a transformer.

Protection is required if: $d > d_c$

where:

    d   is the conventional length in km of the supply line of the considered structure
        with a maximum value of 1 km

    d_c is the critical length in km, equal to $1/N_g$ for level of consequences (iv)
        and equal to $2/N_g$ for level of consequences (v),
        where N_g is the frequency of flashes per km^2 per year.

If this calculation indicates that a surge protective device (SPD) is required, the protection level of these protective devices shall not be higher than the level of overvoltage Category II given in Table 44.3.

443.2.5   *Deleted by BS 7671:2008 Amendment No 1*

443.2.6   Where required or otherwise specified in accordance with this section, overvoltage protective devices shall be located as close as possible to the origin of the installation.

TABLE 44.3 – Required minimum impulse withstand voltage, U_W

Nominal voltage of the installation V	Required minimum impulse withstand voltage kV[1]			
	Category IV (equipment with very high impulse voltage)	Category III (equipment with high impulse voltage)	Category II (equipment with normal impulse voltage)	Category I (equipment with reduced impulse voltage)
230/240 277/480	6	4	2.5	1.5
400/690	8	6	4	2.5
1000	12	8	6	4

[1] This impulse withstand voltage is applied between live conductors and PE.

TABLE 44.4 – Examples of various impulse category equipment

Category	Example
I	Equipment intended to be connected to the fixed electrical installation where protection against transient overvoltage is external to the equipment, either in the fixed installation or between the fixed installation and the equipment. Examples of equipment are household appliances, portable tools and similar loads intended to be connected to circuits in which measures have been taken to limit transient overvoltages.
II	Equipment intended to be connected to the fixed electrical installation e.g. household appliances, portable tools and similar loads, the protective means being either within or external to the equipment.
III	Equipment which is part of the fixed electrical installation and other equipment where a high degree of availability is expected, e.g. distribution boards, circuit-breakers, wiring systems, and equipment for industrial uses, stationary motors with permanent connection to the fixed installation.
IV	Equipment to be used at or in the proximity of the origin of the electrical installation upstream of the main distribution board, e.g. electricity meter, primary overcurrent device, ripple control unit.

444 MEASURES AGAINST ELECTROMAGNETIC DISTURBANCES

444.1 Scope

This section provides basic requirements and recommendations to enable the avoidance and reduction of electromagnetic disturbances.

The designer of the electrical installation shall consider the measures described in this section for reducing electromagnetic disturbances on electrical equipment.

Electromagnetic disturbances can disturb or damage information technology systems or information technology equipment as well as equipment with electronic components or circuits. Currents due to lightning, switching operations, short-circuits and other electromagnetic phenomena might cause overvoltages and electromagnetic interference.

These effects are potentially more severe:

   (i)   where large metal loops exist

   (ii) where different electrical wiring systems are installed in common routes, e.g. for power supply and for signalling and/or data communication cables connecting information technology equipment within a building.

The value of the induced voltage depends on the rate of change (dI/dt) of the interference current and on the size of the loop.

Power cables carrying large currents with a high rate of change of current (dI/dt) (e.g. the starting current of lifts or currents controlled by rectifiers) can induce overvoltages in cables of information technology systems, which can influence or damage information technology equipment or similar electrical equipment.

In or near rooms for medical use, electromagnetic disturbances associated with electrical installations can interfere with medical electrical equipment.

The requirements and recommendations given in this section can have an influence on the overall design of the building including its structural aspects.

The requirements of the following standards shall be applied where appropriate:

   (iii) BS 6701: Telecommunications equipment and telecommunications cabling. Specification for installation, operation and maintenance

   (iv)   BS EN 50310: Application of equipotential bonding and earthing in buildings with information technology equipment

   (v)   BS EN 50174 Series: Information technology. Cabling installation.

   (vi)   BS EN 61000-5-2: Electromagnetic compatibility (EMC). Installation and mitigation guidelines. Earthing and cabling

444.2   Not used

444.3   Not used

444.4   Electromagnetic disturbances

444.4.1   Sources of electromagnetic disturbances

Consideration shall be given to the location of the sources of electromagnetic disturbances relative to the positioning of other equipment. Potential sources of electromagnetic disturbances within an installation typically include:

   (i)   switching devices for inductive loads

   (ii)   electric motors

   (iii) fluorescent lighting

   (iv)   welding machines

   (v)   rectifiers

   (vi)   choppers

   (vii) frequency convertors/regulators including Variable Speed Drives (VSDs)

  (viii) lifts

   (ix)   transformers

   (x)   switchgear

   (xi)   power distribution busbars.

NOTE: For further information refer to the BS EN 50174 series of standards.

444.4.2   Measures to reduce EMI

444.4.2.1   The following measures shall be considered, where appropriate, in order to reduce the effects of electromagnetic interference:

(i)   Where screened signal or data cables are used, care should be taken to limit the fault current from power systems flowing through the screens and cores of signal cables, or data cables, which are earthed. Additional conductors may be necessary, e.g. a bypass conductor for screen reinforcement, see Figure 44.3

Fig 44.3 – Bypass conductor for screen reinforcement to provide a common equipotential bonding system

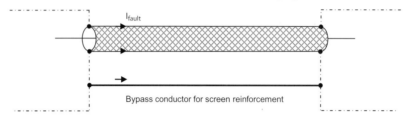

(ii)   The use of surge protective devices and/or filters to improve electromagnetic compatibility with regard to conducted electromagnetic phenomena for electrical equipment sensitive to electromagnetic disturbances

(iii)   The installation of power cables (i.e. line, neutral and any protective earth conductors) close together in order to minimize cable loop areas

(iv)   The separation of power and signal cables

(v)   The installation of an equipotential bonding network, see Regulation 444.5.3.

444.4.3   TN system

To minimize electromagnetic disturbances, the following requirements shall be met:

444.4.3.1   A PEN conductor shall not be used downstream of the origin of the installation.

NOTE: In Great Britain, Regulation 8(4) of the Electricity Safety, Quality and Continuity Regulations 2002 prohibits the use of PEN conductors in consumers' installations.

444.4.3.2   The installation shall have separate neutral and protective conductors downstream of the origin of the installation; see Figure 44.4.

Fig 44.4 – Avoidance of neutral conductor currents in a bonded structure by using an installation forming part of a TN-S system from the origin of the public supply up to and including the final circuit within a building

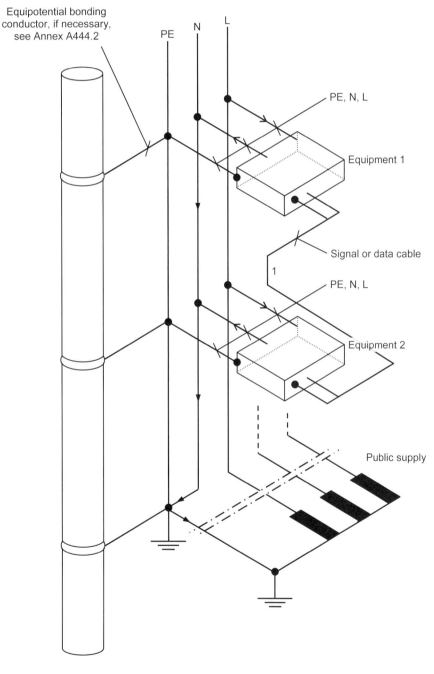

Key :
   1 Loops of limited area formed by signal or data cables.

444.4.3.3   Where the complete low voltage installation including the transformer is operated only by the user, an installation forming part of a TN-S system shall be installed; see Figure 44.5.

Fig 44.5 – Avoidance of neutral conductor currents in a bonded structure by using an installation forming part of a TN-S system downstream of a consumer's private supply transformer

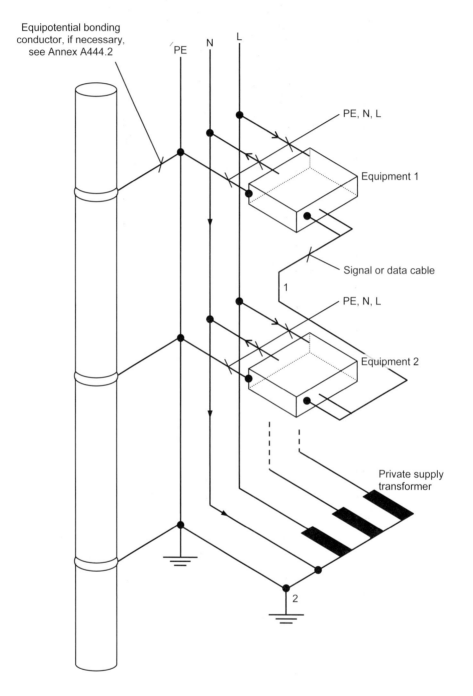

Key :

1 Loops of limited area formed by signal or data cables.

2 The point of neutral earthing may be made at the transformer or the main LV switchgear.

444.4.4   TT system

In an installation forming part of a TT system, such as that shown in Figure 44.6, consideration shall be given to overvoltages which might exist between live parts and extraneous-conductive-parts where the extraneous-conductive-parts of different buildings are connected to different earth electrodes.

The use of an isolating transformer to provide a TN-S system shall be considered.

Fig 44.6 – Installation forming part of a TT system within a building installation

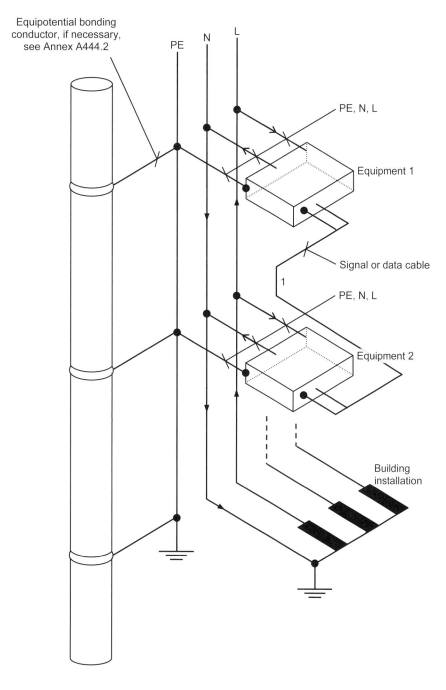

Key :

    1 Loops of limited area formed by signal or data cables.

Where screened signal cables or data cables are common to several buildings supplied from an installation forming part of a TT system, the use of a bypass equipotential bonding conductor (see Figure 44.7) or single-point bonding shall be considered. The bypass conductor shall have a minimum cross-sectional area of 16 mm^2 copper or equivalent, the equivalent cross-sectional area being selected in accordance with Regulation 544.1.

Where the live conductors of the supply into any of the buildings exceed 35 mm^2 in cross-sectional area the bypass conductor shall have a minimum cross-sectional area in accordance with Table 54.8.

Fig 44.7 – Example of a substitute or bypass equipotential bonding conductor in an installation forming part of a TT system

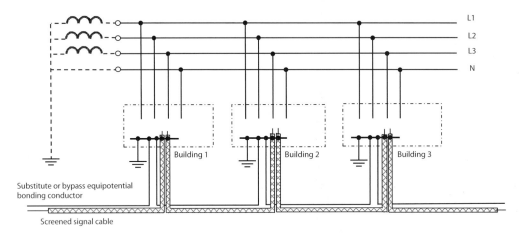

If consent according to the last paragraph of Regulation 411.3.1.2 cannot be obtained, it is the responsibility of the owner or operator of the cable to avoid any danger due to the exclusion of those cables from the connection to the main equipotential bonding.

444.4.5   Not used

444.4.6   Multiple-source TN or TT power supplies

For TN or TT multiple-source power supplies to an installation, the system shall be earthed at one point only.

For a TN system, to avoid having the neutral current flowing through the protective conductor, a single point of connection only shall be made as illustrated in Figure 44.8.

Fig 44.8 – TN multiple-source power supply with single connection between PEN and earth

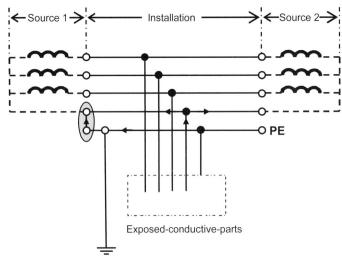

NOTE 1: Where multiple earthing of the star points of the sources of supply is applied, neutral conductor currents might flow back to the relevant star point, not only via the neutral conductor, but also via the protective conductor as shown in Figure 44.9. For this reason the sum of the partial currents flowing in the installation is no longer zero and a stray magnetic field is created, similar to that of a single conductor cable.

NOTE 2: In the case of a single conductor cable carrying a.c. current, a circular electromagnetic field is generated around the conductor that might interfere with electronic equipment. Harmonic currents produce similar electromagnetic fields but they attenuate more rapidly than those produced by the fundamental current.

Fig 44.9 – TN multiple-source power supply with unsuitable multiple connection between PEN and earth

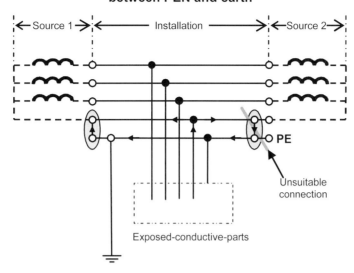

444.4.7 Transfer of supply

In an installation forming part of a TN system the transfer from one supply to an alternative supply shall be by means of a multipole switching device which switches the line conductors and the neutral conductor, if any.

NOTE: This method prevents electromagnetic fields due to stray currents in the main supply system of an installation.

444.4.8 Not used

444.4.9 Separate buildings

Where different buildings have separate equipotential bonding systems, metal-free optical fibre cables or other non-conducting systems are preferred for signal and data transmission, e.g. microwave signal transformer for isolation in accordance with BS EN 61558-2-1, 2-4, 2-6, 2-15 and BS EN 60950-1.

444.4.10 Inside buildings

Within a building, the requirements and recommendations of the following standards shall be applied for control, signalling and communication circuits:

 (i) BS EN 50174-1: Information technology – Cabling installation: Installation specification and quality assurance

 (ii) BS EN 50174-2: Information technology – Cabling installation: Installation planning and practices inside buildings

(iii) BS EN 50310: Application of equipotential bonding and earthing in buildings with information technology equipment.

444.5 Earthing and equipotential bonding

444.5.1 Interconnection of earth electrodes

444.5.1.1 Within a single building

All protective and functional earthing conductors of an installation within a building shall be connected to the main earthing terminal, as required by Regulation 542.4.1, except where this is precluded by the requirements of legislation or Part 7.

444.5.1.2 Between buildings

For communication and data exchange between several buildings, the requirements of Regulation 542.1.3.3 apply to both the protective and functional earthing requirements.

NOTE: Where interconnection of the earth electrodes is not possible or practicable, it is recommended that separation of communications networks is applied, for example, by using optical or radio links.

444.5.2 Equipotential bonding networks

The structure selected for these conductors shall be appropriate for the installation:

(i) Metal sheaths, screens or armouring of cables shall be bonded to the common bonding network (CBN) unless such bonding is required to be omitted for safety reasons

(ii) Where screened signal or data cables are earthed, care shall be taken to limit the fault current from power systems flowing through the screens and cores of signal cables or data cables

(iii) The impedance of equipotential bonding connections intended to carry functional earth currents having high frequency components shall be as low as practicable and this should be achieved by the use of multiple, separated bonds that are as short as possible

> NOTE: Where bonds of up to 1 metre long are used, their inductive reactance and impedance of route can be reduced by choosing a conductive braid or a bonding strap/strip (with a width to thickness ratio of at least 5:1 and a length to width ratio no greater than 5:1).

(iv) Where a lightning protection system is installed, reference shall be made to BS EN 62305.

444.5.3 Sizing and installation of copper bonding ring network conductors

Equipotential bonding designed as a bonding ring network shall have the following minimum nominal dimensions:

(i) Flat cross-section: 25 mm x 3 mm

(ii) Round diameter: 8 mm.

Bare conductors shall be protected against corrosion at their supports and on their passage through walls.

444.5.3.1 Parts to be connected to the equipotential bonding network

The following parts shall be connected to the equipotential bonding network:

(i) Metallic containment, conductive screens, conductive sheaths or armouring of data transmission cables or of information technology equipment

(ii) Functional earthing conductors of antenna systems

(iii) Conductors of the earthed pole of a d.c. supply for information technology equipment

(iv) Functional earthing conductors

(v) Protective conductors.

444.5.4 Not used

444.5.5 Not used

444.5.6 Not used

444.5.7 Earthing arrangements and equipotential bonding of information technology installations for functional purposes

444.5.7.1 Earthing busbar

Where an earthing busbar is required for functional purposes, consideration shall be given to extending the main earthing terminal of the building by using one or more earthing busbars. This enables information technology installations to be connected to the main earthing terminal by the shortest practicable route from any point in the building. Where the earthing busbar is erected to support the equipotential bonding network of a significant amount of information technology equipment in a building, consideration shall be given to the installation of a bonding ring conductor or common mesh bonding network; see Annex A444 Figure A444.2.

Consideration shall be given to the need for accessibility of the earthing busbar throughout its length and to the protection of bare conductors to prevent corrosion.

444.5.7.2   Cross-sectional area of the earthing busbar

For installations connected to a supply having a capacity of 200 A per phase or more, the cross-sectional area of the earthing busbar shall be not less than 50 mm^2 copper and shall be selected in accordance with Regulation 444.5.2(iii).

For supplies having a capacity of less than 200 A per phase the earthing busbar shall be selected in accordance with Table 54.8.

Where the earthing busbar is used as part of a d.c. return current path, its cross-sectional area shall be selected according to the expected d.c. return currents.

444.6   Segregation of circuits

444.6.1   General

Cables that are used at voltage Band II (low voltage) and cables that are used at voltage Band I (extra-low voltage) which share the same cable management system or the same route, shall be installed according to the requirements of Regulations 528.1 and 528.2. Circuits of the same voltage band might also require segregation or separation.

Electrical safety and electromagnetic compatibility might produce different segregation or separation requirements. The design shall meet both requirements.

444.6.2   Equipment

The minimum distance between information technology cables and discharge, neon and mercury vapour (or other high-intensity discharge) lamps shall be 130 mm. In this regard, low energy lamps (cfl) are to be considered as gas discharge sources. Data wiring racks and electrical equipment shall always be separated.

Annex A444 (Informative)
Measures against electromagnetic disturbances

A444.1   Structures for the network of equipotential conductors and earthing conductors

For dwellings, where normally a limited amount of electronic equipment is in use, a protective conductor network in the form of a star network might be acceptable.

For commercial and industrial buildings and similar buildings containing multiple electronic applications, a common equipotential bonding system is useful in order to comply with the EMC requirements of different types of equipment.

The four basic structures described as follows might be used, depending on the importance and vulnerability of equipment.

NOTE: For further information, the methodology referred to in BS EN 50310 (The application of equipotential bonding and earthing in buildings with information technology equipment) is generally applicable.

A444.1.1   Protective conductors in a star network

This type of network is applicable to small installations associated with dwellings, small commercial buildings, etc., and from a general point of view to equipment that is not interconnected by signal cables; see Figure A444.1.

Fig A444.1 – Example of protective conductors in star network

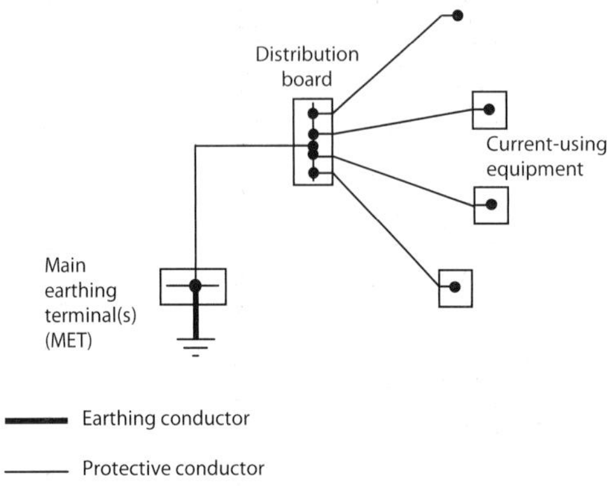

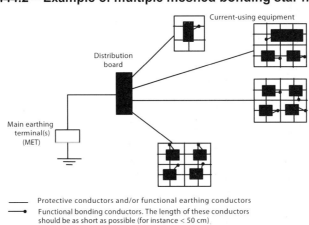

A444.1.2   Multiple meshed bonding star network

This type of network is applicable to small installations with different small groups of interconnected communicating equipment. It enables the local dispersion of currents caused by electromagnetic interference; see Figure A444.2.

Fig A444.2 – Example of multiple meshed bonding star network

A444.1.3   Common meshed bonding star network

This type of network is applicable to installations with a high density of communicating equipment corresponding to critical applications; see Figure A444.3. It is suitable for protection of private automatic branch exchange equipment (PABX) and centralized data processing systems.

A meshed equipotential bonding network is enhanced by the existing metallic structure of the building. It is supplemented by conductors forming the square mesh.

The mesh size depends on the selected level of protection against lightning, on the immunity level of the equipment and on the frequencies used for data transmission.

Mesh size should be adapted to the dimensions of the installation to be protected and should be in accordance with the recommendations of BS EN 50310. Where concerns exist, the mesh size should be adapted to the dimensions of the installation to be protected, but should not exceed 2 m x 2 m in areas where equipment susceptible to electromagnetic environmental interferences is installed.

NOTE: The mesh size refers to the dimensions of square spaces enclosed by the conductors forming the mesh.

In some cases, parts of this network may be meshed more closely in order to meet specific requirements.

Fig A444.3 – Example of a common meshed bonding star network

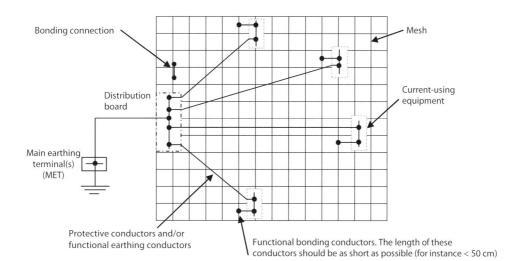

A444.1.4 Protective conductors connected to a bonding ring conductor

An equipotential bonding network in the form of a bonding ring conductor (BRC) is shown in Figure A444.4 on the top floor of the structure. The BRC should preferably be made of copper, bare or insulated, and installed in such a manner that it remains accessible everywhere, e.g. by mounting on a cable tray, in a metallic conduit (see BS EN 61386 series), employing a surface mounted method of installation or cable trunking. All protective and functional earthing conductors may be connected to the BRC.

A444.2   Equipotential bonding networks in buildings with several floors

For buildings with several floors, it is recommended that, on each floor, an equipotential bonding system be installed; see Figure A444.4 for examples of bonding networks in common use; each floor is a type of network. The bonding systems of the different floors should be interconnected, at least twice, by conductors selected in accordance with the requirements of Chapter 54.

Fig A444.4 – Example of equipotential bonding networks in a structure without a lightning protection system

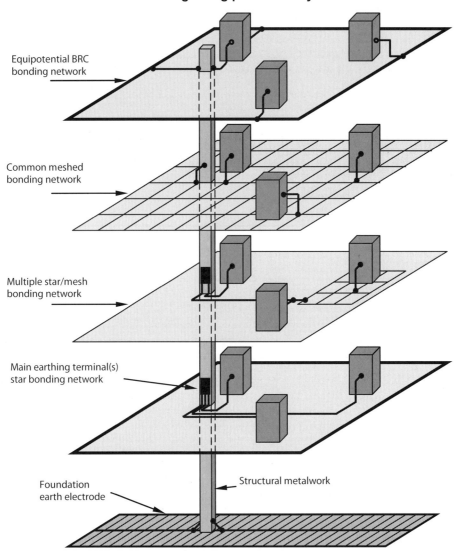

Equipotential BRC bonding network

Common meshed bonding network

Multiple star/mesh bonding network

Main earthing terminal(s) star bonding network

Foundation earth electrode

Structural metalwork

A444.3   Installations containing a high density of interconnected equipment

In severe electromagnetic environments, it is recommended that the common meshed bonding star network described in A444.1.3 be adopted.

A444.4 Design guidelines for segregation of circuits

Where both the specification of the information technology cable and its intended application is known, the requirements of BS EN 50174-2 and BS EN 50174-3 are appropriate.

BS EN 50174 series standards contain requirements and recommendations for the installation of information technology cabling which support a range of applications delivering the following services:

(i) ICT (information communication technologies) e.g. local area networks

(ii) BCT (broadcast communication technologies) e.g. audio-visual, television

(iii) CCCB (command control and communications in buildings) e.g. building automation

(iv) PMCA (process monitoring, control and automation) e.g. industrial networks (Fieldbus).

Where the specification and/or intended application of the information technology cable is not available, then the cable separation distance between the power and IT cables should be a minimum of 200 mm in free air.

This distance can be reduced if a screened power cable, a metallic barrier, or containment system is used as illustrated in Table A444.1.

TABLE A444.1 – Summary of minimum separation distances where the specification and/or the intended application of the information technology cable is not available

These recommendations of segregation are based upon the following assumptions:

(i) The electromagnetic environment complies with the levels defined in the BS EN 61000-6 series of standards for conducted and radiated disturbances (e.g. mains power cabling)

(ii) The LV supply is non-deformed but has high-frequency content consistent with the switching and operation of connected equipment in accordance with the BS EN 61000-6 series of standards

> **NOTE:** "Deformed" LV power supplies and the use of other equipment lie outside the scope of this standard and might require additional engineering practices.

(iii) The total design current in the LV circuits does not exceed 600 A

(iv) Balanced information technology/telecommunications cables have electromagnetic immunity performance in accordance with BS EN 50288 series standards for Category 5 and above

(v) Coaxial information technology/telecommunications cables have electromagnetic immunity performance in accordance with BS EN 50177-4-1 standard for Category BCT-C

(vi) The applications supported by the cabling are designed to operate using the information technology cabling installed or to be installed.

TABLE A444.2 – Summary of minimum separation distances where the specification and/or the intended application of the information technology cable is not available

Containment applied to the mains power cabling		
No containment or open metallic containment A[1]	Perforated open metallic containment B[2]	Solid metallic containment C[3]
200 mm	150 mm	0 mm

NOTE 1: Screening performance (DC-100MHz) equivalent to welded mesh steel basket of mesh size 50 mm x 100 mm (excluding ladders). This screening performance is also achieved with steel tray (duct without cover) of less than 1.0 mm wall thickness and more than 20% equally distributed perforated area.

No part of the cable within the containment should be less than 10 mm below the top of the barrier.

NOTE 2: Screening performance (DC-100 MHz) equivalent to steel tray (duct without cover) of 1.0 mm wall thickness and no more than 20% equally distributed perforated area. This screening performance is also achieved with screened power cables that do not meet the performance defined in Note 1.

No part of the cable within the containment shall be less than 10 mm below the top of the barrier.

NOTE 3: Screening performance (DC-100 MHz) equivalent to a fully enclosed steel containment system having a minimum wall thickness of 1.5 mm. Separation specified is in addition to that provided by any divider/barrier.

NOTE: Zero segregation in the Table references additional segregation/separation for EMC over and above the requirements for safety. Safety considerations must always take precedence over EMC requirements.

Where the above conditions do not apply, further guidance may be obtained from the Chartered Institute of Building Services Engineers (CIBSE), Applications Manual AM7 'Information Technology and Building'.

Additional areas of concern are expressed in Regulation 444.4.1.

The minimum separation between the information technology cables and mains power cables includes all allowances for cable movement between their fixing points or other restraints (see example in Figure A444.5).

Fig A444.5 – Example of cable separation distance

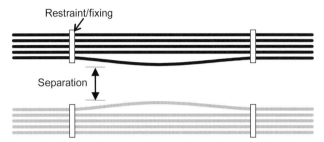

The minimum separation requirement applies in three dimensions. However, where information technology cables and mains power cables are required to cross and required minimum separation cannot be maintained then maintaining the angle of their crossing at 90 degrees on either side of the crossing for a distance no less than the applicable minimum separation requirement will minimize any electromagnetic disturbances.

A444.5   Conditions for zero segregation

See BS EN 50174.

445 PROTECTION AGAINST UNDERVOLTAGE

445.1 General requirements

445.1.1 Suitable precautions shall be taken where a reduction in voltage, or loss and subsequent restoration of voltage, could cause danger. Provisions for a circuit supplying a motor shall comply with Regulation 552.1.3.

Where current-using equipment or any other part of the installation may be damaged by a drop in voltage and it is verified that such damage is unlikely to cause danger, one of the following arrangements shall be adopted:

(i) Suitable precautions against the damage foreseen shall be provided

(ii) It shall be verified, in consultation with the person or body responsible for the operation and maintenance of the installation, that the damage foreseen is an acceptable risk.

445.1.2 A suitable time delay may be incorporated in the operation of an undervoltage protective device if the operation of the equipment to which the protection relates allows without danger a brief reduction or loss of voltage.

445.1.3 Any delay in the opening or reclosing of a contactor shall not impede instantaneous disconnection by a control device or a protective device.

445.1.4 The characteristics of an undervoltage protective device shall be compatible with the requirements for starting and use of the equipment to which the protection relates, as stated in the appropriate British or Harmonized Standard.

445.1.5 Where the reclosure of a protective device is likely to cause danger, the reclosure shall not be automatic.

PART 5

SELECTION AND ERECTION OF EQUIPMENT

CONTENTS

CHAPTER 51
COMMON RULES

CONTENTS

CHAPTER 51

COMMON RULES

510 INTRODUCTION

510.1 General

This chapter deals with the selection of equipment and its erection. It provides common rules for compliance with measures of protection for safety, requirements for proper functioning for intended use of the installation, and requirements appropriate to the external influences.

510.2 Not used

510.3

Every item of equipment shall be selected and erected so as to allow compliance with the regulations stated in this chapter and the relevant regulations in other parts of BS 7671 and shall take account of manufacturers' instructions.

511 COMPLIANCE WITH STANDARDS

511.1

Every item of equipment shall comply with the relevant requirements of the applicable British or Harmonized Standard, appropriate to the intended use of the equipment. The edition of the standard shall be the current edition, with those amendments pertaining at a date to be agreed by the parties to the contract concerned (see Appendix 1).

Alternatively, if equipment complying with a foreign national standard based on an IEC Standard is to be used, the designer or other person responsible for specifying the installation shall verify that any differences between that standard and the corresponding British or Harmonized Standard will not result in a lesser degree of safety than that afforded by compliance with the British or Harmonized Standard.

511.2

Where equipment to be used is not covered by a British or Harmonized Standard or is used outside the scope of its standard, the designer or other person responsible for specifying the installation shall confirm that the equipment provides at least the same degree of safety as that afforded by compliance with the Regulations.

512 OPERATIONAL CONDITIONS AND EXTERNAL INFLUENCES

512.1 Operational conditions

512.1.1 Voltage

Every item of equipment shall be suitable for the nominal voltage (U_0) of the installation or the part of the installation concerned, where necessary taking account of the highest and/or lowest voltage likely to occur in normal service. In an IT system, equipment shall be insulated for the nominal voltage between lines.

512.1.2 Current

Every item of equipment shall be suitable for:
  (i) the design current, taking into account any capacitive and inductive effects, and
  (ii) the current likely to flow in abnormal conditions for such periods of time as are determined by the characteristics of the protective devices concerned.

512.1.3 Frequency

If frequency has an influence on the characteristics of the equipment, the rated frequency of the equipment shall correspond to the nominal frequency of the supply to the circuit concerned.

512.1.4 Power

Every item of equipment selected on the basis of its power characteristics shall be suitable for the duty demanded of the equipment.

512.1.5 Compatibility

Every item of equipment shall be selected and erected so that it will neither cause harmful effects to other equipment nor impair the supply during normal service including switching operations.

Switchgear, protective devices, accessories and other types of equipment shall not be connected to conductors intended to operate at a temperature exceeding 70 °C at the equipment in normal service unless the equipment manufacturer has confirmed that the equipment is suitable for such conditions, or the conductor size shall be chosen based on the current ratings for 70° C cables of a similar construction. See also Regulation 523.1 and Table 4A3.

512.1.6 Impulse withstand voltage

Equipment shall be selected so that its impulse voltage withstand is at least equal to the required minimum impulse withstand voltage according to the overvoltage category at the point of installation as defined in Section 443.

512.2 External influences

512.2.1 Equipment shall be of a design appropriate to the situation in which it is to be used or its mode of installation shall take account of the conditions likely to be encountered.

512.2.2 If the equipment does not, by its construction, have the characteristics relevant to the external influences of its location, it may nevertheless be used on condition that it is provided with appropriate additional protection in the erection of the installation. Such protection shall not adversely affect the operation of the equipment thus protected.

512.2.3 Where different external influences occur simultaneously, they may have independent or mutual effects and the degree of protection shall be provided accordingly.

512.2.4 The selection of equipment according to external influences is necessary not only for proper functioning, but also to ensure the reliability of the measures of protection for safety complying with these Regulations generally. Measures of protection afforded by the construction of equipment are valid only for the given conditions of external influence if the corresponding equipment specification tests are made in these conditions of external influence.

NOTE 1: For the purpose of these Regulations, the following classes of external influence are conventionally regarded as normal:

AA Ambient temperature	AA4
AB Atmospheric humidity	AB4
Other environmental conditions (AC to AS)	XX1 of each parameter
Utilisation and construction of buildings (B and C)	{ XX1 of each parameter, except XX2 for the parameter BC

NOTE 2: The word "normal" appearing in the third column of the table in Appendix 5 signifies that the requirement must generally satisfy applicable standards.

513 ACCESSIBILITY

513.1 Except for a joint in cables where Section 526 allows such a joint to be inaccessible, every item of equipment shall be arranged so as to facilitate its operation, inspection and maintenance and access to each connection. Such facility shall not be significantly impaired by mounting equipment in an enclosure or a compartment.

514 IDENTIFICATION AND NOTICES

514.1 General

514.1.1 Except where there is no possibility of confusion, a label or other suitable means of identification shall be provided to indicate the purpose of each item of switchgear and controlgear. Where the operator cannot observe the operation of switchgear and controlgear and where this might lead to danger, a suitable indicator complying, where applicable, with BS EN 60073 and BS EN 60447, shall be fixed in a position visible to the operator.

514.1.2 So far as is reasonably practicable, wiring shall be so arranged or marked that it can be identified for inspection, testing, repair or alteration of the installation.

514.1.3   Except where there is no possibility of confusion, unambiguous marking shall be provided at the interface between conductors identified in accordance with these Regulations and conductors identified to previous versions of the Regulations.

NOTE: Appendix 7 gives guidance on how this can be achieved.

514.2   Deleted by BS 7671:2008 Amendment No 1

514.3   Identification of conductors

514.3.1   Except where identification is not required by Regulation 514.6, cores of cables shall be identified by:

(i)   colour as required by Regulation 514.4 and/or

(ii)   lettering and/or numbering as required by Regulation 514.5.

514.3.2   Every core of a cable shall be identifiable at its terminations and preferably throughout its length. Binding and sleeves for identification purposes shall comply with BS 3858 where appropriate.

514.3.3   Deleted by BS 7671:2008 Amendment No 1

514.4   Identification of conductors by colour

514.4.1   Neutral or midpoint conductor

Where a circuit includes a neutral or midpoint conductor identified by colour, the colour used shall be blue.

514.4.2   Protective conductor

The bi-colour combination green-and-yellow shall be used exclusively for identification of a protective conductor and this combination shall not be used for any other purpose. In this combination one of the colours shall cover at least 30 % and at most 70 % of the surface being coloured, while the other colour shall cover the remainder of the surface.

Single-core cables that are coloured green-and-yellow throughout their length shall only be used as a protective conductor and shall not be overmarked at their terminations, except as permitted by Regulation 514.4.3.

A bare conductor or busbar used as a protective conductor shall be identified, where necessary, by equal green-and-yellow stripes, each not less than 15 mm and not more than 100 mm wide, close together, either throughout the length of the conductor or in each compartment and unit and at each accessible position.

514.4.3   PEN conductor

A PEN conductor shall, when insulated, be marked by one of the following methods:

(i)   Green-and-yellow throughout its length with, in addition, blue markings at the terminations

(ii)   Blue throughout its length with, in addition, green-and-yellow markings at the terminations.

514.4.4   Other conductors

Other conductors shall be identified by colour in accordance with Table 51.

514.4.5   The single colour green shall not be used.

514.4.6   Bare conductors

A bare conductor shall be identified, where necessary, by the application of tape, sleeve or disc of the appropriate colour prescribed in Table 51 or by painting with such a colour.

514.5   Identification of conductors by letters and/or numbers

514.5.1   The lettering or numbering system applies to identification of individual conductors and of conductors in a group. The identification shall be clearly legible and durable. All numerals shall be in strong contrast to the colour of the insulation. The identification shall be given in letters or Arabic numerals. In order to avoid confusion, unattached numerals 6 and 9 shall be underlined.

514.5.2   Protective conductor

Conductors with green-and-yellow colour identification shall not be numbered other than for the purpose of circuit identification.

514.5.3   Alphanumeric

The preferred alphanumeric system is described in Table 51.

514.5.4   Numeric

Conductors may be identified by numbers, the number 0 being reserved for the neutral or midpoint conductor.

514.6   Omission of identification by colour or marking

514.6.1   Identification by colour or marking is not required for:

  (i)   concentric conductors of cables

  (ii)   metal sheath or armour of cables when used as a protective conductor

  (iii)   bare conductors where permanent identification is not practicable

  (iv)   extraneous-conductive-parts used as a protective conductor

  (v)   exposed-conductive-parts used as a protective conductor.

TABLE 51 – Identification of conductors

Function	Alphanumeric	Colour
Protective conductors		Green-and-yellow
Functional earthing conductor		Cream
a.c. power circuit[1]		
Line of single-phase circuit	L	Brown
Neutral of single- or three-phase circuit	N	Blue
Line 1 of three-phase a.c. circuit	L1	Brown
Line 2 of three-phase a.c. circuit	L2	Black
Line 3 of three-phase a.c. circuit	L3	Grey
Two-wire unearthed d.c. power circuit		
Positive of two-wire circuit	L+	Brown
Negative of two-wire circuit	L-	Grey
Two-wire earthed d.c. power circuit		
Positive (of negative earthed) circuit	L+	Brown
Negative (of negative earthed) circuit[2]	M	Blue
Positive (of positive earthed) circuit[2]	M	Blue
Negative (of positive earthed) circuit	L-	Grey
Three-wire d.c. power circuit		
Outer positive of two-wire circuit derived from three-wire system	L+	Brown
Outer negative of two-wire circuit derived from three-wire system	L-	Grey
Positive of three-wire circuit	L+	Brown
Mid-wire of three-wire circuit[2][3]	M	Blue
Negative of three-wire circuit	L-	Grey
Control circuits, ELV and other applications		
Line conductor	L	Brown, Black, Red, Orange, Yellow, Violet, Grey, White, Pink or Turquoise
Neutral or mid-wire[4]	N or M	Blue

NOTES:

  [1]   Power circuits include lighting circuits.

  [2]   M identifies either the mid-wire of a three-wire d.c. circuit, or the earthed conductor of a two-wire earthed d.c. circuit.

  [3]   Only the middle wire of three-wire circuits may be earthed.

  [4]   An earthed PELV conductor is blue.

514.7   *Not used*

514.8   Identification of a protective device

514.8.1   A protective device shall be arranged and identified so that the circuit protected may be easily recognized.

514.9   Diagrams and documentation

514.9.1   A legible diagram, chart or table or equivalent form of information shall be provided indicating in particular:

   (i)   the type and composition of each circuit (points of utilisation served, number and size of conductors, type of wiring), and

  (ii)   the method used for compliance with Regulation 410.3.2, and

 (iii)   the information necessary for the identification of each device performing the functions of protection, isolation and switching, and its location, and

 (iv)   any circuit or equipment vulnerable to a typical test.

For simple installations the foregoing information may be given in a schedule. A durable copy of the schedule relating to a distribution board shall be provided within or adjacent to each distribution board.

514.9.2   Any symbol used shall comply with IEC 60617.

514.10   Warning notice: voltage

514.10.1   Every item of equipment or enclosure within which a nominal voltage exceeding 230 volts to earth exists and where the presence of such a voltage would not normally be expected, shall be so arranged that before access is gained to a live part, a warning of the maximum voltage present is clearly visible.

514.11   Warning notice: isolation

514.11.1   A notice of such durable material as to be likely to remain easily legible throughout the life of the installation in accordance with Regulation 537.2.1.3, shall be fixed in each position where there are live parts which are not capable of being isolated by a single device. The location of each disconnector (isolator) shall be indicated unless there is no possibility of confusion.

514.12   Notices: periodic inspection and testing

514.12.1   A notice of such durable material as to be likely to remain easily legible throughout the life of the installation, shall be fixed in a prominent position at or near the origin of every installation upon completion of the work carried out in accordance with Chapter 61 or 62. The notice shall be inscribed in indelible characters not smaller than those illustrated here and shall read as follows:

IMPORTANT

This installation should be periodically inspected and tested and a report on its condition obtained, as prescribed in the IET Wiring Regulations BS 7671 Requirements for Electrical Installations.

Date of last inspection ....

Recommended date of next inspection ....

514.12.2   Where an installation incorporates an RCD a notice shall be fixed in a prominent position at or near the origin of the installation. The notice shall be in indelible characters not smaller than those illustrated here and shall read as follows:

> This installation, or part of it, is protected by a device which automatically switches off the supply if an earth fault develops. Test quarterly by pressing the button marked 'T' or 'Test'. The device should switch off the supply and should then be switched on to restore the supply. If the device does not switch off the supply when the button is pressed, seek expert advice.

NOTE: Testing frequencies of RCDs in temporary installations may need increasing.

514.13   Warning notices: earthing and bonding connections

514.13.1   A durable label to BS 951 with the words "Safety Electrical Connection – Do Not Remove" shall be permanently fixed in a visible position at or near:

  (i)   the point of connection of every earthing conductor to an earth electrode, and

  (ii)   the point of connection of every bonding conductor to an extraneous-conductive-part, and

  (iii)   the main earth terminal, where separate from main switchgear.

514.13.2   Where Regulation 418.2.5 or 418.3 applies, the warning notice specified shall be durably marked in legible type not smaller than that illustrated here and shall read as follows:

> The protective bonding conductors associated with the electrical installation in this location MUST NOT BE CONNECTED TO EARTH.
>
> Equipment having exposed-conductive-parts connected to earth must not be brought into this location.

514.14   Warning notice: non-standard colours

514.14.1   If wiring additions or alterations are made to an installation such that some of the wiring complies with Regulation 514.4 but there is also wiring to previous versions of these Regulations, a warning notice shall be affixed at or near the appropriate distribution board with the following wording:

> CAUTION
>
> This installation has wiring colours to two versions of BS 7671. Great care should be taken before undertaking extension, alteration or repair that all conductors are correctly identified.

514.15 Warning notice: alternative supplies

514.15.1 Where an installation includes alternative or additional sources of supply, warning notices shall be affixed at the following locations in the installation:

(i) At the origin of the installation

(ii) At the meter position, if remote from the origin

(iii) At the consumer unit or distribution board to which the alternative or additional sources are connected

(iv) At all points of isolation of all sources of supply.

The warning notice shall have the following wording:

WARNING – MULTIPLE SUPPLIES

Isolate all electrical supplies before carrying out work.

Isolate the mains supply at ....

Isolate the alternative supplies at ....

514.16 Notice: high protective conductor current

See Regulation 543.7.1.105.

515 PREVENTION OF MUTUAL DETRIMENTAL INFLUENCE

515.1 Prevention of mutual detrimental influence

Electrical equipment shall be selected and erected so as to avoid any harmful influence between the electrical installation and any non-electrical installations envisaged.

NOTE: For EMC see Sections 332 and 444.

515.2 Where equipment carrying current of different types or at different voltages is grouped in a common assembly (such as a switchboard, a cubicle or a control desk or box), all the equipment belonging to any one type of current or any one voltage shall be effectively segregated wherever necessary to avoid mutual detrimental influence.

The immunity levels of equipment shall be chosen taking into account the electromagnetic disturbances that can occur when connected and erected as for normal use, and taking into account the intended level of continuity of service necessary for the application. See the specific equipment standard or the relevant part of BS EN 61000 series.

515.3 Deleted by BS 7671:2008 Amendment No 1

CHAPTER 52

SELECTION AND ERECTION OF WIRING SYSTEMS

CONTENTS

CHAPTER 52

SELECTION AND ERECTION OF WIRING SYSTEMS

520   INTRODUCTION

520.1   Scope

This Chapter deals with the selection and erection of wiring systems.

NOTE: These regulations also apply in general to protective conductors. Chapter 54 contains further requirements for those conductors.

520.2   Not used

520.3   Not used

520.4   General

Consideration shall be given to the application of the fundamental principles of Chapter 13 as it applies to:

(i)   cables and conductors

(ii)   their connections, terminations and/or jointing

(iii)   their associated supports or suspensions, and

(iv)   their enclosure or methods of protection against external influences.

521   TYPES OF WIRING SYSTEM

The requirements of Regulations 521.1 to 3 do not apply to busbar and powertrack systems covered by Regulation 521.4.

521.1   The installation method of a wiring system in relation to the type of conductor or cable used shall be in accordance with Table 4A1 of Appendix 4, provided the external influences are taken into account according to Section 522.

521.2   The installation method of a wiring system in relation to the situation concerned shall be in accordance with Table 4A2 of Appendix 4. Other methods of installation of cables and conductors not included in Table 4A2 are permitted provided that they fulfil the requirements of this chapter.

521.3   Examples of wiring systems, excluding systems covered by Regulation 521.4, are shown in Table 4A2. Table 4A2 gives examples of installation methods of cables including reference method for obtaining current-carrying capacity where it is considered that the same current-carrying capacities can safely be used. It is not implied that such methods must be employed or that other methods are prohibited.

521.100   Prefabricated wiring systems intended for permanent connection in fixed installations incorporating installation couplers conforming to BS EN 61535, shall comply with BS 8488.

521.4   Busbar trunking systems and powertrack systems

A busbar trunking system shall comply with BS EN 60439-2 and a powertrack system shall comply with the appropriate part of the BS EN 61534 series. A busbar trunking system or a powertrack system shall be installed in accordance with the manufacturer's instructions taking account of external influences. See also Appendix 8.

521.5   A.C. circuits: electromagnetic effects

521.5.1   Ferromagnetic enclosures: electromagnetic effects

The conductors of an a.c. circuit installed in a ferromagnetic enclosure shall be arranged so that all line conductors and the neutral conductor, if any, and the appropriate protective conductor are contained within the same enclosure.

Where such conductors enter a ferrous enclosure, they shall be arranged such that the conductors are only collectively surrounded by ferromagnetic material.

These requirements do not preclude the use of an additional protective conductor in parallel with the steel wire armouring of a cable where such is required to comply with the requirements of the appropriate regulations in Chapters 41 and 54. It is permitted for such an additional protective conductor to enter the ferrous enclosure individually.

521.5.2   Single-core cables armoured with steel wire or steel tape shall not be used for an a.c. circuit.

NOTE: The steel wire or steel tape armour of a single-core cable is regarded as a ferromagnetic enclosure. For single-core armoured cables, the use of aluminium armour may be considered.

521.5.100   Electromechanical stresses

Every conductor or cable shall have adequate strength and be so installed as to withstand the electromechanical forces that may be caused by any current, including fault current, it may have to carry in service.

521.6   Conduit, ducting, trunking, tray and ladder systems

Two or more circuits are allowed in the same conduit, ducting or trunking system provided the requirements of Section 528 are met.

Cable conduits shall comply with the appropriate part of the BS EN 61386 series, cable trunking or ducting shall comply with the appropriate part of the BS EN 50085 series and cable tray and ladder systems shall comply with BS EN 61537.

521.7   Multicore cables: two or more circuits

Two or more circuits are allowed in the same cable provided the requirements of Section 528 are met.

521.8   Circuit arrangements

521.8.1   Each part of a circuit shall be arranged such that the conductors are not distributed over different multicore cables, conduits, ducting systems, trunking systems or tray or ladder systems.

This requirement need not be met where a number of multicore cables, forming one circuit, are installed in parallel. Where multicore cables are installed in parallel each cable shall contain one conductor of each line.

521.8.2   The line and neutral conductors of each final circuit shall be electrically separate from those of every other final circuit, so as to prevent the indirect energizing of a final circuit intended to be isolated.

521.8.3   Where two or more circuits are terminated in a single junction box this shall comply with BS EN 60670-22.

521.9   Use of flexible cables

521.9.1   A flexible cable shall be used for fixed wiring only where the relevant provisions of the Regulations are met. Flexible cables used for fixed wiring shall be of the heavy duty type unless the risk of damage during installation and service, due to impact or other mechanical stresses, is low or has been minimized or protection against mechanical damage is provided.

NOTE: Descriptions of light, normal and heavy duty types are given in BS 7540.

521.9.2   Equipment that is intended to be moved in use shall be connected by flexible cables, except equipment supplied by contact rails.

521.9.3   Stationary equipment which is moved temporarily for the purposes of connecting, cleaning etc., e.g. cookers or flush-mounting units for installations in false floors, shall be connected with flexible cable. If the equipment is not subject to vibration then non-flexible cables may be used.

521.10   Installation of cables

521.10.1   Non-sheathed cables for fixed wiring shall be enclosed in conduit, ducting or trunking. This requirement does not apply to a protective conductor complying with Section 543.

Non-sheathed cables are permitted if the cable trunking system provides at least the degree of protection IPXXD or IP4X, and if the cover can only be removed by means of a tool or a deliberate action.

521.10.100   A bare live conductor shall be installed on insulators.

522 SELECTION AND ERECTION OF WIRING SYSTEMS IN RELATION TO EXTERNAL INFLUENCES

The installation method selected shall be such that protection against the expected external influences is ensured in all appropriate parts of the wiring system. Particular care shall be taken at changes in direction and where wiring enters into equipment.

NOTE: The external influences categorized in Appendix 5 which are of significance to wiring systems are included in this section.

522.1 Ambient temperature (AA)

522.1.1 A wiring system shall be selected and erected so as to be suitable for the highest and the lowest local ambient temperatures and to ensure that the limiting temperature in normal operation (see Table 52.1) and the limiting temperature in case of a fault (see Table 43.1) will not be exceeded.

522.1.2 Wiring system components, including cables and wiring accessories, shall only be installed or handled at temperatures within the limits stated in the relevant product specification or as given by the manufacturer.

522.2 External heat sources

522.2.1 In order to avoid the effects of heat from external sources, one or more of the following methods or an equally effective method shall be used to protect a wiring system:

(i) Shielding

(ii) Placing sufficiently far from the source of heat

(iii) Selecting a system with due regard for the additional temperature rise which may occur

(iv) Local reinforcement or substitution of insulating material.

NOTE: Heat from external sources may be radiated, conducted or convected, e.g.
- from hot water systems
- from plant, appliances and luminaires
- from a manufacturing process
- through heat conducting materials
- from solar gain of the wiring system or its surrounding medium.

522.2.100 Parts of a cable within an accessory, appliance or luminaire shall be suitable for the temperatures likely to be encountered, as determined in accordance with Regulation 522.1.1, or shall be provided with additional insulation suitable for those temperatures.

522.3 Presence of water (AD) or high humidity (AB)

522.3.1 A wiring system shall be selected and erected so that no damage is caused by condensation or ingress of water during installation, use and maintenance. The completed wiring system shall comply with the IP degree of protection relevant to the particular location.

NOTE: Special considerations apply to wiring systems liable to frequent splashing, immersion or submersion.

522.3.2 Where water may collect or condensation may form in a wiring system, provision shall be made for its escape.

522.3.3 Where a wiring system may be subjected to waves (AD6), protection against mechanical damage shall be afforded by one or more of the methods of Regulations 522.6 to 8.

522.4 Presence of solid foreign bodies (AE)

522.4.1 A wiring system shall be selected and erected so as to minimize the danger arising from the ingress of solid foreign bodies. The completed wiring system shall comply with the IP degree of protection relevant to the particular location.

522.4.2 In a location where dust in significant quantity is present (AE4), additional precautions shall be taken to prevent the accumulation of dust or other substances in quantities which could adversely affect the heat dissipation from the wiring system.

NOTE: A wiring system which facilitates the removal of dust may be necessary (see Section 529).

522.5 Presence of corrosive or polluting substances (AF)

522.5.1 Where the presence of corrosive or polluting substances, including water, is likely to give rise to corrosion or deterioration, parts of the wiring system likely to be affected shall be suitably protected or manufactured from a material resistant to such substances.

NOTE: Suitable protection for application during erection may include protective tapes, paints or grease.

522.5.2 Dissimilar metals liable to initiate electrolytic action shall not be placed in contact with each other, unless special arrangements are made to avoid the consequences of such contact.

522.5.3 Materials liable to cause mutual or individual deterioration or hazardous degradation shall not be placed in contact with each other.

522.6 Impact (AG)

522.6.1 Wiring systems shall be selected and erected so as to minimize the damage arising from mechanical stress, e.g. by impact, abrasion, penetration, tension or compression during installation, use or maintenance.

522.6.2 In a fixed installation where impacts of medium severity (AG2) or high severity (AG3) can occur protection shall be afforded by:

(i) the mechanical characteristics of the wiring system, or

(ii) the location selected, or

(iii) the provision of additional local or general protection against mechanical damage, or

(iv) any combination of the above.

NOTE: Examples are areas where the floor is likely to be penetrated and areas used by forklift trucks.

522.6.3 *Not used*

522.6.4 The degree of protection of electrical equipment shall be maintained after installation of the cables and conductors.

522.6.100 A cable installed under a floor or above a ceiling shall be run in such a position that it is not liable to be damaged by contact with the floor or the ceiling or their fixings. A cable passing through a joist within a floor or ceiling construction or through a ceiling support (e.g. under floorboards), shall:

(i) be at least 50 mm measured vertically from the top, or bottom as appropriate, of the joist or batten, or

(ii) incorporate an earthed metallic covering which complies with the requirements of these Regulations for a protective conductor of the circuit concerned, the cable complying with BS 5467, BS 6724, BS 7846, BS EN 60702-1 or BS 8436, or

(iii) be enclosed in earthed conduit complying with BS EN 61386-21 and satisfying the requirements of these Regulations for a protective conductor, or

(iv) be enclosed in earthed trunking or ducting complying with BS EN 50085-2-1 and satisfying the requirements of these Regulations for a protective conductor, or

(v) be mechanically protected against damage sufficient to prevent penetration of the cable by nails, screws and the like, or

(vi) form part of a SELV or PELV circuit meeting the requirements of Regulation 414.4.

522.6.101 A cable concealed in a wall or partition at a depth of less than 50 mm from a surface of the wall or partition shall:

(i) incorporate an earthed metallic covering which complies with the requirements of these Regulations for a protective conductor of the circuit concerned, the cable complying with BS 5467, BS 6724, BS 7846, BS EN 60702-1 or BS 8436, or

(ii) be enclosed in earthed conduit complying with BS EN 61386-21 and satisfying the requirements of these Regulations for a protective conductor, or

(iii) be enclosed in earthed trunking or ducting complying with BS EN 50085-2-1 and satisfying the requirements of these Regulations for a protective conductor, or

(iv) be mechanically protected against damage sufficient to prevent penetration of the cable by nails, screws and the like, or

(v) be installed in a zone within 150 mm from the top of the wall or partition or within 150 mm of an angle formed by two adjoining walls or partitions. Where the cable is connected to a point, accessory or switchgear on any surface of the wall or partition, the cable may be installed in a zone either horizontally or vertically, to the point, accessory or switchgear. Where the location of the accessory, point or switchgear can be determined from the reverse side, a zone formed on one side of a wall of 100 mm thickness or less or partition of 100 mm thickness or less extends to the reverse side, or

(vi) form part of a SELV or PELV circuit meeting the requirements of Regulation 414.4.

522.6.102   Where Regulation 522.6.101 applies and the installation is not intended to be under the supervision of a skilled or instructed person, a cable installed in accordance with Regulation 522.6.101(v), and not also complying with Regulation 522.6.101(i), (ii), (iii) or (iv), shall be provided with additional protection by means of an RCD having the characteristics specified in Regulation 415.1.1.

522.6.103   Irrespective of the depth of the cable from a surface of the wall or partition, in an installation not intended to be under the supervision of a skilled or instructed person, a cable concealed in a wall or partition the internal construction of which includes metallic parts, other than metallic fixings such as nails, screws and the like, shall:

(i) incorporate an earthed metallic covering which complies with the requirements of these Regulations for a protective conductor of the circuit concerned, the cable complying with BS 5467, BS 6346, BS 6724, BS 7846, BS EN 60702-1 or BS 8436, or

(ii) be enclosed in earthed conduit complying with BS EN 61386-21 and satisfying the requirements of these Regulations for a protective conductor, or

(iii) be enclosed in earthed trunking or ducting complying with BS EN 50085-2-1 and satisfying the requirements of these Regulations for a protective conductor, or

(iv) be mechanically protected sufficiently to avoid damage to the cable during construction of the wall or partition and during installation of the cable, or

(v) form part of a SELV or PELV circuit meeting the requirements of Regulation 414.4, or

(vi) be provided with additional protection by means of an RCD having the characteristics specified in Regulation 415.1.1.

For a cable installed at a depth of 50 mm or less from the surface of a wall or partition the requirements of Regulation 522.6.101 shall also apply.

522.7   Vibration (AH)

522.7.1   A wiring system supported by or fixed to a structure or equipment subject to vibration of medium severity (AH2) or high severity (AH3) shall be suitable for such conditions, particularly where cables and cable connections are concerned.

522.7.2   For the fixed installation of suspended current-using equipment, e.g. luminaires, connection shall be made by cable with flexible cores. Where no vibration or movement can be expected, cable with non-flexible cores may be used.

522.8   Other mechanical stresses (AJ)

522.8.1   A wiring system shall be selected and erected to avoid during installation, use or maintenance, damage to the sheath or insulation of cables and their terminations. The use of any lubricants that can have a detrimental effect on the cable or wiring system are not permitted.

522.8.2   Where buried in the structure, a conduit system or cable ducting system, other than a pre-wired conduit assembly specifically designed for the installation, shall be completely erected between access points before any cable is drawn in.

522.8.3   The radius of every bend in a wiring system shall be such that conductors or cables do not suffer damage and terminations are not stressed.

522.8.4   Where the conductors or cables are not supported continuously due to the method of installation, they shall be supported by suitable means at appropriate intervals in such a manner that the conductors or cables do not suffer damage by their own weight.

522.8.5   Every cable or conductor shall be supported in such a way that it is not exposed to undue mechanical strain and so that there is no appreciable mechanical strain on the terminations of the conductors, account being taken of mechanical strain imposed by the supported weight of the cable or conductor itself.

522.8.6   A wiring system intended for the drawing in or out of conductors or cables shall have adequate means of access to allow this operation.

522.8.7   A wiring system buried in a floor shall be sufficiently protected to prevent damage caused by the intended use of the floor.

522.8.8   *Not used*

522.8.9   *Not used*

522.8.10   Except where installed in a conduit or duct which provides equivalent protection against mechanical damage, a cable buried in the ground shall incorporate an earthed armour or metal sheath or both, suitable for use as a protective conductor. The location of buried cables shall be marked by cable covers or a suitable marking tape. Buried conduits and ducts shall be suitably identified. Buried cables, conduits and ducts shall be at a sufficient depth to avoid being damaged by any reasonably foreseeable disturbance of the ground.

NOTE: BS EN 61386-24 is the standard for underground conduits.

522.8.11   Cable supports and enclosures shall not have sharp edges liable to damage the wiring system.

522.8.12   A cable or conductors shall not be damaged by the means of fixing.

522.8.13   Cables, busbars and other electrical conductors which pass across expansion joints shall be so selected or erected that anticipated movement does not cause damage to the electrical equipment.

522.8.14   No wiring system shall penetrate an element of building construction which is intended to be load bearing unless the integrity of the load-bearing element can be assured after such penetration.

522.9   Presence of flora and/or mould growth (AK)

522.9.1   Where the conditions experienced or expected constitute a hazard (AK2), the wiring system shall be selected accordingly or special protective measures shall be adopted.

NOTE 1: An installation method which facilitates the removal of such growths may be necessary (see Section 529).

NOTE 2: Possible preventive measures are closed types of installation (conduit or channel), maintaining distances to plants and regular cleaning of the relevant wiring system.

522.10   Presence of fauna (AL)

522.10.1   Where conditions experienced or expected constitute a hazard (AL2), the wiring system shall be selected accordingly or special protective measures shall be adopted, for example, by:

   (i)   the mechanical characteristics of the wiring system, or

   (ii)   the location selected, or

   (iii) the provision of additional local or general protection against mechanical damage, or

   (iv)   any combination of the above.

522.11   Solar radiation (AN) and ultraviolet radiation

522.11.1   Where significant solar radiation (AN2) or ultraviolet radiation is experienced or expected, a wiring system suitable for the conditions shall be selected and erected or adequate shielding shall be provided. Special precautions may need to be taken for equipment subject to ionising radiation.

NOTE: See also Regulation 522.2.1 dealing with temperature rise.

522.12   Seismic effects (AP)

522.12.1   The wiring system shall be selected and erected with due regard to the seismic hazards of the location of the installation.

522.12.2   Where the seismic hazards experienced are low severity (AP2) or higher, particular attention shall be paid to the following:

   (i)   The fixing of wiring systems to the building structure

   (ii)   The connections between the fixed wiring and all items of essential equipment, e.g. safety services, shall be selected for their flexible quality.

522.13   Movement of air (AR)

522.13.1   See Regulation 522.7, Vibration (AH), and Regulation 522.8, Other mechanical stresses (AJ).

522.14   Nature of processed or stored materials (BE)

522.14.1   See Section 527, Selection and erection of wiring systems to minimize the spread of fire and Section 422, Precautions where particular risks of fire exist.

522.15   Building design (CB)

522.15.1   Where risks due to structural movement exist (CB3), the cable support and protection system employed shall be capable of permitting relative movement so that conductors and cables are not subjected to excessive mechanical stress.

522.15.2   For a flexible structure or a structure intended to move (CB4), a flexible wiring system shall be used.

523   CURRENT-CARRYING CAPACITIES OF CABLES

523.1   The current, including any harmonic current, to be carried by any conductor for sustained periods during normal operation shall be such that the appropriate temperature limit specified in Table 52.1 is not exceeded. The value of current shall be selected in accordance with Regulation 523.2, or determined in accordance with Regulation 523.3.

TABLE 52.1 – Maximum operating temperatures for types of cable insulation

Type of insulation	Temperature limit[a]
Thermoplastic	70 °C at the conductor
Thermosetting	90 °C at the conductor [b]
Mineral (Thermoplastic covered or bare exposed to touch)	70 °C at the sheath
Mineral (bare not exposed to touch and not in contact with combustible material)	105 °C at the sheath [b, c]

[a] The maximum permissible conductor temperatures given in Table 52.1 on which the tabulated current-carrying capacities given in Appendix 4 are based, have been taken from IEC 60502-1 and BS EN 60702-1 and are shown on these tables in Appendix 4.

[b] Where a conductor operates at a temperature exceeding 70 °C it shall be ascertained that the equipment connected to the conductor is suitable for the resulting temperature at the connection.

[c] For mineral insulated cables, higher operating temperatures may be permissible dependent upon the temperature rating of the cable, its terminations, the environmental conditions and other external influences.

NOTE: For the temperature limits for other types of insulation, refer to cable specification or manufacturer.

523.2   The requirement of Regulation 523.1 is considered to be satisfied if the current for non-sheathed and sheathed cables does not exceed the appropriate values selected from the tables of current-carrying capacity given in Appendix 4 with reference to Table 4A2, subject to any necessary rating factors.

NOTE: The current-carrying capacities given in the Tables are provided for guidance. It is recognized that there will be some tolerance in the current-carrying capacities depending on the environmental conditions and the precise construction of the cables.

523.3   The appropriate value of current-carrying capacity may also be determined as described in BS 7769 series (some parts of the BS 7769 series are now numbered BS IEC 60287 series, eventually all parts will be renumbered), or by test, or by calculation using a recognized method, provided that the method is stated. Where appropriate, account shall be taken of the characteristics of the load and, for buried cables, the effective thermal resistance of the soil.

523.4   The ambient temperature shall be considered to be the temperature of the surrounding medium when the non-sheathed or sheathed cable(s) under consideration are not loaded.

523.5   Groups containing more than one circuit

The group rating factors, see Tables 4C1 to 4C6 of Appendix 4, are applicable to groups of non-sheathed or sheathed cables having the same maximum operating temperature.

For groups containing non-sheathed or sheathed cables having different maximum operating temperatures, the current-carrying capacity of all the non-sheathed or sheathed cables in the group shall be based on the lowest maximum operating temperature of any cable in the group together with the appropriate group rating factor.

If, due to known operating conditions, a non-sheathed or sheathed cable is expected to carry a current not greater than 30 % of its grouped current-carrying capacity, it may be ignored for the purpose of obtaining the rating factor for the rest of the group.

523.6 Number of loaded conductors

523.6.1 The number of conductors to be considered in a circuit are those carrying load current. Where conductors in polyphase circuits carry balanced currents, the associated neutral conductor need not be taken into consideration. Under these conditions a four-core cable is given the same current-carrying capacity as a three-core cable having the same conductor cross-sectional area for each line conductor. The neutral conductor shall be considered as a loaded conductor in the case of the presence of third harmonic current or multiples of the third harmonic presenting a total harmonic distortion greater than 15% of the fundamental line current.

523.6.2 Where the neutral conductor in a multicore cable carries current as a result of an imbalance in the line currents, the temperature rise due to the neutral current is offset by the reduction in the heat generated by one or more of the line conductors. In this case the conductor size shall be chosen on the basis of the highest line current.

In all cases the neutral conductor shall have a cross-sectional area adequate to afford compliance with Regulation 523.1.

523.6.3 Where the neutral conductor carries current without a corresponding reduction in load of the line conductors, the neutral conductor shall be taken into account in ascertaining the current-carrying capacity of the circuit. Such currents may be caused by a significant harmonic current in three-phase circuits. If the total harmonic distortion due to third harmonic current or multiples of the third harmonic is greater than 15 % of the fundamental line current the neutral conductor shall not be smaller than the line conductors. Thermal effects due to the presence of third harmonic or multiples of third harmonic currents and the corresponding rating factors for higher harmonic currents are given in Appendix 4, section 5.5.

523.6.4 Conductors which serve the purpose of protective conductors only are not to be taken into consideration. PEN conductors shall be taken into consideration in the same way as neutral conductors.

523.6.100 The tabulated current-carrying capacities in Appendix 4 are based on the fundamental frequency only and do not take account of the effect of harmonics.

523.7 Conductors in parallel

Where two or more live conductors or PEN conductors are connected in parallel in a system, either:

  (i) measures shall be taken to achieve equal load current sharing between them

  This requirement is considered to be fulfilled if the conductors are of the same material, have the same cross-sectional area, are approximately the same length and have no branch circuits along their length, and either:

  (a) the conductors in parallel are multicore cables or twisted single-core cables or non-sheathed cables, or

  (b) the conductors in parallel are non-twisted single-core cables or non-sheathed cables in trefoil or flat formation and where the cross-sectional area is greater than 50 mm^2 in copper or 70 mm^2 in aluminium, the special configuration necessary for such formations is adopted. These configurations consist of suitable groupings and spacings of the different lines or poles

or (ii) special consideration shall be given to the load current sharing to meet the requirements of Regulation 523.1.

This regulation does not preclude the use of ring final circuits with or without spur connections.

Where adequate current sharing is not possible or where four or more conductors have to be connected in parallel consideration shall be given to the use of busbar trunking.

523.8 Variation of installation conditions along a route

Where the heat dissipation differs from one part of a route to another, the current-carrying capacity at each part of the route shall be appropriate for that part of the route.

523.9 Cables in thermal insulation

A cable should preferably not be installed in a location where it is liable to be covered by thermal insulation. Where a cable is to be run in a space to which thermal insulation is likely to be applied it shall, wherever practicable, be fixed in a position such that it will not be covered by the thermal insulation. Where fixing in such a position is impracticable the cross-sectional area of the cable shall be selected to meet the requirements of Chapter 43. Where necessary, the nature of the load (e.g. cyclic) and diversity shall be taken into account.

For a cable installed in a thermally insulated wall or above a thermally insulated ceiling, the cable being in contact with a thermally conductive surface on one side, current-carrying capacities are tabulated in Appendix 4.

For a single cable likely to be totally surrounded by thermally insulating material over a length of 0.5 m or more, the current-carrying capacity shall be taken, in the absence of more precise information, as 0.5 times the current-carrying capacity for that cable clipped direct to a surface and open (Reference Method C).

Where a cable is to be totally surrounded by thermal insulation for less than 0.5 m the current-carrying capacity of the cable shall be reduced appropriately depending on the size of cable, length in insulation and thermal properties of the insulation. The derating factors in Table 52.2 are appropriate to conductor sizes up to 10 mm^2 in thermal insulation having a thermal conductivity (λ) greater than 0.04 Wm^{-1}K^{-1}.

TABLE 52.2 – Cable surrounded by thermal insulation

Length in insulation (mm)	Derating factor
50	0.88
100	0.78
200	0.63
400	0.51

523.100   Armoured single-core cables

The metallic sheaths and/or non-magnetic armour of single-core cables in the same circuit shall normally be bonded together at both ends of their run (solid bonding). Alternatively, the sheaths or armour of such cables having conductors of cross-sectional area exceeding 50 mm^2 and a non-conducting outer sheath may be bonded together at one point in their run (single point bonding) with suitable insulation at the unbonded ends, in which case the length of the cables from the bonding point shall be limited so that voltages from sheaths and/or armour to Earth:

(i)   do not cause corrosion when the cables are carrying their full load current, for example by limiting the voltage to 25 V, and

(ii)   do not cause danger or damage to property when the cables are carrying short-circuit current.

524   CROSS-SECTIONAL AREAS OF CONDUCTORS

524.1   The cross-sectional area of each conductor in a circuit shall be not less than the values given in Table 52.3, except as provided for extra-low voltage lighting installations according to Regulation 559.11.5.2.

524.2   Neutral conductors

524.2.1   The neutral conductor, if any, shall have a cross-sectional area not less than that of the line conductor:

(i)   in single-phase, two-wire circuits, whatever the cross-sectional area

(ii)   in polyphase and single-phase three-wire circuits, where the size of the line conductors is less than or equal to 16 mm^2 for copper, or 25 mm^2 for aluminium

(iii) in circuits where it is required according to Regulation 523.6.3.

524.2.2   If the total harmonic content due to triplen harmonics is greater than 33% of the fundamental line current, an increase in the cross-sectional area of the neutral conductor may be required (see Regulation 523.6.3 and Appendix 4, section 5.5).

524.2.3   For a polyphase circuit where each line conductor has a cross-sectional area greater than 16 mm^2 for copper or 25 mm^2 for aluminium, the neutral conductor is permitted to have a smaller cross-sectional area than that of the line conductors providing the following conditions are simultaneously fulfilled:

(i)   The expected maximum current including harmonics, if any, in the neutral conductor during normal service is not greater than the current-carrying capacity of the reduced cross-sectional area of the neutral conductor, and

> NOTE: The load carried by the circuit under normal service conditions should be practically equally distributed between the lines.

(ii)   the neutral conductor is protected against overcurrents according to Regulation 431.2, and

(iii) the size of the neutral conductor is at least equal to 16 mm^2 for copper or 25 mm^2 for aluminium, account being taken of Regulation 523.6.3.

TABLE 52.3 – Minimum cross-sectional area of conductors

Type of wiring system	Use of the circuit	Conductor	
		Material	Cross-sectional area mm^2
Non-sheathed and sheathed cables	Lighting circuits	Copper	1.0
		Aluminium	16
	Power circuits	Copper	1.5
		Aluminium	16
	Signalling and control circuits	Copper	0.5 (see Note 1)
Bare conductors	Power circuits	Copper	10
		Aluminium	16
	Signalling and control circuits	Copper	4
Non-sheathed and sheathed flexible cables	For a specific appliance	Copper	As specified in the product standard
	For any other application		0.75 [a]
	Extra-low voltage circuits for special applications (see Note 2)		0.75

NOTE 1: In information technology, signalling and control circuits intended for electronic equipment a minimum cross-sectional area of 0.1 mm^2 is permitted.

NOTE 2: For special requirements for ELV lighting see Section 559.

NOTE 3: Connectors used to terminate aluminium conductors shall be tested and approved for this specific use.

[a] In multicore flexible cables containing seven or more cores, Note 1 applies.

525 VOLTAGE DROP IN CONSUMERS' INSTALLATIONS

525.1 In the absence of any other consideration, under normal service conditions the voltage at the terminals of any fixed current-using equipment shall be greater than the lower limit corresponding to the product standard relevant to the equipment.

525.100 Where fixed current-using equipment is not the subject of a product standard the voltage at the terminals shall be such as not to impair the safe functioning of that equipment.

525.101 The above requirements are deemed to be satisfied if the voltage drop between the origin of the installation (usually the supply terminals) and a socket-outlet or the terminals of fixed current-using equipment does not exceed that stated in Appendix 4 section 6.4.

525.102 A greater voltage drop than stated in Appendix 4 section 6.4 may be accepted for a motor during starting periods and for other equipment with high inrush currents, provided that it is verified that the voltage variations are within the limits specified in the relevant product standard for the equipment or, in the absence of a product standard, in accordance with the manufacturer's recommendations.

526 ELECTRICAL CONNECTIONS

526.1 Every connection between conductors or between a conductor and other equipment shall provide durable electrical continuity and adequate mechanical strength and protection.

NOTE: See Regulation 522.8 – Other mechanical stresses.

526.2 The selection of the means of connection shall take account of, as appropriate:

(i) the material of the conductor and its insulation

(ii) the number and shape of the wires forming the conductor

(iii) the cross-sectional area of the conductor

(iv) the number of conductors to be connected together

(v) the temperature attained at the terminals in normal service such that the effectiveness of the insulation of the conductors connected to them is not impaired

(vi) the provision of adequate locking arrangements in situations subject to vibration or thermal cycling.

Where a soldered connection is used the design shall take account of creep, mechanical stress and temperature rise under fault conditions.

NOTE 1: Applicable standards include BS EN 60947-7, the BS EN 60998 series and BS EN 61535.

NOTE 2: Terminals without the marking R (only rigid conductor), F (only flexible conductor), S or Sol (only solid conductor) are suitable for the connection of all types of conductors.

526.3   Every connection shall be accessible for inspection, testing and maintenance, except for the following:

(i)   A joint designed to be buried in the ground

(ii)   A compound-filled or encapsulated joint

(iii)   A connection between a cold tail and the heating element as in ceiling heating, floor heating or a trace heating system

(iv)   A joint made by welding, soldering, brazing or appropriate compression tool

(v)   Joints or connections made in equipment by the manufacturer of the product and not intended to be inspected or maintained

(vi)   Equipment complying with BS 5733 for a maintenance-free accessory and marked with the symbol and installed in accordance with the manufacturer's instructions.

526.4   Where necessary, precautions shall be taken so that the temperature attained by a connection in normal service shall not impair the effectiveness of the insulation of the conductors connected to it or any insulating material used to support the connection. Where a cable is to be connected to a bare conductor or busbar its type of insulation and/or sheath shall be suitable for the maximum operating temperature of the bare conductor or busbar.

526.5   Every termination and joint in a live conductor or a PEN conductor shall be made within one of the following or a combination thereof:

(i)   A suitable accessory complying with the appropriate product standard

(ii)   An equipment enclosure complying with the appropriate product standard

(iii)   An enclosure partially formed or completed with building material which is non-combustible when tested to BS 476-4.

526.6   There shall be no appreciable mechanical strain on the connections of conductors.

526.7   Where a connection is made in an enclosure the enclosure shall provide adequate mechanical protection and protection against relevant external influences.

526.8   Cores of sheathed cables from which the sheath has been removed and non-sheathed cables at the termination of conduit, ducting or trunking shall be enclosed as required by Regulation 526.5.

526.9   **Connection of multiwire, fine wire and very fine wire conductors**

526.9.1   In order to avoid inappropriate separation or spreading of individual wires of multiwire, fine wire or very fine wire conductors, suitable terminals shall be used or the conductor ends shall be suitably treated.

526.9.2   Soldering (tinning) of the whole conductor end of multiwire, fine wire and very fine wire conductors is not permitted if screw terminals are used.

526.9.3   Soldered (tinned) conductor ends on fine wire and very fine wire conductors are not permissible at connection and junction points which are subject in service to a relative movement between the soldered and the non-soldered part of the conductor.

527 SELECTION AND ERECTION OF WIRING SYSTEMS TO MINIMIZE THE SPREAD OF FIRE

527.1 Precautions within a fire-segregated compartment

527.1.1 The risk of spread of fire shall be minimized by the selection of appropriate materials and erection in accordance with Section 527.

527.1.2 A wiring system shall be installed so that the general building structural performance and fire safety are not reduced.

527.1.3 Cables complying with, at least, the requirements of BS EN 60332-1-2 may be installed without special precautions.

In installations where particular risk is identified, cables shall meet the flame propagation requirements given in the relevant part of the BS EN 60332-3 series.

527.1.4 Cables not complying with the flame propagation requirements of BS EN 60332-1-2 shall be limited to short lengths for connection of appliances to the permanent wiring system and shall not pass from one fire-segregated compartment to another.

527.1.5 Products having the necessary resistance to flame propagation as specified in the BS EN 61386 series, the appropriate part of BS EN 50085 series, BS EN 50086, BS EN 60439-2, BS EN 61534 series, BS EN 61537 or BS EN 60570 may be installed without special precautions. Other products complying with standards having similar requirements for resistance to flame propagation may be installed without special precautions.

527.1.6 Parts of wiring systems other than cables which do not comply, as a minimum, with the flame propagation requirements as specified in the BS EN 61386 series, the appropriate part of BS EN 50085 series, BS EN 50086, BS EN 60439-2, BS EN 61534 series or BS EN 61537 but which comply in all other respects with the requirements of their respective product standard shall, if used, be completely enclosed in suitable non-combustible building materials.

527.2 Sealing of wiring system penetrations

527.2.1 Where a wiring system passes through elements of building construction such as floors, walls, roofs, ceilings, partitions or cavity barriers, the openings remaining after passage of the wiring system shall be sealed according to the degree of fire-resistance (if any) prescribed for the respective element of building construction before penetration.

This requirement is satisfied if the sealing of the wiring system concerned has passed a relevant type test meeting the requirements of Regulation 527.2.3.

527.2.1.1 During the erection of a wiring system temporary sealing arrangements shall be provided as appropriate.

527.2.1.2 During alteration work, sealing which has been disturbed shall be reinstated as soon as practicable.

527.2.2 A wiring system such as a conduit system, cable ducting system, cable trunking system, busbar or busbar trunking system which penetrates elements of building construction having specified fire-resistance shall be internally sealed to the degree of fire-resistance of the respective element before penetration as well as being externally sealed as required by Regulation 527.2.1.

This requirement is satisfied if the sealing of the wiring system concerned has passed a relevant type test meeting the requirements of Regulation 527.2.3.

527.2.3 A conduit system, cable trunking system or cable ducting system classified as non-flame propagating according to the relevant product standard and having a maximum internal cross-sectional area of 710 mm^2 need not be internally sealed provided that:

(i) the system satisfies the test of BS EN 60529 for IP33, and

(ii) any termination of the system in one of the compartments, separated by the building construction being penetrated, satisfies the test of BS EN 60529 for IP33.

527.2.4   Any sealing arrangement intended to satisfy Regulation 527.2.1 or 527.2.1.1 shall resist external influences to the same degree as the wiring system with which it is used and, in addition, it shall meet all of the following requirements:

   (i)   It shall be resistant to the products of combustion to the same extent as the elements of building construction which have been penetrated

  (ii)   It shall provide the same degree of protection from water penetration as that required for the building construction element in which it has been installed

 (iii)   It shall be compatible with the material of the wiring system with which it is in contact

 (iv)   It shall permit thermal movement of the wiring system without reduction of the sealing quality

  (v)   It shall be of adequate mechanical stability to withstand the stresses which may arise through damage to the support of the wiring system due to fire.

The seal and the wiring system shall be protected from dripping water which may travel along the wiring system or which may otherwise collect around the seal unless the materials used in the seal are all resistant to moisture when finally assembled for use.

NOTE: This regulation may be satisfied if:
- either cable cleats, cable ties or cable supports are installed within 750 mm of the seal, and are able to withstand the mechanical loads expected following the collapse of the supports on the fire side of the seal to the extent that no strain is transferred to the seal, or
- the design of the sealing system itself provides adequate support.

528   PROXIMITY OF WIRING SYSTEMS TO OTHER SERVICES

528.1   Proximity to electrical services

Except where one of the following methods is adopted, neither a voltage Band I nor a voltage Band II circuit shall be contained in the same wiring system as a circuit of nominal voltage exceeding that of low voltage, and a Band I circuit shall not be contained in the same wiring system as a Band II circuit:

   (i)   Every cable or conductor is insulated for the highest voltage present

  (ii)   Each conductor of a multicore cable is insulated for the highest voltage present in the cable

 (iii)   The cables are insulated for their system voltage and installed in a separate compartment of a cable ducting or cable trunking system

 (iv)   The cables are installed on a cable tray system where physical separation is provided by a partition

  (v)   A separate conduit, trunking or ducting system is employed

 (vi)   For a multicore cable, the cores of the Band I circuit are separated from the cores of the Band II circuit by an earthed metal screen of equivalent current-carrying capacity to that of the largest core of a Band II circuit.

For SELV and PELV systems the requirements of Regulation 414.4 shall apply.

NOTE 1: In the case of proximity of wiring systems and lightning protection systems, BS EN 62305 should be considered.

NOTE 2: Requirements for separation and segregation in relation to safety services are given in BS 5266 and BS 5839.

528.2   Proximity of communications cables

In the event of crossing or proximity of underground telecommunication cables and underground power cables, a minimum clearance of 100 mm shall be maintained, or the requirements according to (i) or (ii) shall be fulfilled:

   (i)   A fire-retardant partition shall be provided between the cables, e.g. bricks, cable protecting caps (clay, concrete), shaped blocks (concrete), protective cable conduit or troughs made of fire-retardant materials

  (ii)   For crossings, mechanical protection between the cables shall be provided, e.g. cable conduit, concrete cable protecting caps or shaped blocks.

NOTE 1: Special considerations of electrical interference, both electromagnetic and electrostatic, may apply to telecommunication circuits, data transfer circuits and the like.

NOTE 2: Segregation requirements for communications services are given in BS 6701 and BS EN 50174 series.

528.3 Proximity to non-electrical services

528.3.1   A wiring system shall not be installed in the vicinity of services which produce heat, smoke or fumes likely to be detrimental to the wiring, unless it is protected from harmful effects by shielding arranged so as not to affect the dissipation of heat from the wiring.

In areas not specifically designed for the installation of cables, e.g. service shafts and cavities, the cables shall be laid so that they are not exposed to any harmful influence by the normal operation of the adjacent installations (e.g. gas, water or steam lines).

528.3.2   Where a wiring system is routed below services liable to cause condensation (such as water, steam or gas services), precautions shall be taken to protect the wiring system from deleterious effects.

528.3.3   Where an electrical service is to be installed in proximity to one or more non-electrical services it shall be so arranged that any foreseeable operation carried out on the other services will not cause damage to the electrical service or the converse.

NOTE: This may be achieved by:
    (i)   suitable spacing between the services, or
    (ii)   the use of mechanical or thermal shielding.

528.3.4   Where an electrical service is located in close proximity to one or more non-electrical services, both the following conditions shall be met:

(i)   The wiring system shall be suitably protected against the hazards likely to arise from the presence of the other services in normal use

(ii)   Fault protection shall be afforded in accordance with the requirements of Section 411.

NOTE: The requirements for segregation between low pressure gas systems and electrical equipment are given in BS 6891.

528.3.5   No cable shall be run in a lift or hoist shaft unless it forms part of the lift installation as defined in BS EN 81-1 series.

529   SELECTION AND ERECTION OF WIRING SYSTEMS IN RELATION TO MAINTAINABILITY, INCLUDING CLEANING

529.1   With regard to maintainability, reference shall be made to Regulation 132.12.

529.2   Where it is necessary to remove any protective measure in order to carry out maintenance, provision shall be made so that the protective measure can be reinstated without reduction of the degree of protection originally intended.

529.3   Provision shall be made for safe and adequate access to all parts of a wiring system which may require maintenance.

NOTE: In some situations, it may be necessary to provide permanent means of access by ladders, walkways, etc.

CHAPTER 53

PROTECTION, ISOLATION, SWITCHING, CONTROL AND MONITORING

CONTENTS

CHAPTER 53
PROTECTION, ISOLATION, SWITCHING, CONTROL
AND MONITORING

530 INTRODUCTION

530.1 Scope

This chapter provides general requirements for protection, isolation, switching, control and monitoring with the requirements for selection and erection of the devices provided to fulfil such functions, including requirements to provide compliance with measures of protection for safety.

530.2 Not used

530.3 General and common requirements

Equipment shall be selected and installed to provide for the safety and proper functioning for the intended use of the installation. Equipment installed shall be appropriate to the external influences foreseen.

530.3.1 In multiphase circuits the moving contacts of all poles of a multipole device shall be so coupled mechanically that they make and break substantially together, except:

  (i) that contacts solely intended for the neutral may close before and open after the other contacts

  (ii) in accordance with Regulation 543.3.3(ii).

NOTE 1: Switching devices, contactors, circuit-breakers, RCDs, isolating switches, control and protective switching devices for equipment (CPS), etc., complying with their relevant standard (see Table 53.4) fulfil this requirement.

NOTE 2: The requirement to make and break substantially together may not apply to control and auxiliary contacts.

530.3.2 Except as provided in Regulation 537.2.2.5, in multiphase circuits an independently operated single-pole switching device or protective device shall not be inserted in the neutral conductor. In single-phase circuits an independently operated single-pole switching or protective device shall not be inserted in the neutral conductor alone.

530.3.3 A device embodying more than one function shall comply with all the requirements of this chapter appropriate to each separate function.

530.3.4 For an installation with a 230 V single-phase supply rated up to 100 A that is under the control of ordinary persons, switchgear and controlgear assemblies shall either comply with BS EN 60439-3 and Regulation 432.1 or be a consumer unit incorporating components and protective devices specified by the manufacturer complying with BS EN 60439-3, including the conditional short-circuit test described in Annex ZA of BS EN 60439-3.

530.3.5 An auto-reclosing device for protection, isolation, switching or control may be installed only in an installation intended to be under the supervision of skilled or instructed persons and intended to be inspected and tested by competent persons.

An auto-reclosing device shall not be used to meet the requirements of Regulations 411.3.3, 415.1, 522.6.102 or 522.6.103 unless the device is of a type that automatically verifies the insulation resistance in the part of the installation it controls is satisfactory before the device recloses.

An auto-reclosing device shall be of a type that satisfies all the following requirements:

  (i) The automatic reclose function of the device cannot be engaged after the device is manually switched to the off (or open) position

  (ii) There is a time delay before the first automatic reclosure

  (iii) The number of consecutive automatic reclosures is limited (to 3, for example).

A warning notice shall be clearly displayed on or near the device, indicating that no work should be carried out on the part of the installation controlled by the device unless the automatic reclose function of the device is disengaged, the device is manually switched off, and that part of the installation is securely isolated from the supply.

530.4 Fixing of equipment

530.4.1 Except where specifically designed for direct connection to flexible wiring, equipment shall be fixed according to the manufacturer's instructions in such a way that connections between wiring and equipment shall not be subject to undue stress or strain resulting from the normal use of the equipment.

530.4.2 Unenclosed equipment shall be mounted in a suitable mounting box or enclosure in compliance with the relevant part of BS EN 60670, BS EN 62208 or other relevant standard. Socket-outlets, connection units, plate switches and similar accessories shall be fitted to a mounting box complying with BS 4662 or BS 5733 and with the relevant part of BS EN 60670.

530.4.3 Wherever equipment is fixed on or in cable trunking, skirting trunking or in mouldings it shall not be fixed on covers which can be removed inadvertently.

531 DEVICES FOR FAULT PROTECTION BY AUTOMATIC DISCONNECTION OF SUPPLY

531.1 Overcurrent protective devices

531.1.1 TN system

In a TN system, overcurrent protective devices where used as devices for fault protection shall be selected and erected in order to comply with the requirements specified in Chapter 41.

531.1.2 TT system

In a TT system, overcurrent protective devices where used as devices for fault protection shall be selected and erected in order to comply with the requirements specified in Chapter 41.

531.1.3 IT system

Overcurrent protective devices where used as devices for fault protection, in the event of a second fault, shall comply with:

(i) Regulation 531.1.1, taking into account the requirements of Regulation 411.6.4(i), where exposed-conductive-parts are interconnected

(ii) Regulation 531.1.2, taking into account the requirements of Regulation 411.6.4(ii), where exposed-conductive-parts are earthed in groups or individually.

Overcurrent protective devices used in IT systems shall be suitable for line-to-line voltage applications for operation in case of a second insulation fault.

In an IT system, in the event of a second fault, an overcurrent protective device shall disconnect all corresponding live conductors, including the neutral conductor, if any (see also Regulation 431.2.2 of Chapter 43).

531.2 RCDs

531.2.1 An RCD shall be capable of disconnecting all the line conductors of the circuit at substantially the same time.

531.2.2 The magnetic circuit of the transformer of an RCD shall enclose all the live conductors of the protected circuit. The associated protective conductor shall be outside the magnetic circuit.

531.2.3 The rated residual operating current of the protective device shall comply with the requirements of Section 411 as appropriate to the type of system earthing.

531.2.4 An RCD shall be so selected and the electrical circuits so subdivided that any protective conductor current which may be expected to occur during normal operation of the connected load(s) will be unlikely to cause unnecessary tripping of the device.

531.2.5 The use of an RCD associated with a circuit normally expected to have a protective conductor shall not be considered sufficient for fault protection if there is no such conductor, even if the rated residual operating current ($I_{\Delta n}$) of the device does not exceed 30 mA.

531.2.6   An RCD which is powered from an independent auxiliary source and which does not operate automatically in the case of failure of the auxiliary source shall be used only if one of the following conditions is fulfilled:

(i)   Fault protection is maintained even in the case of failure of the auxiliary source

(ii) The device is incorporated in an installation intended to be supervised by a skilled or instructed person and inspected and tested by a competent person.

531.2.7   An RCD shall be located so that its operation will not be impaired by magnetic fields caused by other equipment.

531.2.8   Where an RCD is used for fault protection with, but separately from, an overcurrent protective device, it shall be verified that the residual current operated device is capable of withstanding, without damage, the thermal and mechanical stresses to which it is likely to be subjected in the case of a fault occurring on the load side of the point at which it is installed.

531.2.9   Where, for compliance with the requirements of the regulations for fault protection or otherwise to prevent danger, two or more RCDs are in series, and where discrimination in their operation is necessary to prevent danger, the characteristics of the devices shall be such that the intended discrimination is achieved.

NOTE: In such cases the downstream RCD may need to disconnect all live conductors.

531.2.10   Where an RCD may be operated by a person other than a skilled or instructed person, it shall be designed or installed so that it is not possible to modify or adjust the setting or the calibration of its rated residual operating current ($I_{\Delta n}$) or time delay mechanism without a deliberate act involving the use of either a key or a tool and resulting in a visible indication of its setting or calibration.

531.3   RCDs in a TN system

531.3.1   In a TN system, where, for certain equipment in a certain part of the installation, the requirement of Regulation 411.4.5 cannot be satisfied, that part may be protected by an RCD.

The exposed-conductive-parts of that part of the installation shall be connected to the TN earthing system protective conductor or to a separate earth electrode which affords an impedance appropriate to the operating current of the RCD.

In this latter case the circuit shall be treated as a TT system and Regulations 411.5.1 to 3 apply.

531.4   RCDs in a TT system

531.4.1   If an installation which is part of a TT system is protected by a single RCD, this shall be placed at the origin of the installation unless the part of the installation between the origin and the device complies with the requirements for protection by the use of Class II equipment or equivalent insulation (Section 412). Where there is more than one origin this requirement applies to each origin.

531.5   RCDs in an IT system

531.5.1   Where protection is provided by an RCD and disconnection following a first fault is not envisaged, the non-operating residual current of the device shall be at least equal to the current which circulates on the first fault to Earth of negligible impedance affecting a line conductor.

531.6   Insulation monitoring devices

531.6.1   An insulation monitoring device shall be so designed or installed that it shall be possible to modify the setting only by the use of a key or a tool.

532   DEVICES FOR PROTECTION AGAINST THE RISK OF FIRE

532.1   Where, in accordance with the requirements of Regulation 422.3.9, it is necessary to limit the consequence of fault currents in a wiring system from the point of view of fire risk, the circuit shall be either:

(i)   protected by an RCD complying with Regulation 531.2 for fault protection, and

  – the RCD shall be installed at the origin of the circuit to be protected, and

  – the RCD shall switch all live conductors, and

  – the rated residual operating current of the RCD shall not exceed 300 mA

or (ii)   continuously monitored by an insulation monitoring device(s) complying with Regulation 538.1 and which initiates an alarm on the occurrence of an insulation fault.

NOTE 1: A fault location system complying with Regulation 538.2 which is able to locate the faulty circuit may be helpful.

NOTE 2: For locations having a risk of explosion see BS EN 60079-10, BS EN 60079-14, BS EN 61241-10 and BS EN 61241-14.

NOTE 3: The following methods which are additional to the requirements may also be employed:

  - devices intended to provide protection from the effects of arc faults

  - devices intended to provide protection in case of overheating

  - optically operated devices that provide signalling to another device intended to break the circuit

  - smoke detection devices that provide signalling to another device intended to break the circuit.

533   DEVICES FOR PROTECTION AGAINST OVERCURRENT

533.1   General requirements

A device for protection against overcurrent shall comply with one or more of the following:

  - BS 88 series
  - BS 646
  - BS 1362
  - BS 3036
  - BS EN 60898-1 and -2
  - BS EN 60947-2 and -3
  - BS EN 60947-4-1, -6-1 and –6-2
  - BS EN 61009-1.

The use of another device is not precluded provided that its time/current characteristics provide a level of protection not less than that given by the devices listed above.

For every fuse and circuit-breaker there shall be provided on or adjacent to it an indication of its intended rated current as appropriate to the circuit it protects. For a semi-enclosed fuse, the intended rated current to be indicated is the value to be selected in accordance with Regulation 533.1.1.3.

533.1.1   Fuses

533.1.1.1   A fuse base shall be arranged so as to exclude the possibility of the fuse carrier making contact between conductive parts belonging to two adjacent fuse bases.

A fuse base using screw-in fuses shall be connected so that the centre contact is connected to the conductor from the supply and the shell contact is connected to the conductor to the load.

533.1.1.2   Fuses having fuse links likely to be removed or replaced by persons other than skilled or instructed persons shall be of a type which complies with BS 88-3, BS 3036 or BS 1362. Such a fuse link shall either:

(i)   have marked on or adjacent to it an indication of the type of fuse link intended to be used, or

(ii)   be of a type such that there is no possibility of inadvertent replacement by a fuse link having the intended rated current but a higher fusing factor than that intended.

NOTE: In multiphase systems additional measures may be needed, e.g. an all-pole switch on the supply side, in order to prevent the risk of unintentional contact with live parts on the load side.

Fuses or combination units having fuse links likely to be removed and replaced only by skilled or instructed persons shall be installed in such a manner that it is ensured that the fuse links can be removed or replaced without unintentional contact with live parts.

533.1.1.3   A fuse shall preferably be of the cartridge type. Where a semi-enclosed fuse is selected, it shall be fitted with an element in accordance with the manufacturer's instructions, if any. In the absence of such instructions, it shall be fitted with a single element of tinned copper wire of the appropriate diameter specified in Table 53.1.

TABLE 53.1 – Sizes of tinned copper wire for use in semi-enclosed fuses

Rated current of fuse element (A)	Nominal diameter of wire (mm)
3	0.15
5	0.2
10	0.35
15	0.5
20	0.6
25	0.75
30	0.85
45	1.25
60	1.53
80	1.8
100	2.0

533.1.2   Circuit-breakers

Where a circuit-breaker may be operated by a person other than a skilled or instructed person, it shall be designed or installed so that it is not possible to modify the setting or the calibration of its overcurrent release without a deliberate act involving the use of either a key or a tool and resulting in a visible indication of its setting or calibration.

Where a screw-in type circuit-breaker is used in a fuse base, the requirements of Regulation 533.1.1.1 also apply.

533.2   Selection of devices for overload protection of wiring systems

533.2.1   The rated current (or current setting) of the protective device shall be chosen in accordance with Regulation 433.1.

In certain cases, to avoid unintentional operation, the peak current values of the loads may have to be taken into consideration.

In the case of a cyclic load, the values of I_n and I_2 shall be chosen on the basis of values of I_b and I_z for the thermally equivalent constant load

where:

I_b is the current for which the circuit is designed

I_z is the continuous current-carrying capacity of the cable

I_n is the rated current of the protective device

I_2 is the current ensuring effective operation of the overload protective device within the conventional time as stated in the product standard.

533.2.2   Additional requirements for protection against overload when harmonic currents are present

When selecting an overload protective device to comply with Regulation 433.1, account shall be taken of harmonic currents in accordance with Regulation 431.2.3.

533.3   Selection of devices for protection of wiring systems against fault current

The application of the regulations of Chapter 43 shall take into account minimum and maximum fault current conditions, so as to ensure the highest energy let-through is taken into account.

Where the standard covering a protective device specifies both a rated service short-circuit breaking capacity and a rated ultimate short-circuit breaking capacity, it is acceptable to select the protective device on the basis of the ultimate short-circuit breaking capacity for the maximum fault current conditions. Operational circumstances may, however, make it desirable to select the protective device on the service short-circuit breaking capacity, e.g. where a protective device is placed at the origin of the installation.

Where the short-circuit breaking capacity of the protective device is lower than the maximum prospective short-circuit or earth fault current that is expected at its point of installation, it is necessary to comply with the requirements of the last paragraph of Regulation 536.1 and Regulation 536.5.

534   DEVICES FOR PROTECTION AGAINST OVERVOLTAGE

For further information see Appendix 16.

534.1   Scope and object

This section contains provisions for the application of voltage limitation to obtain insulation coordination in the cases described in Section 443, BS EN 60664-1, BS EN 62305-4 and BS EN 61643 (CLC/TS 61643-12). Protection against malfunction of electrical and electronic equipment due to overvoltages is not covered in this section. Protective measures against malfunction of such equipment are detailed in the BS EN 61643 series (CLC/TS 61643-12). BS EN 62305-4 and CLC/TS 61643-12 deal with the protection against the effects of direct lightning strokes or strokes near to the supply system. Both documents describe the selection and the application of surge protective devices (SPDs) according to the lightning protection zones (LPZ) concept. The LPZ concept describes the installation of Type 1, Type 2 and Type 3 SPDs.

SPDs, specific isolating transformers, filters or a combination of these may be used for protection against overvoltages.

These requirements are for the selection and erection of:

(i)   SPDs for electrical installations of buildings to obtain a limitation of transient overvoltages of atmospheric origin transmitted via the supply distribution system and against switching overvoltages

(ii)   SPDs for the protection against transient overvoltages caused by direct lightning strokes or lightning strokes in the vicinity of buildings protected by a lightning protection system.

These requirements do not take into account surge protective components which may be incorporated in the appliances connected to the installation. The presence of such components may modify the behaviour of the main surge protective device of the installation and may need an additional coordination.

These requirements also cover protection against overcurrent due to SPD failure and the consequences of that failure.

NOTE 1: For example, failure can be due to mains supply faults or due to the SPD reaching the end of its life.

These requirements apply to a.c. power circuits.

NOTE 2: Overvoltages of atmospheric origin and electrical switching events can affect metallic data, signal and telecommunication lines. Protection measures for these systems are detailed within CLC/TS 61643-22.

534.2   Selection and erection of surge protective devices

534.2.1   Use of SPDs

Where required by Section 443 or otherwise specified, SPDs shall be installed:

(i)   near the origin of an installation, or

(ii)   in the main distribution assembly nearest the origin of an installation.

NOTE 1: Annex E of CLC/TS 61643-12 provides examples of application of the risk analysis as described in Section 443.

When two or more SPDs are used on the same conductor, they shall be coordinated and as such are referred to as coordinated SPDs. In accordance with BS EN 62305-4, protection of electrical and electronic systems within structures requires coordinated SPDs.

Where required and in accordance with BS EN 62305-4 or otherwise specified, SPDs shall be installed at the origin of an installation.

NOTE 2: Depending on the voltage stress, Type 1 or Type 2 SPDs may be used at the origin whilst Type 2 or Type 3 are also suited for location close to the protected equipment to further protect against switching transients generated within the building.

NOTE 3: Type 1 SPDs are often referred to as equipotential bonding SPDs and are fitted at the origin of the installation to specifically prevent dangerous sparking which could lead to fire or electric shock hazards. A lightning protection system which only employs equipotential bonding SPDs provides no effective protection against failure of sensitive electrical and electronic systems.

Additional SPDs are required to protect sensitive and critical equipment (for example hospital equipment and fire/security alarm systems). Such SPDs shall be coordinated with the SPDs installed upstream and be installed as close as practicable to the equipment to be protected. Where SPDs are required in accordance with Section 443 or BS EN 62305-4, consideration shall be given to the installation of SPDs on other incoming networks (such as metallic telecommunication and signalling services).

534.2.2   Connection of SPDs

Where required by Regulation 534.2.1, SPDs at or near the origin of the installation shall be connected between specific conductors according to Table 53.2.

Where the equipment to be protected has a sufficient overvoltage withstand or is located close to the main distribution board, one SPD assembly may be sufficient if it provides sufficient modes of protection.

NOTE 1: Sensitive electronic equipment requires both common mode protection (conductors with respect to earth e.g. line and earth, neutral and earth) as well as differential mode protection (between live conductors, e.g. line and neutral).

The SPD assembly shall be installed as close as possible to the origin of the installation and shall have sufficient surge withstand capability for this location.

SPDs at or near the origin of the installation shall be connected at least between the following points:

(i)   If there is no direct connection between the neutral conductor and protective conductor at or near the origin of the installation:

between each line conductor and either the main earthing terminal or the main protective conductor, and between the neutral conductor and either the main earthing terminal or the protective conductor, whichever route is the shorter - Connection Type 1 (CT 1), see Figure 53.1

or alternatively,

between each line conductor and the neutral conductor and between the neutral conductor and either the main earthing terminal or the protective conductor, whichever route is the shorter – Connection Type 2 (CT 2) see Figure 53.2.

NOTE 2: If a line conductor is earthed, it is considered to be equivalent to a neutral conductor for the application of this regulation.

NOTE 3: It is recommended that the power and signalling cabling enter the structure to be protected in close proximity and are bonded together at the main earthing terminal.

Fig 53.1 – Connection Type 1 (CT 1)

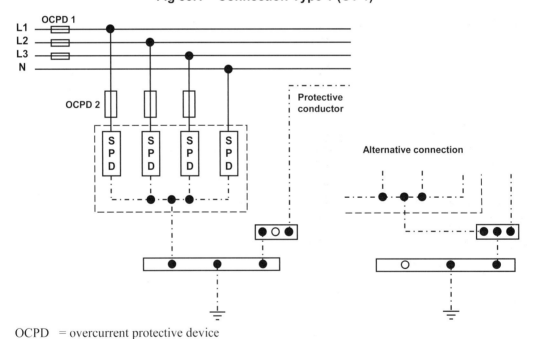

OCPD = overcurrent protective device

Fig 53.2 – Connection Type 2 (CT 2)

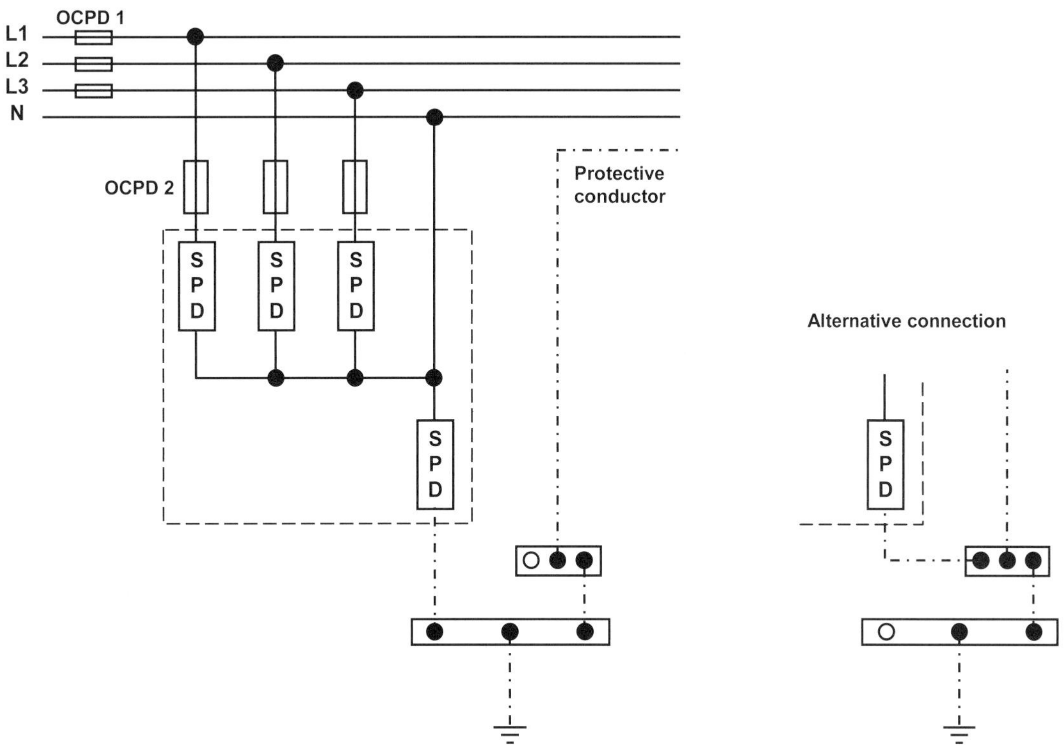

TABLE 53.2 – Modes of protection for various LV systems

SPDs connected between:	TN-C-S, TN-S or TT Installation in accordance with		IT without distributed neutral
	CT 1	CT 2	
Each line conductor and neutral conductor	+	•	N/A
Each line conductor and PE conductor	•	N/A	•
Neutral conductor and PE conductor	•	•	N/A
Each line conductor and PEN conductor	N/A	N/A	N/A
Line conductors	+	+	+

• = SPD required,   N/A = not applicable,   + = optional, in addition,   CT = Connection Type

NOTE 1: If more than one SPD is connected on the same conductor, it is necessary to ensure co-ordination between them.

NOTE 2: The number of modes of protection depends on the type of equipment to be protected (for example if the equipment is not connected to earth, line or neutral to earth protection may not be necessary), the withstand of the equipment in accordance with each mode of protection, the electrical system structure and earthing and the characteristics of the incoming surge. For example, protection between line/neutral and protective conductor or between line and neutral are generally sufficient, and protection between line and line is not used generally.

Consideration shall be given to the installation of additional SPDs at other locations. Sensitive and critical equipment requires protection in both common and differential modes to ensure further protection against switching transients.

534.2.3 Selection of surge protective devices (SPDs)

SPDs shall be selected in accordance with the following regulations.

534.2.3.1 Selection with regard to voltage protection level (U_p)

534.2.3.1.1 The impulse withstand voltage of the equipment (or impulse immunity of critical equipment) to be protected and the nominal voltage of the system shall be considered in selecting the preferred voltage protection level (U_p) value of the SPD.

SPDs shall be selected to provide a voltage protection level lower than the impulse withstand capability of the equipment or, in some cases where the continuous operation of the equipment is critical, lower than the impulse immunity of the equipment.

SPDs with lower voltage protection levels provide better protection. However, consideration shall be given to the effects of temporary overvoltages on an SPD (or built-in surge protective component) having an excessively low protection level. SPDs shall be selected which will protect equipment from failure, remain operational during surge activity and withstand most temporary overvoltage conditions.

The voltage protection level (U_p) of SPDs shall be selected in accordance with impulse withstand voltage Category II of Table 44.3. If BS EN 62305-2 requires SPDs for the protection against overvoltages caused by direct lightning strokes, the protection level of these SPDs shall also be selected in accordance with impulse withstand voltage Category II of Table 44.3. In installations operating at 230/400 V, the protection level of the installed SPD shall not exceed 2.5 kV, as the SPD's connecting leads have additional inductive voltage drop across them (see Regulation 534.2.9). It may therefore be necessary to select an SPD with a lower voltage protection level.

If the distance between the SPD and equipment to be protected (protective distance) is greater than 10 m, oscillations could lead to a voltage at the equipment terminals of up to twice the SPD's voltage protection level. Consideration shall be given to the provision of additional coordinated SPDs, closer to the equipment, or the selection of SPDs with a lower voltage protection level.

534.2.3.1.2 Where SPDs are connected as in Figure 53.2 the voltage protection level (U_p) required shall be met by the serial combination of the two SPDs between line and neutral and between neutral and the protective conductor.

Where the voltage protection level required cannot be obtained with a single SPD, additional SPDs, which are coordinated, shall be installed to ensure the required voltage protection level.

To protect sensitive and critical equipment, consideration shall be given to the installation of additional coordinated SPDs selected in accordance with impulse withstand voltage Category I (1.5 kV) of Table 44.3.

NOTE 1: Annex D of CLC/TS 61643-12 provides application examples of selecting SPDs and choice of protection level U_p.

534.2.3.2 Selection with regard to continuous operating voltage (U_c)

The maximum continuous operating voltage U_c of SPDs shall be equal to or higher than that required by Table 53.3.

TABLE 53.3 – Minimum required U_c of the SPD dependent on supply system configuration

SPDs connected between:	TN-C-S, TN-S or TT	IT without distributed neutral
Line conductor and neutral conductor	$1.1\ U_{\alpha\ spd}$	N/A
Each line conductor and PE conductor	$1.1\ U_{\alpha\ spd}$	Line to line voltage (Note 2)
Neutral conductor and PE conductor	$U_{\alpha\ spd}$ (Note 2)	N/A
Each line conductor and PEN conductor	N/A	N/A
N/A = not applicable		
NOTE 1: $U_{\alpha\ spd}$ is the nominal a.c. rms line voltage of the low voltage system to Earth.		
NOTE 2: These values are related to worst-case fault conditions, therefore the tolerance of 10 % is not taken into account.		
NOTE 3: In extended IT systems, higher values of U_c may be necessary.		

534.2.3.3   Selection with regard to temporary overvoltages (TOVs)

SPDs shall be selected and erected in accordance with manufacturers' instructions.

NOTE: The loss of neutral is not covered by these requirements. Although there is currently no specific test in BS EN 61643-11, SPDs are expected to fail safely.

534.2.3.4   Selection with regard to nominal discharge current (I_{nspd}) and impulse current (I_{imp})

534.2.3.4.1 SPDs shall be selected according to their withstand capability, as classified in BS EN 61643-11 for power systems and in BS EN 61643-21 for telecommunication systems.

The nominal discharge current I_{nspd} of the SPD shall be not less than 5 kA with a waveform characteristic 8/20 for each mode of protection.

534.2.3.4.2 For Connection Type 2 (CT 2), the nominal discharge current I_{nspd} for the SPD connected between the neutral conductor and the protective conductor shall be not less than 20 kA with a waveform characteristic 8/20 for three-phase systems and 10 kA with a waveform characteristic 8/20 for single-phase systems.

Type 1 SPDs shall be installed where a structural lightning protection system is fitted or the installation is supplied by an overhead line at risk of direct lightning strike.

The lightning impulse current I_{imp} according to BS EN 61643-11 shall be calculated according to BS EN 62305-4.

Where Type 1 SPDs are required, the value of I_{imp} shall be not less than 12.5 kA for each mode of protection, if I_{imp} cannot be calculated.

534.2.3.4.3 For Connection Type 2 (CT 2) installations, the lightning impulse current I_{imp} for the SPD connected between the neutral conductor and the protective conductor shall be calculated in accordance with BS EN 62305-4. If the current value cannot be established, the value of I_{imp} shall be not less than 50 kA for three-phase systems and 25 kA for single-phase systems.

534.2.3.4.4 When a single SPD is used for protection according to both BS EN 62305-4 and Section 443, the rating of I_{nspd} and of I_{imp} shall comply with the requirements of Regulation 534.2.3.4.3.

534.2.3.5   Selection with regard to the prospective fault current and the follow current interrupt rating

The short-circuit withstand of the combination SPD and overcurrent protective device (OCPD), as stated by the SPD manufacturer shall be equal to or higher than the maximum prospective fault current expected at the point of installation.

NOTE 1: The OCPD may be either internal or external to the SPD.

When a follow current interrupt rating is declared by the manufacturer, it shall be equal to or higher than the prospective line to neutral fault current at the point of installation.

NOTE 2: Follow current interrupt rating is only applicable to certain voltage switching devices such as spark gaps and gas discharge tubes.

SPDs connected between the neutral conductor and the protective conductor in TT or TN systems, which allow a power frequency follow current after operation (e.g. spark gaps) shall have a follow current interrupt rating greater than or equal to 100 A.

In IT systems, the follow current interrupt rating for SPDs connected between the neutral connector and the protective conductor shall be the same as for SPDs connected between line and neutral.

534.2.3.6   Co-ordination of SPDs

SPDs shall be selected and erected such as to ensure coordination in operation.

534.2.4   Protection against overcurrent and consequences of SPD's end of life

Protection against SPD short-circuits is provided by OCPDs (see Figures 53.1 and 53.2) which shall be selected according to the recommended ratings given in the manufacturer's SPD instructions.

OCPD2 may be omitted if the characteristics of OCPD1 meet the requirements given in the manufacturer's instructions.

534.2.5 Fault protection integrity

Fault protection shall remain effective in the protected installation even in case of failures of SPDs.

In TN systems, automatic disconnection of supply shall be obtained by correct operation of the overcurrent protective device on the supply side of the SPD.

In TT systems this shall be obtained by either:

   (i)   the installation of SPDs on the load side of an RCD (see Figure 16A.2 in Appendix 16), or

  (ii)   the installation of SPDs on the supply side of an RCD (see Figure 16A.3 in Appendix 16).

Owing to the possibility of the failure of an SPD between neutral and protective conductors,

  (iii)   the conditions of Regulation 411.4.1 shall be met, and

  (iv)   the SPD shall be installed in accordance with Regulation 534.2.2, Connection Type 2 (CT 2).

In IT systems, no additional measure is required.

534.2.6 SPD installation in conjunction with RCDs

Where SPDs are installed in accordance with Regulation 534.2.1 and are on the load side of a residual current device, an RCD having an immunity to surge currents of at least 3 kA 8/20, shall be used.

NOTE 1: S-type RCDs satisfy this requirement.

NOTE 2: In the case of surge current higher than 3 kA 8/20, the RCD may trip causing interruption of the power supply.

534.2.7 *Not used*

534.2.8 SPD status indication

Indication that the SPD no longer provides overvoltage protection shall be provided by a status indicator local to the SPD itself and/or remote, to provide electrical, visual or audible alarms.

534.2.9 Critical length of connecting conductors

To gain maximum protection the supply conductors shall be kept as short as possible, to minimize additive inductive voltage drops across the conductors. Current loops shall be avoided (see Figure 53.3). The total lead length *(a + b)* should preferably not exceed 0.5 m, but shall in no case exceed 1.0 m. Where an SPD is fitted in-line (see Figure 53.4), the protective conductor (length *c*) should not exceed 0.5 m in length, and in no case shall exceed 1.0 m.

NOTE: To keep the connections of SPDs as short and inductance as low as possible, SPDs may be connected to the main earthing terminal or to the protective conductor, e.g. via the metallic enclosures of the assembly being connected to the protective conductor.

Fig 53.3 – Example of a shunt installation of SPDs at or near the origin of the electrical installation

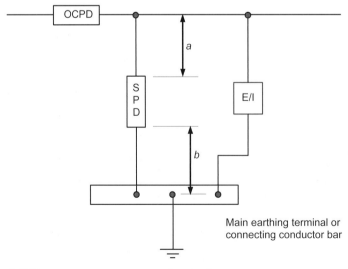

OCPD   = overcurrent protective device
SPD   = surge protective device
E/I   = equipment or installation to be protected against overvoltages

**Fig 53.4 – Example of an in-line installation of SPDs
at or near the origin of the electrical installation**

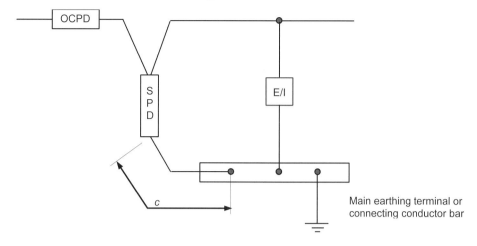

OCPD = overcurrent protective device
SPD = surge protective device
E/I = equipment or installation to be protected against overvoltages
NOTE: The line terminal on the SPD must be capable of carrying the load current.

534.2.10   Cross-sectional area of connecting conductors

The connecting conductors of SPDs shall either:

(i)   have a cross-sectional area of not less than 4 mm^2 copper (or equivalent) if the cross-sectional area of the line conductors is greater than or equal to 4 mm^2, or

(ii)   have a cross-sectional area not less than that of the line conductors, where the line conductors have a cross-sectional area less than 4 mm^2.

Where there is a structural lightning protection system, the minimum cross-sectional area for Type 1 SPDs shall be 16 mm^2 copper, or equivalent.

535   DEVICES FOR PROTECTION AGAINST UNDERVOLTAGE

A device for protection against undervoltage shall be selected and erected so as to allow compliance with the requirements of Section 445.

536   CO-ORDINATION OF PROTECTIVE DEVICES

536.1   General

Where co-ordination of series protective devices is necessary to prevent danger and where required for proper functioning of the installation, consideration shall be given to selectivity and/or any necessary back-up protection.

Selectivity between protective devices depends on the co-ordination of the operating characteristics of two or more protective devices such that, on the incidence of fault currents within stated limits, only the device intended to operate within these limits does so (see Regulation 536.5).

The rated breaking capacity of a protective device shall be not less than the maximum prospective short-circuit or earth fault current at the point at which the device is installed unless back-up protection is provided. A lower breaking capacity is permitted if another protective device (a back-up protective device) having the necessary breaking capacity is installed on the supply side and the characteristics of the devices are suitably co-ordinated such that the energy let-through of the upstream device does not exceed that which can be withstood without damage by the downstream device (see Chapter 43).

NOTE: Back-up protection on the load side of a protective device is acceptable only if the risk of a fault in the circuit between the two protective devices in series is negligible.

536.2   Selectivity between overcurrent protective devices

Where selectivity between overcurrent protective devices is necessary to prevent danger and where required for proper functioning of the installation, the manufacturer's instructions shall be taken into account.

536.3 Selectivity between RCDs

Where selectivity between RCDs is necessary to prevent danger and where required for proper functioning of the installation, the manufacturer's instructions shall be taken into account.

536.4 Back-up protection

Where necessary to prevent danger and where required for proper functioning of the installation, back-up protection shall be provided according to the manufacturer's information.

NOTE: Technical data for the selection of protective devices for the purpose of back-up protection are published by the manufacturer of the device to be protected.

536.5 Co-ordination of switching devices with overcurrent protective devices

536.5.1 A switching device shall be protected against overcurrent.

A switching device without integral overcurrent protection shall be co-ordinated with an appropriate overcurrent protective device.

The rated current and the characteristics of the protective device against fault current shall be in accordance with those stated by the manufacturer of the switching device.

536.5.2 Motor starters shall be co-ordinated with the appropriate fault current protective device according to BS EN 60947-4-1.

537 ISOLATION AND SWITCHING

537.1 General

This section provides requirements for:

(i) non-automatic local and remote isolation and switching measures for the prevention or removal of dangers associated with electrical installations or electrically-powered equipment and machines, and

(ii) functional switching and control.

537.1.1 According to the intended function(s), every device provided for isolation or switching shall comply with the relevant requirements of this section.

NOTE: Table 53.4 provides information on selection.

TABLE 53.4 – Guidance on the selection of protective, isolation and switching devices

Device	Standard	Isolation[4]	Emergency switching[2]	Functional switching
Switching device	BS EN 50428	No	No	Yes
	BS EN 60669-1	No	Yes	Yes
	BS EN 60669-2-1	No	No	Yes
	BS EN 60669-2-2	No	Yes	Yes
	BS EN 60669-2-3	No	Yes	Yes
	BS EN 60669-2-4	Yes[3]	Yes	Yes
	BS EN 60947-3	Yes[1,3]	Yes	Yes
	BS EN 60947-5-1	No	Yes	Yes
Contactor	BS EN 60947-4-1	Yes[1,3]	Yes	Yes
	BS EN 61095	No	No	Yes
Circuit-breaker	BS EN 60898	Yes[3]	Yes	Yes
	BS EN 60947-2	Yes[1,3]	Yes	Yes
	BS EN 61009-1	Yes[3]	Yes	Yes
RCD	BS EN 60947-2	Yes[1,3]	Yes	Yes
	BS EN 61008-1	Yes[3]	Yes	Yes
	BS EN 61009-1	Yes[3]	Yes	Yes
Isolating switch	BS EN 60669-2-4	Yes[3]	Yes	Yes
	BS EN 60947-3	Yes[1,3]	Yes	Yes
Plug and socket-outlet (≤ 32 A)	BS EN 60309	Yes[3]	No	Yes
	IEC 60884	Yes[3]	No	Yes
	IEC 60906	Yes[3]	No	Yes
Plug and socket-outlet (> 32 A)	BS EN 60309	Yes[3]	No	No
Device for the connection of luminaire	BS EN 61995-1 BS IEC 61995-1	Yes[3]	No	No
Control and protective switching device for equipment (CPS)	BS EN 60947-6-1	Yes[1,3]	Yes	Yes
	BS EN 60947-6-2	Yes[1,3]	Yes	Yes
Fuse	BS 88 series	Yes	No	No
Device with semiconductors	BS EN 50428	No	No	Yes
	BS EN 60669-2-1	No	No	Yes
Luminaire Supporting Coupler	BS 6972	Yes[3]	No	No
Plug and unswitched socket-outlet	BS 1363-1	Yes[3]	No	Yes
	BS 1363-2	Yes[3]	No	Yes
Plug and switched socket-outlet	BS 1363-1	Yes[3]	No	Yes
	BS 1363-2	Yes[3]	No	Yes
Plug and socket-outlet	BS 5733	Yes[3]	No	Yes
Switched fused connection unit	BS 1363-4	Yes[3]	Yes	Yes
Unswitched fused connection unit	BS 1363-4	Yes[3] (Removal of fuse link)	No	No
Fuse	BS 1362	Yes	No	No
Cooker Control Unit switch	BS 4177	Yes[3]	Yes	Yes

Yes = Function provided, No = Function not provided

[1] Function provided if the device is suitable and marked with the symbol for isolation (see BS EN 60617 identity number S00288). ⎯╱├⎯

[2] See Regulation 537.4.2.5

[3] Device is suitable for on-load isolation, i.e. disconnection whilst carrying load current.

[4] In an installation forming part of a TT or IT system, isolation requires disconnection of all the live conductors. See Regulation 537.2.2.1.

NOTE 1: An entry of (1,3) means that the device is suitable for on-load isolation only if it is marked with the symbol for **on-load** isolation ⎯╱o├⎯

NOTE 2: In the above table, the functions provided by the devices for isolation and switching are summarized, together with the indication of the relevant product standards.

537.1.2   Combined protective and neutral (PEN) conductors shall not be isolated or switched except as permitted by Regulation 543.3.3.

Except as required by Regulation 537.1.4, in a TN-S or TN-C-S system the neutral conductor need not be isolated or switched where it can be regarded as being reliably connected to Earth by a suitably low impedance. For supplies which are provided in accordance with the Electricity Safety, Quality and Continuity Regulations 2002, the supply neutral conductor (PEN or N) is considered to be connected to Earth by a suitably low impedance.

537.1.3   Each installation shall have provision for disconnection from the supply.

Where the distributor provides a means of disconnection complying with Chapter 53 at the origin of the installation and agrees that it may be used as the means of isolation for the part of the installation between the origin and the main linked switch or circuit-breaker required by Regulation 537.1.4, the requirement for isolation and switching of that part of the installation is satisfied.

NOTE: A cut-out fuse may be withdrawn only by a person authorized to do so by the distributor.

537.1.4   A main linked switch or linked circuit-breaker shall be provided as near as practicable to the origin of every installation as a means of switching the supply on load and as a means of isolation.

A main switch intended for operation by ordinary persons, e.g. of a household or similar installation, shall interrupt both live conductors of a single-phase supply.

537.1.5   Where an installation is supplied from more than one source of energy, one of which requires a means of earthing independent of the means of earthing of other sources and it is necessary to ensure that not more than one means of earthing is applied at any time, a switch may be inserted in the connection between the neutral point and the means of earthing, provided that the switch is a linked switch arranged to disconnect and connect the earthing conductor for the appropriate source, at substantially the same time as the related live conductors.

537.1.6   Where an installation is supplied from more than one source, a main switch shall be provided for each source of supply and a durable warning notice shall be permanently fixed in such a position that any person seeking to operate any of these main switches will be warned of the need to operate all such switches to achieve isolation of the installation. Alternatively, a suitable interlock system shall be provided. (See also Regulation 514.15.1.)

537.2   Isolation

537.2.1   General

537.2.1.1   Every circuit shall be capable of being isolated from each of the live supply conductors. In a TN-S or TN-C-S system, it is not necessary to isolate or switch the neutral conductor where it is regarded as being reliably connected to Earth by a suitably low impedance.

Provision may be made for isolation of a group of circuits by a common means, if the service conditions allow this.

537.2.1.2   Suitable means shall be provided to prevent any equipment from being inadvertently or unintentionally energized.

537.2.1.3   Where an installation or an item of equipment or enclosure contains live parts connected to more than one supply, a durable warning notice shall be placed in such a position that any person before gaining access to live parts, will be warned of the need to isolate those parts from the various supplies unless an interlocking arrangement is provided to ensure that all the circuits concerned are isolated.

537.2.1.4   Where necessary, suitable means shall be provided for the discharge of stored electrical energy.

537.2.1.5   Where an isolating device for a particular circuit is placed remotely from the equipment to be isolated, provision shall be made so that the means of isolation can be secured in the open position. Where this provision takes the form of a lock or removable handle, the key or handle shall be non-interchangeable with any other used for a similar purpose within the premises.

537.2.1.6   For every electric discharge lighting installation having an open circuit voltage exceeding low voltage, one or more of the following means shall be provided for the isolation of every self-contained luminaire, or of every circuit supplying luminaires at a voltage exceeding low voltage:

   (i)   An interlock on a self-contained luminaire, so arranged that before access can be had to live parts the supply is automatically disconnected, such means being additional to the switch normally used for controlling the circuit

   (ii)   An effective local means for the isolation of the circuit from the supply, such means being additional to the switch normally used for controlling the circuit

   (iii)   A switch having a lock or removable handle, or a distribution board which can be locked, in either case complying with Regulation 537.2.1.5.

537.2.1.7   Provision shall be made for disconnecting the neutral conductor. Where this is a joint it shall be such that it is in an accessible position, can only be disconnected by means of a tool, is mechanically strong and will reliably maintain electrical continuity.

537.2.2   Devices for isolation

537.2.2.1   A device for isolation shall isolate all live supply conductors from the circuit concerned, subject to the provisions of Regulation 537.1.2.

A device suitable for isolation shall be selected according to the requirements which are based on the overvoltage category applicable at the point of installation.

Equipment of overvoltage Categories I and II, according to BS EN 60664-1, shall not be used for isolation.

NOTE: Except for the plug of a plug and socket-outlet identified in Table 53.4 as suitable for isolation, equipment of overvoltage Categories I and II should not be used for isolation.

Semiconductor devices shall not be used as isolating devices.

Equipment used for isolation shall comply with Regulations 537.2.1.1 and 537.2.2.2 to 6.

537.2.2.2   The position of the contacts or other means of isolation shall be either externally visible or clearly and reliably indicated. An indication of the isolated position shall occur only when the specified isolation has been obtained in each pole.

537.2.2.3   A device for isolation shall be designed and/or installed so as to prevent unintentional or inadvertent closure.

537.2.2.4   Provision shall be made for securing off-load isolating devices against inadvertent or unauthorized opening.

NOTE: This may be achieved by locating the device in a lockable space or enclosure or by padlocking. Alternatively, the off-load device may be interlocked with a load-breaking one.

Where a link is inserted in the neutral conductor, the link shall comply with either or both of the following requirements:

(i)   It cannot be removed without the use of a tool

(ii)   It is accessible to skilled persons only.

537.2.2.5   Means of isolation shall preferably be provided by a multipole switching device which disconnects all applicable poles of the relevant supply but single-pole devices situated adjacent to each other are not excluded.

537.2.2.6   Each device used for isolation shall be clearly identified by position or durable marking to indicate the installation or circuit it isolates.

537.3   Switching off for mechanical maintenance

537.3.1   General

Where electrically powered equipment is within the scope of BS EN 60204, the requirements for switching off for mechanical maintenance of that standard apply.

537.3.1.1   Means of switching off for mechanical maintenance shall be provided where mechanical maintenance may involve a risk of physical injury.

NOTE 1: Electrically powered mechanical equipment may include rotating machines as well as heating elements and electromagnetic equipment.

NOTE 2: Systems powered by other means, e.g. pneumatic, hydraulic or steam, are not covered by these regulations. In such cases, switching off any associated supply of electricity may not be a sufficient measure.

537.3.1.2   Suitable means shall be provided to prevent electrically powered equipment from becoming unintentionally reactivated during mechanical maintenance, unless the means of switching off is continuously under the control of any person performing such maintenance.

537.3.2 Devices for switching off for mechanical maintenance

537.3.2.1 Where practicable, a device for switching off for mechanical maintenance shall be inserted in the main supply circuit.

Where a switch is provided for this purpose, it shall be capable of cutting off the full load current of the relevant part of the installation.

Interruption of a circuit for the control of mechanical movement is permitted only where a condition equivalent to the direct interruption of the main supply is provided by one of the following:

(i) Supplementary safeguards, such as mechanical restrainers

(ii) Compliance with the requirements of a British or Harmonized Standard specification for the control devices used.

NOTE: Switching off for mechanical maintenance may be achieved, for example, by one of the following:
- multipole switch
- circuit-breaker
- control and protective switching device (CPS)
- control switch operating a contactor
- plug and socket-outlet.

537.3.2.2 A device for switching off for mechanical maintenance or a control switch for such a device shall require manual operation.

The open position of the contacts of the device shall be visible or be clearly and reliably indicated.

NOTE: The indication required by this regulation may be achieved by the use of the symbols 'O' and 'I' to indicate the open and closed positions respectively.

537.3.2.3 A device for switching off for mechanical maintenance shall be designed and/or installed so as to prevent inadvertent or unintentional switching on.

537.3.2.4 A device for switching off for mechanical maintenance shall be so placed and durably marked so as to be readily identifiable and convenient for the intended use.

537.3.2.5 Where a switch is used as a device for switching off for mechanical maintenance, the switch need not necessarily interrupt the neutral conductor.

537.3.2.6 A plug and socket-outlet or similar device of rating not exceeding 16 A may be used as a device for switching off for mechanical maintenance.

537.4 Emergency switching

537.4.1 General

Where electrically powered equipment is within the scope of BS EN 60204, the requirements for emergency switching of that standard apply.

NOTE: Emergency switching may be emergency switching on or emergency switching off.

537.4.1.1 Means shall be provided for emergency switching of any part of an installation where it may be necessary to control the supply to remove an unexpected danger.

537.4.1.2 Except as provided in Regulation 537.1.2, where a risk of electric shock is involved the emergency switching device shall be an isolating device and shall interrupt all live conductors.

Except as required by Regulation 537.1.4, where the neutral conductor can be regarded as being reliably connected to Earth in a TN-S or TN-C-S system the neutral conductor need not be isolated or switched.

537.4.1.3 Means for emergency switching shall act as directly as possible on the appropriate supply conductors.

The arrangement shall be such that one single action only will interrupt the appropriate supply.

537.4.1.4 The arrangement of the emergency switching shall be such that its operation does not introduce a further danger or interfere with the complete operation necessary to remove the danger.

537.4.2 Devices for emergency switching

537.4.2.1 A device for emergency switching shall be capable of breaking the full load current of the relevant part(s) of the installation taking account of stalled motor currents where appropriate.

537.4.2.2 Means for emergency switching may consist of:

(i) one switching device capable of directly cutting off the appropriate supply, or

(ii) a combination of equipment activated by a single action for the purpose of cutting off the appropriate supply.

NOTE 1: Emergency switching may be achieved, for example, by means of:
- a switch in the main circuit
- pushbuttons and the like in the control (auxiliary) circuit.

NOTE 2: Emergency stopping may include the retention of supply for electric braking facilities.

537.4.2.3 Hand-operated switching devices for direct interruption of the main circuit shall be selected where practicable.

A device such as a circuit-breaker or a contactor operated by remote control shall open on de-energisation of the coil, or another technique of suitable reliability shall be employed.

537.4.2.4 The means of operating (handle, push-button, etc.) a device for emergency switching shall be clearly identified, preferably by colour. If a colour is used, this shall be red with a contrasting background.

537.4.2.5 The means of operation shall be readily accessible at places where a danger might occur and, where appropriate, at any additional remote position from which that danger can be removed.

537.4.2.6 The means of operation shall be capable of latching or being restrained in the 'off' or 'stop' position, unless both the means of operation for emergency switching and for re-energizing are under the control of the same person.

The release of an emergency switching device shall not re-energize the relevant part of the installation.

537.4.2.7 A device for emergency switching shall be so placed and durably marked so as to be readily identifiable and convenient for the intended use.

537.4.2.8 A plug and socket-outlet or similar device shall not be selected as a device for emergency switching.

537.5 Functional switching (control)

537.5.1 General

537.5.1.1 A functional switching device shall be provided for each part of a circuit which may require to be controlled independently of other parts of the installation.

537.5.1.2 Functional switching devices need not necessarily control all live conductors of a circuit. Switching of the neutral shall be in compliance with Regulation 530.3.2.

537.5.1.3 In general, all current-using equipment requiring control shall be controlled by an appropriate functional switching device.

A single functional switching device may control two or more items of equipment intended to operate simultaneously.

537.5.1.4 Functional switching devices ensuring the changeover of supply from alternative sources shall affect all live conductors and shall not be capable of putting the sources in parallel, unless the installation is specifically designed for this condition.

In these cases, no provision shall be made for isolation of the PEN or protective conductors.

537.5.2 Functional switching devices

537.5.2.1 Functional switching devices shall be suitable for the most onerous duty they are intended to perform.

537.5.2.2 Functional switching devices may control the current without necessarily opening the corresponding poles.

NOTE 1: Semiconductor switching devices are examples of devices capable of interrupting the current in the circuit but not opening the corresponding poles.

537.5.2.3 Off-load isolators (disconnectors), fuses and links shall not be used for functional switching.

537.5.3   Control circuits (auxiliary circuits)

A control circuit shall be designed, arranged and protected to limit dangers resulting from a fault between the control circuit and other conductive parts liable to cause malfunction (e.g. inadvertent operation) of the controlled equipment.

537.5.4   Motor control

Where electrically powered equipment is within the scope of BS EN 60204, the requirements for motor control of that standard apply.

537.5.4.1 Motor control circuits shall be designed so as to prevent any motor from restarting automatically after a stoppage due to a fall in or loss of voltage, if such starting is liable to cause danger.

537.5.4.2 Where reverse-current braking of a motor is provided, provision shall be made for the avoidance of reversal of the direction of rotation at the end of braking if such reversal may cause danger.

537.5.4.3 Where safety depends on the direction of rotation of a motor, provision shall be made for the prevention of reverse operation due to, for example, a reversal of phases.

537.6   Firefighter's switches

537.6.1 A firefighter's switch shall be provided in the low voltage circuit supplying:

(i) exterior electrical installations operating at a voltage exceeding low voltage, and

(ii) interior discharge lighting installations operating at a voltage exceeding low voltage.

For the purpose of this regulation, an installation in a covered market, arcade or shopping mall is considered to be an exterior installation. A temporary installation in a permanent building used for exhibitions is considered not to be an exterior installation.

This requirement does not apply to a portable discharge lighting luminaire or to a sign of rating not exceeding 100 W and fed from a readily accessible socket-outlet.

537.6.2 Every exterior installation covered by Regulation 537.6.1 in each single premises shall wherever practicable be controlled by a single firefighter's switch. Similarly, every internal installation covered by Regulation 537.6.1 in each single premises shall be controlled by a single firefighter's switch independent of the switch for any exterior installation.

537.6.3 Every firefighter's switch provided for compliance with Regulation 537.6.1 shall comply with all the relevant requirements of the following items (i) to (iv) and any requirements of the local fire authority:

(i) For an exterior installation, the switch shall be outside the building and adjacent to the equipment, or alternatively a notice indicating the position of the switch shall be placed adjacent to the equipment and a notice shall be fixed near the switch so as to render it clearly distinguishable

(ii) For an interior installation, the switch shall be in the main entrance to the building or in another position to be agreed with the local fire authority

(iii) The switch shall be placed in a conspicuous position, reasonably accessible to firefighters and, except where otherwise agreed with the local fire authority, at not more than 2.75 m from the ground or the standing beneath the switch

(iv) Where more than one switch is installed on any one building, each switch shall be clearly marked to indicate the installation or part of the installation which it controls.

537.6.4 A firefighter's switch provided for compliance with Regulations 537.6.1 to 3 shall:

(i) be coloured red and have fixed on or near it a permanent durable nameplate marked with the words 'FIREFIGHTER'S SWITCH', the plate being of minimum size 150 mm by 100 mm, and having lettering easily legible from a distance appropriate to the site conditions but not less than 36 point, and

(ii) have its ON and OFF positions clearly indicated by lettering legible to a person standing at the intended site, with the OFF position at the top, and

(iii) be provided with a device to prevent the switch being inadvertently returned to the ON position, and

(iv) be arranged to facilitate operation by a firefighter.

538 MONITORING

NOTE: Monitoring is a function intended to observe the operation of a system or part of a system to verify correct functioning or detect incorrect functioning by measuring system variables and comparing the measured values with specified values.

538.1 Insulation monitoring devices (IMDs) for IT system

An IMD is intended to be permanently connected to an IT system and to continuously monitor the insulation resistance of the complete system (secondary side of the power supply and the complete installation supplied by this power supply) to which it is connected.

NOTE: An IMD is not intended to provide protection against electric shock.

538.1.1 In accordance with the requirement of Regulation 411.6.3.1, an IMD shall be installed in an IT system.

The IMD shall be in accordance with BS EN 61557-8, unless it complies with Regulation 538.1.4.

Instructions shall be provided indicating that when the IMD detects a fault to Earth, the fault shall be located and eliminated, in order to restore normal operating conditions as soon as possible.

538.1.2 Installation of insulation monitoring devices

The 'line' terminal(s) of the IMD shall be connected as close as practicable to the origin of the system to either:

(i) the neutral point of the power supply, or

(ii) an artificial neutral point with impedances connected to the line conductors, or

(iii) a line conductor or two or more line conductors.

Where the IMD is connected between one line and Earth, it shall be suitable to withstand at least the line-to-line voltage between its 'line' terminal and its 'earth' terminal.

NOTE: This voltage appears across these two terminals in the case of a single insulation fault on another line conductor.

For d.c. installations, the 'line' terminal(s) of the IMD shall be connected either directly to the midpoint, if any, or to one or all of the supply conductors.

The 'earth' or 'functional earth' terminal of the IMD shall be connected to the main earth terminal of the installation.

The supply circuit of the IMD shall be connected either to the installation on the same circuit of the connecting point of the 'line' terminal and as close as possible to the origin of the system, or to an auxiliary supply.

The connecting point to the installation shall be selected in such a way that the IMD is able to monitor the insulation of the installation under all operating conditions.

Where the installation is supplied from more than one power supply, connected in parallel, one IMD per supply shall be used, provided they are interlocked in such a way that only one IMD remains connected to the system. All other IMDs monitor the disconnected power supply enabling the reconnection of this supply without any pre-existing insulation fault.

538.1.3 Adjustment of the insulation monitoring device

An IMD is designed to indicate any important reduction of the insulation level of the system in order to find the cause before a second insulation fault occurs, thus avoiding any power supply interruption. Consequently, the IMD shall be set to a lower value corresponding to the normal insulation of the system when operating normally with the maximum of loads connected.

IMDs, installed in locations where persons other than instructed persons or skilled persons have access to their use, shall be designed or installed in such a way that it shall be impossible to modify the settings, except by the use of a key, a tool or a password.

538.1.4   Passive insulation monitoring devices

In some particular d.c. IT two-conductor installations, a passive IMD that does not inject current into the system may be used, provided that:

   (i)   the insulation of all live distributed conductors is monitored, and

   (ii)   all exposed-conductive-parts of the installation are interconnected, and

   (iii)   circuit conductors are selected and installed so as to reduce the risk of an earth fault to a minimum.

538.2   Equipment for insulation fault location in an IT system

Equipment for insulation fault location shall be in accordance with BS EN 61557-9. Where an IT system has been selected for continuity of service, it is recommended to combine the IMD with devices enabling the fault to be located while the circuit is operating. The function of the devices is to indicate the faulty circuit when the insulation monitoring device has detected an insulation fault.

538.3   Installation of equipment for insulation fault location in an IT system

An IMD shall be used on a circuit comprising safety equipment which is normally de-energized by a switching means disconnecting all live poles and which is only energized in the event of an emergency, provided that the IMD is automatically deactivated whenever the safety equipment is activated.

NOTE 1: The above arrangement is intended to ensure that the safety equipment is allowed to work without interruption of supply during the emergency.

The reduction of the insulation level shall be indicated locally by a visual or an audible signal with the choice of remote indication.

The IMD shall be connected between Earth and a live conductor of the monitored equipment.

The measuring circuit shall be automatically disconnected when the equipment is energized.

NOTE 2: If the equipment is disconnected from the installation during the off-load insulation measuring process, the insulation levels to be measured are generally very high. The alarm threshold should be above 300 kΩ.

538.4   Residual current monitor (RCM)

A residual current monitor permanently monitors any leakage current in the downstream installation or part of it.

RCMs for use in a.c. systems shall comply with BS EN 62020.

Where an RCM is used in an a.c. IT system, it is recommended to use a directionally discriminating RCM in order to avoid inopportune signalling of leakage current where high leakage capacitances are liable to exist downstream from the point of installation of the RCM.

NOTE: An RCM is not intended to provide protection against electric shock.

538.4.1   In supply systems, RCMs may be installed to reduce the risk of operation of the protective device in the event of excessive leakage current of the installation or the connected appliances.

The RCM is intended to alert the user of the installation before the protective device is activated.

Where an RCD is installed upstream of the RCM, it is recommended that the RCM has a rated residual operating current not exceeding a third of that of the RCD.

In all cases, the RCM shall have a rated residual operating current not higher than the first fault current level intended to be detected.

538.4.2   In an IT system where interruption of the supply in case of a first insulation fault to Earth is not required or not permitted, an RCM may be installed to facilitate the location of a fault. It is recommended to install the RCM at the beginning of the outgoing circuits.

CHAPTER 54

EARTHING ARRANGEMENTS AND PROTECTIVE CONDUCTORS

CONTENTS

CHAPTER 54
EARTHING ARRANGEMENTS AND PROTECTIVE CONDUCTORS

541 GENERAL

541.1 Every means of earthing and every protective conductor shall be selected and erected so as to satisfy the requirements of the Regulations.

541.2 The earthing system of the installation may be subdivided, in which case each part thus divided shall comply with the requirements of this chapter.

541.3 Where there is also a lightning protection system, reference shall be made to BS EN 62305.

542 EARTHING ARRANGEMENTS

542.1 General requirements

542.1.1 The earthing arrangements may be used jointly or separately for protective and functional purposes, according to the requirements of the installation.

542.1.100 The main earthing terminal shall be connected with Earth by one of the methods described in Regulations 542.1.2.1 to 3, as appropriate to the type of system of which the installation is to form a part and in compliance with Regulations 542.1.3.1 and 542.1.3.2.

NOTE: Refer to Part 2 and Appendix 9 for definitions of systems.

542.1.2 Supply arrangements

542.1.2.1 For a TN-S system, means shall be provided for the main earthing terminal of the installation to be connected to the earthed point of the source of energy. Part of the connection may be formed by the distributor's lines and equipment.

542.1.2.2 For a TN-C-S system, where protective multiple earthing is provided, means shall be provided for the main earthing terminal of the installation to be connected by the distributor to the neutral of the source of energy.

542.1.2.3 For a TT or IT system, the main earthing terminal shall be connected via an earthing conductor to an earth electrode complying with Regulation 542.2.

542.1.2.4 Where the supply to an installation is at high voltage, protection against faults between the high voltage supply and earth shall be provided in accordance with Section 442.

542.1.3 Installation earthing arrangements

542.1.3.1 The earthing arrangements shall be such that:

(i) the value of impedance from the consumer's main earthing terminal to the earthed point of the supply for TN systems, or to Earth for TT and IT systems, is in accordance with the protective and functional requirements of the installation, and considered to be continuously effective, and

(ii) earth fault currents and protective conductor currents which may occur are carried without danger, particularly from thermal, thermomechanical and electromechanical stresses, and

(iii) they are adequately robust or have additional mechanical protection appropriate to the assessed conditions of external influence.

542.1.3.2 Precautions shall be taken against the risk of damage to other metallic parts through electrolysis.

542.1.3.3 Where a number of installations have separate earthing arrangements, any protective conductors common to any of these installations shall either be capable of carrying the maximum fault current likely to flow through them or be earthed within one installation only and insulated from the earthing arrangements of any other installation. In the latter circumstances, if the protective conductor forms part of a cable, the protective conductor shall be earthed only in the installation containing the associated protective device.

542.2   Earth electrodes

542.2.1   The design used for, and the construction of, an earth electrode shall be such as to withstand damage and to take account of possible increase in resistance due to corrosion.

542.2.2   Not used

542.2.3   Suitable earth electrodes shall be used. The following types of earth electrode are recognized for the purposes of the Regulations:

- (i)   Earth rods or pipes
- (ii)   Earth tapes or wires
- (iii)   Earth plates
- (iv)   Underground structural metalwork embedded in foundations
- (v)   Welded metal reinforcement of concrete (except pre-stressed concrete) embedded in the ground
- (vi)   Lead sheaths and other metal coverings of cables, where not precluded by Regulation 542.2.5
- (vii)   other suitable underground metalwork.

NOTE: Further information on earth electrodes can be found in BS 7430.

542.2.4   The type and embedded depth of an earth electrode shall be such that soil drying and freezing will not increase its resistance above the required value.

542.2.5   The use, as an earth electrode, of the lead sheath or other metal covering of a cable shall be subject to all of the following conditions:

- (i)   Adequate precautions to prevent excessive deterioration by corrosion
- (ii)   The sheath or covering shall be in effective contact with Earth
- (iii)   The consent of the owner of the cable shall be obtained
- (iv)   Arrangements shall exist for the owner of the electrical installation to be warned of any proposed change to the cable which might affect its suitability as an earth electrode.

542.2.6   A metallic pipe for gases or flammable liquids shall not be used as an earth electrode. The metallic pipe of a water utility supply shall not be used as an earth electrode. Other metallic water supply pipework shall not be used as an earth electrode unless precautions are taken against its removal and it has been considered for such a use.

542.2.7   An earth electrode shall not consist of a metal object immersed in water.

542.3   Earthing conductors

542.3.1   Every earthing conductor shall comply with Section 543 and, where PME conditions apply, shall meet the requirements of Regulation 544.1.1 for the cross-sectional area of a main protective bonding conductor. In addition, where buried in the ground, the earthing conductor shall have a cross-sectional area not less than that stated in Table 54.1. For a tape or strip conductor, the thickness shall be such as to withstand mechanical damage and corrosion.

NOTE: For further information see BS 7430.

TABLE 54.1 –
Minimum cross-sectional area of a buried earthing conductor

	Protected against mechanical damage	Not protected against mechanical damage
Protected against corrosion by a sheath	2.5 mm^2 copper 10 mm^2 steel	16 mm^2 copper 16 mm^2 coated steel
Not protected against corrosion	25 mm^2 copper 50 mm^2 steel	

542.3.2   The connection of an earthing conductor to an earth electrode or other means of earthing shall be soundly made and be electrically and mechanically satisfactory, and labelled in accordance with Regulation 514.13.1. It shall be suitably protected against corrosion.

542.4   Main earthing terminals or bars

542.4.1   In every installation a main earthing terminal shall be provided to connect the following to the earthing conductor:

(i)   The circuit protective conductors

(ii)   The protective bonding conductors

(iii) Functional earthing conductors (if required)

(iv) Lightning protection system bonding conductor, if any (see Regulation 411.3.1.2).

542.4.2   To facilitate measurement of the resistance of the earthing arrangements, means shall be provided in an accessible position for disconnecting the earthing conductor. Such means may conveniently be combined with the main earthing terminal or bar. Any joint shall be capable of disconnection only by means of a tool.

543   PROTECTIVE CONDUCTORS

543.1   Cross-sectional areas

543.1.1   The cross-sectional area of every protective conductor, other than a protective bonding conductor, shall be:

(i)   calculated in accordance with Regulation 543.1.3, or

(ii)   selected in accordance with Regulation 543.1.4.

Calculation in accordance with Regulation 543.1.3 is necessary if the choice of cross-sectional area of line conductors has been determined by considerations of short-circuit current and if the earth fault current is expected to be less than the short-circuit current.

If the protective conductor:

(iii) is not an integral part of a cable, or

(iv) is not formed by conduit, ducting or trunking, or

(v)   is not contained in an enclosure formed by a wiring system,

the cross-sectional area shall be not less than 2.5 mm^2 copper equivalent if protection against mechanical damage is provided, and 4 mm^2 copper equivalent if mechanical protection is not provided (see also Regulation 543.3.1).

For a protective conductor buried in the ground Regulation 542.3.1 for earthing conductors also applies. The cross-sectional area of a protective bonding conductor shall comply with Section 544.

543.1.2   Where a protective conductor is common to two or more circuits, its cross-sectional area shall be:

(i)   calculated in accordance with Regulation 543.1.3 for the most onerous of the values of fault current and operating time encountered in each of the various circuits, or

(ii)   selected in accordance with Regulation 543.1.4 so as to correspond to the cross-sectional area of the largest line conductor of the circuits.

543.1.3   The cross-sectional area, where calculated, shall be not less than the value determined by the following formula or shall be obtained by reference to BS 7454:

$$S = \frac{\sqrt{I^2 t}}{k}$$

NOTE: This equation is an adiabatic equation and is applicable for disconnection times not exceeding 5s.

where:

S   is the nominal cross-sectional area of the conductor in mm^2

I   is the value in amperes (rms for a.c.) of fault current for a fault of negligible impedance, which can flow through the associated protective device, due account being taken of the current limiting effect of the circuit impedances and the limiting capability ($I^2 t$) of that protective device

t   is the operating time of the disconnecting device in seconds corresponding to the fault current I amperes

k   is a factor taking account of the resistivity, temperature coefficient and heat capacity of the conductor material, and the appropriate initial and final temperatures.

Values of k for protective conductors in various use or service are as given in Tables 54.2 to 6. The values are based on the initial and final temperatures indicated in each table.

Where the application of the formula produces a non-standard size, a conductor having the nearest larger standard cross-sectional area shall be used.

TABLE 54.2 –
Values of k for insulated protective conductor not incorporated in a cable and not bunched with cables, or for separate bare protective conductor in contact with cable covering but not bunched with cables where the assumed initial temperature is 30 °C

Material of conductor	Insulation of protective conductor or cable covering		
	70 °C thermoplastic	90 °C thermoplastic	90 °C thermosetting
Copper	143/133*	143/133*	176
Aluminium	95/88*	95/88*	116
Steel	52	52	64
Assumed initial temperature	30 °C	30 °C	30 °C
Final temperature	160 °C/140 °C*	160 °C/140 °C*	250 °C

* Above 300 mm²

TABLE 54.3 –
Values of k for protective conductor incorporated in a cable or bunched with cables, where the assumed initial temperature is 70 °C or greater

Material of conductor	Insulation material		
	70 °C thermoplastic	90 °C thermoplastic	90 °C thermosetting
Copper	115/103*	100/86*	143
Aluminium	76/68*	66/57*	94
Assumed initial temperature	70 °C	90 °C	90 °C
Final temperature	160 °C/140 °C*	160 °C/140 °C*	250 °C

* Above 300 mm²

TABLE 54.4 –
Values of k for protective conductor as a sheath or armour of a cable

Material of conductor	Insulation material		
	70 °C thermoplastic	90 °C thermoplastic	90 °C thermosetting
Aluminium	93	85	85
Steel	51	46	46
Lead	26	23	23
Assumed initial temperature	60 °C	80 °C	80 °C
Final temperature	200 °C	200 °C	200 °C

TABLE 54.5 –
Values of k for steel conduit, ducting and trunking as the protective conductor

Material of protective conductor	Insulation material		
	70 °C thermoplastic	90 °C thermoplastic	90 °C thermosetting
Steel conduit, ducting and trunking	47	44	58
Assumed initial temperature	50 °C	60 °C	60 °C
Final temperature	160 °C	160 °C	250 °C

TABLE 54.6 –
Values of k for bare conductor where there is
no risk of damage to any neighbouring material by the temperatures indicated
The temperatures indicated are valid only where they do not impair the quality of the connections

Material of conductor	Conditions		
	Visible and in restricted areas	Normal conditions	Fire risk
Copper	228	159	138
Aluminium	125	105	91
Steel	82	58	50
Assumed initial temperature	30 °C	30 °C	30 °C
Final temperature			
Copper conductors	500 °C	200 °C	150 °C
Aluminium conductors	300 °C	200 °C	150 °C
Steel conductors	500 °C	200 °C	150 °C

543.1.4   Where it is desired not to calculate the minimum cross-sectional area of a protective conductor in accordance with Regulation 543.1.3, the cross-sectional area may be determined in accordance with Table 54.7.

Where the application of Table 54.7 produces a non-standard size, a conductor having the nearest larger standard cross-sectional area shall be used.

TABLE 54.7 –
Minimum cross-sectional area of protective conductor
in relation to the cross-sectional area of associated line conductor

Cross-sectional area of line conductor S	Minimum cross-sectional area of the corresponding protective conductor	
	If the protective conductor is of the same material as the line conductor	If the protective conductor is not of the same material as the line conductor
(mm^2)	(mm^2)	(mm^2)
$S \leq 16$	S	$\dfrac{k_1}{k_2} \times S$
$16 < S \leq 35$	16	$\dfrac{k_1}{k_2} \times 16$
$S > 35$	$\dfrac{S}{2}$	$\dfrac{k_1}{k_2} \times \dfrac{S}{2}$

where:

k_1 is the value of k for the line conductor, selected from Table 43.1 in Chapter 43 according to the materials of both conductor and insulation.

k_2 is the value of k for the protective conductor, selected from Tables 54.2 to 6, as applicable.

543.2 Types of protective conductor

543.2.1 A protective conductor may consist of one or more of the following:

(i) A single-core cable

(ii) A conductor in a cable

(iii) An insulated or bare conductor in a common enclosure with insulated live conductors

(iv) A fixed bare or insulated conductor

(v) A metal covering, for example, the sheath, screen or armouring of a cable

(vi) A metal conduit, metallic cable management system or other enclosure or electrically continuous support system for conductors

(vii) an extraneous-conductive-part complying with Regulation 543.2.6.

543.2.2 Where a metal enclosure or frame of a low voltage switchgear or controlgear assembly or busbar trunking system is used as a protective conductor, it shall satisfy the following three requirements:

(i) Its electrical continuity shall be assured, either by construction or by suitable connection, in such a way as to be protected against mechanical, chemical or electrochemical deterioration

(ii) Its cross-sectional area shall be at least equal to that resulting from the application of Regulation 543.1, or verified by test in accordance with BS EN 60439-1

(iii) It shall permit the connection of other protective conductors at every predetermined tap-off point.

543.2.3 A gas pipe, an oil pipe, flexible or pliable conduit, support wires or other flexible metallic parts, or constructional parts subject to mechanical stress in normal service, shall not be selected as a protective conductor.

543.2.4 A protective conductor of the types described in items (i) to (iv) of Regulation 543.2.1 and of cross-sectional area 10 mm^2 or less, shall be of copper.

543.2.5 The metal covering including the sheath (bare or insulated) of a cable, in particular the sheath of a mineral insulated cable, trunking and ducting for electrical purposes and metal conduit, may be used as a protective conductor for the associated circuit, if it satisfies both requirements of items (i) and (ii) of Regulation 543.2.2.

543.2.6 Except as prohibited in Regulation 543.2.3, an extraneous-conductive-part may be used as a protective conductor if it satisfies all the following requirements:

(i) Electrical continuity shall be assured, either by construction or by suitable connection, in such a way as to be protected against mechanical, chemical or electrochemical deterioration

(ii) The cross-sectional area shall be at least equal to that resulting from the application of Regulation 543.1.1

(iii) Unless compensatory measures are provided, precautions shall be taken against its removal

(iv) It has been considered for such a use and, if necessary, suitably adapted.

543.2.7 Where the protective conductor is formed by metal conduit, trunking or ducting or the metal sheath and/or armour of a cable, the earthing terminal of each accessory shall be connected by a separate protective conductor to an earthing terminal incorporated in the associated box or other enclosure.

543.2.8 *Deleted by BS 7671:2008 Amendment No 1 (content moved to 543.3.5)*

543.2.9 Except where the circuit protective conductor is formed by a metal covering or enclosure containing all of the conductors of the ring, the circuit protective conductor of every ring final circuit shall also be run in the form of a ring having both ends connected to the earthing terminal at the origin of the circuit.

543.2.10 A separate metal enclosure for cable shall not be used as a PEN conductor.

543.3 Preservation of electrical continuity of protective conductors

543.3.1 A protective conductor shall be suitably protected against mechanical and chemical deterioration and electrodynamic effects.

543.3.2 Every connection and joint shall be accessible for inspection, testing and maintenance except as provided by Regulation 526.3.

543.3.100 A protective conductor having a cross-sectional area up to and including 6 mm^2 shall be protected throughout by a covering at least equivalent to that provided by the insulation of a single-core non-sheathed cable of appropriate size having a voltage rating of at least 450/750 V, except for the following:

(i) A protective conductor forming part of a multicore cable

(ii) A metal conduit, metallic cable management system or other enclosure or electrically continuous support system for conductors, where used as a protective conductor.

Where the sheath of a cable incorporating an uninsulated protective conductor of cross-sectional area up to and including 6 mm^2 is removed adjacent to joints and terminations, the protective conductor shall be protected by insulating sleeving complying with BS EN 60684 series.

543.3.3 A switching device shall not be inserted in a protective conductor except for the following:

(i) as permitted by Regulation 537.1.5

(ii) multipole linked switching or plug-in devices in which the protective conductor circuit shall not be interrupted before the live conductors and shall be re-established not later than when the live conductors are reconnected.

Joints intended to be disconnected for test purposes are permitted in a protective conductor circuit.

543.3.4 Where electrical monitoring of earthing is used, no dedicated devices (e.g. operating sensors, coils) shall be connected in series with the protective conductor (see BS 4444).

543.3.5 An exposed-conductive-part of equipment shall not be used to form a protective conductor for other equipment except as provided by Regulations 543.2.1, 543.2.2 and 543.2.5.

543.3.6 Every joint in metallic conduit shall be mechanically and electrically continuous.

543.4 Combined protective and neutral (PEN) conductors

543.4.1 PEN conductors shall not be used within an installation except as permitted by Regulation 543.4.2.

NOTE: In Great Britain, Regulation 8(4) of the Electricity Safety, Quality and Continuity Regulations 2002 prohibits the use of PEN conductors in consumers' installations.

543.4.2 The provisions of Regulations 543.4.3 to 8 may be applied only:

(i) where any necessary authorisation for use of a PEN conductor has been obtained and where the installation complies with the conditions for that authorisation, or

(ii) where the installation is supplied by a privately owned transformer or convertor in such a way that there is no metallic connection (except for the earthing connection) with the distributor's network, or

(iii) where the supply is obtained from a private generating plant.

543.4.3 If, from any point of the installation, the neutral and protective functions are provided by separate conductors, those conductors shall not then be reconnected together beyond that point. At the point of separation, separate terminals or bars shall be provided for the protective and neutral conductors. The PEN conductor shall be connected to the terminals or bar intended for the protective earthing conductor and the neutral conductor. The conductance of the terminal link or bar shall be not less than that specified in Regulation 543.4.5.

543.4.4 The outer conductor of a concentric cable shall not be common to more than one circuit. This requirement does not preclude the use of a twin or multicore cable to serve a number of points contained within one final circuit.

543.4.5   The conductance of the outer conductor of a concentric cable (measured at a temperature of 20 °C) shall:

(i)   for a single-core cable, be not less than that of the internal conductor

(ii)   for a multicore cable serving a number of points contained within one final circuit or having the internal conductors connected in parallel, be not less than that of the internal conductors connected in parallel.

543.4.6   At every joint in the outer conductor of a concentric cable and at a termination, the continuity of that joint shall be supplemented by a conductor additional to any means used for sealing and clamping the outer conductor. The conductance of the additional conductor shall be not less than that specified in Regulation 543.4.5 for the outer conductor.

543.4.7   No means of isolation or switching shall be inserted in the outer conductor of a concentric cable.

543.4.8   Excepting a cable to BS EN 60702-1 installed in accordance with manufacturers' instructions, the PEN conductor of every cable shall be insulated or have an insulating covering suitable for the highest voltage to which it may be subjected.

543.4.9   *Moved by BS 7671:2008 Amendment No 1 to 543.4.3*

543.4.100   For a fixed installation, a conductor of a cable not subject to flexing and having a cross-sectional area not less than 10 mm^2 for copper or 16 mm^2 for aluminium may serve as a PEN conductor provided that the part of the installation concerned is not supplied through an RCD.

543.5   Earthing arrangements for combined protective and functional purposes

543.5.1   Where earthing for combined protective and functional purposes is required, the requirements for protective measures shall take precedence.

543.6   Earthing arrangements for protective purposes

543.6.1   Where overcurrent protective devices are used for fault protection, the protective conductor shall be incorporated in the same wiring system as the live conductors or in their immediate proximity.

543.7   Earthing requirements for the installation of equipment having high protective conductor currents

543.7.1   General

543.7.1.101   Equipment having a protective conductor current exceeding 3.5 mA but not exceeding 10 mA, shall be either permanently connected to the fixed wiring of the installation without the use of a plug and socket-outlet or connected by means of a plug and socket-outlet complying with BS EN 60309-2.

543.7.1.102   Equipment having a protective conductor current exceeding 10 mA shall be connected to the supply by one of the following methods:

(i)   Permanently connected to the wiring of the installation, with the protective conductor selected in accordance with Regulation 543.7.1.103. The permanent connection to the wiring may be by means of a flexible cable

(ii)   A flexible cable with a plug and socket-outlet complying with BS EN 60309-2, provided that either:

   (a) the protective conductor of the associated flexible cable is of a cross-sectional area not less than 2.5mm^2 for plugs rated at 16A and not less than 4 mm^2 for plugs rated above 16A, or

   (b) the protective conductor of the associated flexible cable is of a cross-sectional area not less than that of the line conductor

(iii)   A protective conductor complying with Section 543 with an earth monitoring system to BS 4444 installed which, in the event of a continuity fault occurring in the protective conductor, automatically disconnects the supply to the equipment.

543.7.1.103 The wiring of every final circuit and distribution circuit intended to supply one or more items of equipment, such that the total protective conductor current is likely to exceed 10 mA, shall have a high integrity protective connection complying with one or more of the following:

(i)   A single protective conductor having a cross-sectional area of not less than 10 mm², complying with the requirements of Regulations 543.2 and 543.3

(ii) A single copper protective conductor having a cross-sectional area of not less than 4 mm², complying with the requirements of Regulations 543.2 and 543.3, the protective conductor being enclosed to provide additional protection against mechanical damage, for example, within a flexible conduit

(iii) Two individual protective conductors, each complying with the requirements of Section 543. The two protective conductors may be of different types e.g. a metallic conduit together with an additional conductor of a cable enclosed in the same conduit.

Where the two individual protective conductors are both incorporated in a multicore cable, the total cross-sectional area of all the conductors including the live conductors shall be not less than 10 mm². One of the protective conductors may be formed by the metallic sheath, armour or wire braid screen incorporated in the construction of the cable and complying with Regulation 543.2.5

(iv) An earth monitoring system to BS 4444 may be installed which, in the event of a continuity fault occurring in the protective conductor, automatically disconnects the supply to the equipment

(v)   Connection of the equipment to the supply by means of a double-wound transformer or equivalent unit, such as a motor-alternator set, the protective conductor of the incoming supply being connected to the exposed-conductive-parts of the equipment and to a point of the secondary winding of the transformer or equivalent device. The protective conductor(s) between the equipment and the transformer or equivalent device shall comply with one of the arrangements described in (i) to (iv) above.

543.7.1.104 Where two protective conductors are used in accordance with Regulation 543.7.1.103 (iii), the ends of the protective conductors shall be terminated independently of each other at all connection points throughout the circuit, e.g. the distribution board, junction boxes and socket-outlets. This requires an accessory to be provided with two separate earth terminals.

543.7.1.105 At the distribution board information shall be provided indicating those circuits having a high protective conductor current. This information shall be positioned so as to be visible to a person who is modifying or extending the circuit.

543.7.2   Socket-outlet final circuits

543.7.2.101 For a final circuit with a number of socket-outlets or connection units intended to supply two or more items of equipment, where it is known or reasonably to be expected that the total protective conductor current in normal service will exceed 10 mA, the circuit shall be provided with a high integrity protective conductor connection complying with the requirements of Regulation 543.7.1. The following arrangements of the final circuit are acceptable:

(i)   A ring final circuit with a ring protective conductor. Spurs, if provided, require high integrity protective conductor connections complying with the requirements of Regulation 543.7.1

(ii) A radial final circuit with a single protective conductor:

(a) the protective conductor being connected as a ring, or

(b) a separate protective conductor being provided at the final socket-outlet by connection to the metal conduit or ducting, or

(c) where two or more similar radial circuits supply socket-outlets in adjacent areas and are fed from the same distribution board, have identical means of short-circuit and overcurrent protection and circuit protective conductors of the same cross-sectional area, then a second protective conductor may be provided at the final socket-outlet on one circuit by connection to the protective conductor of the adjacent circuit

(iii) Other circuits complying with the requirements of Regulation 543.7.1.

544 PROTECTIVE BONDING CONDUCTORS

544.1 Main protective bonding conductors

544.1.1 Except where PME conditions apply, a main protective bonding conductor shall have a cross-sectional area not less than half the cross-sectional area required for the earthing conductor of the installation and not less than 6 mm^2. The cross-sectional area need not exceed 25 mm^2 if the bonding conductor is of copper or a cross-sectional area affording equivalent conductance in other metals.

Except for highway power supplies and street furniture, where PME conditions apply the main protective bonding conductor shall be selected in accordance with the neutral conductor of the supply and Table 54.8.

Where an installation has more than one source of supply to which PME conditions apply, a main protective bonding conductor shall be selected according to the largest neutral conductor of the supply.

TABLE 54.8 –
Minimum cross-sectional area of the main protective bonding conductor
in relation to the neutral of the supply

NOTE: Local distributor's network conditions may require a larger conductor.

Copper equivalent cross-sectional area of the supply neutral conductor	Minimum copper equivalent[*] cross-sectional area of the main protective bonding conductor
35 mm^2 or less	10 mm^2
over 35 mm^2 up to 50 mm^2	16 mm^2
over 50 mm^2 up to 95 mm^2	25 mm^2
over 95 mm^2 up to 150 mm^2	35 mm^2
over 150 mm^2	50 mm^2

[*] The minimum copper equivalent cross-sectional area is given by a copper bonding conductor of the tabulated cross-sectional area or a bonding conductor of another metal affording equivalent conductance.

544.1.2 The main equipotential bonding connection to any gas, water or other service shall be made as near as practicable to the point of entry of that service into the premises. Where there is an insulating section or insert at that point, or there is a meter, the connection shall be made to the consumer's hard metal pipework and before any branch pipework. Where practicable the connection shall be made within 600 mm of the meter outlet union or at the point of entry to the building if the meter is external.

544.2 Supplementary bonding conductors

544.2.1 A supplementary bonding conductor connecting two exposed-conductive-parts shall have a conductance, if sheathed or otherwise provided with mechanical protection, not less than that of the smaller protective conductor connected to the exposed-conductive-parts. If mechanical protection is not provided, its cross-sectional area shall be not less than 4 mm^2.

544.2.2 A supplementary bonding conductor connecting an exposed-conductive-part to an extraneous-conductive-part shall have a conductance, if sheathed or otherwise provided with mechanical protection, not less than half that of the protective conductor connected to the exposed-conductive-part. If mechanical protection is not provided, its cross-sectional area shall be not less than 4 mm^2.

544.2.3 A supplementary bonding conductor connecting two extraneous-conductive-parts shall have a cross-sectional area not less than 2.5 mm^2 if sheathed or otherwise provided with mechanical protection or 4 mm^2 if mechanical protection is not provided, except that where one of the extraneous-conductive-parts is connected to an exposed-conductive-part in compliance with Regulation 544.2.2, that regulation shall apply also to the conductor connecting the two extraneous-conductive-parts.

544.2.4 Except where Regulation 544.2.5 applies, supplementary bonding shall be provided by a supplementary conductor, a conductive part of a permanent and reliable nature, or by a combination of these.

544.2.5 Where supplementary bonding is to be applied to a fixed appliance which is supplied via a short length of flexible cable from an adjacent connection unit or other accessory, incorporating a flex outlet, the circuit protective conductor within the flexible cable shall be deemed to provide the supplementary bonding connection to the exposed-conductive-parts of the appliance, from the earthing terminal in the connection unit or other accessory.

CHAPTER 55

OTHER EQUIPMENT

CONTENTS

CHAPTER 55

OTHER EQUIPMENT

551   **LOW VOLTAGE GENERATING SETS**

551.1   **Scope**

This section applies to low voltage and extra-low voltage installations which incorporate generating sets intended to supply, either continuously or occasionally, all or part of the installation. Requirements are included for installations with the following supply arrangements:

- (i)   Supply to an installation which is not connected to a system for distribution of electricity to the public

- (ii)   Supply to an installation as an alternative to a system for distribution of electricity to the public

- (iii)   Supply to an installation in parallel with a system for distribution of electricity to the public

- (iv)   Appropriate combinations of the above.

This section does not apply to self-contained items of extra-low voltage electrical equipment which incorporate both the source of energy and the energy-using load and for which a specific product standard exists that includes the requirements for electrical safety.

NOTE: Where a generating set with an output not exceeding 16 A is to be connected in parallel with a system for distribution of electricity to the public, procedures for informing the electricity distributor are given in the ESQCR (2002). In addition to the ESQCR requirements, where a generating set with an output exceeding 16 A is to be connected in parallel with a system for distribution of electricity to the public, requirements of the electricity distributor should be ascertained before the generating set is connected. Requirements of the distributor for the connection of units rated up to 16A are given in BS EN 50438.

551.1.1   Generating sets with the following power sources are considered:

- (i)   Combustion engines
- (ii)   Turbines
- (iii)   Electric motors
- (iv)   Photovoltaic cells
- (v)   Electrochemical accumulators
- (vi)   Other suitable sources.

551.1.2   Generating sets with the following electrical characteristics are considered:

- (i)   Mains-excited and separately excited synchronous generators
- (ii)   Mains-excited and self-excited asynchronous generators
- (iii)   Mains-commutated and self-commutated static convertors with or without bypass facilities.

551.1.3   The use of generating sets for the following purposes is considered:

- (i)   Supply to permanent installations
- (ii)   Supply to temporary installations
- (iii)   Supply to mobile equipment which is not connected to a permanent fixed installation
- (iv)   Supply to mobile units (Section 717 also applies).

551.2   **General requirements**

551.2.1   The means of excitation and commutation shall be appropriate for the intended use of the generating set and the safety and proper functioning of other sources of supply shall not be impaired by the generating set.

551.2.2   The prospective short-circuit current and prospective earth fault current shall be assessed for each source of supply or combination of sources which can operate independently of other sources or combinations. The short-circuit rating of protective devices within the installation and, where appropriate, connected to a system for distribution of electricity to the public, shall not be exceeded for any of the intended methods of operation of the sources.

551.2.3   Where the generating set is intended to provide a supply to an installation which is not connected to a system for distribution of electricity to the public or to provide a supply as a switched alternative to such a system, the capacity and operating characteristics of the generating set shall be such that danger or damage to equipment does not arise after the connection or disconnection of any intended load as a result of the deviation of the voltage or frequency from the intended operating range. Means shall be provided to automatically disconnect such parts of the installation as may be necessary if the capacity of the generating set is exceeded.

NOTE 1: Consideration should be given to the intended duty cycle and size of individual connected loads as a proportion of the capacity of the generating set and to the starting characteristics of any connected electric motors.

NOTE 2: Consideration should be given to the power factor specified for protective devices in the installation.

NOTE 3: The installation of a generating set within an existing building or installation may change the conditions of external influence for the installation (see Part 3), for example by the introduction of moving parts, parts at high temperature or by the presence of flammable fluids and noxious gases, etc.

551.2.4   Provision for isolation shall meet the requirements of Section 537 for each source or combination of sources of supply.

551.3   Protective measure: Extra-low voltage provided by SELV or PELV

551.3.1   Additional requirements for SELV and PELV where the installation is supplied from more than one source

Where a SELV or PELV system may be supplied by more than one source, the requirements of Regulation 414.3 shall apply to each source. Where one or more of the sources is earthed, the requirements for PELV systems in Regulation 414.4 shall apply.

If one or more of the sources does not meet the requirements of Regulation 414.3, the system shall be treated as a FELV system and the requirements of Regulation 411.7 shall apply.

551.3.2   Where it is necessary to maintain the supply to an extra-low voltage system following the loss of one or more sources of supply, each source of supply or combination of sources of supply which can operate independently of other sources or combinations shall be capable of supplying the intended load of the extra-low voltage system. Provisions shall be made so that loss of the low voltage supply to an extra-low voltage source does not lead to danger or damage to other extra-low voltage equipment.

NOTE: Such provisions may be necessary in supplies for safety services (see Chapter 56).

551.4   Fault protection

551.4.1   Fault protection shall be provided for the installation in respect of each source of supply or combination of sources of supply that can operate independently of other sources or combinations of sources.

The fault protective provisions shall be selected or precautions shall be taken to ensure that where fault protective provisions are achieved in different ways within the same installation or part of an installation according to the active sources of supply, no influence shall occur or conditions arise that could impair the effectiveness of the fault protective provisions.

NOTE: This might, for example, require the use of a transformer providing electrical separation between parts of the installation using different earthing systems.

551.4.2   The generating set shall be connected so that any provision within the installation for protection by RCDs in accordance with Chapter 41 remains effective for every intended combination of sources of supply.

NOTE: Connection of live parts of the generator with Earth may affect the protective measure.

551.4.3   Protection by automatic disconnection of supply

551.4.3.1   Protection by automatic disconnection of supply shall be provided in accordance with Section 411, except as modified for particular cases by Regulation 551.4.3.2, 551.4.3.3 or 551.4.4.

551.4.3.2   Additional requirements for installations where the generating set provides a switched alternative to the system for distribution of electricity to the public (standby systems)

551.4.3.2.1   Protection by automatic disconnection of supply shall not rely upon the connection to the earthed point of the system for distribution of electricity to the public when the generator is operating as a switched alternative to a TN system. A suitable means of earthing shall be provided.

551.4.3.3 Additional requirements for installations incorporating static convertors

551.4.3.3.1 Where fault protection for parts of the installation supplied by the static convertor relies upon the automatic closure of the bypass switch and the operation of protective devices on the supply side of the bypass switch is not within the time required by Section 411, supplementary equipotential bonding shall be provided between simultaneously accessible exposed-conductive-parts and extraneous-conductive-parts on the load side of the static convertor in accordance with Regulation 415.2.

The resistance (R) of the supplementary protective bonding conductor required between simultaneously accessible exposed-conductive parts and extraneous-conductive-parts shall fulfil the following condition:

$$R \leq 50/I_a$$

where I_a is the maximum earth fault current which can be supplied by the static convertor when the bypass switch is closed.

NOTE: Where such equipment is intended to operate in parallel with a system for distribution of electricity to the public, the requirements of Regulation 551.7 also apply.

551.4.3.3.2 Precautions shall be taken or equipment shall be selected so that the correct operation of protective devices is not impaired by d.c. currents generated by a static convertor or by the presence of filters.

551.4.3.3.3 A means of isolation shall be installed on both sides of a static convertor. This requirement does not apply on the power source side of a static convertor which is integrated in the same enclosure as the power source.

551.4.4 Additional requirements for protection by automatic disconnection where the installation and generating set are not permanently fixed

This regulation applies to portable generating sets and to generating sets which are intended to be moved to unspecified locations for temporary or short-term use. Such generating sets may be part of an installation which is subject to similar use. This regulation does not apply to permanent fixed installations.

NOTE: For suitable connection arrangements see BS EN 60309 series.

551.4.4.1 Between separate items of equipment, protective conductors shall be provided which are part of a suitable cable and which comply with Table 54.7.

All protective conductors shall comply with Chapter 54.

551.4.4.2 In a TN, TT or IT system, protection by means of one or more RCDs with a rated residual operating current of not more than 30 mA shall be installed in accordance with Regulation 415.1 to protect every circuit.

NOTE: In an IT system, an RCD may not operate unless one of the earth faults is on a part of the system on the supply side of the device.

551.5 Protection against overcurrent

551.5.1 Where overcurrent protection of the generating set is required, it shall be located as near as practicable to the generator terminals.

NOTE: The contribution to the prospective short-circuit current by a generating set may be time-dependent and may be much less than the contribution made by a system for distribution of electricity to the public.

551.5.2 Where a generating set is intended to operate in parallel with a system for distribution of electricity to the public, or where two or more generating sets may operate in parallel, circulating harmonic currents shall be limited so that the thermal rating of conductors is not exceeded.

The effects of circulating harmonic currents may be limited by one or more of the following:

   (i) The selection of generating sets with compensated windings

  (ii) The provision of a suitable impedance in the connection to the generator star points

 (iii) The provision of switches which interrupt the circulatory circuit but which are interlocked so that at all times fault protection is not impaired

 (iv) The provision of filtering equipment

  (v) Other suitable means.

NOTE: Consideration should be given to the maximum voltage which may be produced across an impedance connected to limit circulating harmonic currents.

551.6 Additional requirements for installations where the generating set provides a supply as a switched alternative to the system for distribution of electricity to the public (standby systems)

551.6.1 Precautions complying with the relevant requirements of Section 537 for isolation shall be taken so that the generator cannot operate in parallel with the system for distribution of electricity to the public. Suitable precautions may include one or more of the following:

(i) An electrical, mechanical or electromechanical interlock between the operating mechanisms or control circuits of the changeover switching devices

(ii) A system of locks with a single transferable key

(iii) A three-position break-before-make changeover switch

(iv) An automatic changeover switching device with a suitable interlock

(v) Other means providing equivalent security of operation.

551.6.2 For a TN-S system where the neutral is not isolated, any RCD shall be positioned to avoid incorrect operation due to the existence of any parallel neutral-earth path.

NOTE: It may be desirable in a TN system to disconnect the neutral of the installation from the neutral or PEN of the system for distribution of electricity to the public to avoid disturbances such as induced voltage surges caused by lightning.

551.7 Additional requirements for installations where the generating set may operate in parallel with other sources including systems for distribution of electricity to the public

551.7.1 Where a generating set is used as an additional source of supply in parallel with another source, protection against thermal effects in accordance with Chapter 42 and protection against overcurrent in accordance with Chapter 43 shall remain effective in all situations.

551.7.2 A generating set used as an additional source of supply in parallel with another source shall be installed:

– on the supply side of all the protective devices for the final circuits of the installation, or

– on the load side of all the protective devices for a final circuit of the installation, but in this case all the following additional requirements shall be fulfilled:

(i) The conductors of the final circuit shall meet the following requirement:

$$I_z \geq I_n + I_g$$

where:

I_z is the current-carrying capacity of the final circuit conductors

I_n is the rated current of the protective device of the final circuit

I_g is the rated output current of the generating set

(ii) A generating set shall not be connected to a final circuit by means of a plug and socket

(iii) A residual current device providing additional protection of the final circuit in accordance with Regulation 415.1 shall disconnect all live conductors including the neutral conductor

(iv) The line and neutral conductors of the final circuit and of the generating set shall not be connected to Earth

(v) Unless the device providing automatic disconnection of the final circuit in accordance with Regulation 411.3.2 disconnects the line and neutral conductors, it shall be verified that the combination of the disconnection time of the protective device for the final circuit and the time taken for the output voltage of the generating set to reduce to 50 V or less is not greater than the disconnection time required by Regulation 411.3.2 for the final circuit.

This regulation does not apply to an uninterruptible power supply provided to supply specific items of current-using equipment within the final circuit to which it is connected.

551.7.3 In selecting and using a generating set to run in parallel with the system for distribution of electricity to the public, care shall be taken to avoid adverse effects to that system and to other installations in respect of power factor, voltage changes, harmonic distortion, unbalance, starting, synchronizing or voltage fluctuation effects. Where synchronization is necessary, the use of an automatic synchronizing system which considers frequency, phase and voltage is to be preferred.

551.7.4   Means of automatic switching shall be provided to disconnect the generating set from the system for distribution of electricity to the public in the event of loss of that supply or deviation of the voltage or frequency at the supply terminals from declared values.

For a generating set with an output exceeding 16 A, the type of protection and the sensitivity and operating times depend upon the protection of the system for distribution of electricity to the public and the number of generating sets connected and shall be agreed by the distributor. For a generating set with an output not exceeding 16 A, the settings shall comply with BS EN 50438.

In the case of the presence of a static convertor, the means of switching shall be provided on the load side of the static convertor.

551.7.5   Means shall be provided to prevent the connection of a generating set to the system for distribution of electricity to the public in the event of loss of that supply or deviation of the voltage or frequency at the supply terminals from values required by Regulation 551.7.4.

NOTE: For a generating set with an output not exceeding 16 A intended to operate in parallel with a system for distribution of electricity to the public the requirements are given in BS EN 50438.

551.7.6   Means shall be provided to enable the generating set to be isolated from the system for distribution of electricity to the public. For a generating set with an output exceeding 16 A, the accessibility of this means of isolation shall comply with national rules and distribution system operator requirements. For a generating set with an output not exceeding 16 A, the accessibility of this means of isolation shall comply with BS EN 50438.

551.7.7   Where a generating set may operate as a switched alternative to the system for distribution of electricity to the public, the installation shall also comply with Regulation 551.6.

551.8   Requirements for installations incorporating stationary batteries

551.8.1   Stationary batteries shall be installed so that they are accessible only to skilled or instructed persons.

NOTE: This generally requires the battery to be installed in a secure location or, for smaller batteries, a secure enclosure.

The location or enclosure shall be adequately ventilated.

551.8.2   Battery connections shall have basic protection by insulation or enclosures or shall be arranged so that two bare conductive parts having between them a potential difference exceeding 120 volts cannot be inadvertently touched simultaneously.

552   ROTATING MACHINES

NOTE: See also Regulation 537.5.4 Motor control.

552.1   Rotating machines

552.1.1   All equipment, including cable, of every circuit carrying the starting, accelerating and load currents of a motor shall be suitable for a current at least equal to the full-load current rating of the motor when rated in accordance with the appropriate British or Harmonized Standard. Where the motor is intended for intermittent duty and for frequent starting and stopping, account shall be taken of any cumulative effects of the starting or braking currents upon the temperature rise of the equipment of the circuit.

552.1.2   Every electric motor having a rating exceeding 0.37 kW shall be provided with control equipment incorporating means of protection against overload of the motor. This requirement does not apply to a motor incorporated in an item of current-using equipment complying as a whole with an appropriate British or Harmonized Standard.

552.1.3   Except where failure to start after a brief interruption would be likely to cause greater danger, every motor shall be provided with means to prevent automatic restarting after a stoppage due to a drop in voltage or failure of supply, where unexpected restarting of the motor might cause danger. These requirements do not preclude arrangements for starting a motor at intervals by an automatic control device, where other adequate precautions are taken against danger from unexpected restarting.

553   ACCESSORIES

553.1   Plugs and socket-outlets

553.1.1   Every plug and socket-outlet shall comply with all the requirements of items (i) and (ii) below and, in addition, with the appropriate requirements of Regulations 553.1.2 to 553.2.2:

(i)   Except for SELV circuits, it shall not be possible for any pin of a plug to make contact with any live contact of its associated socket-outlet while any other pin of the plug is completely exposed

(ii)   It shall not be possible for any pin of a plug to make contact with any live contact of any socket-outlet within the same installation other than the type of socket-outlet for which the plug is designed.

553.1.2   Except for SELV or a special circuit from Regulation 553.1.5, every plug and socket-outlet shall be of the non-reversible type, with provision for the connection of a protective conductor.

553.1.3   Except where Regulation 553.1.5 applies, in a low voltage circuit every plug and socket-outlet shall conform with the applicable British Standard listed in Table 55.1.

TABLE 55.1 –
Plugs and socket-outlets for low voltage circuits

Type of plug and socket-outlet	Rating (amperes)	Applicable British Standard
Fused plugs and shuttered socket-outlets, 2-pole and earth, for a.c.	13	BS 1363 (fuses to BS 1362)
Plugs, fused or non-fused, and socket-outlets, 2-pole and earth	2, 5, 15, 30	BS 546 (fuses, if any, to BS 646)
Plugs and socket-outlets (industrial type)	16, 32, 63, 125	BS EN 60309-2

553.1.100   Every socket-outlet for household and similar use shall be of the shuttered type and, for an a.c. installation, shall preferably be of a type complying with BS 1363.

553.1.5   A plug and socket-outlet not complying with BS 1363, BS 546 or BS EN 60309-2, may be used in single-phase a.c. or two-wire d.c. circuits operating at a nominal voltage not exceeding 250 volts for:

(i)   the connection of an electric clock, provided that the plug and socket-outlet are designed specifically for that purpose, and that each plug incorporates a fuse of rating not exceeding 3 amperes complying with BS 646 or BS 1362 as appropriate

(ii)   the connection of an electric shaver, provided that the socket-outlet is either incorporated in a shaver supply unit complying with BS EN 61558-2-5 or, in a room other than a bathroom, is a type complying with BS 4573

(iii) a circuit having special characteristics such that danger would otherwise arise or it is necessary to distinguish the function of the circuit.

553.1.6   A socket-outlet on a wall or similar structure shall be mounted at a height above the floor or any working surface to minimize the risk of mechanical damage to the socket-outlet or to an associated plug and its flexible cable which might be caused during insertion, use or withdrawal of the plug.

553.1.7   Where mobile equipment is likely to be used, provision shall be made so that the equipment can be fed from an adjacent and conveniently accessible socket-outlet, taking account of the length of flexible cable normally fitted to portable appliances and luminaires.

553.2   Cable couplers

553.2.1   Except for a SELV or a Class II circuit, a cable coupler shall comply where appropriate with BS 6991, BS EN 61535, BS EN 60309-2 or BS EN 60320-1, shall be non-reversible and shall have provision for the connection of a protective conductor.

553.2.2   A cable coupler shall be arranged so that the connector of the coupler is fitted at the end of the cable remote from the supply.

554 CURRENT-USING EQUIPMENT

554.1 Electrode water heaters and boilers

554.1.1   Every electrode water heater and electrode boiler shall be connected to an a.c. system only, and shall be selected and erected in accordance with the appropriate requirements of this section.

554.1.2   The supply to the electrode water heater or electrode boiler shall be controlled by a linked circuit-breaker arranged to disconnect the supply from all electrodes simultaneously and provided with an overcurrent protective device in each conductor feeding an electrode.

554.1.3   The earthing of the electrode water heater or electrode boiler shall comply with the requirements of Chapter 54 and, in addition, the shell of the electrode water heater or electrode boiler shall be bonded to the metallic sheath and armour, if any, of the incoming supply cable. The protective conductor shall be connected to the shell of the electrode water heater or electrode boiler and shall comply with Regulation 543.1.1.

554.1.4   Where an electrode water heater or electrode boiler is directly connected to a supply at a voltage exceeding low voltage, the installation shall include an RCD arranged to disconnect the supply from the electrodes on the occurrence of a sustained earth leakage current in excess of 10 % of the rated current of the electrode water heater or electrode boiler under normal conditions of operation, except that if in any instance a higher value is essential to ensure stability of operation of the electrode water heater or electrode boiler, the value may be increased to a maximum of 15 %. A time delay may be incorporated in the device to prevent unnecessary operation in the event of imbalance of short duration.

554.1.5   Where an electrode water heater or electrode boiler is connected to a three-phase low voltage supply, the shell of the electrode water heater or electrode boiler shall be connected to the neutral of the supply as well as to the earthing conductor. The current-carrying capacity of the neutral conductor shall be not less than that of the largest line conductor connected to the equipment.

554.1.6   Except as provided by Regulation 554.1.7, where the supply to an electrode water heater or electrode boiler is single-phase and one electrode is connected to a neutral conductor earthed by the distributor, the shell of the electrode water heater or electrode boiler shall be connected to the neutral of the supply as well as to the earthing conductor.

554.1.7   Where the electrode water heater or electrode boiler is not piped to a water supply or in physical contact with any earthed metal, and where the electrodes and the water in contact with the electrodes are so shielded in insulating material that they cannot be touched while the electrodes are live, a fuse in the line conductor may be substituted for the circuit-breaker required under Regulation 554.1.2 and the shell of the electrode water heater or electrode boiler need not be connected to the neutral of the supply.

554.2 Heaters for liquids or other substances having immersed heating elements

554.2.1   Every heater for liquid or other substance shall incorporate or be provided with an automatic device to prevent a dangerous rise in temperature.

554.3 Water heaters having immersed and uninsulated heating elements

554.3.1   Every single-phase water heater or boiler having an uninsulated heating element immersed in the water shall comply with the requirements of Regulations 554.3.2 and 554.3.3. This type of water heater or boiler is deemed not to be an electrode water heater or electrode boiler.

554.3.2   All metal parts of the heater or boiler which are in contact with the water (other than current-carrying parts) shall be solidly and metallically connected to a metal water pipe through which the water supply to the heater or boiler is provided, and that water pipe shall be connected to the main earthing terminal by means independent of the circuit protective conductor.

554.3.3   The heater or boiler shall be permanently connected to the electricity supply through a double-pole linked switch which is either separate from and within easy reach of the heater or boiler or is incorporated therein, and the wiring from the heater or boiler shall be connected directly to that switch without the use of a plug and socket-outlet; in addition, where the heater or boiler is installed in a room containing a fixed bath, the switch shall comply with Section 701.

554.3.4   Before a heater or boiler of the type referred to in Regulation 554.3.1 is connected, the installer shall confirm that no single-pole switch, non-linked circuit-breaker or fuse is fitted in the neutral conductor in any part of the circuit between the heater or boiler and the origin of the installation.

554.4 Heating conductors and cables

NOTE: For electric floor and ceiling heating systems in buildings the requirements of Section 753 must also be met.

554.4.1 Where a heating cable is required to pass through, or be in close proximity to, material which presents a fire hazard, the cable shall be enclosed in material having the ignitability characteristic 'P' as specified in BS 476-12 and shall be adequately protected from any mechanical damage reasonably foreseeable during installation and use.

554.4.2 A heating cable intended for laying directly in soil, concrete, cement screed or other material used for road and building construction shall be:

(i) capable of withstanding mechanical damage under the conditions that can reasonably be expected to prevail during its installation, and

(ii) constructed of material that will be resistant to damage from dampness and/or corrosion under normal conditions of service.

554.4.3 A heating cable laid directly in soil, a road or the structure of a building shall be installed so that it:

(i) is completely embedded in the substance it is intended to heat, and

(ii) does not suffer damage in the event of movement normally to be expected in it or the substance in which it is embedded, and

(iii) complies in all respects with the manufacturer's instructions and recommendations.

554.4.4 The load of every floor-warming cable under operation shall be limited to a value such that the manufacturer's stated conductor temperature is not exceeded. Other factors can limit the maximum temperature at which the cable can be run, such as the temperature rating of any terminations or accessories, and any material with which it is in contact.

554.5 *Deleted by BS 7671:2008 Amendment No 1*

555 TRANSFORMERS

555.1 Autotransformers and step-up transformers

555.1.1 Where an autotransformer is connected to a circuit having a neutral conductor, the common terminal of the winding shall be connected to the neutral conductor.

555.1.2 A step-up autotransformer shall not be connected to an IT system.

555.1.3 Where a step-up transformer is used, a linked switch shall be provided for disconnecting the transformer from all live conductors of the supply.

559 LUMINAIRES AND LIGHTING INSTALLATIONS

559.1 Scope

This section applies to the selection and erection of luminaires and lighting installations intended to be part of the fixed installation and to highway power supplies and street furniture.

Particular requirements are given for:

(i) fixed outdoor lighting installations

(ii) extra-low voltage lighting installations supplied from a source with a maximum rated voltage of 50 V a.c. or 120 V d.c.

(iii) lighting for display stands.

NOTE 1: For lighting installations in special locations, refer to Part 7.

The requirements of this section do not apply to:

(iv) high voltage signs supplied at low voltage (such as neon tubes)

(v) signs and luminous discharge tube installations operating from a no-load rated output voltage exceeding 1 kV but not exceeding 10 kV (BS EN 50107).

NOTE 2: The requirements for high voltage signs are given in BS 559 and the BS EN 50107 series.

559.2 *Not used*

559.3 Outdoor lighting installations

An outdoor lighting installation comprises one or more luminaires, a wiring system and accessories.

The following are included:

(i) Lighting installations such as those for roads, parks, car parks, gardens, places open to the public, sporting areas, illumination of monuments and floodlighting

(ii) Other lighting arrangements in places such as telephone kiosks, bus shelters, advertising panels and town plans

(iii) Road signs.

The following are excluded:

(iv) Equipment of the owner or operator of a system for distribution of electricity to the public

(v) Temporary festoon lighting

(vi) Luminaires fixed to the outside of a building and supplied directly from the internal wiring of that building

(vii) Road traffic signal systems.

559.4 General requirements for installations

NOTE: See Table 55.2 for an explanation of the symbols used in luminaires, in controlgear for luminaires and in the installation of luminaires.

559.4.1 Every luminaire shall comply with the relevant standard for manufacture and test of that luminaire and shall be selected and erected in accordance with the manufacturer's instructions.

559.4.2 For the purposes of this section, luminaires without transformers or convertors but which are fitted with extra-low voltage lamps connected in series shall be considered as low voltage equipment not extra-low voltage equipment.

559.4.3 Where a luminaire is installed in a pelmet, there shall be no adverse effects due to the presence or operation of curtains or blinds.

559.4.4 A track system for luminaires shall comply with the requirements of BS EN 60570.

559.5 Protection against fire

559.5.1 General

In the selection and erection of a luminaire the thermal effects of radiant and convected energy on the surroundings shall be taken into account, including:

(i) the maximum permissible power dissipated by the lamps

(ii) the fire-resistance of adjacent material
- at the point of installation, and
- in the thermally affected areas

(iii) the minimum distance to combustible materials, including material in the path of a spotlight beam.

559.6 Wiring systems

559.6.1 Common rules

559.6.1.1 Connection to the fixed wiring

At each fixed lighting point one of the following shall be used:

(i) A ceiling rose to BS 67

(ii) A luminaire supporting coupler to BS 6972 or BS 7001

(iii) A batten lampholder or a pendant set to BS EN 60598

(iv) A luminaire to BS EN 60598

(v) A suitable socket-outlet to BS 1363-2, BS 546 or BS EN 60309-2

(vi) A plug-in lighting distribution unit to BS 5733

(vii) A connection unit to BS 1363-4

(viii) Appropriate terminals enclosed in a box complying with the relevant part of BS EN 60670 series or BS 4662

(ix) A device for connecting a luminaire (DCL) outlet according to IEC 61995-1.

NOTE: In suspended ceilings one plug-in lighting distribution unit may be used for a number of luminaires.

559.6.1.2   A ceiling rose or lampholder for a filament lamp shall not be installed in any circuit operating at a voltage normally exceeding 250 volts.

559.6.1.3   A ceiling rose shall not be used for the attachment of more than one outgoing flexible cable unless it is specially designed for multiple pendants.

559.6.1.4   Luminaire supporting couplers are designed specifically for the mechanical support and electrical connection of luminaires and shall not be used for the connection of any other equipment.

559.6.1.5   Adequate means to fix luminaires shall be provided.

The fixing means may be mechanical accessories (e.g. hooks or screws), boxes or enclosures which are able to support luminaires or supporting devices for connecting a luminaire.

In places where the fixing means is intended to support a pendant luminaire, the fixing means shall be capable of carrying a mass of not less than 5 kg. If the mass of the luminaire is greater than 5 kg, the installer shall ensure that the fixing means is capable of supporting the mass of the pendant luminaire.

The installation of the fixing means shall be in accordance with the manufacturer's instructions.

The weight of luminaires and their eventual accessories shall be compatible with the mechanical capability of the ceiling or suspended ceiling or supporting structure where installed.

Any flexible cable between the fixing means and the luminaire shall be installed so that any expected stresses in the conductors, terminals and terminations will not impair the safety of the installation. (See also Table 4F3A of Appendix 4.)

559.6.1.6   Lighting circuits incorporating B15, B22, E14, E27 or E40 lampholders shall be protected by an overcurrent protective device of maximum rating 16A.

559.6.1.7   Bayonet lampholders B15 and B22 shall comply with BS EN 61184 and shall have the temperature rating T2 described in that standard.

559.6.1.8   In circuits of a TN or TT system, except for E14 and E27 lampholders complying with BS EN 60238, the outer contact of every Edison screw or single centre bayonet cap type lampholder shall be connected to the neutral conductor. This regulation also applies to track mounted systems.

559.6.1.9   A lighting installation shall be appropriately controlled.

NOTE: See Table 53.4 for guidance on the selection of suitable protective, isolation and switching devices.

559.6.1.100   Consideration shall be given to the provision of the neutral conductor, at each switch position, to facilitate the installation of electronic switching devices.

559.6.2   Through wiring

559.6.2.1   The installation of through wiring in a luminaire is only permitted if the luminaire is designed for such wiring.

559.6.2.2   A cable for through wiring shall be selected in accordance with the temperature information on the luminaire or on the manufacturer's instruction sheet, if any, as follows:

  (i)   For a luminaire complying with BS EN 60598 but with temperature marking, cables suitable for the marked temperature shall be used

  (ii) Unless specified in the manufacturer's instructions, for a luminaire complying with BS EN 60598 but with no temperature marking, heat-resistant cables are not required

  (iii) In the absence of information, heat-resistant cables and/or insulated conductors of type H05S-U, H05S-K, H05SJ-K, H05SS-K (BS 6007) or equivalent shall be used.

559.6.2.3   Groups of luminaires divided between the three line conductors of a three-phase system with only one common neutral conductor shall be provided with at least one device that simultaneously disconnects all line conductors.

559.7 Independent lamp controlgear, e.g. ballasts

Only independent lamp controlgear marked as suitable for independent use, according to the relevant standard, shall be used external to a luminaire.

Only the following are permitted to be mounted on flammable surfaces:

(i) A "class P" thermally protected ballast(s)/transformer(s).

(ii) A temperature declared thermally protected ballast(s)/transformer(s) with a marked value equal to or below 130 °C.

NOTE: For an explanation of symbols used see Table 55.2.

559.8 Compensation capacitors

Compensation capacitors having a total capacitance exceeding 0.5 μF shall only be used in conjunction with discharge resistors. Capacitors and their marking shall be in accordance with BS EN 61048.

This requirement does not apply to capacitors forming part of the equipment.

559.9 Stroboscopic effect

In the case of lighting for premises where machines with moving parts are in operation, consideration shall be given to stroboscopic effects which can give a misleading impression of moving parts being stationary. Such effects may be avoided by selecting luminaires with suitable lamp controlgear, such as high frequency controlgear, or by distributing lighting loads across all the phases of a three-phase supply.

559.10 Requirements for outdoor lighting installations, highway power supplies and street furniture

559.10.1 Protective measures: Placing out of reach and obstacles

The protective measures of placing out of reach and obstacles shall not be used.

Except where the maintenance of equipment is to be restricted to skilled persons who are specially trained, where items of street furniture are within 1.5 m of a low voltage overhead line, basic protection of the low voltage overhead line shall be provided by means other than placing out of reach.

559.10.2 Protective measures: Non-conducting location and earth-free local equipotential bonding

The protective measures non-conducting location and earth-free local equipotential bonding shall not be used.

559.10.3 Protective measure: Automatic disconnection of supply

559.10.3.1 Where the protective measure automatic disconnection of supply is used:

(i) all live parts of electrical equipment shall be protected by insulation or by barriers or enclosures providing basic protection. A door in street furniture, used for access to electrical equipment, shall not be used as a barrier or an enclosure

(ii) for every accessible enclosure live parts shall only be accessible with a key or a tool, unless the enclosure is in a location where only skilled or instructed persons have access

(iii) a door giving access to electrical equipment and located less than 2.50 m above ground level shall be locked with a key or shall require the use of a tool for access. In addition, basic protection shall be provided when the door is open either by the use of equipment having at least a degree of protection IPXXB or IP2X by construction or by installation, or by installing a barrier or an enclosure giving the same degree of protection

(iv) for a luminaire at a height of less than 2.80 m above ground level, access to the light source shall only be possible after removing a barrier or an enclosure requiring the use of a tool

(v) for an outdoor lighting installation, a metallic structure (such as a fence, grid etc.), which is in the proximity of but is not part of the outdoor lighting installation need not be connected to the main earthing terminal.

559.10.3.2 Lighting in places such as telephone kiosks, bus shelters and town plans shall be provided with additional protection by an RCD having the characteristics specified in Regulation 415.1.1.

559.10.3.3 A maximum disconnection time of 5 s shall apply to all circuits feeding fixed equipment used in highway power supplies for compliance with Regulation 411.3.2.3 (TN system) or 411.3.2.4 (TT system).

559.10.3.4 The earthing conductor of a street electrical fixture shall have a minimum copper equivalent cross-sectional area not less than that of the supply neutral conductor at that point or not less than 6 mm^2, whichever is the smaller.

559.10.4   Protective measure: Double or reinforced insulation

For an outdoor lighting installation, where the protective measure for the whole installation is by double or reinforced insulation, no protective conductor shall be provided and the conductive parts of the lighting column shall not be intentionally connected to the earthing system.

559.10.5   External influences

559.10.5.1   Classification of external influences

The following classes are generally recommended:

   (i)   Ambient temperature: AA2 and AA4 (from -40 °C to +40 °C)

   (ii)   Climatic conditions: AB2 and AB4 (relative humidity between 5 % and 100 %).

The classes given for the following external influences are minimum requirements:

   (iii)   Presence of water: AD3 (sprays)

   (iv)   Presence of foreign bodies: AE2 (small objects).

559.10.5.2 Electrical equipment shall have, by construction or by installation, a degree of protection of at least IP33.

559.10.6   Devices for isolation and switching

559.10.6.1 Where it is intended that isolation and switching is carried out only by instructed persons and subject to suitable provisions being made so that precautions can be taken to prevent any equipment from being inadvertently or unintentionally energized, for TN systems, the means of switching the supply on load and the means of isolation is permitted to be provided by a suitably rated fuse carrier.

559.10.6.2 Where the distributor's cut-out is used as the means of isolation of a highway power supply the approval of the distributor shall be obtained.

559.10.7   Warning notices

559.10.7.1 The requirements for notices for:

   (i)   periodic inspection and testing (Regulation 514.12.1) and

   (ii)   the testing of RCDs (Regulation 514.12.2)

need not be applied where the installation is subject to a programmed inspection and testing procedure.

559.10.7.2 On every temporary supply unit there shall be an externally mounted durable label stating the maximum sustained current to be supplied from that unit.

559.11   Requirements for extra-low voltage lighting installations

559.11.1   Protective measure: FELV

The protective measure FELV shall not be used.

559.11.2   Protective measure: SELV

An extra-low voltage luminaire without provision for the connection of a protective conductor shall be installed only as part of a SELV system.

559.11.3   Transformers and convertors

559.11.3.1 A safety isolating transformer for an extra-low voltage lighting installation shall comply with BS EN 61558-2-6 and shall meet at least one of the following requirements:

   (i)   The transformer shall be protected on the primary side by a protective device complying with the requirements of Regulation 559.11.4.2

   (ii)   The transformer shall be short-circuit proof (both inherently and non-inherently).

NOTE: For an explanation of symbols used see Table 55.2.

559.11.3.2 An electronic convertor for an extra-low voltage lighting installation shall comply with BS EN 61347-2-2.

NOTE: For an explanation of symbols used see Table 55.2.

559.11.4 Fire risk due to short-circuit

559.11.4.1 Where both the live circuit conductors are uninsulated, either:

(i) they shall be provided with a protective device complying with the requirements of Regulation 559.11.4.2, or

(ii) the system shall comply with BS EN 60598-2-23.

559.11.4.2 A device providing protection against the risk of fire in accordance with Regulation 559.11.4.1 shall meet the following requirements:

(i) The device shall continuously monitor the power demand of the luminaires

(ii) The device shall automatically disconnect the supply circuit within 0.3 s in the case of a short-circuit or failure which causes a power increase of more than 60 W

(iii) The device shall provide automatic disconnection while the supply circuit is operating with reduced power (for example, by gating control or a regulating process or a lamp failure) if there is a failure which causes a power increase of more than 60 W

(iv) The device shall provide automatic disconnection upon connection of the supply circuit if there is a failure which causes a power increase of more than 60 W

(v) The device shall be fail-safe.

NOTE: Account needs to be taken of starting currents.

559.11.5 Wiring systems

559.11.5.1 Metallic structural parts of buildings, for example, pipe systems or parts of furniture, shall not be used as live conductors.

559.11.5.2 The minimum cross-sectional area of the extra-low voltage conductors shall be:

(i) 1.5 mm^2 copper, but in the case of flexible cables with a maximum length of 3 m a cross-sectional area of 1 mm^2 copper may be used

(ii) 4 mm^2 copper in the case of suspended flexible cables or insulated conductors, for mechanical reasons

(iii) 4 mm^2 copper in the case of composite cables consisting of braided tinned copper outer sheath, having a material of high tensile strength inner core.

559.11.5.3 Bare conductors

If the nominal voltage does not exceed 25 V a.c. or 60 V d.c., bare conductors may be used providing that the extra-low voltage lighting installation complies with all the following requirements:

(i) The lighting installation shall be designed, and installed or enclosed in such a way that the risk of a short-circuit is reduced to a minimum

(ii) The conductors used shall have a cross-sectional area of at least 4 mm^2, for mechanical reasons

(iii) The conductors shall not be placed directly on combustible material.

For suspended bare conductors, at least one conductor and its terminals shall be insulated for that part of the circuit between the transformer and the short-circuit protective device to prevent a short-circuit.

559.11.6 Suspended systems

Suspension devices for extra-low voltage luminaires, including supporting conductors, shall be capable of carrying five times the mass of the luminaires (including their lamps) intended to be supported, but not less than 5 kg.

Terminations and connections of conductors shall be made by screw terminals or screwless clamping devices complying with BS EN 60998-2-1 or BS EN 60998-2-2.

Insulation piercing connectors and termination wires which rely on counterweights hung over suspended conductors to maintain the electrical connection shall not be used.

The suspended system shall be fixed to walls or ceilings by insulated distance cleats and shall be continuously accessible throughout the route.

TABLE 55.2 – Explanation of symbols used in luminaires, in controlgear for luminaires and in the installation of luminaires

BS EN 60598-1:2004		BS EN 60598-1:2008	
F (triangle)	Luminaire suitable for direct mounting on normally flammable surfaces	flames symbol (recessed, crossed)	Recessed luminaire not suitable for direct mounting on normally flammable surfaces
F (triangle, crossed out)	Luminaire suitable for direct mounting on non-combustible surfaces only	flames symbol (surface)	Surface mounted luminaire not suitable for direct mounting on normally flammable surfaces
F (triangle with line above)	Luminaire suitable for direct mounting in/on normally flammable surfaces when thermally insulating material may cover the luminaire.	crossed symbol	Luminaire not suitable for covering with thermally insulating material

NOTE   Luminaires suitable for direct mounting on normally flammable surfaces may be marked with the symbol shown according to BS EN 60598-1:2004.

With the publication of BS EN 60598-1 Ed. 7, luminaires suitable for direct mounting on normally flammable surfaces have no special marking and only luminaires not suitable for mounting on normally flammable surfaces are marked with a symbol (see annex N.4 of BS EN 60598-1:2008 for further explanations).

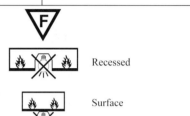

F (triangle)

flames symbol (recessed)   Recessed

flames symbol (surface)   Surface

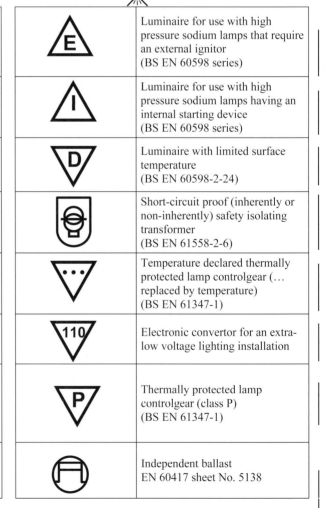

BS EN 60598-1:2004		BS EN 60598-1:2008	
t ..... °C (cable symbol)	Use of heat-resistant supply cables, interconnecting cables, or external wiring (number of conductors of cable is optional) (BS EN 60598 series)	E (triangle)	Luminaire for use with high pressure sodium lamps that require an external ignitor (BS EN 60598 series)
bowl mirror lamp symbol	Luminaire designed for use with bowl mirror lamps (BS EN 60598 series)	I (triangle)	Luminaire for use with high pressure sodium lamps having an internal starting device (BS EN 60598 series)
ta .... °C	Rated maximum ambient temperature (BS EN 60598 series)	D (inverted triangle)	Luminaire with limited surface temperature (BS EN 60598-2-24)
COOL BEAM (crossed)	Warning against the use of cool-beam lamps (BS EN 60598 series)	transformer symbol	Short-circuit proof (inherently or non-inherently) safety isolating transformer (BS EN 61558-2-6)
⊂ – – m symbol	Minimum distance from lighted objects (metres) (BS EN 60598 series)	••• (inverted triangle)	Temperature declared thermally protected lamp controlgear (... replaced by temperature) (BS EN 61347-1)
hammer symbol	Rough service luminaire (BS EN 60598 series)	110 (inverted triangle)	Electronic convertor for an extra-low voltage lighting installation
cracked screen symbol	Replace any cracked protective screen (BS EN 60598 series)	P (inverted triangle)	Thermally protected lamp controlgear (class P) (BS EN 61347-1)
self-shielded lamp symbol	Luminaire designed for use with self-shielded tungsten halogen lamps or self-shielded metal halide lamps only (BS EN 60598 series)	ballast symbol	Independent ballast EN 60417 sheet No. 5138

CHAPTER 56

SAFETY SERVICES

CONTENTS

CHAPTER 56

SAFETY SERVICES

560.1   SCOPE

This chapter covers general requirements for safety services, selection and erection of electrical supply systems for safety services and electrical safety sources. Standby electrical supply systems are outside the scope of this chapter. This chapter does not apply to installations in hazardous areas (BE3), for which requirements are given in BS EN 60079-14.

NOTE: Examples of safety services include (this list is not exhaustive):

- Emergency lighting
- Fire pumps
- Fire rescue service lifts
- Fire detection and alarm systems
- CO detection and alarm systems
- Fire evacuation systems
- Smoke ventilation systems
- Fire services communication systems
- Essential medical systems
- Industrial safety systems.

560.2   *Not used*

560.3   *Not used*

560.4 CLASSIFICATION

560.4.1   An electrical safety service supply is either:

(i) a non-automatic supply, the starting of which is initiated by an operator, or

(ii) an automatic supply, the starting of which is independent of an operator.

An automatic supply is classified as follows, according to the maximum changeover time:

(iii) No-break: an automatic supply which can ensure a continuous supply within specified conditions during the period of transition, for example as regards variations in voltage and frequency

(iv) Very short break: an automatic supply available within 0.15 s

(v) Short break: an automatic supply available within 0.5 s

(vi) Normal break: an automatic supply available within 5 s

(vii) Medium break: an automatic supply available within 15 s

(viii) Long break: an automatic supply available in more than 15 s.

560.4.2   The essential equipment for safety services shall be compatible with the changeover time in order to maintain the specified operation.

560.5 GENERAL

560.5.1   Safety services may be required to operate at all relevant times including during mains and local supply failure and through fire conditions. To meet this requirement specific sources, equipment, circuits and wiring are necessary. Some applications also have particular requirements, as in Regulations 560.5.2 and 560.5.3.

560.5.2   For safety services required to operate in fire conditions, the following two conditions shall be fulfilled:

(i) An electrical source for safety supply shall be so selected as to maintain a supply of adequate duration, and

(ii) all equipment of safety services shall be so provided, either by construction or by erection, with protection ensuring fire-resistance of adequate duration.

NOTE: The safety source is generally additional to the normal source. The normal source is, for example, the public supply network.

560.5.3   Where automatic disconnection of supply is used as a protective measure against electric shock, non-disconnection on the first fault is preferred. In IT systems, continuous insulation monitoring devices shall be provided which give an audible and visual indication in the event of a first fault.

560.5.4   A failure in the control or bus system of a normal installation shall not adversely affect the function of safety services.

560.6 ELECTRICAL SOURCES FOR SAFETY SERVICES

560.6.1   The following electrical sources for safety services are recognized:

(i) Storage batteries

(ii) Primary cells

(iii) Generator sets independent of the normal supply

(iv) A separate feeder of the supply network that is effectively independent of the normal feeder.

560.6.2   Safety sources for safety services shall be installed as fixed equipment and in such a manner that they cannot be adversely affected by failure of the normal source.

560.6.3   Safety sources shall be installed in a suitable location and be accessible only to skilled or instructed persons (BA5 or BA4).

560.6.4   The location of the safety source shall be properly and adequately ventilated so that exhaust gases, smoke or fumes from the safety source cannot penetrate areas occupied by persons.

560.6.5   Separated independent feeders from a distributor's network shall not serve as electrical sources for safety services unless assurance can be obtained that the two supplies are unlikely to fail concurrently.

560.6.6   The safety source shall have sufficient capability to supply its related safety service.

560.6.7   A safety source may, in addition, be used for purposes other than safety services, provided the availability for safety services is not thereby impaired. A fault occurring in a circuit for purposes other than safety services shall not cause the interruption of any circuit for safety services.

560.6.8   Special requirements for safety services having sources not capable of operation in parallel

560.6.8.1   Adequate precautions shall be taken to avoid the paralleling of sources.

NOTE: This may be achieved by mechanical interlocking.

560.6.8.2   Short-circuit protection and fault protection shall be ensured for each source.

560.6.9   Special requirements for safety services having sources capable of operation in parallel

Short-circuit protection and fault protection shall be ensured when the installation is supplied separately by either of the two sources or by both in parallel.

NOTE 1:   The parallel operation of a private source with the public supply network is subject to authorisation by the distribution network operator (DNO). This may require special devices, for example to prevent reverse power. Refer also to Section 551 and ENA publications Engineering Recommendation G.59/1 and ETR 113.

NOTE 2:   Precautions may be necessary to limit current circulation in the connection between the neutral points of the sources, in particular the effect of triplen harmonics.

560.6.10   Central power supply sources

Batteries shall be of vented or valve-regulated maintenance-free type and shall be of heavy duty industrial design, for example cells complying with BS EN 60623 or the appropriate part of the BS EN 60896 series.

NOTE: The minimum design life of the batteries at 20 °C should be 10 years.

560.6.11   Low power supply sources

The power output of a low power supply system is limited to 500 W for 3-hour duration and 1500 W for a 1-hour duration. Batteries shall be of heavy duty industrial design, for example cells complying with BS EN 60623 or the appropriate part of the BS EN 60896 series are suitable.

NOTE: The minimum design life of the batteries at 20 °C should be 5 years.

560.6.12   Uninterruptible power supply sources (UPS)

Where an uninterruptible power supply is used, it shall:

  (i)   be able to operate distribution circuit protective devices, and

  (ii)   be able to start the safety devices when it is operating in the emergency condition from the inverter supplied by the battery, and

  (iii)   comply with the requirements of Regulation 560.6.10, and

  (iv)   comply with BS EN 62040-1 and BS EN 62040-3, as applicable.

560.6.13   Generator sets for safety services

Where a generating set is used as a safety source, it shall comply with BS 7698-12.

560.6.14   Monitoring of safety sources

The condition of the source for safety services (ready for operation, under fault conditions, feeding from the source for safety services) shall be monitored.

560.7   CIRCUITS OF SAFETY SERVICES

560.7.1   Circuits of safety services shall be independent of other circuits.

NOTE: This means that any electrical fault or any intervention or modification in one system must not affect the correct functioning of the other. This may necessitate separation by fire-resistant materials or different routes or enclosures.

560.7.2   Circuits of safety services shall not pass through locations exposed to fire risk (BE2) unless they are fire-resistant. The circuits shall not, in any case, pass through zones exposed to explosion risk (BE3).

NOTE: Where practicable, the passage of circuits through locations presenting a fire risk should be avoided.

560.7.3   In accordance with Regulation 433.3.3, protection against overload may be omitted where the loss of supply may cause a greater hazard. Where protection against overload is omitted, the occurrence of an overload shall be monitored.

560.7.4   Overcurrent protective devices shall be selected and erected so as to avoid an overcurrent in one circuit impairing the correct operation of circuits of safety services.

560.7.5   Switchgear and controlgear shall be clearly identified and grouped in locations accessible only to skilled or instructed persons (BA5 or BA4).

560.7.6   In equipment supplied by two different circuits, a fault occurring in one circuit shall not impair the protection against electric shock or the correct operation of the other circuit. Such equipment shall be connected to the protective conductors of both circuits, if necessary.

560.7.7   Safety circuit cables, other than metallic screened, fire-resistant cables, shall be adequately and reliably separated by distance or by barriers from other circuit cables, including other safety circuit cables.

NOTE: For battery cables, special requirements may apply.

560.7.8   Circuits for safety services, with the exception of wiring for fire and rescue service lift supply cables and wiring for lifts with special requirements, shall not be installed in lift shafts or other flue-like openings.

NOTE: While fire-resistant cables will survive most fires, if they are located in an unstopped vertical shaft the upward air draught in a fire can generate excessive temperatures above the capabilities of the cable so they should be avoided as a route for safety systems.

560.7.9   In addition to a general schematic diagram, full details of all electrical safety sources shall be given. The information shall be maintained adjacent to the distribution board. A single-line diagram is sufficient.

560.7.10   Drawing(s) of the electrical safety installations shall be available showing the exact location of:

(i)   all electrical equipment and distribution boards, with equipment designations

(ii)   safety equipment with final circuit designation and particulars and purpose of the equipment

(iii)   special switching and monitoring equipment for the safety power supply (e.g. area switches, visual or acoustic warning equipment).

560.7.11   A list of all the current-using equipment permanently connected to the safety power supply, indicating the nominal electrical power, rated nominal voltage, current and starting current, together with its duration, shall be available.

NOTE: This information may be included in the circuit diagrams.

560.7.12   Operating instructions for safety equipment and electrical safety services shall be available. They shall take into account all the particulars of the installation.

560.8   WIRING SYSTEMS

560.8.1   One or more of the following wiring systems shall be utilised for safety services required to operate in fire conditions:

(i)   Mineral insulated cable systems complying with BS EN 60702-1 and BS EN 60702-2 and BS EN 60332-1-2

(ii)   Fire-resistant cables complying with IEC 60331-1 or IEC 60331-2 or IEC 60331-3 and with BS EN 60332-1-2

(iii)   Fire-resistant cables complying with test requirements of BS EN 50200 or BS 8434 or BS 8491, appropriate for the cable size and with BS EN 60332-1-2

(iv)   A wiring system maintaining the necessary fire and mechanical protection.

The wiring system selected shall meet the requirements of the relevant code of practice appropriate to the application and shall be mounted and installed in such a way that the circuit integrity will not be impaired during a fire.

NOTE 1: The codes of practice in the UK for emergency lighting, fire alarms and life safety and fire fighting applications are:

(i)   BS 5266 Part 1: Emergency lighting - Code of practice for the emergency lighting of premises.

(ii)   BS 5839 Part 1: Fire detection and fire alarm systems for buildings – Code of practice for system design, installation, commissioning and maintenance.

(iii)   BS 8519: Code of practice for the selection and installation of fire–resistant cables and systems for life safety and fire fighting applications.

NOTE 2: BS 5266, BS 5839 and BS 8519 specifies cables to BS EN 60702-1, BS 7629-1 and BS 7846 as being suitable when appropriately selected for the application.

NOTE 3: Examples of a system maintaining the necessary fire and mechanical protection could be:

    (i)   constructional enclosures to maintain fire and mechanical protection, or

    (ii)   wiring systems in separate fire compartments.

560.8.2   Wiring for control and bus systems of safety services shall be in accordance with the same requirements as the wiring which is to be used for the safety services. This does not apply to circuits that do not adversely affect the operation of the safety equipment.

560.8.3   *Deleted by BS 7671:2008 Amendment No 1 and reserved for future use*

560.8.4   Circuits for safety services which can be supplied by direct current shall be provided with two-pole overcurrent protection mechanisms.

560.8.5   Switchgear and controlgear used for both a.c. and d.c. supply sources shall be suitable for both a.c. and d.c. operation.

560.9   EMERGENCY LIGHTING SYSTEMS

Emergency lighting systems shall comply with the relevant parts of BS 5266 series.

560.10   FIRE DETECTION AND FIRE ALARM SYSTEMS

Fire detection and fire alarm systems shall comply with the relevant parts of BS 5839 series.

PART 6

INSPECTION AND TESTING

CONTENTS

CHAPTER 61
INITIAL VERIFICATION

610 GENERAL

610.1 Every installation shall, during erection and on completion before being put into service, be inspected and tested to verify, so far as is reasonably practicable, that the requirements of the Regulations have been met.

Precautions shall be taken to avoid danger to persons and livestock, and to avoid damage to property and installed equipment, during inspection and testing.

610.2 The result of the assessment of the fundamental principles, Section 131, the general characteristics required by Sections 311 to 313, together with the information required by Regulation 514.9.1, shall be made available to the person or persons carrying out the inspection and testing.

610.3 The verification shall include comparison of the results with relevant criteria to confirm that the requirements of the Regulations have been met.

610.4 For an addition or alteration to an existing installation, it shall be verified that the addition or alteration complies with the Regulations and does not impair the safety of the existing installation.

610.5 The verification shall be made by a competent person.

610.6 On completion of the verification, according to Regulations 610.1 to 5, a certificate shall be prepared.

611 INSPECTION

611.1 Inspection shall precede testing and shall normally be done with that part of the installation under inspection disconnected from the supply.

611.2 The inspection shall be made to verify that the installed electrical equipment is:

  (i) in compliance with the requirements of Section 511 (this may be ascertained by mark or by certification furnished by the installer or the manufacturer), and

 (ii) correctly selected and erected in accordance with the Regulations, taking into account manufacturers' instructions, and

(iii) not visibly damaged or defective so as to impair safety.

611.3 The inspection shall include at least the checking of the following items where relevant, including as appropriate all particular requirements for special installations or locations (Part 7):

  (i) Connection of conductors

 (ii) Identification of conductors

(iii) Routing of cables in safe zones, or protection against mechanical damage, in compliance with Section 522

 (iv) Selection of conductors for current-carrying capacity and voltage drop, in accordance with the design

  (v) Connection of single-pole devices for protection or switching in line conductors only

 (vi) Correct connection of accessories and equipment

(vii) Presence of fire barriers, suitable seals and protection against thermal effects

(viii) Methods of protection against electric shock

        (a) both basic protection and fault protection, i.e.:
            - SELV
            - PELV
            - Double insulation
            - Reinforced insulation

        (b) basic protection (including measurement of distances, where appropriate), i.e.:
            - protection by insulation of live parts
            - protection by a barrier or an enclosure
            - protection by obstacles
            - protection by placing out of reach

(c) fault protection:

  (i) automatic disconnection of supply

    confirmed for presence and sized in accordance with the design

- earthing conductor
- circuit protective conductors
- protective bonding conductors
- supplementary bonding conductors
- earthing arrangements for combined protective and functional purposes

    presence of adequate arrangements for alternative sources(s), where applicable

    FELV

    choice and setting of protective and monitoring devices (for fault and/or overcurrent protection)

  (ii) non-conducting location (including measurement of distances, where appropriate)

    absence of protective conductors

  (iii) earth-free local equipotential bonding

    presence of earth-free protective bonding conductors

  (iv) electrical separation

(d) additional protection

(ix) Prevention of mutual detrimental influence

(x) Presence of appropriate devices for isolation and switching correctly located

(xi) Presence of undervoltage protective devices

(xii) Labelling of protective devices, switches and terminals

(xiii) Selection of equipment and protective measures appropriate to external influences

(xiv) Adequacy of access to switchgear and equipment

(xv) Presence of danger notices and other warning signs

(xvi) Presence of diagrams, instructions and similar information

(xvii) Erection methods.

612 TESTING

612.1 General

The tests of Regulations 612.2 to 13, where relevant, shall be carried out and the results compared with relevant criteria.

Measuring instruments and monitoring equipment and methods shall be chosen in accordance with the relevant parts of BS EN 61557. If other measuring equipment is used, it shall provide no less degree of performance and safety.

When undertaking testing in a potentially explosive atmosphere, appropriate safety precautions in accordance with BS EN 60079-17 and BS EN 61241-17 are necessary.

The tests of Regulations 612.2 to 6, where relevant, shall be carried out in that order before the installation is energized. Where the installation incorporates an earth electrode, the test of Regulation 612.7 shall also be carried out before the installation is energized.

If any test indicates a failure to comply, that test and any preceding test, the results of which may have been influenced by the fault indicated, shall be repeated after the fault has been rectified.

Some methods of test are described in IET Guidance Note 3, Inspection & Testing, published by the Institution of Engineering and Technology. Other methods are not precluded provided they give valid results.

612.2　Continuity of conductors

612.2.1　Continuity of protective conductors, including main and supplementary equipotential bonding

A continuity test shall be made. It is recommended that the test be carried out with a supply having a no-load voltage between 4 V and 24 V, d.c. or a.c., and a short-circuit current of not less than 200 mA.

612.2.2　Continuity of ring final circuit conductors

A test shall be made to verify the continuity of each conductor, including the protective conductor, of every ring final circuit.

612.3　Insulation resistance

612.3.1　The insulation resistance shall be measured between live conductors and between live conductors and the protective conductor connected to the earthing arrangement. Where appropriate during this measurement, line and neutral conductors may be connected together.

612.3.2　The insulation resistance measured with the test voltages indicated in Table 61 shall be considered satisfactory if the main switchboard and each distribution circuit tested separately, with all its final circuits connected but with current-using equipment disconnected, has an insulation resistance not less than the appropriate value given in Table 61.

NOTE :　More stringent requirements are applicable for the wiring of fire detection and fire alarm systems in buildings, see BS 5839-1.

TABLE 61 – Minimum values of insulation resistance

Circuit nominal voltage (V)	Test voltage d.c. (V)	Minimum insulation resistance (MΩ)
SELV and PELV	250	0.5
Up to and including 500 V with the exception of the above systems	500	1.0
Above 500 V	1000	1.0

Table 61 shall be applied when verifying insulation resistance between non-earthed protective conductors and Earth.

Where surge protective devices (SPDs) or other equipment are likely to influence the verification test, or be damaged, such equipment shall be disconnected before carrying out the insulation resistance test. Where it is not reasonably practicable to disconnect such equipment (e.g. fixed socket-outlet incorporating an SPD), the test voltage for the particular circuit may be reduced to 250 V d.c., but the insulation resistance shall have a value of at least 1 MΩ.

NOTE 1:　In locations exposed to fire hazard, a measurement of the insulation resistance between the live conductors should be applied. In practice, it may be necessary to carry out this measurement during erection of the installation and before connection of the equipment.

NOTE 2:　Insulation resistance values are usually much higher than those of Table 61.

612.3.3　Where the circuit includes electronic devices which are likely to influence the results or be damaged, only a measurement between the live conductors connected together and the earthing arrangement shall be made.

NOTE:　Additional precautions, such as disconnection, may be necessary to avoid damage to electronic devices.

612.4　Protection by SELV, PELV or by electrical separation

It shall be verified that the separation of circuits is in accordance with Regulation 612.4.1 in the case of protection by SELV, Regulation 612.4.2 in the case of protection by PELV and Regulation 612.4.3 in the case of protection by electrical separation. The resistance values obtained in the tests of Regulations 612.4.1 to 3 shall be at least that of the circuit with the highest voltage present in accordance with Table 61.

612.4.1 Protection by SELV

The separation of the live parts from those of other circuits and from Earth, according to Section 414, shall be confirmed by a measurement of the insulation resistance. The resistance values obtained shall be in accordance with Table 61.

612.4.2 Protection by PELV

The separation of the live parts from other circuits, according to Section 414, shall be confirmed by a measurement of the insulation resistance. The resistance values obtained shall be in accordance with Table 61.

612.4.3 Protection by electrical separation

The separation of the live parts from those of other circuits and from Earth, according to Section 413, shall be confirmed by a measurement of the insulation resistance. The resistance values obtained shall be in accordance with Table 61. In the case of electrical separation with more than one item of current-using equipment, either by measurement or by calculation, it shall be verified that in case of two coincidental faults with negligible impedance between different line conductors and either the protective bonding conductor or exposed-conductive-parts connected to it, at least one of the faulty circuits shall be disconnected. The disconnection time shall be in accordance with that for the protective measure automatic disconnection of supply in a TN system.

612.4.4 Functional extra-low voltage circuits

Functional extra-low voltage circuits shall meet all the test requirements for low voltage circuits.

612.4.5 Basic protection by a barrier or an enclosure provided during erection

Where basic protection is intended to be afforded by a barrier or an enclosure provided during erection in accordance with Regulation 416.2, it shall be verified by test that each barrier or enclosure affords a degree of protection not less than IPXXB or IP2X, or IPXXD or IP4X as appropriate, where that regulation so requires.

612.5 Insulation resistance/impedance of floors and walls

612.5.1 Where it is necessary to comply with the requirements of Regulation 418.1, at least three measurements shall be made in the same location, one of these measurements being approximately 1 m from any accessible extraneous-conductive-part in the location. The other two measurements shall be made at greater distances. The measurement of resistance/impedance of insulating floors and walls is carried out with the system voltage to Earth at nominal frequency. The above series of measurements shall be repeated for each relevant surface of the location.

NOTE: Further information on measurement of the insulation resistance/impedance of floors and walls can be found in Appendix 13.

612.5.2 Any insulation or insulating arrangement of extraneous-conductive-parts intended to satisfy Regulation 418.1.4(iii):

(i) when tested at 500 V d.c. shall be not less than 1 megohm, and

(ii) shall be able to withstand a test voltage of at least 2 kV a.c. rms, and

(iii) shall not pass a leakage current exceeding 1 mA in normal conditions of use.

612.6 Polarity

A test of polarity shall be made and it shall be verified that:

(i) every fuse and single-pole control and protective device is connected in the line conductor only, and

(ii) except for E14 and E27 lampholders to BS EN 60238, in circuits having an earthed neutral conductor, centre contact bayonet and Edison screw lampholders have the outer or screwed contacts connected to the neutral conductor, and

(iii) wiring has been correctly connected to socket-outlets and similar accessories.

612.7 Earth electrode resistance

Where the earthing system incorporates an earth electrode as part of the installation, the electrode resistance to Earth shall be measured.

NOTE: Where a measurement of R_A is not practicable the measured value of external earth fault loop impedance may be used.

612.8   Protection by automatic disconnection of the supply

Where RCDs are applied also for protection against fire, the verification of the conditions for protection by automatic disconnection of the supply may be considered as satisfying the relevant requirements of Chapter 42.

612.8.1   General

The verification of the effectiveness of the measures for fault protection by automatic disconnection of supply is effected as follows:

a) TN system

Compliance with Regulation 411.4 shall be verified by:

1) measurement of the earth fault loop impedance (see Regulation 612.9)

2) verification of the characteristics and/or the effectiveness of the associated protective device. This verification shall be made:

   –   for overcurrent protective devices, by visual inspection (i.e. short-time or instantaneous tripping setting for circuit-breakers, current rating and type for fuses)

   –   for RCDs, by visual inspection and test.
       The effectiveness of automatic disconnection of supply by RCDs shall be verified using suitable test equipment according to BS EN 61557-6 (see Regulation 612.1) to confirm that the relevant requirements in Chapter 41 are met.

   The disconnection times required by Chapter 41 shall be verified.

b) TT system

Compliance with Regulation 411.5 shall be verified by:

1) measurement of the resistance of the earth electrode for exposed-conductive-parts of the installation (see Regulation 612.7)

   NOTE: Where a measurement of R_A is not practicable the measured value of external earth fault loop impedance may be used.

2) verification of the characteristics and/or effectiveness of the associated protective device. This verification shall be made:

   –   for overcurrent protective devices, by visual inspection (i.e. short-time or instantaneous tripping setting for circuit-breakers, current rating and type for fuses)

   –   for RCDs, by visual inspection and test.
       The effectiveness of disconnection of supply by RCDs shall be verified using suitable test equipment according to BS EN 61557-6 (see Regulation 612.1) to confirm that the relevant requirements in Chapter 41 are met.

   The disconnection times required by Chapter 41 shall be verified.

c) IT system

Compliance with Regulation 411.6 shall be verified by calculation or measurement of the current I_d in case of a first fault at the line conductor or at the neutral.

NOTE 1: The measurement is made only if calculation is not possible because all the parameters are not known. Precautions are to be taken while making the measurement in order to avoid the danger due to a double fault.

Where conditions that are similar to conditions of a TT system occur, in the event of a second fault in another circuit (see Regulation 411.6.4(ii)), verification is made as for a TT system (see point **b)** of this regulation).

Where conditions that are similar to conditions of a TN system occur, in the event of a second fault in another circuit (see Regulation 411.6.4(i)), verification is made as for a TN system (see point **a)** of this regulation).

NOTE 2: During measurement of the fault loop impedance, it is necessary to establish a connection of negligible impedance between the neutral point of the system and the protective conductor preferably at the origin of the installation or, where this is not acceptable, at the point of measurement.

612.9   Earth fault loop impedance

Where protective measures are used which require a knowledge of earth fault loop impedance, the relevant impedances shall be measured, or determined by an alternative method.

NOTE: Further information on measurement of earth fault loop impedance can be found in Appendix 14.

612.10   Additional protection

The verification of the effectiveness of the measures applied for additional protection is fulfilled by visual inspection and test. Where RCDs are required for additional protection, the effectiveness of automatic disconnection of supply by RCDs shall be verified using suitable test equipment according to BS EN 61557-6 (see Regulation 612.1) to confirm that the relevant requirements in Chapter 41 are met.

612.11   Prospective fault current

The prospective short-circuit current and prospective earth fault current shall be measured, calculated or determined by another method, at the origin and at other relevant points in the installation.

612.12   Check of phase sequence

For multiphase circuits, it shall be verified that the phase sequence is maintained.

612.13   Functional testing

612.13.1   Where fault protection and/or additional protection is to be provided by an RCD, the effectiveness of any test facility incorporated in the device shall be verified.

612.13.2   Equipment, such as switchgear and controlgear assemblies, drives, controls and interlocks, shall be subjected to a functional test to show that it is properly mounted, adjusted and installed in accordance with the relevant requirements of these Regulations.

612.14   Verification of voltage drop

Where required to verify compliance with Section 525, the following options may be used:

(i)   The voltage drop may be evaluated by measuring the circuit impedance

(ii)   The voltage drop may be evaluated by using calculations, for example, by diagrams or graphs showing maximum cable length v load current for different conductor cross-sectional areas with different percentage voltage drops for specific nominal voltages, conductor temperatures and wiring systems.

NOTE: Verification of voltage drop is not normally required during initial verification.

CHAPTER 62
PERIODIC INSPECTION AND TESTING

621 GENERAL

621.1 Where required, periodic inspection and testing of every electrical installation shall be carried out in accordance with Regulations 621.2 to 5 in order to determine, so far as is reasonably practicable, whether the installation is in a satisfactory condition for continued service. Wherever possible, the documentation arising from the initial certification and any previous periodic inspection and testing shall be taken into account. Where no previous documentation is available, investigation of the electrical installation shall be undertaken prior to carrying out the periodic inspection and testing.

621.2 Periodic inspection comprising a detailed examination of the installation shall be carried out without dismantling, or with partial dismantling as required, supplemented by appropriate tests from Chapter 61 to show that the requirements for disconnection times, as set out in Chapter 41 for protective devices, are complied with, to provide for:

(i) safety of persons and livestock against the effects of electric shock and burns

(ii) protection against damage to property by fire and heat arising from an installation defect

(iii) confirmation that the installation is not damaged or deteriorated so as to impair safety

(iv) the identification of installation defects and departures from the requirements of these Regulations that may give rise to danger.

NOTE: A generic list of examples of items requiring inspection is given in Appendix 6.

621.3 Precautions shall be taken to ensure that the periodic inspection and testing shall not cause danger to persons or livestock and shall not cause damage to property and equipment even if the circuit is defective. Measuring instruments and monitoring equipment and methods shall be chosen in accordance with relevant parts of BS EN 61557. If other measuring equipment is used, it shall provide no less degree of performance and safety.

621.4 The extent and results of the periodic inspection and testing of an installation, or any part of an installation, shall be recorded.

621.5 Periodic inspection and testing shall be undertaken by a competent person.

622 FREQUENCY OF INSPECTION AND TESTING

622.1 The frequency of periodic inspection and testing of an installation shall be determined having regard to the type of installation and equipment, its use and operation, the frequency and quality of maintenance and the external influences to which it is subjected. The results and recommendations of the previous report, if any, shall be taken into account.

622.2 In the case of an installation under an effective management system for preventive maintenance in normal use, periodic inspection and testing may be replaced by an adequate regime of continuous monitoring and maintenance of the installation and all its constituent equipment by skilled persons, competent in such work. Appropriate records shall be kept.

CHAPTER 63
CERTIFICATION AND REPORTING

631 GENERAL

631.1 Upon completion of the verification of a new installation or changes to an existing installation, an Electrical Installation Certificate, based on the model given in Appendix 6, shall be provided. Such documentation shall include details of the extent of the installation covered by the Certificate, together with a record of the inspection, the results of testing and a recommendation for the interval until the first periodic inspection.

631.2 Upon completion of the periodic inspection and testing of an existing installation, an Electrical Installation Condition Report, based on the model given in Appendix 6, shall be provided. Such documentation shall include details of the extent of the installation and limitations of the inspection and testing covered by the Report, together with records of inspection, the results of testing and a recommendation for the interval until the next periodic inspection.

631.3   Where minor electrical installation work does not include the provision of a new circuit, a Minor Electrical Installation Works Certificate, based on the model given in Appendix 6, may be provided for each circuit altered or extended as an alternative to an Electrical Installation Certificate.

631.4   Electrical Installation Certificates, Electrical Installation Condition Reports and Minor Electrical Installation Works Certificates shall be compiled and signed or otherwise authenticated by a competent person or persons.

631.5   Electrical Installation Certificates, Electrical Installation Condition Reports and Minor Electrical Installation Works Certificates may be produced in any durable medium, including written and electronic media. Regardless of the media used for original certificates, reports or their copies, their authenticity and integrity shall be verified by a reliable process or method. The process or method shall also verify that any copy is a true copy of the original.

632   INITIAL VERIFICATION

632.1   Following the initial verification required by Chapter 61, an Electrical Installation Certificate, together with schedules of inspection and schedules of test results, shall be given to the person ordering the work. These schedules shall be based on the models given in Appendix 6.

632.2   The schedule of test results shall identify every circuit, including its related protective device(s), and shall record the results of the appropriate tests and measurements detailed in Chapter 61.

632.3   The person or persons responsible for the design, construction, inspection and testing of the installation shall, as appropriate, give to the person ordering the work a Certificate which takes account of their respective responsibilities for the safety of that installation, together with the schedules described in Regulation 632.1.

632.4   Defects or omissions revealed during inspection and testing of the installation work covered by the Certificate shall be made good before the Certificate is issued.

633   ADDITIONS AND ALTERATIONS

633.1   The requirements of Sections 631 and 632 for the issue of an Electrical Installation Certificate or a Minor Electrical Installation Works Certificate shall apply to all the work of the additions or alterations.

633.2   The contractor or other person responsible for the new work, or a person authorized to act on their behalf, shall record on the Electrical Installation Certificate or the Minor Electrical Installation Works Certificate, any defects found, so far as is reasonably practicable, in the existing installation.

634   PERIODIC INSPECTION AND TESTING

634.1   Following the periodic inspection and testing described in Chapter 62, an Electrical Installation Condition Report, together with schedules of inspection and schedules of test results, shall be given by the person carrying out the inspection, or a person authorized to act on their behalf, to the person ordering the inspection. These schedules shall be based on the models given in Appendix 6. The schedule of test results shall record the results of the appropriate tests required of Chapter 61.

634.2   Any damage, deterioration, defects, dangerous conditions and non-compliance with the requirements of the Regulations, which may give rise to danger, together with any significant limitations of the inspection and testing, including their reasons, shall be recorded.

PART 7

SPECIAL INSTALLATIONS OR LOCATIONS

PARTICULAR REQUIREMENTS

CONTENTS

SECTION 700
GENERAL

700   GENERAL

The particular requirements for each section (special installation or location) in Part 7 supplement or modify the general requirements contained in other parts of the Regulations.

The absence of reference to the exclusion of a part, a chapter, a section or a regulation means that the corresponding general regulations are applicable.

The number appearing after a section number generally refers to the corresponding chapter, section or regulation within Parts 1 to 6. The numbering does not, therefore, necessarily follow sequentially and new numbers have been added as required. Numbering of figures and tables takes the number of the section followed by a sequential number.

SECTION 701

LOCATIONS CONTAINING A BATH OR SHOWER

701.1   Scope

The particular requirements of this section apply to the electrical installations in locations containing a fixed bath (bath tub, birthing pool) or shower and to the surrounding zones as described in these regulations.

These regulations do not apply to emergency facilities, e.g. emergency showers used in industrial areas or laboratories.

701.3   Assessment of general characteristics

701.32   Classification of external influences

701.32.1   General

When applying this section, the zones specified in Regulations 701.32.2 to 4 shall be taken into account. For fixed prefabricated bath or shower units, the zones are applied to the situation when the bath or shower basin is in its usable configuration(s).

Horizontal or inclined ceilings, walls with or without windows, doors, floors and fixed partitions may be taken into account where these effectively limit the extent of locations containing a bath or shower as well as their zones. Where the dimensions of fixed partitions are smaller than the dimensions of the relevant zones, e.g. partitions having a height lower than 2.25 m, the minimum distance in the horizontal and vertical directions shall be taken into account (see Figures 701.1 and 701.2).

For electrical equipment in parts of walls or ceilings limiting the zones specified in Regulations 701.32.2 to 4, but being part of the surface of that wall or ceiling, the requirements for the respective zone apply.

701.32.2   Description of zone 0

Zone 0 is the interior of the bath tub or shower basin (see Figures 701.1 and 2).

For showers without a basin, the height of zone 0 is 0.10 m and its surface extent has the same horizontal extent as zone 1 (see Figure 701.2).

701.32.3   Description of zone 1

Zone 1 is limited by:

(i)   the finished floor level and the horizontal plane corresponding to the highest fixed shower head or water outlet or the horizontal plane lying 2.25 m above the finished floor level, whichever is higher

(ii)   the vertical surface:

a) circumscribing the bath tub or shower basin (see Figure 701.1)

b) at a distance of 1.20 m from the centre point of the fixed water outlet on the wall or ceiling for showers without a basin (see Figure 701.1(e) and (f)).

Zone 1 does not include zone 0.

The space under the bath tub or shower basin is considered to be zone 1. However, if the space under the bath tub or shower basin is only accessible with a tool, it is considered to be outside the zones.

701.32.4   Description of zone 2

Zone 2 is limited by:

(i)   the finished floor level and the horizontal plane corresponding to the highest fixed shower head or water outlet or the horizontal plane lying 2.25 m above the finished floor level, whichever is higher

(ii)   the vertical surface at the boundary of zone 1 and the parallel vertical surface at a distance of 0.60 m from the zone 1 border (see Figure 701.1).

For showers without a basin, there is no zone 2 but an increased zone 1 is provided by the horizontal dimension of 1.20 m mentioned in Regulation 701.32.3(ii)b) (see Figure 701.1(e) and (f)).

701.41   Protection for safety: protection against electric shock

701.410.3   General requirements

701.410.3.5 The protective measures of obstacles and placing out of reach (Section 417) are not permitted.

701.410.3.6 The protective measures of non-conducting location (Regulation 418.1) and earth-free local equipotential bonding (Regulation 418.2) are not permitted.

701.411.3.3 Additional protection by RCDs

Additional protection shall be provided for all low voltage circuits of the location, by the use of one or more RCDs having the characteristics specified in Regulation 415.1.1.

NOTE: See also Regulations 314.1(iv) and 531.2.4 concerning the avoidance of unwanted tripping.

701.413   Protective measure: Electrical separation

Protection by electrical separation shall only be used for:

   (i)   circuits supplying one item of current-using equipment, or

   (ii)   one single socket-outlet.

For electric floor heating systems, see Regulation 701.753.

701.414   Protective measure: Extra-low voltage provided by SELV or PELV

701.414.4.5 Requirements for SELV and PELV circuits

Where SELV or PELV is used, whatever the nominal voltage, basic protection for equipment in zones 0, 1 and 2 shall be provided by:

   (i)   basic insulation complying with Regulation 416.1, or

   (ii)   barriers or enclosures complying with Regulation 416.2.

701.415   Additional protection

701.415.2   Supplementary equipotential bonding

Local supplementary equipotential bonding according to Regulation 415.2 shall be established connecting together the terminals of the protective conductor of each circuit supplying Class I and Class II equipment to the accessible extraneous-conductive-parts, within a room containing a bath or shower, including the following:

   (i)   metallic pipes supplying services and metallic waste pipes (e.g. water, gas)

   (ii)   metallic central heating pipes and air conditioning systems

   (iii) accessible metallic structural parts of the building (metallic door architraves, window frames and similar parts are not considered to be extraneous-conductive-parts unless they are connected to metallic structural parts of the building).

Supplementary equipotential bonding may be installed outside or inside rooms containing a bath or shower, preferably close to the point of entry of extraneous-conductive-parts into such rooms.

Where the location containing a bath or shower is in a building with a protective equipotential bonding system in accordance with Regulation 411.3.1.2, supplementary equipotential bonding may be omitted where all of the following conditions are met:

   (iv)   All final circuits of the location comply with the requirements for automatic disconnection according to Regulation 411.3.2

   (v)   All final circuits of the location have additional protection by means of an RCD in accordance with Regulation 701.411.3.3

   (vi)   All extraneous-conductive-parts of the location are effectively connected to the protective equipotential bonding according to Regulation 411.3.1.2.

NOTE: The effectiveness of the connection of extraneous-conductive-parts in the location to the main earthing terminal may be assessed, where necessary, by the application of Regulation 415.2.2.

701.5   Selection and erection of equipment

701.512.2   External influences

Installed electrical equipment shall have at least the following degrees of protection:

(i)   In zone 0: IPX7

(ii)   In zones 1 and 2: IPX4.

This requirement does not apply to shaver supply units complying with BS EN 61558-2-5 installed in zone 2 and located where direct spray from showers is unlikely.

Electrical equipment exposed to water jets, e.g. for cleaning purposes, shall have a degree of protection of at least IPX5.

701.512.3   Erection of switchgear, controlgear and accessories according to external influences

The following requirements do not apply to switches and controls which are incorporated in fixed current-using equipment suitable for use in that zone or to insulating pull cords of cord operated switches.

In zone 0:

   switchgear or accessories shall not be installed.

In zone 1:

   only switches of SELV circuits supplied at a nominal voltage not exceeding 12 V a.c. rms or 30 V ripple-free d.c. shall be installed, the safety source being installed outside zones 0, 1 and 2.

In zone 2:

   switchgear, accessories incorporating switches or socket-outlets shall not be installed with the exception of:

   (i) switches and socket-outlets of SELV circuits, the safety source being installed outside zones 0, 1 and 2, and

   (ii) shaver supply units complying with BS EN 61558-2-5.

Except for SELV socket-outlets complying with Section 414 and shaver supply units complying with BS EN 61558-2-5, socket-outlets are prohibited within a distance of 3 m horizontally from the boundary of zone 1.

701.55   Current-using equipment

In zone 0, current-using equipment shall only be installed provided that all the following requirements are met:

(i)   The equipment complies with the relevant standard and is suitable for use in that zone according to the manufacturer's instructions for use and mounting

(ii)   The equipment is fixed and permanently connected

(iii) The equipment is protected by SELV at a nominal voltage not exceeding 12 V a.c. rms or 30 V ripple-free d.c., the safety source being installed outside zones 0, 1 and 2.

In zone 1, only the following fixed and permanently connected current-using equipment shall be installed, provided it is suitable for installation in zone 1 according to the manufacturer's instructions:

(iv)   Whirlpool units

(v)   Electric showers

(vi)   Shower pumps

(vii) Equipment protected by SELV or PELV at a nominal voltage not exceeding 25 V a.c. rms or 60 V ripple-free d.c., the safety source being installed outside zones 0, 1 and 2

(viii) Ventilation equipment

(ix)   Towel rails

(x)   Water heating appliances

(xi)   Luminaires.

701.753 Electric floor heating systems

For electric floor heating systems, only heating cables according to relevant product standards or thin sheet flexible heating elements according to the relevant equipment standard shall be erected provided that they have either a metal sheath or a metal enclosure or a fine mesh metallic grid. The fine mesh metallic grid, metal sheath or metal enclosure shall be connected to the protective conductor of the supply circuit. Compliance with the latter requirement is not required if the protective measure SELV is provided for the floor heating system.

For electric floor heating systems the protective measure "protection by electrical separation" is not permitted.

Fig 701.1 – Examples of zone dimensions (plan)
NOT TO SCALE (See Regulation 701.32 for definitions of zones)

a) Bath tub

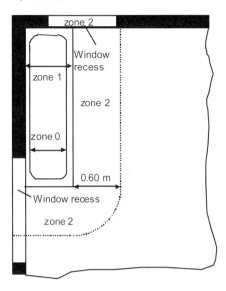

b) Bath tub, with permanent fixed partition

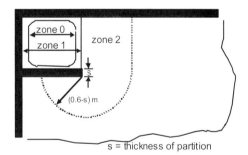

c) Shower basin

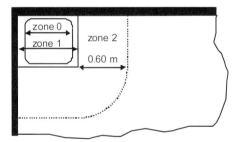

d) Shower basin with permanent fixed partition

e) Shower, without basin

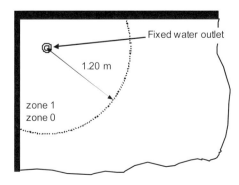

f) Shower, without basin, but with permanent fixed partition

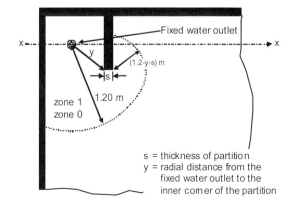

Fig 701.2 – Examples of zone dimensions (elevation)
NOT TO SCALE (See Regulation 701.32 for definitions of zones)

a) Bath tub

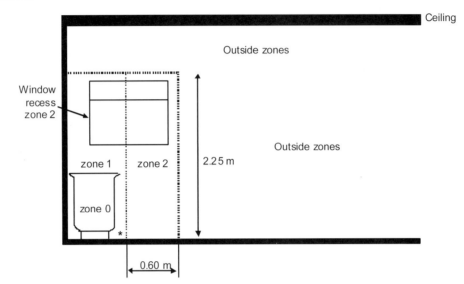

c) Shower basin

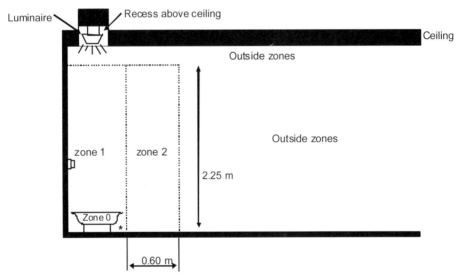

f) Shower without basin, but with permanent fixed partition

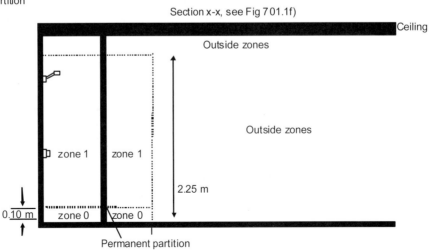

* Zone 1 if the space is accessible without the use of a tool.
Spaces under the bath accessible only with the use of a tool are outside the zones.

SECTION 702

SWIMMING POOLS AND OTHER BASINS

702.1 Scope, object and fundamental principles

702.11 Scope

The particular requirements of this section apply to the basins of swimming pools, the basins of fountains and the basins of paddling pools. The particular requirements also apply to the surrounding zones of these basins.

In these areas, in normal use, the risk of electric shock is increased by a reduction in body resistance and contact of the body with Earth potential. Swimming pools within the scope of an equipment standard are outside the scope of these regulations.

Except for areas especially designed as swimming pools, the requirements of this section do not apply to natural waters, lakes in gravel pits, coastal areas and the like.

702.3 Assessment of general characteristics

702.32 Classification of external influences

These requirements are based on the dimensions of three zones (examples are given in Figures 702.1 to 4).

Zones 1 and 2 may be limited by fixed partitions having a minimum height of 2.5 m.

(i) Zone 0
This zone is the interior of the basin of the swimming pool or fountain including any recesses in its walls or floors, basins for foot cleaning and waterjets or waterfalls and the space below them.

(ii) Zone 1
This zone is limited by:
- zone 0
- a vertical plane 2 m from the rim of the basin
- the floor or surface expected to be occupied by persons
- the horizontal plane 2.5 m above the floor or the surface expected to be occupied by persons.

Where the swimming pool or fountain contains diving boards, springboards, starting blocks, chutes or other components expected to be occupied by persons, zone 1 comprises the zone limited by:
- a vertical plane situated 1.5 m from the periphery of the diving boards, springboards, starting blocks, chutes and other components such as accessible sculptures, viewing bays and decorative basins
- the horizontal plane 2.5 m above the highest surface expected to be occupied by persons.

(iii) Zone 2
This zone is limited by:
- the vertical plane external to zone 1 and a parallel plane 1.5 m from the former
- the floor or surface expected to be occupied by persons
- the horizontal plane 2.5 m above the floor or surface expected to be occupied by persons.

There is no zone 2 for fountains.

702.4 Protection for safety

702.410.3 General requirements

702.410.3.4 Application of protective measures against electric shock

702.410.3.4.1 Zones 0 and 1

Except for fountains as stated in Regulation 702.410.3.4.2, in zone 0 only protection by SELV at a nominal voltage not exceeding 12 V a.c. rms or 30 V ripple-free d.c. is permitted, the source for SELV being installed outside zones 0, 1 and 2.

Except for fountains as stated in Regulation 702.410.3.4.2, in zone 1 only protection by SELV at a nominal voltage not exceeding 25 V a.c. rms or 60 V ripple-free d.c. is permitted, the source for SELV being installed outside zones 0, 1 and 2.

Equipment for use in the interior of basins which is only intended to be in operation when people are not inside zone 0 shall be supplied by a circuit protected by:

(i) SELV (Section 414), the source for SELV being installed outside zones 0, 1 and 2. However, it is permitted to install the source for SELV in zone 2 if its supply circuit is protected by an RCD having the characteristics specified in Regulation 415.1.1, or

(ii) automatic disconnection of the supply (Section 411), using an RCD having the characteristics specified in Regulation 415.1.1, or

(iii) electrical separation (Section 413), the source for electrical separation supplying only one item of current-using equipment and being installed outside zones 0, 1 and 2. However, it is permitted to install the source in zone 2 if its supply circuit is protected by an RCD having the characteristics specified in Regulation 415.1.1.

The socket-outlet of a circuit supplying such equipment and the control device of such equipment shall have a notice in order to warn the user that this equipment shall be used only when the swimming pool is not occupied by persons.

702.410.3.4.2   Zones 0 and 1 of fountains

In zones 0 and 1, one or more of the following protective measures shall be employed:

(i) SELV (Section 414), the source for SELV being installed outside zones 0 and 1

(ii) Automatic disconnection of supply (Section 411), using an RCD having the characteristics specified in Regulation 415.1.1

(iii) Electrical separation (Section 413), the source for electrical separation supplying only one item of current-using equipment and being installed outside zones 0 and 1.

702.410.3.4.3   Zone 2

One or more of the following protective measures shall be employed:

(i) SELV (Section 414), the source for SELV being installed outside zones 0, 1 and 2. However, it is permitted to install the source for SELV in zone 2 if its supply circuit is protected by an RCD having the characteristics specified in Regulation 415.1.1

(ii) Automatic disconnection of supply (Section 411), using an RCD having the characteristics specified in Regulation 415.1.1

> **NOTE:** Where a PME earthing facility is used as the means of earthing for the electrical installation of a swimming pool or other basin, it is recommended that an earth mat or earth electrode of suitably low resistance, e.g. 20 ohms or less, be installed and connected to the protective equipotential bonding.

(iii) Electrical separation (Section 413), the source for electrical separation supplying only one item of current-using equipment and being installed outside zones 0, 1 and 2. However, it is permitted to install the source in zone 2 if its supply circuit is protected by an RCD having the characteristics specified in Regulation 415.1.1.

There is no zone 2 for fountains.

702.410.3.5 The protective measures of obstacles and placing out of reach (Section 417) are not permitted.

702.410.3.6 The protective measures of non-conducting location (Regulation 418.1) and earth-free local equipotential bonding (Regulation 418.2) are not permitted.

702.414   Protective measure: Extra-low voltage provided by SELV or PELV

702.414.4   Requirements for SELV and PELV circuits

702.414.4.5 Where SELV is used, whatever the nominal voltage, basic protection shall be provided by:

(i) basic insulation complying with Regulation 416.1, or

(ii) barriers or enclosures complying with Regulation 416.2.

702.415 Additional protection

702.415.2 Additional protection: Supplementary equipotential bonding

All extraneous-conductive-parts in zones 0, 1 and 2 shall be connected by supplementary protective bonding conductors to the protective conductors of exposed-conductive-parts of equipment situated in these zones, in accordance with Regulation 415.2.

NOTE: The connection with the protective conductor may be provided in the proximity of the location, e.g. in an accessory or in a local distribution board.

702.5 Selection and erection of equipment

702.51 Common rules

702.512 Operational conditions and external influences

702.512.2 External influences

Electrical equipment shall have at least the following degree of protection according to BS EN 60529:

(i) zone 0: IPX8

(ii) zone 1: IPX4, IPX5 where water jets are likely to occur for cleaning purposes

(iii) zone 2: IPX2 for indoor locations, IPX4 for outdoor locations, IPX5 where water jets are likely to occur for cleaning purposes.

702.52 Wiring systems

702.520 General

The following regulations apply to surface wiring systems and to wiring systems embedded in the walls, ceilings or floors at a depth not exceeding 50 mm.

702.522 Selection and erection in relation to external influences

702.522.21 Erection according to the zones

In zones 0, 1 and 2, any metallic sheath or metallic covering of a wiring system shall be connected to the supplementary equipotential bonding.

NOTE: Cables should preferably be installed in conduits made of insulating material.

702.522.22 Limitation of wiring systems according to the zones

In zones 0 and 1, a wiring system shall be limited to that necessary to supply equipment situated in these zones.

702.522.23 Additional requirements for the wiring of fountains

For a fountain, the following additional requirements shall be met:

(i) A cable for electrical equipment in zone 0 shall be installed as far outside the basin rim as is reasonably practicable and run to the electrical equipment inside zone 0 by the shortest practicable route

(ii) In zone 1, a cable shall be selected, installed and provided with mechanical protection to medium severity (AG2) and the relevant submersion in water depth (AD8). The cable type H07RN8-F (BS 7919) is suitable up to a depth of 10 m of water. For depths of water greater than 10 m the cable manufacturer shall be consulted.

702.522.24 Junction boxes

A junction box shall not be installed in zones 0 or 1, but in the case of SELV circuits it is permitted to install junction boxes in zone 1.

702.53 Switchgear and controlgear

In zones 0 or 1, switchgear or controlgear shall not be installed.

In zones 0 or 1, a socket-outlet shall not be installed.

In zone 2, a socket-outlet or a switch is permitted only if the supply circuit is protected by one of the following protective measures:

(i) SELV (Section 414), the source of SELV being installed outside zones 0, 1 and 2. However, it is permitted to install the source of SELV in zone 2 if its supply circuit is protected by an RCD having the characteristics specified in Regulation 415.1.1

(ii) Automatic disconnection of supply (Section 411), using an RCD having the characteristics specified in Regulation 415.1.1

(iii) Electrical separation (Section 413), the source for electrical separation supplying only one item of current-using equipment, or one socket-outlet, and being installed outside zones 0, 1 and 2. However, it is permitted to install the source in zone 2 if its supply circuit is protected by an RCD having the characteristics specified in Regulation 415.1.1.

For a swimming pool where it is not possible to locate a socket-outlet or switch outside zone 1, a socket-outlet or switch, preferably having a non-conductive cover or coverplate, is permitted in zone 1 if it is installed at least 1.25 m horizontally from the border of zone 0, is placed at least 0.3 m above the floor, and is protected by:

(iv) SELV (Section 414), at a nominal voltage not exceeding 25 V a.c. rms or 60 V ripple-free d.c., the source for SELV being installed outside zones 0 and 1, or

(v) automatic disconnection of supply (Section 411), using an RCD having the characteristics specified in Regulation 415.1.1, or

(vi) electrical separation (Section 413) for a supply to only one item of current-using equipment, the source for electrical separation being installed outside zones 0 and 1.

702.55 Other equipment

702.55.1 Current-using equipment of swimming pools

In zones 0 and 1, it is only permitted to install fixed current-using equipment specifically designed for use in a swimming pool, in accordance with the requirements of Regulations 702.55.2 and 702.55.4.

Equipment which is intended to be in operation only when people are outside zone 0 may be used in all zones provided that it is supplied by a circuit protected according to Regulation 702.410.3.4.

It is permitted to install an electric heating unit embedded in the floor, provided that it:

(i) is protected by SELV (Section 414), the source of SELV being installed outside zones 0, 1 and 2. However, it is permitted to install the source of SELV in zone 2 if its supply circuit is protected by an RCD having the characteristics specified in Regulation 415.1.1, or

(ii) incorporates an earthed metallic sheath connected to the supplementary equipotential bonding specified in Regulation 702.415.2 and its supply circuit is additionally protected by an RCD having the characteristics specified in Regulation 415.1.1, or

(iii) is covered by an embedded earthed metallic grid connected to the supplementary equipotential bonding specified in Regulation 702.415.2 and its supply circuit is additionally protected by an RCD having the characteristics specified in Regulation 415.1.1.

702.55.2 Underwater luminaires for swimming pools

A luminaire for use in the water or in contact with the water shall be fixed and shall comply with BS EN 60598-2-18.

Underwater lighting located behind watertight portholes, and serviced from behind, shall comply with the appropriate part of BS EN 60598 and be installed in such a way that no intentional or unintentional conductive connection between any exposed-conductive-part of the underwater luminaires and any conductive parts of the portholes can occur.

702.55.3 Electrical equipment of fountains

Electrical equipment in zones 0 or 1 shall be provided with mechanical protection to medium severity (AG2), e.g. by use of mesh glass or by grids which can only be removed by the use of a tool.

A luminaire installed in zones 0 or 1 shall be fixed and shall comply with BS EN 60598-2-18.

An electric pump shall comply with the requirements of BS EN 60335-2-41.

702.55.4 Special requirements for the installation of electrical equipment in zone 1 of swimming pools and other basins

Fixed equipment designed for use in swimming pools and other basins (e.g. filtration systems, jet stream pumps) and supplied at low voltage is permitted in zone 1, subject to all the following requirements being met:

(i) The equipment shall be located inside an insulating enclosure providing at least Class II or equivalent insulation and providing protection against mechanical impact of medium severity (AG2)

This regulation applies irrespective of the classification of the equipment.

(ii) The equipment shall only be accessible via a hatch (or a door) by means of a key or a tool. The opening of the hatch (or door) shall disconnect all live conductors. The supply cable and the main disconnecting means shall be installed in a way which provides protection of Class II or equivalent insulation

(iii) The supply circuit of the equipment shall be protected by:

– SELV at a nominal voltage not exceeding 25 V a.c. rms or 60 V ripple-free d.c., the source of SELV being installed outside zones 0, 1 and 2, or

– an RCD having the characteristics specified in Regulation 415.1.1, or

– electrical separation (Section 413), the source for electrical separation supplying a single fixed item of current-using equipment and being installed outside zones 0, 1 and 2.

For swimming pools where there is no zone 2, lighting equipment supplied by other than a SELV source at 12 V a.c. rms or 30 V ripple-free d.c. may be installed in zone 1 on a wall or on a ceiling, provided that the following requirements are fulfilled:

- The circuit is protected by automatic disconnection of the supply and additional protection is provided by an RCD having the characteristics specified in Regulation 415.1.1

- The height from the floor is at least 2 m above the lower limit of zone 1.

In addition, every luminaire shall have an enclosure providing Class II or equivalent insulation and providing protection against mechanical impact of medium severity.

Fig 702.1 – zone dimensions for swimming pools and paddling pools

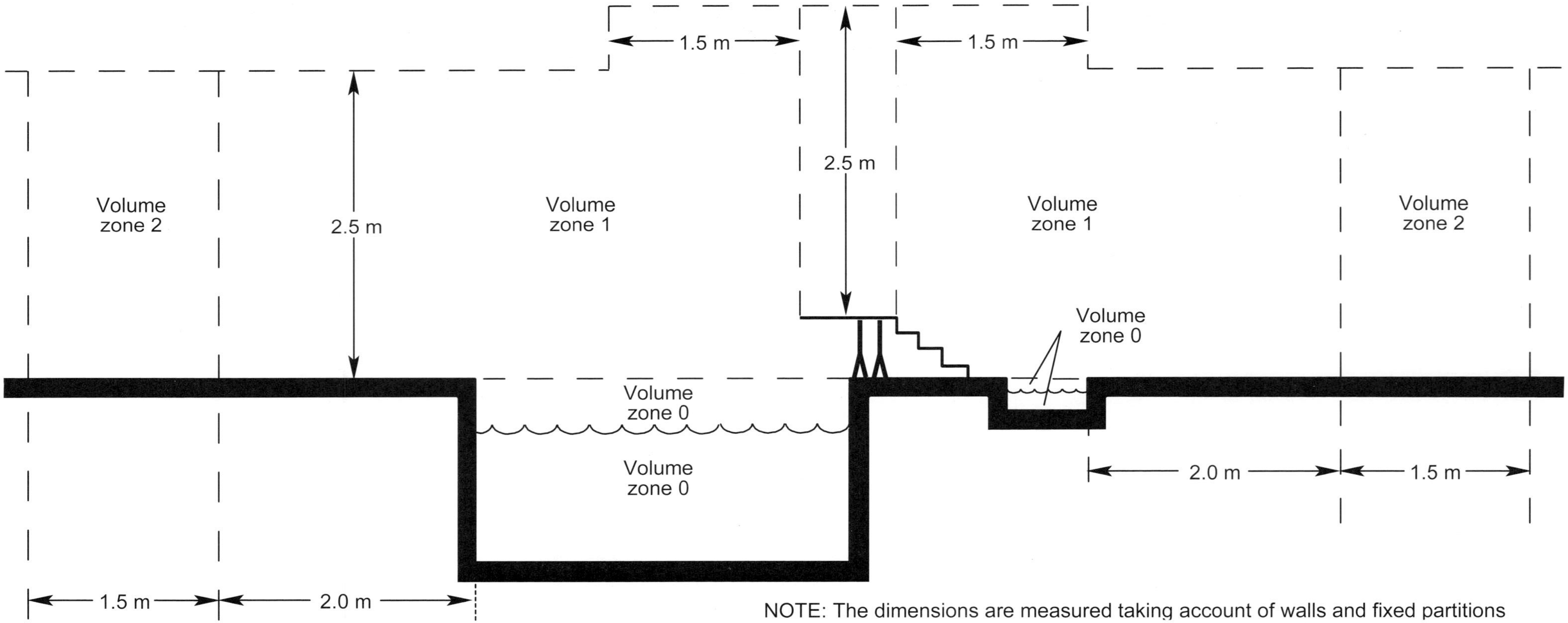

Volume zone 2

Volume zone 1

Volume zone 0

Volume zone 1

Volume zone 0

Volume zone 0

Volume zone 2

1.5 m

2.0 m

1.5 m

2.5 m

1.5 m

2.5 m

2.0 m

1.5 m

NOTE: The dimensions are measured taking account of walls and fixed partitions

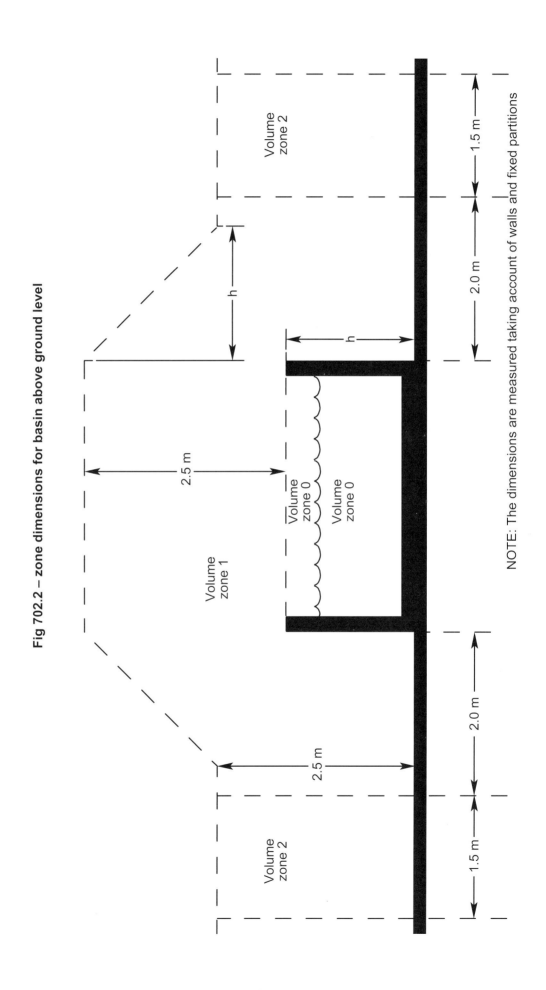

Fig 702.2 – zone dimensions for basin above ground level

Volume zone 2

Volume zone 2

Volume zone 1

Volume zone 0

Volume zone 0

2.5 m

2.5 m

h

h

1.5 m

2.0 m

2.0 m

1.5 m

NOTE: The dimensions are measured taking account of walls and fixed partitions

Fig 702.3 – Example of zone dimensions (plan) with fixed partitions of height at least 2.5 m

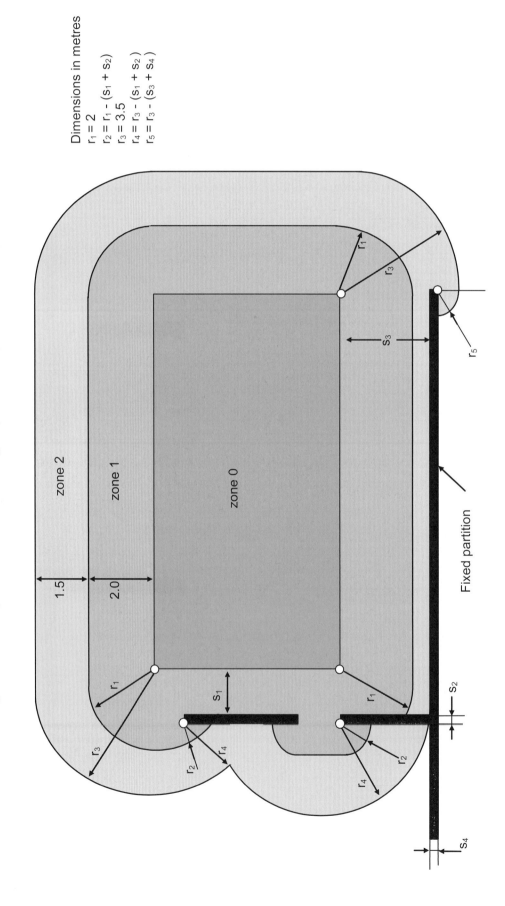

Dimensions in metres
$r_1 = 2$
$r_2 = r_1 - (s_1 + s_2)$
$r_3 = 3.5$
$r_4 = r_3 - (s_1 + s_2)$
$r_5 = r_3 - (s_3 + s_4)$

Fig 702.4 – Example of determination of the zones of a fountain

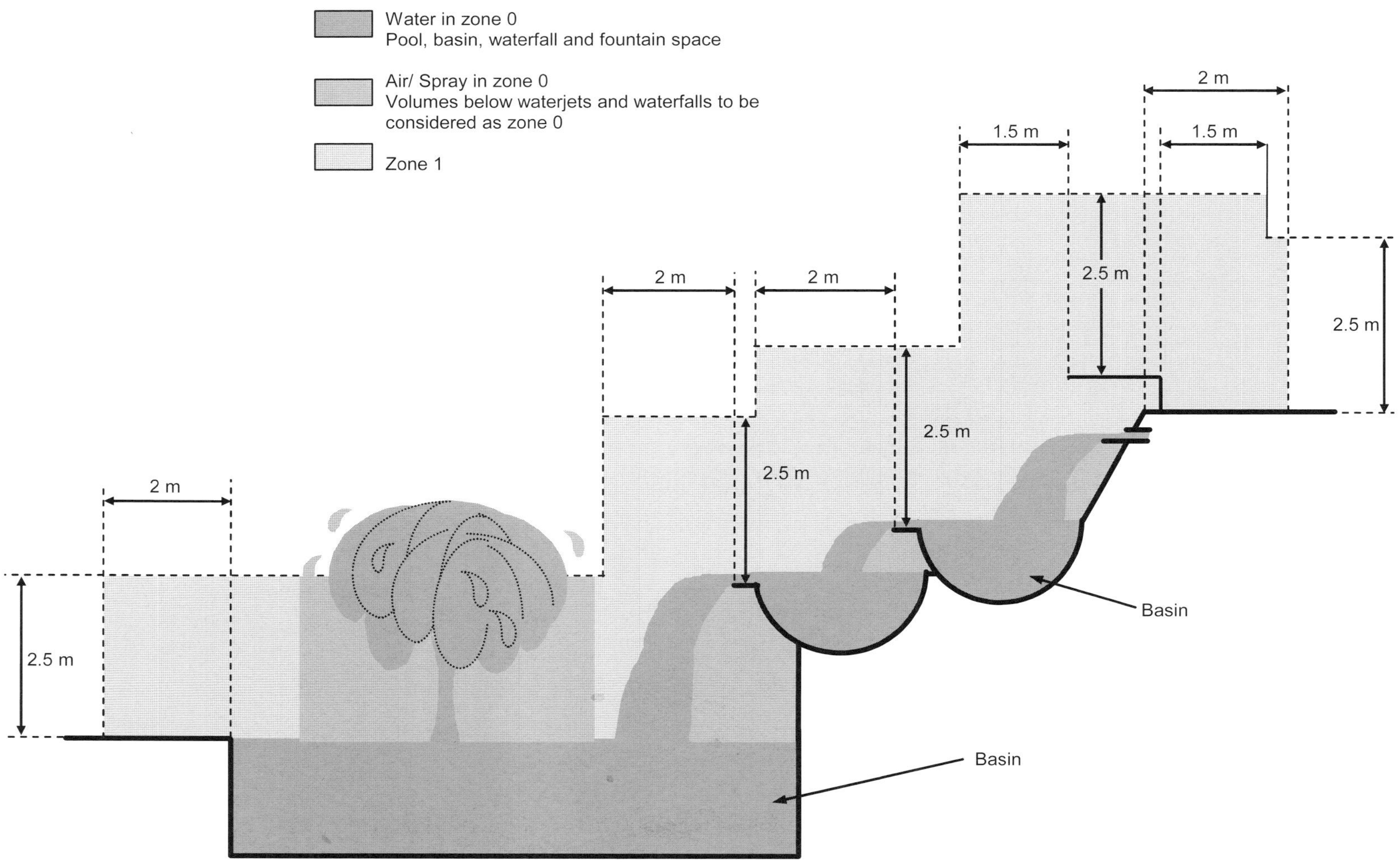

Water in zone 0
Pool, basin, waterfall and fountain space

Air/ Spray in zone 0
Volumes below waterjets and waterfalls to be considered as zone 0

Zone 1

Basin

Basin

2.5 m

2 m

1.5 m

2.5 m

1.5 m

2.5 m

2 m

2.5 m

2 m

2 m

2.5 m

SECTION 703

ROOMS AND CABINS CONTAINING SAUNA HEATERS

703.1 Scope

The particular requirements of this section apply to:

(i) sauna cabins erected on site, e.g. in a location or in a room

(ii) the room where the sauna heater is, or the sauna heating appliances are installed. In this case the whole room is considered as the sauna.

The requirements of this section do not apply to prefabricated sauna cabins complying with a relevant equipment standard.

Where facilities such as showers, etc. are installed, the requirements of Section 701 also apply.

703.3 Assessment of general characteristics

703.32 General

When applying these regulations, the zones specified in Regulations 703.32.1 to 3 shall be taken into account (see also Figure 703).

703.32.1 Description of zone 1

Zone 1 is the volume containing the sauna heater, limited by the floor, the cold side of the thermal insulation of the ceiling and a vertical surface circumscribing the sauna heater at a distance 0.5 m from the surface of the heater. If the sauna heater is located closer than 0.5 m to a wall, then zone 1 is limited by the cold side of the thermal insulation of that wall.

703.32.2 Description of zone 2

Zone 2 is the volume outside zone 1, limited by the floor, the cold side of the thermal insulation of the walls and a horizontal surface located 1.0 m above the floor.

703.32.3 Description of zone 3

Zone 3 is the volume outside zone 1, limited by the cold side of the thermal insulation of the ceiling and walls and a horizontal surface located 1.0 m above the floor.

703.41 Protection against electric shock

703.410.3 General requirements

703.410.3.5 The protective measures of obstacles and placing out of reach (Section 417) are not permitted.

703.410.3.6 The protective measures of non-conducting location (Regulation 418.1) and earth-free local equipotential bonding (Regulation 418.2) are not permitted.

703.411.3.3 Additional protection by RCDs

Additional protection shall be provided for all circuits of the sauna, by the use of one or more RCDs having the characteristics specified in Regulation 415.1.1. RCD protection need not be provided for the sauna heater unless such protection is recommended by the manufacturer.

703.414 Protective measure: Extra-low voltage provided by SELV or PELV

703.414.4.5 Where SELV or PELV is used, whatever the nominal voltage, basic protection shall be provided by:

(i) basic insulation complying with Regulation 416.1, or

(ii) barriers or enclosures complying with Regulation 416.2.

703.51   Selection and erection of equipment: Common rules

703.512.2   External influences

The equipment shall have a degree of protection of at least IPX4.

If cleaning by use of water jets may be reasonably expected, electrical equipment shall have a degree of protection of at least IPX5.

Three zones are defined as shown in Figure 703:

(i)   In zone 1: only the sauna heater and equipment belonging to the sauna heater shall be installed

(ii) In zone 2: there is no special requirement concerning heat-resistance of equipment

(iii) In zone 3: the equipment shall withstand a minimum temperature of 125 °C and the insulation and sheaths of cables shall withstand a minimum temperature of 170 °C (see also Regulation 703.52 for wiring).

703.52   Selection and erection of equipment: Wiring systems

The wiring system should be preferably installed outside the zones, i.e. on the cold side of the thermal insulation. If the wiring system is installed on the warm side of the thermal insulation in zones 1 or 3, it shall be heat-resisting. Metallic sheaths and metallic conduits shall not be accessible in normal use.

703.53   Selection and erection of equipment: Isolation, switching, control and accessories

703.537.5   Switchgear and controlgear which forms part of the sauna heater equipment or of other fixed equipment installed in zone 2, may be installed within the sauna room or cabin in accordance with the manufacturer's instructions. Other switchgear and controlgear, e.g. for lighting, shall be placed outside the sauna room or cabin. Socket-outlets shall not be installed within the location containing the sauna heater.

703.55   Other equipment

Sauna heating appliances shall comply with BS EN 60335-2-53 and be installed in accordance with the manufacturer's instructions.

Fig 703 – zone dimensions for a sauna

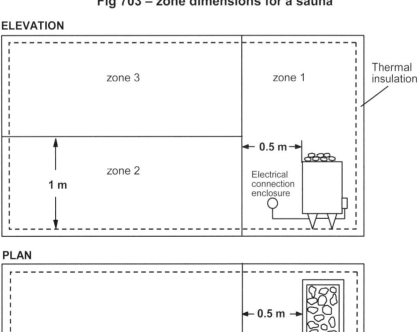

SECTION 704

CONSTRUCTION AND DEMOLITION SITE INSTALLATIONS

704.0   This section should be read in conjunction with BS 7375.

704.1   Scope

704.1.1   The particular requirements of this section apply to temporary installations for construction and demolition sites during the period of the construction or demolition work, including, for example, the following:

(i)   Construction work of new buildings

(ii)   Repair, alteration, extension, demolition of existing buildings or parts of existing buildings

(iii)   Engineering works

(iv)   Earthworks

(v)   Work of similar nature.

The requirements apply to fixed or movable installations.

The regulations do not apply to:

(vi)   installations covered by the IEC 60621 series 2, where equipment of a similar nature to that used in surface mining applications is involved

(vii)   installations in administrative locations of construction sites (e.g. offices, cloakrooms, meeting rooms, canteens, restaurants, dormitories, toilets), where the general requirements of Parts 1 to 6 apply.

704.1.2   For special situations, further particular requirements apply, e.g. Section 706 for conducting locations with restricted movement.

704.313   Supplies

704.313.3   Equipment shall be identified with and be compatible with the particular supply from which it is energized and shall contain only components connected to one and the same installation, except for control or signalling circuits and inputs from standby supplies.

704.4   Protection for safety

704.41   Protection against electric shock

704.410.3   General requirements

704.410.3.5   The protective measures of obstacles and placing out of reach (Section 417) are not permitted.

704.410.3.6   The protective measures of non-conducting location (Regulation 418.1) and earth-free local equipotential bonding (Regulation 418.2) are not permitted.

704.410.3.10   A circuit supplying a socket-outlet with a rated current up to and including 32 A and any other circuit supplying hand-held electrical equipment with rated current up to and including 32A shall be protected by:

(i)   reduced low voltage (Regulation 411.8), or

(ii)   automatic disconnection of supply (Section 411) with additional protection provided by an RCD having the characteristics specified in Regulation 415.1.1, or

(iii)   electrical separation of circuits (Section 413), each socket-outlet and item of hand-held electrical equipment being supplied by an individual transformer or by a separate winding of a transformer, or

(iv)   SELV or PELV (Section 414).

Where electrical separation is used special attention should be paid to the requirements of Regulation 413.3.4.

NOTE 1: The reduced low voltage system is strongly preferred for the supply to portable hand lamps for general use and portable hand tools and local lighting up to 2 kW.

NOTE 2: The SELV system is strongly preferred for portable hand lamps in confined or damp locations.

704.411 Protective measure: Automatic disconnection of supply

704.411.3.1

A PME earthing facility shall not be used for the means of earthing for an installation falling within the scope of this section unless all extraneous-conductive-parts are reliably connected to the main earthing terminal in accordance with Regulation 411.3.1.2.

NOTE: If the PME earthing facility is considered for use, see also BS 7375.

704.411.3.2 Automatic disconnection in case of a fault

704.411.3.2.1 For any circuit supplying one or more socket-outlets with a rated current exceeding 32A, Regulations 411.3.2.5 and 411.3.2.6 are not applicable. For any circuit supplying one or more socket-outlets with a rated current exceeding 32A, an RCD having a rated residual operating current not exceeding 500 mA shall be provided to automatically interrupt the supply to the line conductors of a circuit or equipment in the event of a fault of negligible impedance between a line conductor and an exposed-conductive-part or a protective conductor in the circuit or equipment within the disconnection time required in Regulation 411.3.2.3 or 411.3.2.4 as appropriate.

704.414 Protective measure: Extra-low voltage provided by SELV or PELV

704.414.4 Requirements for SELV and PELV circuits

704.414.4.5 Where SELV or PELV is used, whatever the nominal voltage, basic protection shall be provided by:

(i) basic insulation complying with Regulation 416.1, or

(ii) barriers or enclosures complying with Regulation 416.2.

704.5 Selection and erection of equipment

704.51 Common rules

704.511.1 All assemblies on construction and demolition sites for the distribution of electricity shall be in compliance with the requirements of BS EN 60439-4.

A plug or socket-outlet with a rated current equal to or greater than 16 A shall comply with the requirements of BS EN 60309-2.

704.52 Wiring systems

704.522.8.10 Cable shall not be installed across a site road or a walkway unless adequate protection of the cable against mechanical damage is provided.

704.522.8.11 For reduced low voltage systems, low temperature 300/500 V thermoplastic (BS 7919) or equivalent flexible cables shall be used. For applications exceeding reduced low voltage, flexible cable shall be H07RN-F (BS 7919) type or equivalent having 450/750 V rating and resistant to abrasion and water.

704.53 Switchgear and controlgear

704.537.2.2 Devices for isolation

Each Assembly for Construction Sites (ACS) shall incorporate suitable devices for the switching and isolation of the incoming supply.

A device for isolating the incoming supply shall be suitable for securing in the off position (see Regulation 537.2.1.5), for example, by padlock or location of the device inside a lockable enclosure.

Current-using equipment shall be supplied by ACSs, each ACS comprising:

(i) overcurrent protective devices, and

(ii) devices affording fault protection, and

(iii) socket-outlets, if required.

Safety and standby supplies shall be connected by means of devices arranged to prevent interconnection of the different supplies.

SECTION 705

AGRICULTURAL AND HORTICULTURAL PREMISES

705.1   Scope

The particular requirements of this section apply to fixed electrical installations indoors and outdoors in agricultural and horticultural premises. Some of the requirements are also applicable to other locations that are in common buildings belonging to the agricultural and horticultural premises. Where special requirements also apply to residences and other locations in such common buildings this is stated in the text of the relevant regulations.

Rooms, locations and areas for household applications and similar are not covered by this section.

NOTE: Section 705 does not cover electric fence installations. Refer to BS EN 60335-2-76.

705.41   Protection against electric shock

705.410.3   General requirements

705.410.3.5   The protective measures of obstacles and placing out of reach (Section 417) are not permitted.

705.410.3.6   The protective measures of non-conducting location (Regulation 418.1) and earth-free local equipotential bonding (Regulation 418.2) are not permitted.

705.411   Protective measure: Automatic disconnection of supply

705.411.1   General

In circuits, whatever the type of earthing system, the following disconnection devices shall be provided:

(i)   In final circuits supplying socket-outlets with rated current not exceeding 32 A, an RCD having the characteristics specified in Regulation 415.1.1

(ii)   In final circuits supplying socket-outlets with rated current more than 32 A, an RCD with a rated residual operating current not exceeding 100 mA

(iii)   In all other circuits, RCDs with a rated residual operating current not exceeding 300 mA.

705.411.4   TN system

A TN-C system shall not be used. This requirement applies also to residences and other locations belonging to agricultural or horticultural premises according to the definition of "Residences and other… premises" (see Part 2).

705.414   Protective measure: Extra-low voltage provided by SELV or PELV

705.414.4   Requirements for SELV and PELV circuits

705.414.4.5   Where SELV or PELV is used, whatever the nominal voltage, basic protection shall be provided by:

(i)   basic insulation complying with Regulation 416.1, or

(ii)   barriers or enclosures complying with Regulation 416.2.

705.415.2.1 Additional protection: Supplementary equipotential bonding

In locations intended for livestock, supplementary bonding shall connect all exposed-conductive-parts and extraneous-conductive-parts that can be touched by livestock. Where a metal grid is laid in the floor, it shall be included within the supplementary bonding of the location (Figure 705 shows an example of this, other suitable arrangements of a metal grid are not precluded).

Extraneous-conductive-parts in, or on, the floor, e.g. concrete reinforcement in general or reinforcement of cellars for liquid manure, shall be connected to the supplementary equipotential bonding.

It is recommended that spaced floors made of prefabricated concrete elements be part of the supplementary equipotential bonding. The supplementary equipotential bonding and the metal grid, if any, shall be erected so that it is durably protected against mechanical stresses and corrosion.

NOTE: Unless a metal grid is laid in the floor, the use of a PME earthing facility as the means of earthing for the electrical installation is not recommended.

705.42   Protection against thermal effects

705.422   Measures for protection against fire

705.422.6   Electrical heating appliances used for the breeding and rearing of livestock shall comply with BS EN 60335-2-71 and shall be fixed so as to maintain an appropriate distance from livestock and combustible material, to minimize any risks of burns to livestock and of fire. For radiant heaters the clearance shall be not less than 0.5 m or such other clearance as recommended by the manufacturer.

705.422.7   For fire protection purposes, RCDs shall be installed with a rated residual operating current not exceeding 300 mA. RCDs shall disconnect all live conductors. Where improved continuity of service is required, RCDs not protecting socket-outlets shall be of the S type or have a time delay.

NOTE: The protection of the final circuits by RCD required according to Regulation 411.1 is also effective for protection against fire.

705.422.8   In locations where a fire risk exists conductors of circuits supplied at extra-low voltage shall be protected either by barriers or enclosures affording a degree of protection of IPXXD or IP4X or, in addition to their basic insulation, by an enclosure of insulating material.

NOTE: For example, cables of the type H07RN-F (BS 7919) for outdoor use are in compliance with this requirement.

705.51   Selection and erection of equipment: Common rules

705.512   Operational conditions and external influences

705.512.2   External influences

In agricultural or horticultural premises, electrical equipment shall have a minimum degree of protection of IP44, when used under normal conditions. Where equipment of IP44 rating is not available, it shall be placed in an enclosure complying with IP44.

Socket-outlets shall be installed in a position where they are unlikely to come into contact with combustible material.

Where there are conditions of external influences >AD4, >AE3 and/or >AG1, socket-outlets shall be provided with the appropriate protection.

Protection may also be provided by the use of additional enclosures or by installation in building recesses.

These requirements do not apply to residential locations, offices, shops and locations with similar external influences belonging to agricultural and horticultural premises where, for socket-outlets, BS 1363-2 or BS 546 applies.

Where corrosive substances are present, e.g. in dairies or cattle sheds, the electrical equipment shall be adequately protected.

705.513   Accessibility

705.513.2   Accessibility by livestock

Electrical equipment generally shall be inaccessible to livestock. Equipment that is unavoidably accessible to livestock such as equipment for feeding and basins for watering, shall be adequately constructed and installed to avoid damage by, and to minimize the risk of injury to, livestock.

705.514   Identification

705.514.9   Diagrams and documentation

705.514.9.3   The following documentation shall be provided to the user of the installation:

  (i)   A plan indicating the location of all electrical equipment

  (ii)   The routing of all concealed cables

  (iii)   A single-line distribution diagram

  (iv)   An equipotential bonding diagram indicating locations of bonding connections.

705.52 Selection and erection of equipment: Wiring systems

705.522 Selection and erection of wiring systems in relation to external influences

In locations accessible to, and enclosing, livestock, wiring systems shall be erected so that they are inaccessible to livestock or suitably protected against mechanical damage.

Overhead lines shall be insulated.

In areas of agricultural premises where vehicles and mobile agricultural machines are operated, the following methods of installation shall be applied:

    (i) Cables shall be buried in the ground at a depth of at least 0.6 m with added mechanical protection

    (ii) Cables in arable or cultivated ground shall be buried at a depth of at least 1 m

    (iii) Self-supporting suspension cables shall be installed at a height of at least 6 m.

705.522.10   Special attention shall be given to the presence of different kinds of fauna, e.g. rodents.

705.522.16 Conduit systems, cable trunking systems and cable ducting systems

For locations where livestock is kept, external influences shall be classified AF4, and conduits shall have protection against corrosion of at least Class 2 (medium) for indoor use and Class 4 (high protection) outdoors according to BS EN 61386-21.

For locations where the wiring system may be exposed to impact and mechanical shock due to vehicles and mobile agricultural machines, etc, the external influences shall be classified AG3 and:

    (i) conduits shall provide a degree of protection against impact of 5 J according to BS EN 61386-21

    (ii) cable trunking and ducting systems shall provide a degree of protection against impact of 5 J according to BS EN 50085-2-1.

705.53 Selection and erection of equipment: Isolation, switching and control

Only electrical heating appliances with visual indication of the operating position shall be used.

705.537 Isolation and switching

705.537.2 Isolation

The electrical installation of each building or part of a building shall be isolated by a single isolation device according to Chapter 53.

Means of isolation of all live conductors, including the neutral conductor, shall be provided for circuits used occasionally, e.g. during harvest time.

The isolation devices shall be clearly marked according to the part of the installation to which they belong.

Devices for isolation and switching and devices for emergency stopping or emergency switching shall not be erected where they are accessible to livestock or in any position where access may be impeded by livestock.

705.54 Selection and erection of equipment: Earthing arrangements and protective conductors

705.544 Protective bonding conductors

705.544.2 Supplementary bonding conductors

Protective bonding conductors shall be protected against mechanical damage and corrosion, and shall be selected to avoid electrolytic effects.

For example, the following may be used:

    (i) Hot-dip galvanized steel strip with dimensions of at least 30 mm × 3 mm

    (ii) Hot-dip galvanized round steel of at least 8 mm diameter

    (iii) Copper conductor having a minimum cross-sectional area of 4 mm^2.

Other suitable materials may be used.

705.55   Selection and erection of equipment: Other equipment

705.553.1   Socket-outlets

Socket-outlets of agricultural and horticultural premises shall comply with:

  (i) BS EN 60309-1, or

 (ii) BS EN 60309-2 when interchangeability is required, or

(iii) BS 1363 or BS 546 provided the rated current does not exceed 20 A.

705.56   Safety services

705.560.6   Automatic life support for high density livestock rearing

For high density livestock rearing, systems operating for the life support of livestock shall be taken into account as follows:

  (i) Where the supply of food, water, air and/or lighting to livestock is not ensured in the event of power supply failure, a secure source of supply shall be provided, such as an alternative or back-up supply (see also Section 551). For the supply of ventilation and lighting units separate final circuits shall be provided. Such circuits shall only supply electrical equipment necessary for the operation of the ventilation and lighting

 (ii) Discrimination of the main circuits supplying the ventilation shall be ensured in case of any overcurrent and/or short-circuit to Earth

(iii) Where electrically powered ventilation is necessary in an installation one of the following shall be provided:

    a) A standby electrical source ensuring sufficient supply for ventilation equipment, or

      NOTE: A notice should be placed adjacent to the standby electrical source, indicating that it should be tested periodically according to the manufacturer's instructions.

    b) temperature and supply voltage monitoring. This can be achieved by one or more monitoring devices. The device(s) shall provide a visual or audible signal that can be readily observed by the user and shall operate independently from the normal supply.

Fig 705 – Example of supplementary equipotential bonding within a cattle shed

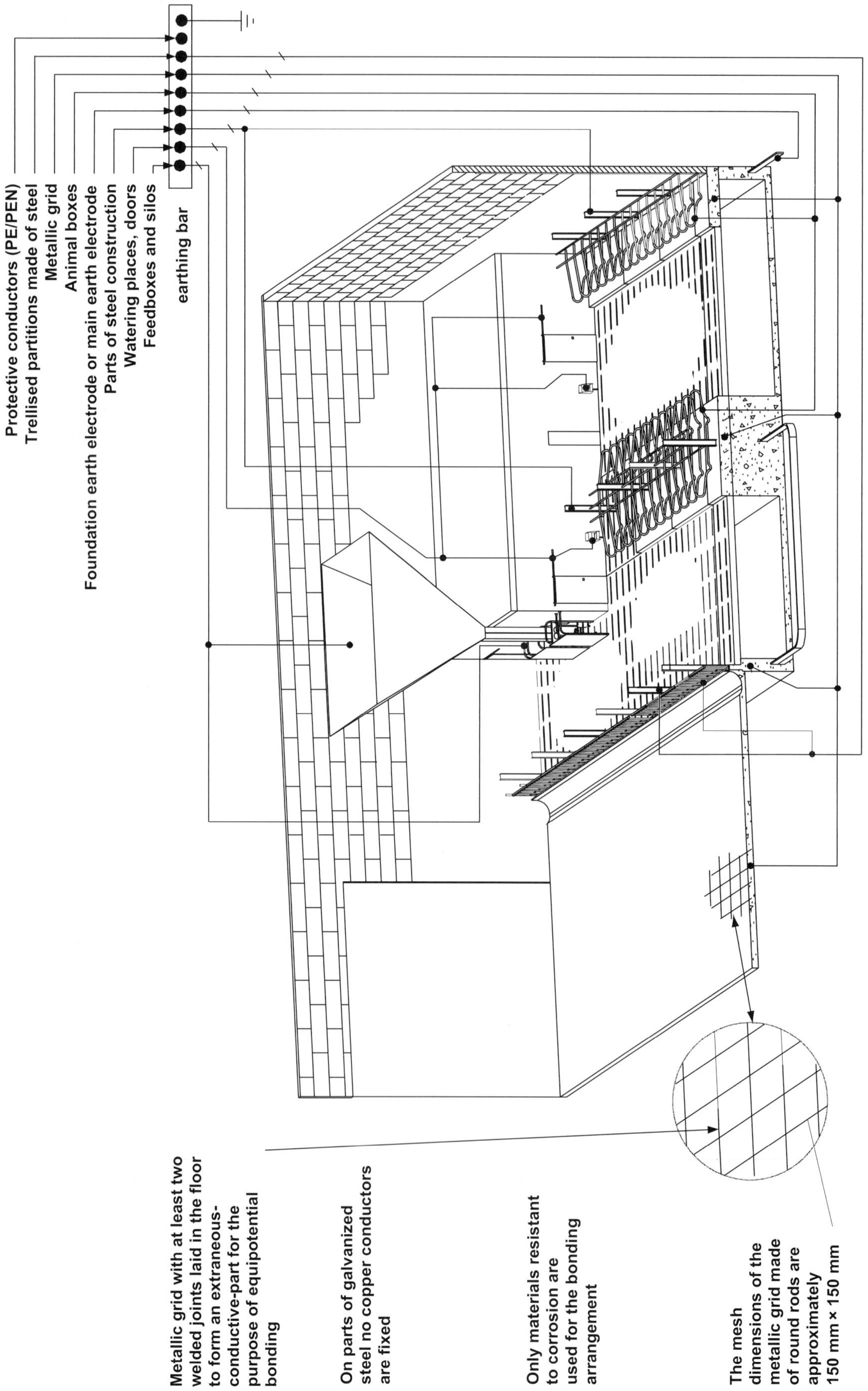

Protective conductors (PE/PEN)
Trellised partitions made of steel
Metallic grid
Animal boxes
Foundation earth electrode or main earth electrode
Parts of steel construction
Watering places, doors
Feedboxes and silos
earthing bar

Metallic grid with at least two welded joints laid in the floor to form an extraneous-conductive-part for the purpose of equipotential bonding

On parts of galvanized steel no copper conductors are fixed

Only materials resistant to corrosion are used for the bonding arrangement

The mesh dimensions of the metallic grid made of round rods are approximately 150 mm × 150 mm

SECTION 706

CONDUCTING LOCATIONS WITH RESTRICTED MOVEMENT

706.1   Scope

The particular requirements of this section apply to:

(i) fixed equipment in conducting locations where movement of persons is restricted by the location, and

(ii) to supplies for mobile equipment for use in such locations.

A conducting location with restricted movement is comprised mainly of metallic or other conductive surrounding parts, within which it is likely that a person will come into contact through a substantial portion of the body with the metallic or other conductive surrounding parts and where the possibility of interrupting this contact is limited.

The particular requirements of this section do not apply to locations which allow a person freedom of bodily movement to work, enter and leave the location without physical constraint. For installation and use of arc welding equipment, see IEC/TS 62081.

This section does not apply to electrical systems as defined in BS 7909 used in structures, sets, mobile units etc as used for public or private events, touring shows, theatrical, radio, TV or film productions and similar activities of the entertainment industry.

706.41   Protection against electric shock

706.410.3.5   The protective measures of obstacles and placing out of reach (Section 417) are not permitted.

706.410.3.10 In a conducting location with restricted movement the following protective measures apply to circuits supplying the following current-using equipment:

(i) For the supply to a hand-held tool or an item of mobile equipment:

    a) electrical separation (Section 413), subject to only one item of equipment being connected to a secondary winding of the transformer, or
        NOTE: The transformer may have two or more secondary windings.

    b) SELV (Section 414).

(ii) For the supply to handlamps:

    a) SELV (Section 414). It is permissible for the SELV circuit to supply a fluorescent luminaire with a built-in step-up transformer with electrically separated windings.

(iii) For the supply to fixed equipment:

    a) automatic disconnection of the supply (Section 411) with supplementary equipotential bonding (Regulation 415.2). The supplementary bonding shall connect exposed-conductive-parts of fixed equipment and the conductive parts of the location, or

    b) by use of Class II equipment or equipment having equivalent insulation (Section 412), provided the supply circuits have additional protection by the use of RCDs having the characteristics specified in Regulation 415.1.1, or

    c) electrical separation (Section 413), subject to only one item of equipment being connected to a secondary winding of the isolating transformer, or

    d) SELV (Section 414), or

    e) PELV (Section 414), where equipotential bonding is provided between all exposed-conductive-parts, all extraneous-conductive-parts inside the conducting location with restricted movement, and the connection of the PELV system to Earth.

706.411   Protective measure: Automatic disconnection of supply

706.411.1   General

706.411.1.1 Only circuits and the protective measures for supplying equipment indicated in Regulation 706.410.3.10 are permitted.

706.411.1.2 If a functional earth is required for certain equipment, for example measuring and control equipment, equipotential bonding shall be provided between all exposed-conductive-parts and extraneous-conductive-parts inside the conducting location with restricted movement and the functional earth.

706.413   Protective measure: Electrical separation

706.413.1.2 The unearthed source shall have simple separation and shall be situated outside the conducting location with restricted movement, unless the source is part of the fixed installation within the conducting location with restricted movement as provided by item (iii) of Regulation 706.410.3.10.

706.414   Protective measure: Extra-low voltage provided by SELV or PELV

706.414.3   Sources for SELV and PELV

706.414.3(ii) A source for SELV or PELV shall be situated outside the conducting location with restricted movement, unless it is part of the fixed installation within the conducting location with restricted movement as provided by item (iii) of Regulation 706.410.3.10.

706.414.4   Requirements for SELV and PELV circuits

706.414.4.5 Where SELV or PELV is used, whatever the nominal voltage, basic protection shall be provided by:

(i)   basic insulation complying with Regulation 416.1, or

(ii)   barriers or enclosures complying with Regulation 416.2.

SECTION 708

ELECTRICAL INSTALLATIONS IN CARAVAN / CAMPING PARKS AND SIMILAR LOCATIONS

NOTE: In order not to mix regulations on different subjects, such as those for electrical installation of caravan parks with those for electrical installation inside caravans, two sections have been created:
Section 708, which concerns electrical installations in caravan parks, camping parks and similar locations and
Section 721, which concerns electrical installations in caravans and motor caravans.

708.1 Scope

The particular requirements of this section apply to that portion of the electrical installation in caravan / camping parks and similar locations providing facilities for supplying leisure accommodation vehicles (including caravans) or tents.

They do not apply to the internal electrical installations of leisure accommodation vehicles or mobile or transportable units.

NOTE 1: For installations in caravans and motor caravans which are operated at 12 V d.c., BS EN 1648-1 and 2 apply.

NOTE 2: For installations in caravans and motor caravans which are operated at voltages other than 12 V d.c., Section 721 applies.

708.3 Assessment of general characteristics

708.313.1.2 The nominal supply voltage of the installation for the supply of leisure accommodation vehicles shall be 230 V a.c. single-phase or 400 V a.c. three-phase.

708.4 Protection for safety

708.41 Protection against electric shock

708.410.3 General requirements

708.410.3.5 The protective measures of obstacles and placing out of reach (Section 417) are not permitted.

708.410.3.6 The protective measures of non-conducting location (Regulation 418.1) and earth-free local equipotential bonding (Regulation 418.2) are not permitted.

708.411.4 TN system

The Electricity Safety, Quality and Continuity Regulations 2002 (ESQCR) prohibit the connection of a PME earthing facility to any metalwork in a leisure accommodation vehicle (including a caravan).

This does not preclude the use of a PME earthing facility as the means of earthing for other purposes, such as to the installations of permanent buildings.

NOTE: The requirements of other sections of Part 7 may also apply.

708.5 Selection and erection of equipment

708.512.2 External influences

Electrical equipment installed outside in caravan parks shall comply at least with the following external influences:

   (i)   Presence of water: AD4 (splashes), IPX4 in accordance with BS EN 60529

  (ii) Presence of foreign solid bodies: AE2 (small objects), IP3X in accordance with BS EN 60529

 (iii) Mechanical stress: AG3 (high severity), IK08 in accordance with BS EN 62262.

708.521.1 Wiring systems in caravan parks

The following wiring systems are suitable for distribution circuits feeding caravan or tent pitch electrical supply equipment:

   (i)   Underground distribution circuits

  (ii) Overhead distribution circuits.

NOTE: The preferred method of supply is by means of underground distribution circuits.

708.521.1.1 Underground distribution circuits

Underground cables shall be buried at a depth of at least 0.6 m and, unless having additional mechanical protection, be placed outside any caravan pitch or away from any surface where tent pegs or ground anchors are expected to be present.

708.521.1.2 Overhead distribution circuits

All overhead conductors shall be insulated.

Poles and other supports for overhead wiring shall be located or protected so that they are unlikely to be damaged by any foreseeable vehicle movement.

Overhead conductors shall be at a height above ground of not less than 6 m in all areas subject to vehicle movement and 3.5 m in all other areas.

708.53   Switchgear and controlgear

708.530.3   Caravan pitch electrical supply equipment

Caravan pitch electrical supply equipment shall be located adjacent to the pitch and not more than 20 m from the connection facility on the leisure accommodation vehicle or tent when on its pitch.

NOTE: Not more than four socket-outlets should be grouped in one location, in order to avoid the supply cable crossing a pitch other than the one intended to be supplied.

708.55   Selection and erection of equipment: Other equipment

708.553.1   Plugs and socket-outlets

708.553.1.8   Each socket-outlet and its enclosure forming part of the caravan pitch electrical supply equipment shall comply with BS EN 60309-2 and meet the degree of protection of at least IP44 in accordance with BS EN 60529.

708.553.1.9   The socket-outlets shall be placed at a height of 0.5 m to 1.5 m from the ground to the lowest part of the socket-outlet. In special cases, due to environmental conditions such as risk of flooding or heavy snowfall, the maximum height is permitted to exceed 1.5 m.

708.553.1.10   The current rating of socket-outlets shall be not less than 16 A. Socket-outlets of higher current ratings shall be provided where greater demands are envisaged.

708.553.1.11   At least one socket-outlet shall be provided for each caravan pitch.

708.553.1.12   Each socket-outlet shall be protected by an individual overcurrent protective device, in accordance with the requirements of Chapter 43.

A fixed connection for supply to each leisure accommodation vehicle (including caravans) or tent shall be protected individually by an overcurrent protective device, in accordance with the requirements of Chapter 43.

708.553.1.13   Socket-outlets shall be protected individually by an RCD having the characteristics specified in Regulation 415.1.1. Devices selected shall disconnect all poles including the neutral.

Final circuits intended for fixed connection for the supply to leisure accommodation vehicles (including caravans) or tents shall be protected individually by an RCD having the characteristics specified in Regulation 415.1.1. Devices selected shall disconnect all poles including the neutral.

708.553.1.14   Socket-outlet protective conductors shall not be connected to a PME earthing facility.

The protective conductor of each socket-outlet shall be connected to an earth electrode and the requirements of Regulation 411.5 for a TT system shall be complied with.

Fig 708 – Example of a 2-pole and protective conductor supply system between the caravan pitch supply equipment and the caravan or motor caravan

NOTE: See also Regulation 721.55.2.6

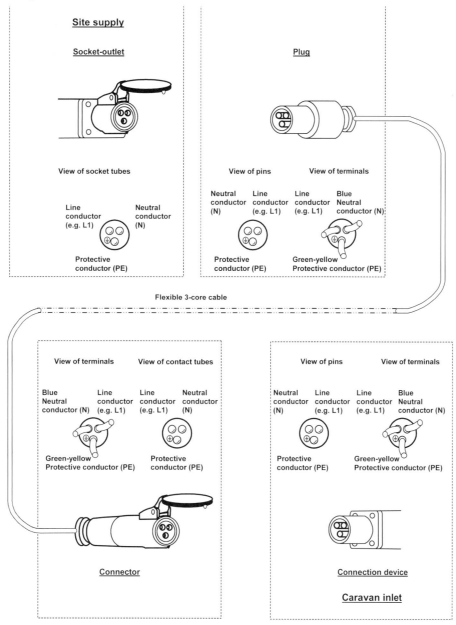

NOTE: Typical requirements for cord extension sets

The means of connection between the caravan pitch socket-outlet and the leisure accommodation vehicle should be an assembly of the following:

- a plug complying with BS EN 60309-2;

- a flexible cable type H05RN-F (BS 7919) or equivalent, with a protective conductor and having the following characteristics:

  - length: 25 m (±2 m)

  - for current rating 16A: minimum cross-sectional area: 2.5 mm^2. For higher current ratings, the cross-sectional area must be chosen so that secure tripping of the overcurrent protective device is achieved at the lowest fault current calculated at the end of the cord extension set

  - conductors to be identified in accordance with Table 51.

- a connector complying with BS EN 60309-2.

SECTION 709

MARINAS AND SIMILAR LOCATIONS

709.1 Scope

The particular requirements of this section are applicable only to circuits intended to supply pleasure craft or houseboats in marinas and similar locations.

NOTE 1: In this section "marina" means "marina and similar locations".

The particular requirements do not apply to the supply to houseboats if they are supplied directly from the public network.

The particular requirements do not apply to the internal electrical installations of pleasure craft or houseboats.

NOTE 2: For electrical installations of pleasure craft see BS EN 60092-507.

NOTE 3: The electrical installations of houseboats shall comply with the general requirements of these Regulations together with the relevant particular requirements of Part 7.

For the remainder of the electrical installation of marinas and similar locations the general requirements of these Regulations together with the relevant particular requirements of Part 7 apply.

709.3 Assessment of general characteristics

709.313 Supplies

709.313.1.2 The nominal supply voltage of the installation for the supply to pleasure craft or houseboats shall be 230 V a.c. single-phase, or 400 V a.c. three-phase.

709.41 Protection against electric shock

709.410.3 General requirements

709.410.3.5 The protective measures of obstacles and placing out of reach (Section 417) are not permitted.

709.410.3.6 The protective measures of non-conducting location (Regulation 418.1) and earth-free local equipotential bonding (Regulation 418.2) are not permitted.

709.411.4 TN system

The Electricity Safety, Quality and Continuity Regulations 2002 (ESQCR) prohibit the connection of a PME earthing facility to any metalwork in a boat.

This does not preclude the use of a PME earthing facility as the means of earthing for other purposes, such as to the installations of permanent buildings.

NOTE: The requirements of other sections of Part 7 may also apply.

709.5 Selection and erection of equipment

709.512 Operational conditions and external influences

709.512.2 External influences

For marinas, particular attention is given in this section to the likelihood of corrosive elements, movement of structures, mechanical damage, presence of flammable fuel and the increased risk of electric shock due to:

   (i)   presence of water

  (ii)   reduction in body resistance

 (iii)   contact of the body with Earth potential.

709.512.2.1.1 Presence of water (AD)

In marinas, equipment installed on or above a jetty, wharf, pier or pontoon shall be selected as follows, according to the external influences which may be present:

   (i)   Water splashes (AD4): IPX4

  (ii)   Water jets (AD5): IPX5

 (iii)   Water waves (AD6): IPX6.

709.512.2.1.2 Presence of solid foreign bodies (AE)

Equipment installed on or above a jetty, wharf, pier or pontoon shall be selected with a degree of protection of at least IP3X in order to protect against the ingress of small objects (AE2).

709.512.2.1.3   Presence of corrosive or polluting substances (AF)

Equipment installed on or above a jetty, wharf, pier or pontoon shall be suitable for use in the presence of atmospheric corrosive or polluting substances (AF2). If hydrocarbons are present, AF3 is applicable.

709.512.2.1.4   Impact (AG)

Equipment installed on or above a jetty, wharf, pier or pontoon shall be protected against mechanical damage (Impact of medium severity AG2). Protection shall be afforded by one or more of the following:
- (i)   The position or location selected to avoid being damaged by any reasonably foreseeable impact
- (ii)   The provision of local or general mechanical protection
- (iii)   Installing equipment complying with a minimum degree of protection for external mechanical impact IK08 (see BS EN 62262).

709.521   Types of wiring system

709.521.1   Wiring systems of marinas

709.521.1.4   The following wiring systems are suitable for distribution circuits of marinas:
- (i)   Underground cables
- (ii)   Overhead cables or overhead insulated conductors
- (iii)   Cables with copper conductors and thermoplastic or elastomeric insulation and sheath installed within an appropriate cable management system taking into account external influences such as movement, impact, corrosion and ambient temperature
- (iv)   Mineral-insulated cables with a PVC protective covering
- (v)   Cables with armouring and serving of thermoplastic or elastomeric material
- (vi)   Other cables and materials that are no less suitable than those listed above.

709.521.1.5   The following wiring systems shall not be used on or above a jetty, wharf, pier or pontoon:
- (i)   Cables in free air suspended from or incorporating a support wire, e.g. as installation methods Nos. 35 and 36 in Table 4A2
- (ii)   Non-sheathed cables in cable management systems
- (iii)   Cables with aluminium conductors
- (iv)   Mineral insulated cables.

709.521.1.6   Cables shall be selected and installed so that mechanical damage due to tidal and other movement of floating structures is prevented.

Cable management systems shall be installed to allow the drainage of water by drainage holes and/or installation of the equipment on an incline.

709.521.1.7 Underground cables

Underground distribution cables shall, unless provided with additional mechanical protection, be buried at a sufficient depth to avoid being damaged, e.g. by heavy vehicle movement.

NOTE: A depth of 0.5 m is generally considered as a minimum depth to fulfil this requirement.

709.521.1.8 Overhead cables or overhead insulated conductors

All overhead conductors shall be insulated.

Poles and other supports for overhead wiring shall be located or protected so that they are unlikely to be damaged by any foreseeable vehicle movement.

Overhead conductors shall be at a height above ground of not less than 6 m in all areas subjected to vehicle movement and 3.5 m in all other areas.

709.531   Devices for fault protection by automatic disconnection of supply

709.531.2   RCDs

Socket–outlets shall be protected individually by an RCD having the characteristics specified in Regulation 415.1.1. Devices selected shall disconnect all poles, including the neutral.

Final circuits intended for fixed connection for the supply to houseboats shall be protected individually by an RCD having the characteristics specified in Regulation 415.1.1. The device selected shall disconnect all poles, including the neutral.

709.533　Devices for protection against overcurrent

Each socket–outlet shall be protected by an individual overcurrent protective device, in accordance with the requirements of Chapter 43.

A fixed connection for supply to each houseboat shall be protected individually by an overcurrent protective device, in accordance with the requirements of Chapter 43.

709.537　Isolation and switching

709.537.2　Isolation

709.537.2.1 General

709.537.2.1.1　At least one means of isolation shall be installed in each distribution cabinet. This switching device shall disconnect all live conductors including the neutral conductor. One isolating switching device for a maximum of four socket-outlets shall be installed.

709.55　Other equipment

709.553.1　Plugs and socket-outlets

709.553.1.8　Socket-outlets shall comply with BS EN 60309-1 above 63 A and BS EN 60309-2 up to 63 A. Every socket-outlet shall meet the degree of protection of at least IP44 or such protection shall be provided by an enclosure.

Where the codes AD5 or AD6 are applicable the degree of protection shall be at least either IPX5 or IPX6 respectively.

709.553.1.9　Every socket-outlet shall be located as close as practicable to the berth to be supplied.

Socket-outlets shall be installed in the distribution board or in separate enclosures.

709.553.1.10　In order to avoid any hazard due to long connection cords, a maximum of four socket–outlets shall be grouped together in one enclosure.

NOTE:　See Figure 709.3 regarding the recommended instruction notice to be placed in marinas adjacent to each group of socket–outlets.

709.553.1.11　One socket-outlet shall supply only one pleasure craft or houseboat.

709.553.1.12　In general, single-phase socket-outlets with rated voltage 200 V – 250 V and rated current 16 A shall be provided.

Where greater demands are envisaged socket-outlets with higher current ratings shall be provided.

709.553.1.13　Socket-outlets shall be placed at a height of not less than 1 m above the highest water level. In the case of floating pontoons or walkways only, this height may be reduced to 300 mm above the highest water level provided that appropriate additional measures are taken to protect against the effects of splashing.

709.553.1.14　Socket-outlet protective conductors shall not be connected to a PME earthing facility.

Figs 709.1 & 2 – Examples of methods of obtaining supply in marinas

Fig 709.1 – Connection to a single-phase mains supply with RCD

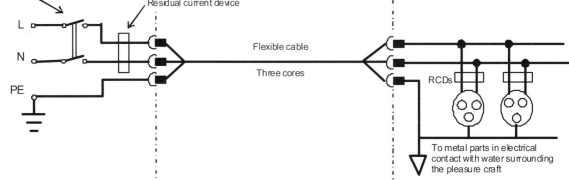

Fig 709.2 – Connection to a three-phase mains supply with RCD

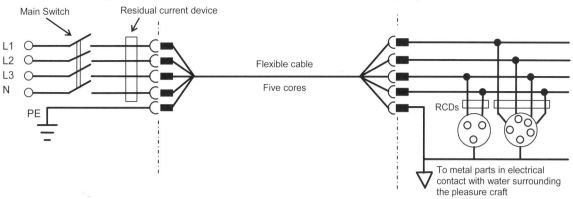

Fig 709.3 – Example of an instruction notice to be placed in marinas

NOTE 1: It is recommended that the marina operator provides every pleasure craft operator who wishes to connect a pleasure craft to an electrical supply with an up-to-date copy of this instruction notice.

NOTE 2: The instruction notice should contain, at least, the following:

INSTRUCTIONS FOR ELECTRICITY SUPPLY

BERTHING INSTRUCTIONS FOR CONNECTION TO SHORE SUPPLY

This marina provides power for use on your pleasure craft with a direct connection to the shore supply which is connected to earth. Unless you have an isolating transformer fitted on board to isolate the electrical system on your craft from the shore supply system, corrosion through electrolysis could damage your craft or surrounding craft.

ON ARRIVAL

(i) Ensure the supply is switched off and disconnect all current-using equipment on the craft, before inserting the craft plug. Connect the flexible cable **firstly** at the pleasure-craft inlet socket and **then** at the marina socket-outlet.

(ii) The supply at this berth is * V, * Hz. The socket-outlet will accommodate a standard marina plug colour * (technically described as BS EN 60309-2, position 6 h).

(iii) For safety reasons, your craft must not be connected to any other socket-outlet than that allocated to you and the internal wiring on your craft must comply with the appropriate standards.

(iv) Every effort must be made to prevent the connecting flexible cable from falling into the water if it should become disengaged. For this purpose, securing hooks are provided alongside socket-outlets for anchorage at a loop of tie cord.

(v) For safety reasons, only one pleasure-craft connecting cable supplying one pleasure craft may be connected to any one socket-outlet.

(vi) The connecting flexible cable must be in one length, without signs of damage, and not contain joints or other means to increase its length.

(vii) The entry of moisture and salt into the pleasure-craft inlet socket may cause a hazard. Examine carefully and clean the plug and socket before connecting the supply.

(viii) It is dangerous to attempt repairs or alterations. If any difficulty arises, contact the marina management.

BEFORE LEAVING

(i) Ensure that the supply is switched off and disconnect all current-using equipment on the craft, before the connecting cable is disconnected and any tie cord loops are unhooked.

(ii) The connecting flexible cable should be disconnected **firstly** from the marina socket-outlet and **then** from the pleasure-craft inlet socket. Any cover that may be provided to protect the inlet from weather should be securely replaced. The connecting flexible cable should be coiled up and stored in a dry location where it will not be damaged.

---

* appropriate figures and colours to be inserted.
nominally 230 V 50 Hz blue - single-phase, and
nominally 400 V 50 Hz red - three-phase

SECTION 710

MEDICAL LOCATIONS

710.1 Scope

This section applies to electrical installations in medical locations to ensure safety of patients and medical staff. These requirements, in the main, refer to hospitals, private clinics, medical and dental practices, healthcare centres and dedicated medical rooms in the workplace.

The requirements of this section do not apply to medical electrical equipment.

This section also applies to electrical installations in locations designed for medical research.

NOTE 1 It may be necessary to modify the existing electrical installation, in accordance with this Standard, when a change of utilization of the location occurs. Special care should be taken where intracardiac procedures are performed in existing installations.

NOTE 2 Where applicable, this Standard can also be used in veterinary clinics.

NOTE 3 For medical electrical equipment and medical electrical systems, refer to the BS EN 60601 series.

NOTE 4 Care should be taken so that other installations do not compromise the level of safety provided by installations meeting the requirements of this section.

NOTE 5 Supporting information on the electrical installations of medical locations can be found in Health Technical Memorandum (HTM) 06-01 (Part A), published by The Department of Health.

710.3 Assessment of general characteristics

In order to determine the classification and Group number of a medical location, it is necessary that the relevant medical staff indicate which medical procedures will take place within the location. Based on the intended use, the appropriate classification for the location shall be determined.

NOTE 1: Classification of a medical location should be related to the type of contact between applied parts and the patient, the threat to the safety of the patient that represents a discontinuity (failure) of the electrical supply, as well as the purpose for which the location is used. (Guidance on the allocation of a Group number and classification of safety services for medical locations is shown in Annex A710.)

NOTE 2: To ensure protection of patients from possible electrical hazards, additional protective measures need to be applied in medical locations. The type and description of these hazards can vary according to the treatment being administered. The purpose for which a location is to be used may justify areas with different classifications (Group 0, 1 or 2) for different medical procedures.

NOTE 3: Applied parts are defined by the particular standards for medical electrical equipment.

NOTE 4: Where a medical location may be used for different medical procedures the requirements of the higher Group classification should be applied; refer to Annex A710.

710.31 Purpose, supplies and structure

710.312.2 Types of system earthing

PEN conductors shall not be used in medical locations and medical buildings downstream of the main distribution board.

NOTE: In Great Britain, Regulation 8(4) of the Electricity Safety, Quality and Continuity Regulations 2002 prohibits the use of PEN conductors in consumers' installations.

710.313 Supplies

710.313.1 General

In medical locations, the distribution system shall be designed and installed to facilitate the automatic changeover from the main distribution network to the electrical safety source feeding essential loads, as required by Regulation 560.5.

710.4 Protection for safety

710.41 Protection against electric shock

710.410.3 General requirements

710.410.3.5 The protective measures of obstacles and placing out of reach (Section 417) are not permitted.

710.410.3.6 The protective measures of non-conducting location (Regulation 418.1), earth-free local equipotential bonding (Regulation 418.2) and electrical separation for the supply of more than one item of current-using equipment (Regulation 418.3), are not permitted.

710.411.3 Requirements for fault protection (protection against indirect contact)

710.411.3.2 Automatic disconnection in case of a fault

710.411.3.2.1   Care shall be taken to ensure that simultaneous use of many items of equipment connected to the same circuit cannot cause unwanted tripping of the residual current protective device (RCD).

In medical locations of Group 1 and Group 2, where RCDs are required, only type A according to BS EN 61008 and BS EN 61009 or type B according to IEC 62423 shall be selected, depending on the possible fault current arising. Type AC RCDs shall not be used.

710.411.3.2.5   In medical locations of Group 1 and Group 2, the following shall apply:

(i)   For TN, TT and IT systems, the voltage presented between simultaneously accessible exposed-conductive-parts and/or extraneous-conductive-parts shall not exceed 25 V a.c. or 60 V d.c.

(ii)   For TN and TT systems, the requirements of Table 710 shall apply.

TABLE 710 – Maximum disconnection times

System	$25\ V < U_0 \leq 50\ V$ s		$50\ V < U_0 \leq 120\ V$ s		$120\ V < U_0 \leq 230\ V$ s		$230\ V < U_0 \leq 400\ V$ s		$U_0 > 400\ V$ s	
	a.c.	d.c.	a.c.	d.c.	a.c.	d.c.	a.c.	d.c.	a.c.	d.c.
TN	5	5	0.3	2	0.3	0.5	0.05	0.06	0.02	0.02
TT	5	5	0.15	0.2	0.05	0.1	0.02	0.06	0.02	0.02

NOTE: In TN systems, 25 V a.c. or 60 V d.c. may be met with protective equipotential bonding, by complying with the disconnection time in accordance with Table 710.

710.411.3.3 Additional protection

Where a medical IT system is used, additional protection by means of an RCD shall not be used.

710.411.4   TN system

In final circuits of Group 1 rated up to 63 A, RCDs having the characteristics specified in Regulation 415.1.1 shall be used.

In TN-S systems, the insulation level of all live conductors shall be monitored.

In medical locations of Group 2 (except for the medical IT system), protection by automatic disconnection of supply by means of RCDs having the characteristics specified in Regulation 415.1.1 shall only be used on the following circuits:

(i)   Circuits for the supply of movements of fixed operating tables, or

(ii)   circuits for X-ray units, or

(iii)   circuits for large equipment with a rated power greater than 5 kVA.

NOTE: The requirement in (ii) above is mainly applicable to mobile X-ray units brought into Group 2 locations.

710.411.5   TT system

In medical locations of Group 1 and Group 2, RCDs shall be used as disconnection devices and the requirements of Regulation 710.411.4 apply.

710.411.6   IT system

710.411.6.3.1   In Group 2 medical locations, an IT system shall be used for final circuits supplying medical electrical equipment and systems intended for life support, surgical applications and other electrical equipment located or that may be moved into the "patient environment", excluding equipment listed in Regulation 710.411.4.

For each group of rooms serving the same function, at least one IT system is necessary. The IT system shall be equipped with an insulation monitoring device (IMD) in accordance with BS EN 61557-8, with the following additional specific requirements:

(i)   A.C. internal impedance shall be $\geq 100\ k\Omega$

(ii)   Internal resistance shall be $\geq 250\ k\Omega$

(iii)   Test voltage shall be $\leq 25\ V$ d.c.

(iv)   Injected current, even under fault conditions, shall be ≤ 1 mA peak

(v)   Indication shall take place at the latest when the insulation resistance has decreased to 50 kΩ. If the response value is adjustable, the lowest possible setpoint value shall be ≥ 50 kΩ. A test device shall be provided

(vi)   Response and alarm-off time shall be ≤ 5 s.

NOTE: An indication is recommended if the protective earth (PE) or wiring connection of the IMD is lost.

For each medical IT system, an acoustic and visual alarm system incorporating the following components shall be arranged at a suitable place so that it can be permanently monitored (audible and visual signals) by the medical staff and, furthermore, is forwarded to the technical staff:

(vii)   A green signal lamp to indicate normal operation

(viii)   A yellow signal lamp which lights when the minimum value set for the insulation resistance is reached. It shall not be possible for this light to be cancelled or disconnected

(ix)   An audible alarm which sounds when the minimum value set for the insulation resistance is reached. This audible alarm may be silenced

(x)   The yellow signal shall go out on removal of the fault and when the normal condition is restored.

Documentation shall be easily readable in the medical location and it shall include:

(xi)   the meaning of each type of signal, and

(xii)   the procedure to be followed in case of an alarm at first fault.

NOTE: For illustration of a typical theatre layout refer to Figure 710.2.

710.411.6.3.2   Monitoring of overload and high temperature for the IT transformer is required.

710.411.6.3.3   Fault location systems which localize insulation faults in any part of the medical IT system may also be installed in addition to an insulation monitoring device.

The insulation fault location system shall be in accordance with BS EN 61557-9.

710.411.7   Functional extra-low voltage (FELV)

In medical locations, functional extra-low voltage (FELV) is not permitted as a method of protection against electric shock.

710.414   Protective measure: Extra-low voltage provided by SELV or PELV

710.414.1   General

When using SELV and/or PELV circuits in medical locations of Group 1 and Group 2, the nominal voltage applied to current-using equipment shall not exceed 25 V. a.c. rms or 60 V ripple-free d.c. Protection by basic insulation of live parts as required by Regulation 416.1 or by barriers or enclosures as required by Regulation 416.2, shall be provided.

710.414.4.1   In medical locations of Group 2, where PELV is used, exposed-conductive-parts of equipment, e.g. operating theatre luminaires, shall be connected to the circuit protective conductor.

710.415.2   Additional protection: Supplementary equipotential bonding

710.415.2.1   In each medical location of Group 1 and Group 2, supplementary equipotential bonding shall be installed and the supplementary bonding conductors shall be connected to the equipotential bonding busbar for the purpose of equalizing potential differences between the following parts, which are located or that may be moved into the "patient environment":

(i)   Protective conductors

(ii)   Extraneous-conductive-parts

(iii)   Screening against electrical interference fields, if installed

(iv)   Connection to conductive floor grids, if installed

(v)   Metal screens of isolating transformers, via the shortest route to the earthing conductor.

Supplementary equipotential bonding connection points for the connection of medical electrical equipment shall be provided in each medical location, as follows:

(vi)   Group 1: one per patient location

(vii)   Group 2: one per medical IT socket-outlet.

NOTE: Fixed conductive non-electrical patient supports such as operating theatre tables, physiotherapy couches and dental chairs should be connected to the equipotential bonding conductor unless they are intended to be isolated from earth.

710.415.2.2 In medical locations of Group 1, the resistance of the protective conductors, including the resistance of the connections, between the terminals for the protective conductor of socket-outlets and of fixed equipment or any extraneous-conductive-parts and the equipotential bonding busbar shall not exceed 0.7 Ω.

In medical locations of Group 2, the resistance of the protective conductors, including the resistance of the connections, between the terminals for the protective conductor of socket-outlets and of fixed equipment or any extraneous-conductive-parts and the equipotential bonding busbar shall not exceed 0.2 Ω.

NOTE: In TN systems, a value of 25 V ac. or 60 V d.c. may be obtained by the provision of a satisfactory value of R_A using protective equipotential bonding in conjunction with circuit protective conductors for the particular circuit.

710.415.2.3 The equipotential bonding busbar shall be located in or near the medical location. Connections shall be so arranged that they are accessible, labelled, clearly visible and can easily be disconnected individually.

NOTE: It is recommended that radial wiring patterns are used to avoid "earth loops" that may cause electromagnetic interference.

710.444 Protection against electromagnetic disturbances

Special considerations have to be made concerning electromagnetic interference (EMI) and electromagnetic compatibility (EMC).

710.5 Selection and erection of equipment

710.51 Common rules

710.51.1 Distribution boards

Distribution boards shall meet the requirements of BS EN 60439 series.

Distribution boards for Group 2 shall be installed in close proximity to the Group 2 medical locations and clearly labelled.

710.512 Operational conditions and external influences

710.512.1 Operational conditions

710.512.1.1 Transformers for IT systems

Transformers shall be in accordance with BS EN 61558-2-15, installed in close proximity to the medical location and with the following additional requirements:

(i) The leakage current of the output winding to earth and the leakage current of the enclosure, when measured in no-load condition and the transformer supplied at rated voltage and rated frequency, shall not exceed 0.5 mA

(ii) At least one single-phase transformer per room or functional group of rooms shall be used to form the IT systems for mobile and fixed equipment and the rated output shall be not less than 0.5 kVA and shall not exceed 10 kVA. Where several transformers are needed to supply equipment in one room, they shall not be connected in parallel

(iii) If the supply of three-phase loads via an IT system is also required, a separate three-phase transformer shall be provided for this purpose.

For monitoring see Regulation 710.411.6.3.1.

Capacitors shall not be used in transformers for medical IT systems.

710.512.1.2 Power supply for medical locations of Group 2

In the event of a first fault to earth, a total loss of supply in Group 2 locations shall be prevented.

710.512.2.1 Explosion risk

Electrical devices, e.g. socket-outlets and switches, installed below any medical-gas outlets for oxidizing or flammable gases shall be located at a distance of at least 0.2 m from the outlet (centre to centre), so as to minimize the risk of ignition of flammable gases.

NOTE: Requirements for medical electrical equipment for use in conjunction with flammable gases and vapours are contained in BS EN 60601.

710.52　Selection and erection of wiring systems

Any wiring system within Group 2 medical locations shall be exclusive for the use of equipment and accessories within those locations.

710.53　Protection, isolation, switching, control and monitoring

710.531.1　Overcurrent protective devices - protection of wiring systems in medical locations of Group 2

Overload current protection shall not be used in either the primary or secondary circuit of the transformer of a medical IT system.

Overcurrent protection against short-circuit and overload current is required for each final circuit.

NOTE:　Overcurrent protective devices (e.g. fuses) may be used in the primary circuit of the transformer for short-circuit protection only.

710.531.2　RCDs

710.531.2.4 Socket-outlets protected by RCDs

For each circuit protected by an RCD having the characteristics specified in Regulation 415.1.1, consideration shall be given to reduce the possibility of unwanted tripping of the RCD due to excessive protective conductor currents produced by equipment in normal operation.

710.537　Isolation and switching
710.537.1　General

Automatic changeover devices shall be arranged so that safe separation between supply lines is maintained. Automatic changeover devices shall comply with BS EN 60947-6-1.

710.55　Other equipment

710.553.1　Socket-outlet circuits in the medical IT system for medical locations of Group 2

Socket-outlets intended to supply medical electrical equipment shall be unswitched.

At each patient's place of treatment, e.g. bedheads, the configuration of socket-outlets shall be as follows:

(i)　each socket-outlet supplied by an individually protected circuit, or

(ii)　several socket-outlets separately supplied by a minimum of two circuits.

Socket-outlets used on medical IT systems, shall be coloured blue and clearly and permanently marked "Medical equipment only".

NOTE:　See also Health Technical Memorandum (HTM) 06-01 (Part A).

710.559　Luminaires and lighting installations

In medical locations of Group 1 and Group 2, at least two different sources of supply shall be provided. One of the sources shall be connected to the electrical supply system for safety services.

710.56　Safety services

In medical locations a power supply for safety purposes is required, which, in accordance with the standard, will energize the installations needed for continuous operation in case of failure of the general power system, for a defined period within a pre-set changeover time.

The safety power supply system shall automatically take over if the voltage of one or more incoming live conductors of the main distribution board of the building with the main power supply has dropped for more than 0.5 s and by more than 10 % in regard to the nominal voltage; refer to Annex A710 Table A710 – List of examples.

710.560.4　Classification

Classification of safety services is given in Regulation 560.4.1.

NOTE:　Safety services provided for locations having differing classifications should meet that classification which gives the highest security of supply.

710.560.5.5 General requirements for safety power supply sources of Group 1 and Group 2

Primary cells are not allowed as safety power sources.

An additional main incoming power supply, from the general power supply, is not regarded as a source of the safety power supply.

The availability (readiness for service) of safety power sources shall be monitored and indicated at a suitable location.

710.560.5.6 In case of a failure of the general power supply source, the power supply for safety services shall be energized to feed the equipment stated in Regulations 710.560.6.1.1 to 3 with electrical energy for a defined period of time and within a predetermined changeover period.

710.560.5.7 Where socket-outlets are supplied from the safety power supply source they shall be readily identifiable according to their safety services classification.

710.560.6 Electrical sources for safety services

710.560.6.1 Detailed requirements for safety power supply services

NOTE: Also refer to Regulation 710.560.5.5.

710.560.6.1.1 Power supply sources with a changeover period less than or equal to 0.5 s

In the event of a voltage failure on one or more line conductors at the distribution board, a safety power supply source shall be used and be capable of providing power for a period of at least 3 h for the following:

(i) Luminaires of operating theatre tables

(ii) Medical electrical equipment containing light sources being essential for the application of the equipment, e.g. endoscopes, including associated essential equipment, e.g. monitors

(iii) Life-supporting medical electrical equipment.

The normal power supply shall be restored within a changeover period not exceeding 0.5 s.

NOTE: The duration of 3 h may be reduced to 1 h if a power source meeting the requirements of Regulation 710.560.6.1.2 is installed.

710.560.6.1.2 Power supply sources with a changeover period less than or equal to 15 s

Equipment meeting the requirements of Regulations 710.560.9.1 and 710.560.11 shall be connected within 15 s to a safety power supply source capable of maintaining it for a minimum period of 24 h, when the voltage of one or more live conductors at the main distribution board for the safety services has decreased by more than 10 % of the nominal value of supply voltage and for a duration greater than 3 s.

710.560.6.1.3 Power supply sources with a changeover period greater than 15 s

Equipment, other than that covered by Regulations 710.560.6.1.1 and 710.560.6.1.2, which is required for the maintenance of hospital services, shall be connected either automatically or manually to a safety power supply source capable of maintaining it for a minimum period of 24 h. This equipment may include, for example:

(i) Sterilization equipment

(ii) Technical building installations, in particular air conditioning, heating and ventilation systems, building services and waste disposal systems

(iii) Cooling equipment

(iv) Catering equipment

(v) Storage battery chargers.

710.560.7 Circuits of safety services

The circuit which connects the power supply source for safety services to the main distribution board shall be considered a safety circuit.

710.560.9   Emergency lighting systems

710.560.9.1 Safety lighting

In the event of mains power failure, the changeover period to the safety services source shall not exceed 15 s. The necessary minimum illuminance shall be provided for the following:

(i) Emergency lighting and exit signs to BS 5266

(ii) Locations for switchgear and controlgear for emergency generating sets, for main distribution boards of the normal power supply and for power supply for safety services

(iii) Rooms in which essential services are intended. In each such room at least one luminaire shall be supplied from the power source for safety services

(iv) Locations of central fire alarm and monitoring systems

(v) Rooms of Group 1 medical locations. In each such room at least one luminaire shall be supplied from the power supply source for safety services

(vi) Rooms of Group 2 medical locations. A minimum of 90 % of the lighting shall be supplied from the power source for safety services.

The luminaires of the escape routes shall be arranged on alternate circuits.

710.560.11   Other services

Other services which may require a safety service supply with a changeover period not exceeding 15 s include, for example, the following:

(i) Selected lifts for firefighters

(ii) Ventilation systems for smoke extraction

(iii) Paging systems

(iv) Medical electrical equipment used in Group 2 medical locations which serves for surgical or other procedures of vital importance. Such equipment will be defined by responsible staff

(v) Electrical equipment of medical gas supply including compressed air, vacuum supply and narcosis (anaesthetics) exhaustion as well as their monitoring devices

(vi) Fire detection and fire alarms to BS 5839

(vii) Fire extinguishing systems.

710.6   Inspection and testing

710.61   Initial verification

The dates and results of each verification shall be recorded.

The tests specified below under items (i), (ii) and (iii) in addition to the requirements of Chapter 61 and HTM 06-01 (Part A) shall be carried out, both prior to commissioning and after alteration or repairs and before re-commissioning:

(i) Complete functional tests of the insulation monitoring devices (IMDs) associated with the medical IT system including insulation failure, transformer high temperature, overload, discontinuity and the acoustic/visual alarms linked to them

(ii) Measurements of leakage current of the output circuit and of the enclosure of the medical IT transformers in no-load condition, Regulation 710.512.1.1(i)

(iii) Measurements to verify that the resistance of the supplementary equipotential bonding is within the limits stipulated by Regulation 710.415.2.2.

710.62   Periodic inspection and testing

NOTE 1: Periodic inspection and testing should be carried out in accordance with Health Technical Memorandum (HTM) 06-01 (Part B) and local Health Authority requirements, if any.

NOTE 2: In addition to the requirements of Chapter 62, the following procedures are recommended at the given intervals:

(i) Annually - Complete functional tests of the insulation monitoring devices (IMDs) associated with the medical IT system including insulation failure, transformer high temperature, overload, discontinuity and the acoustic/visual alarms linked to them

(ii) Every 3 years - Measurements of leakage current of the output circuit and of the enclosure of the medical IT transformers in no-load condition, Regulation 710.512.1.1(i)

(iii) Annually - Measurements to verify that the resistance of the supplementary equipotential bonding is within the limits stipulated by Regulation 710.415.2.2.

Fig 710.1 – Example of patient environment (BS EN 60601)

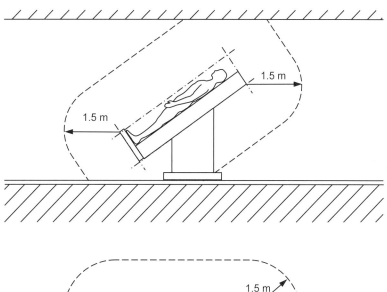

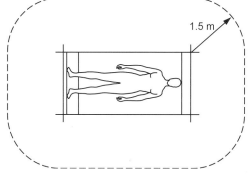

NOTE: The dimensions in the figure show the minimum
extent of the patient environment in a free surrounding.

Fig 710.2 – Typical theatre layout

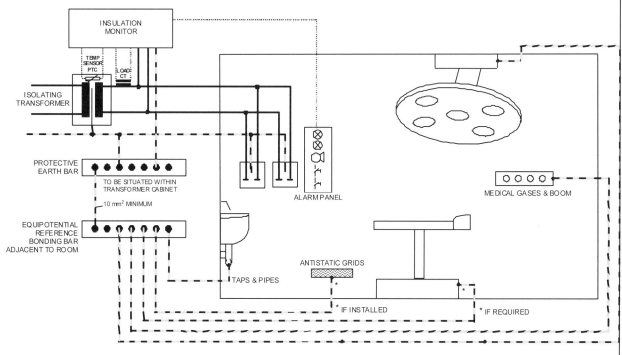

Annex A710 (informative)
Medical locations

Examples for allocation of Group numbers and classification for safety services of medical locations

A definitive list of medical locations showing their assigned Groups is impracticable, as is the use to which locations (rooms) might be put. The accompanying list of examples is provided as a guide only and should be read in conjunction with Regulation 710.3.

TABLE A710 – List of examples

Medical location	Group			Class	
	0	1	2	≤ 0.5 s	> 0.5 s ≤ 15 s
1 Massage room	X	X			X
2 Bedrooms		X			X
3 Delivery room		X		X[a]	X
4 ECG, EEG, EHG room		X			X
5 Endoscopic room		X[b]		X	X[b]
6 Examination or treatment room		X		X	X
7 Urology room		X[b]		X	X[b]
8 Radiological diagnostic and therapy room		X	X	X	X
9 Hydrotherapy room		X			X
10 Physiotherapy room		X			X
11 Anaesthetic area			X	X[a]	X
12 Operating theatre			X	X[a]	X
13 Operating preparation room			X	X[a]	X
14 Operating plaster room			X	X[a]	X
15 Operating recovery room			X	X[a]	X
16 Heart catheterization room			X	X[a]	X
17 Intensive care room			X	X[a]	X
18 Angiographic examination room			X	X[a]	X
19 Haemodialysis room			X		X
20 Magnetic resonance imaging (MRI) room		X	X	X	X
21 Nuclear medicine		X			X
22 Premature baby room			X	X[a]	X
23 Intermediate Care Unit (IMCU)			X	X	X

a   Luminaires and life-support medical electrical equipment which needs power supply within 0.5 s or less.

b   Not being an operating theatre.

SECTION 711

EXHIBITIONS, SHOWS AND STANDS

711.1 Scope

The particular requirements of this section apply to the temporary electrical installations in exhibitions, shows and stands (including mobile and portable displays and equipment) to protect users. Unless specifically stated, this section does not apply to exhibits for which requirements are given in the relevant standards.

This section does not apply to the fixed electrical installation of the building, if any, in which the exhibition, shows or stands may take place.

This section does not apply to electrical systems as defined in BS 7909 used in structures, sets, mobile units etc as used for public or private events, touring shows, theatrical, radio, TV or film productions and similar activities of the entertainment industry.

The requirements of other sections of Part 7 may also apply.

711.3 Assessment of general characteristics

711.313 Supplies

The nominal supply voltage of a temporary electrical installation in an exhibition, show or stand shall not exceed 230/400 V a.c. or 500 V d.c.

711.32 Classification of external influences

The external influence conditions of the particular location where the temporary electrical installation is erected, e.g. the presence of water or mechanical stresses, shall be taken into account.

711.41 Protection against electric shock

711.410.3 General requirements

711.410.3.4 A cable intended to supply temporary structures shall be protected at its origin by an RCD whose rated residual operating current does not exceed 300 mA. This device shall provide a delay by using a device in accordance with BS EN 60947-2, or be of the S-type in accordance with BS EN 61008-1 or BS EN 61009-1 for discrimination with RCDs protecting final circuits.

NOTE: The requirement for additional protection relates to the increased risk of damage to cables in temporary locations.

711.410.3.5 The protective measures of obstacles and placing out of reach (Section 417) are not permitted.

711.410.3.6 The protective measures of non-conducting location (Regulation 418.1) and earth-free local equipotential bonding (Regulation 418.2) are not permitted.

711.411 Protective measure: Automatic disconnection of supply

711.411.3.1.2 Protective equipotential bonding

Structural metallic parts which are accessible from within the stand, vehicle, wagon, caravan or container shall be connected through the main protective bonding conductors to the main earthing terminal within the unit.

711.411.3.3 Additional protection

Each socket-outlet circuit not exceeding 32 A and all final circuits other than for emergency lighting shall be protected by an RCD having the characteristics specified in Regulation 415.1.1.

711.411.4 TN system

Except for a part of an installation within a building, a PME earthing facility shall not be used as the means of earthing for an installation falling within the scope of this section except:

(i)   where the installation is continuously under the supervision of a skilled or instructed person, and

(ii) the suitability and effectiveness of the means of earthing has been confirmed before the connection is made.

711.414   Protective measure: Extra-low voltage provided by SELV or PELV

711.414.4.5 Where SELV or PELV is used, whatever the nominal voltage, basic protection shall be provided by:

(i)   basic insulation complying with Regulation 416.1, or

(ii) by barriers or enclosures complying with Regulation 416.2 and affording a degree of protection of at least IPXXD or IP4X.

711.42   Protection against thermal effects

711.422   Protection against fire

711.422.4.2 Heat generation

Lighting equipment such as incandescent lamps, spotlights and small projectors, and other equipment or appliances with high temperature surfaces, shall be suitably guarded, and installed and located in accordance with the relevant standard.

Showcases and signs shall be constructed of material having adequate heat-resistance, mechanical strength, electrical insulation and ventilation, taking into account the combustibility of exhibits in relation to the heat generation.

Stand installations containing a concentration of electrical equipment, luminaires or lamps liable to generate excessive heat shall not be installed unless adequate ventilation provisions are made, e.g. well ventilated ceiling constructed of incombustible material.

In all cases, the manufacturer's instructions shall be followed.

711.5   Selection and erection of equipment

711.51   Common rules

Switchgear and controlgear shall be placed in closed cabinets which can only be opened by the use of a key or a tool, except for those parts designed and intended to be operated by ordinary persons.

711.52   Wiring systems

Armoured cables or cables protected against mechanical damage shall be used wherever there is a risk of mechanical damage.

Wiring cables shall be copper, have a minimum cross-sectional area of 1.5 mm^2, and shall comply with an appropriate British or Harmonized Standard for either thermoplastic or thermosetting insulated electric cables.

Flexible cables shall not be laid in areas accessible to the public unless they are protected against mechanical damage.

711.521   Types of wiring system

Where no fire alarm system is installed in a building used for exhibitions etc. cable systems shall be either:

(i)   flame retardant to BS EN 60332-1-2 or to a relevant part of the BS EN 60332-3 series, and low smoke to BS EN 61034-2, or

(ii) single-core or multicore unarmoured cables enclosed in metallic or non-metallic conduit or trunking, providing fire protection in accordance with BS EN 61386 series or BS EN 50085 series and providing a degree of protection of at least IP4X.

711.526   Electrical connections

711.526.1   Joints shall not be made in cables except where necessary as a connection into a circuit. Where joints are made, these shall either use connectors in accordance with relevant standards or be in enclosures with a degree of protection of at least IPXXD or IP4X.

Where strain can be transmitted to terminals the connection shall incorporate suitable cable anchorage(s).

711.537.2   Isolation

711.537.2.3 Every separate temporary structure, such as a vehicle, stand or unit, intended to be occupied by one specific user and each distribution circuit supplying outdoor installations shall be provided with its own readily accessible and properly identifiable means of isolation. The means of isolation shall be selected and erected in accordance with Regulation 537.2.

711.55 Other equipment

711.55.1.5 All such equipment shall be so fixed and protected that a focusing or concentration of heat is not likely to cause ignition of any material.

711.55.4 Electric motors

711.55.4.1 Isolation

Where an electric motor might give rise to a hazard, the motor shall be provided with an effective means of isolation on all poles and such means shall be adjacent to the motor which it controls (see BS EN 60204-1).

711.55.6 ELV transformers and electronic convertors

A manual reset protective device shall protect the secondary circuit of each transformer or electronic convertor.

Particular care shall be taken when installing ELV transformers, which shall be mounted out of arm's reach of the public and shall have adequate ventilation. Access by a competent person for testing and by a skilled person competent in such work for maintenance shall be provided.

Electronic convertors shall conform with BS EN 61347-1.

711.55.7 Socket-outlets and plugs

An adequate number of socket-outlets shall be installed to allow user requirements to be met safely.

Where a floor mounted socket-outlet is installed, it shall be adequately protected from accidental ingress of water and have sufficient strength to be able to withstand the expected traffic load.

711.559 Luminaires and lighting installations

711.559.4.2 ELV lighting systems for filament lamps

Extra-low voltage systems for filament lamps shall comply with BS EN 60598-2-23.

711.559.4.3 Lampholders

Insulation piercing lampholders shall not be used unless the cables and lampholders are compatible, and providing the lampholders are non-removable once fitted to the cable.

711.559.4.4 Electric discharge lamp installations

Installations of any luminous tube, sign or lamp as an illuminated unit on a stand, or as an exhibit, with nominal power supply voltage higher than 230/400 V a.c., shall comply with Regulations 711.559.4.4.1 to 3.

711.559.4.4.1 Location

The sign or lamp shall be installed out of arm's reach or shall be adequately protected to reduce the risk of injury to persons.

711.559.4.4.2 Installation

The facia or stand fitting material behind luminous tubes, signs or lamps shall be non-ignitable.

711.559.4.4.3 Emergency switching device

A separate circuit shall be used to supply signs, lamps or exhibits, which shall be controlled by an emergency switch. The switch shall be easily visible, accessible and clearly marked.

711.559.5 Protection against thermal effects

Luminaires mounted below 2.5 m (arm's reach) from floor level or otherwise accessible to accidental contact shall be firmly and adequately fixed, and so sited or guarded as to prevent risk of injury to persons or ignition of materials.

NOTE: In the case of outdoor lighting installations, Section 559 also applies, and a degree of protection of at least IP33 may be required.

711.6 Inspection and testing

The temporary electrical installations of exhibitions, shows and stands shall be inspected and tested on site in accordance with Chapter 61 after each assembly on site.

SECTION 712

SOLAR PHOTOVOLTAIC (PV) POWER SUPPLY SYSTEMS

712.1 Scope

The particular requirements of this section apply to the electrical installations of PV power supply systems including systems with a.c. modules.

NOTE: Requirements for PV power supply systems which are intended for stand-alone operation are under consideration.

712.3 Assessment of general characteristics

712.31 Purpose, supplies and structure

712.312 System earthing

712.312.2 Type of earthing arrangement

Earthing of one of the live conductors of the d.c. side is permitted, if there is at least simple separation between the a.c. side and the d.c. side.

NOTE: Any connections with Earth on the d.c. side should be electrically connected so as to avoid corrosion (see BS 7361-1:1991).

712.4 Protection for safety

712.41 Protection against electric shock

712.410.3 General requirements

PV equipment on the d.c. side shall be considered to be energized, even when the system is disconnected from the a.c. side.

712.410.3.6 The protective measures of non-conducting location (Regulation 418.1) and earth-free local equipotential bonding (Regulation 418.2) are not permitted on the d.c. side.

712.411 Protective measure: Automatic disconnection of supply

712.411.3.2.1.1 On the a.c. side, the PV supply cable shall be connected to the supply side of the protective device for automatic disconnection of circuits supplying current-using equipment.

712.411.3.2.1.2 Where an electrical installation includes a PV power supply system without at least simple separation between the a.c. side and the d.c. side, an RCD installed to provide fault protection by automatic disconnection of supply shall be type B according to IEC 62423.

Where the PV convertor is, by construction, not able to feed d.c. fault currents into the electrical installation, an RCD of type B according to IEC 62423 is not required.

712.412 Protective measure: Double or reinforced insulation

Protection by the use of Class II or equivalent insulation shall preferably be adopted on the d.c. side.

712.414 Protective measure: Extra-low voltage provided by SELV or PELV

712.414.1 General

712.414.1.1 For SELV and PELV systems, $U_{OC\ STC}$ replaces U_0 and shall not exceed 120 V d.c.

712.433 Protection against overload on the d.c. side

712.433.1 Overload protection may be omitted to PV string and PV array cables when the continuous current-carrying capacity of the cable is equal to or greater than 1.25 times $I_{SC\ STC}$ at any location.

712.433.2 Overload protection may be omitted to the PV main cable if the continuous current-carrying capacity is equal to or greater than 1.25 times $I_{SC\ STC}$ of the PV generator.

NOTE: The requirements of Regulations 712.433.1 and 712.433.2 are only relevant for protection of the cables. See also the manufacturer's instructions for protection of PV modules.

712.434 Protection against fault current

712.434.1 The PV supply cable on the a.c. side shall be protected against fault current by an overcurrent protective device installed at the connection to the a.c. mains.

712.444 Protection against electromagnetic disturbances

712.444.4.4 To minimize voltages induced by lightning, the area of all wiring loops shall be as small as possible.

712.5 Selection and erection of equipment

712.51 Common rules

712.511 Compliance with standards

712.511.1   PV modules shall comply with the requirements of the relevant equipment standard, e.g. BS EN 61215 for crystalline PV modules. PV modules of Class II construction or with equivalent insulation are recommended if $U_{oc\,STC}$ of the PV strings exceeds 120 V d.c.

The PV array junction box, PV generator junction box and switchgear assemblies shall be in compliance with BS EN 60439-1.

712.512 Operational conditions and external influences

712.512.1.1   Electrical equipment on the d.c. side shall be suitable for direct voltage and direct current.

PV modules may be connected in series up to the maximum allowed operating voltage of the PV modules ($U_{oc\,STC}$ of the PV strings) and the PV convertor, whichever is lower. Specifications for this equipment shall be obtained from the equipment manufacturer.

If blocking diodes are used, their reverse voltage shall be rated for 2 x $U_{oc\,STC}$ of the PV string. The blocking diodes shall be connected in series with the PV strings.

712.512.2.1   As specified by the manufacturer, the PV modules shall be installed in such a way that there is adequate heat dissipation under conditions of maximum solar radiation for the site.

712.513 Accessibility

712.513.1   The selection and erection of equipment shall facilitate safe maintenance and shall not adversely affect provisions made by the manufacturer of the PV equipment to enable maintenance or service work to be carried out safely.

712.52 Selection and erection of wiring systems

712.522 Selection and erection of wiring systems in relation to external influences

712.522.8.1 PV string cables, PV array cables and PV d.c. main cables shall be selected and erected so as to minimize the risk of earth faults and short-circuits.

NOTE: This may be achieved, for example, by reinforcing the protection of the wiring against external influences by the use of single-core sheathed cables.

712.522.8.3 Wiring systems shall withstand the expected external influences such as wind, ice formation, temperature and solar radiation.

712.53 Isolation, switching and control

712.537 Isolation and switching

712.537.2 Isolation

712.537.2.1.1   To allow maintenance of the PV convertor, means of isolating the PV convertor from the d.c. side and the a.c. side shall be provided.

NOTE: Further requirements with regard to the isolation of a PV installation operating in parallel with the public supply system are given in Regulation 551.7.6.

712.537.2.2 Devices for isolation

712.537.2.2.1   In the selection and erection of devices for isolation and switching to be installed between the PV installation and the public supply, the public supply shall be considered the source and the PV installation shall be considered the load.

712.537.2.2.5   A switch-disconnector shall be provided on the d.c. side of the PV convertor.

712.537.2.2.5.1 All junction boxes (PV generator and PV array boxes) shall carry a warning label indicating that parts inside the boxes may still be live after isolation from the PV convertor.

712.54 Earthing arrangements and protective conductors

Where protective bonding conductors are installed, they shall be parallel to and in as close contact as possible with d.c. cables and a.c. cables and accessories.

Fig 712.1 – PV installation - General schematic - One array

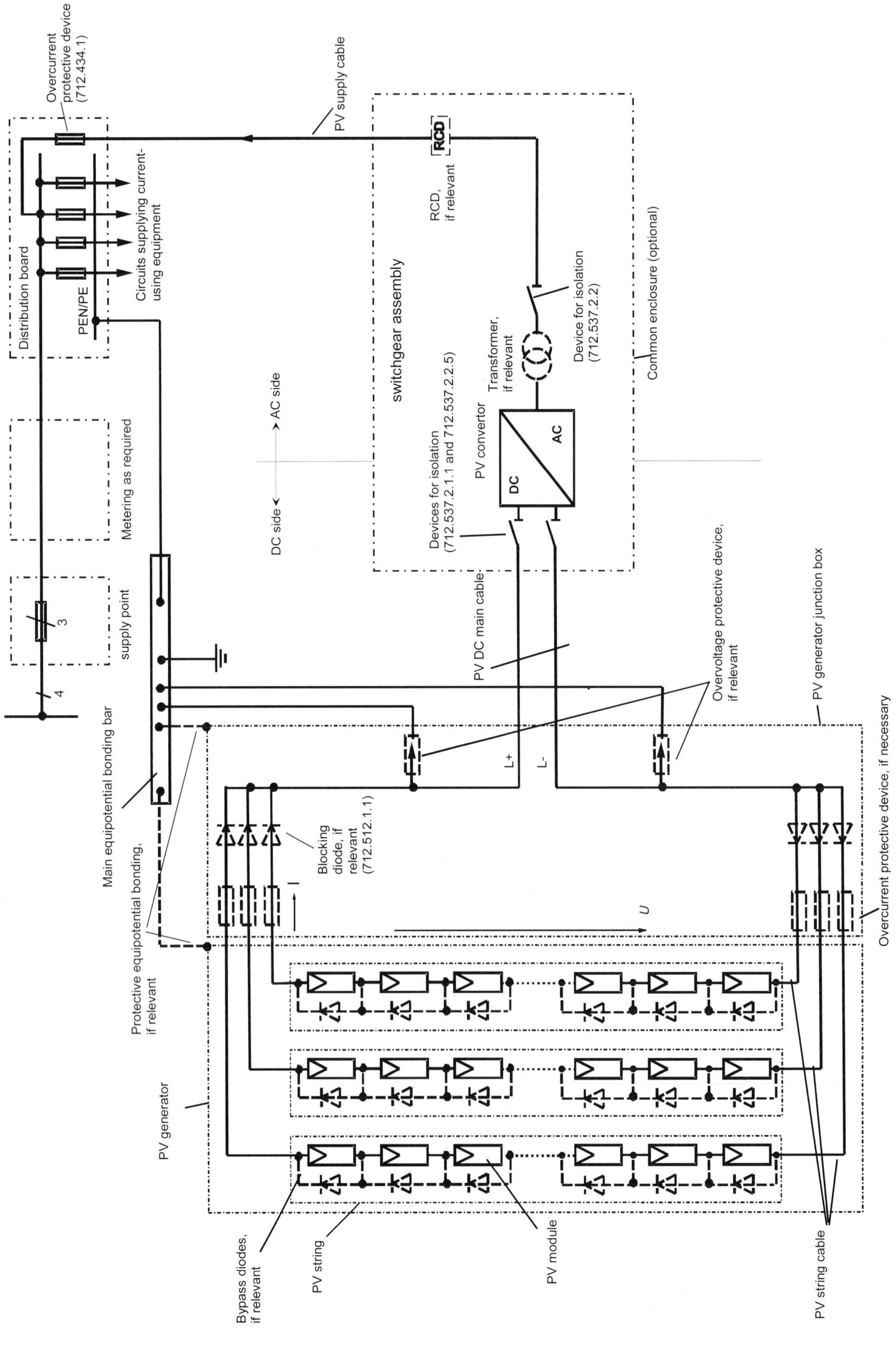

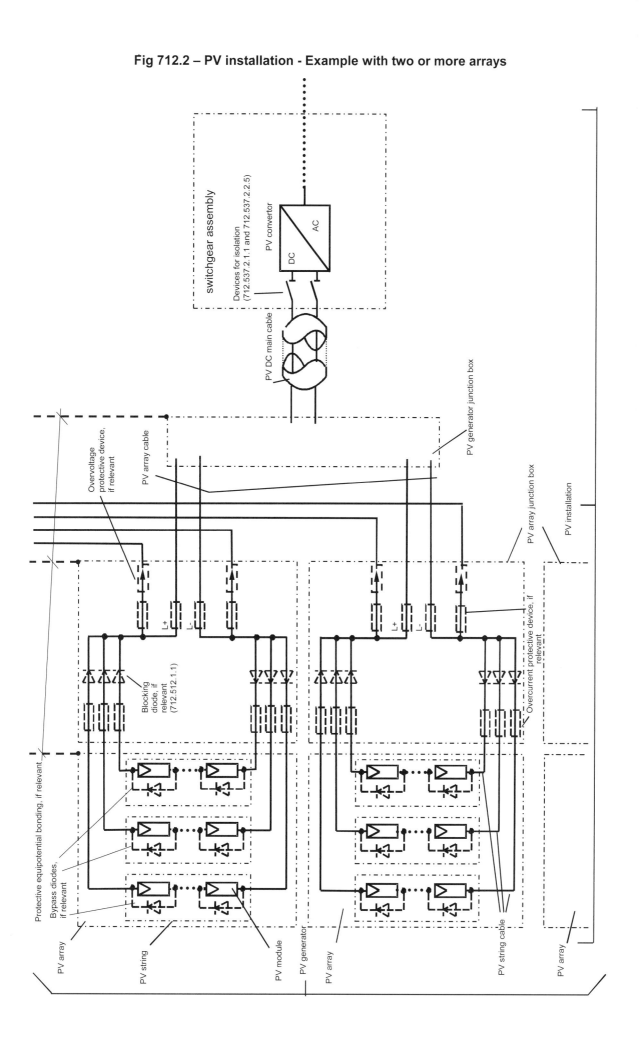

Fig 712.2 – PV installation - Example with two or more arrays

switchgear assembly

Devices for isolation
(712.537.2.1.1 and 712.537.2.2.5)

PV convertor

DC

AC

PV DC main cable

PV generator junction box

Overvoltage protective device, if relevant

PV array cable

PV array junction box

PV installation

Overcurrent protective device, if relevant

L+

L-

Blocking diode, if relevant (712.512.1.1)

Protective equipotential bonding, if relevant

Bypass diodes, if relevant

PV array

PV string

PV module

PV generator

PV array

PV string cable

PV array

SECTION 717

MOBILE OR TRANSPORTABLE UNITS

717.1 Scope

The particular requirements of this section apply to a.c. and d.c. installations for mobile or transportable units.

For the purposes of this section, the term "unit" is intended to mean a vehicle and/or mobile or transportable structure in which all or part of an electrical installation is contained, which is provided with a temporary supply by means of, for example, a plug and socket-outlet.

Units are either:

(i) of the mobile type, e.g. vehicles (self-propelled or towed), or

(ii) of the transportable type, e.g. containers or cabins.

Examples of the units include technical and facilities vehicles for the entertainment industry, medical or health screening services, welfare units, promotion & demonstration, firefighting, workshops, offices, transportable catering units etc.

The requirements are not applicable to:

(iii) generating sets

(iv) marinas and pleasure craft

(v) mobile machinery in accordance with BS EN 60204-1

(vi) caravans to Section 721

(vii) traction equipment of electric vehicles

(viii) electrical equipment required by a vehicle to allow it to be driven safely or used on the highway.

Additional requirements shall be applied where necessary for units including showers, or for medical locations, etc.

NOTE: Guidance on temporary electrical systems for entertainment and related purposes is given in BS 7909.

717.313 Supplies

One or more of the following methods shall be used to supply a unit:

(i) Connection to a low voltage generating set in accordance with Section 551 (see Figures 717.1 and 717.2)

(ii) Connection to a fixed electrical installation in which the protective measures are effective (see Figure 717.3)

(iii) Connection through means providing simple separation, in accordance with Section 413, from a fixed electrical installation (see Figures 717.4, 717.6 and 717.7)

NOTE 1: In cases (i), (ii) and (iii), an earth electrode may be provided where supplies are used external to the vehicle (see Regulation 717.411.4).

NOTE 2: In the case of Figure 717.4, an earth electrode may be necessary for protective purposes (see Regulation 717.411.6.2(ii)).

NOTE 3: Simple separation or electrical separation is appropriate, for example, where information technology equipment is used in the unit or where a reduction of electromagnetic disturbances is necessary, or if high leakage currents are to be expected (use of frequency convertors), and/or if the supply to the unit comes from alternative supply systems (as is the case in disaster management).

The sources, means of connection or separation may be within the unit.

NOTE 4: Where there is a potential hazard due to moving the unit whilst connected to an external installation, it is recommended that the unit is equipped with an electrical interlock, warning, alarm or other appropriate means to reduce the risk.

NOTE 5: For the purpose of this section, power inverters or frequency convertors supplied from the unit's electrical system or an auxiliary system driven by the unit's prime mover are also considered as generating sets.

Power inverters or frequency convertors shall include electrical separation where both the d.c. supply and the a.c. neutral point are earthed.

717.4 Protection for safety

717.41 Protection against electric shock

717.410.3 General requirements

717.410.3.5 The protective measures of obstacles and placing out of reach (Section 417) are not permitted.

717.410.3.6 The protective measure of non-conducting location (Regulation 418.1) is not permitted

The protective measure of earth-free local equipotential bonding (Regulation 418.2) is not recommended.

717.411 Protective measure: Automatic disconnection of supply

717.411.1 General

Automatic disconnection of the supply shall be provided by means of an RCD.

717.411.3.1.2 Protective equipotential bonding

Accessible conductive parts of the unit, such as the chassis, shall be connected through the main protective bonding conductors to the main earthing terminal within the unit. The main protective bonding conductors shall be finely stranded.

NOTE: Cable types H05V-K and H07V-K to BS 6004 are considered appropriate.

717.411.4 TN system

A PME earthing facility shall not be used as the means of earthing for an installation falling within the scope of this section except:

   (i)   where the installation is continuously under the supervision of a skilled or instructed person, and

   (ii)   the suitability and effectiveness of the means of earthing has been confirmed before the connection is made.

717.411.6 IT system

717.411.6.2 An IT system can be provided by:

   (i)   an isolating transformer or a low voltage generating set, with an insulation monitoring device or an insulation fault location system, both without automatic disconnection of the supply in case of the first fault and without a need of connection to an earthing installation (see Figure 717.7); the second fault shall be automatically disconnected by overcurrent protective devices according to Regulation 411.6.4, or

   (ii)   a transformer providing simple separation, e.g. in accordance with BS EN 61558-1, with an RCD and an earth electrode installed to provide automatic disconnection in the case of failure in the transformer providing the simple separation (see Figure 717.4)

717.415 Additional protection

717.415.1   Additional protection by an RCD having the characteristics specified in Regulation 415.1.1, shall be provided for every socket-outlet intended to supply current-using equipment outside the unit, with the exception of socket-outlets which are supplied from circuits with protection by:

   (i)   SELV, or

   (ii)   PELV, or

   (iii)   electrical separation.

717.5 Selection and erection of equipment

717.51 Common requirements

717.514 Identification and notices

A permanent notice of such durable material as to be likely to remain easily legible throughout the life of the installation, shall be fixed to the unit in a prominent position, preferably adjacent to each supply inlet connector. The notice should state in clear and unambiguous terms the following:

   (i)   The type of supplies which may be connected to the unit

   (ii)   The voltage rating of the unit

   (iii)   The number of supplies, phases and their configuration

   (iv)   The on-board earthing arrangement

   (v)   The maximum power requirement of the unit.

717.52   Wiring systems

717.52.1   Flexible cables in accordance with H07RN-F (BS 7919), or cables of equivalent design, having a minimum cross-sectional area of 2.5 mm² copper shall be used for connecting the unit to the supply. The flexible cable shall enter the unit by an insulating inlet in such a way as to minimize the possibility of any insulation damage or fault which might energize the exposed-conductive-parts of the unit.

717.52.2   The wiring system shall be installed using one or more of the following:

(i)   Unsheathed flexible cable with thermoplastic or thermosetting insulation to BS 6004 or BS 7211 installed in conduit in accordance with the appropriate part of the BS EN 61386 series or in trunking or ducting in accordance with the appropriate part of the BS EN 50085 series

(ii)   Sheathed flexible cable with thermoplastic or thermosetting insulated to BS 6004, BS 7211 or BS 7919, if precautionary measures are taken such that no mechanical damage is likely to occur due to any sharp-edged parts or abrasion.

717.528.3   Proximity to non-electrical services

717.528.3.4   No electrical equipment, including wiring systems, except ELV equipment for gas supply control, shall be installed in any gas cylinder storage compartment.

ELV cables and electrical equipment may only be installed within the LPG cylinder compartment if the installation serves the operation of the gas cylinder (e.g. indication of empty gas cylinders) or is for use within the compartment. Such electrical installations and components shall be constructed and installed so that they are not a potential source of ignition.

Where cables have to run through such a compartment, they shall be protected against mechanical damage by installation within a conduit system complying with the appropriate part of the BS EN 61386 series or within a ducting system complying with the appropriate part of the BS EN 50085 series.

Where installed, this conduit or ducting system shall be able to withstand an impact equivalent to AG3 without visible physical damage.

717.55   Other equipment

717.55.1   Connecting devices mounted, accessed or used outside the unit and used to connect the unit to the supply, or supply other equipment, shall comply with BS EN 60309-2 and shall meet the following requirements:

(i)   Connecting devices shall be within an enclosure of insulating material

(ii)   Connecting devices shall afford a degree of protection not less than IP44

(iii)   Enclosures containing the connecting devices shall provide a degree of protection of at least IP55 when no cable connections are made to the unit. When cable connections are made to the unit the enclosure shall provide a degree of protection not less than IP44.

717.55.2   *Moved by BS 7671:2008 Amendment No 1 to 717.55.1(iii).*

717.55.3   Generating sets able to produce voltages other than SELV or PELV, mounted in a mobile unit, shall automatically be switched off in case of an accident to the unit (e.g. event causing the release of airbags). If this requirement is difficult to implement an emergency switch, easily accessible, shall be installed.

Fig 717.1 – Example of connection to low voltage generating set located inside the unit, with or without an earth electrode

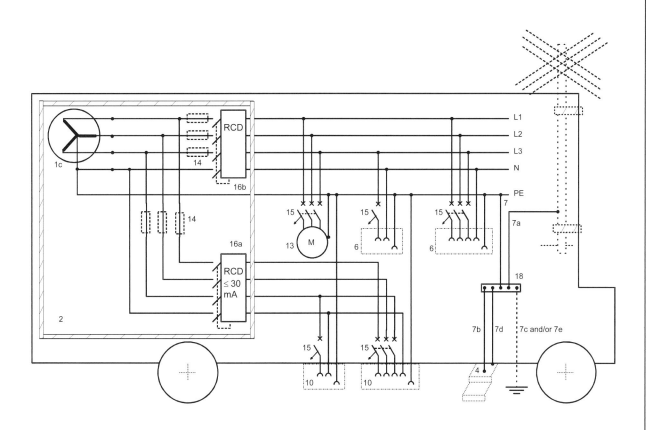

Fig 717.2 – Example of connection to a low voltage generating set located outside the unit, with or without an earth electrode

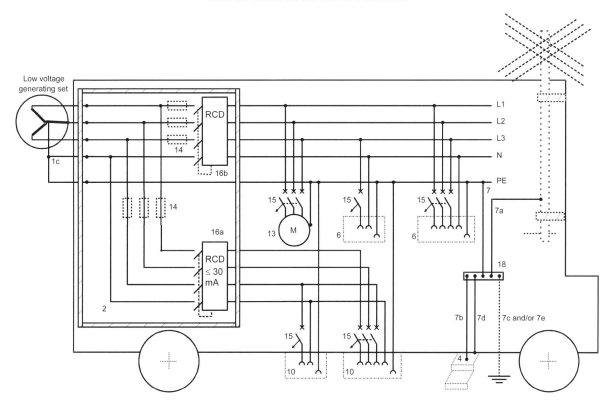

Fig 717.3 – Example of connection to a TN or TT electrical installation, with or without an earth electrode at the unit

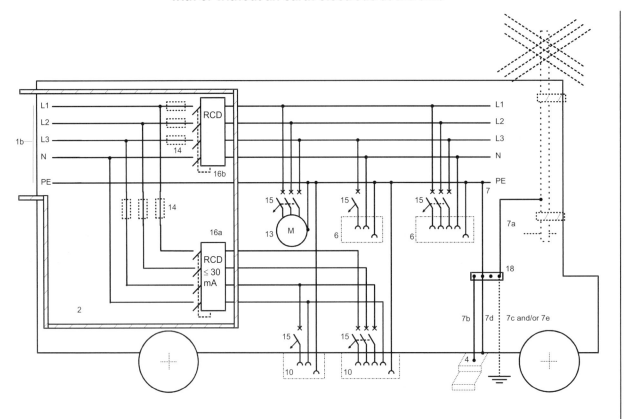

NOTE: Where a PME earthing facility is used, see Regulation 717.411.4.

Fig 717.4 – Example of connection to a fixed electrical installation with any type of earthing system using a simple separation transformer and an internal IT system, with an earth electrode

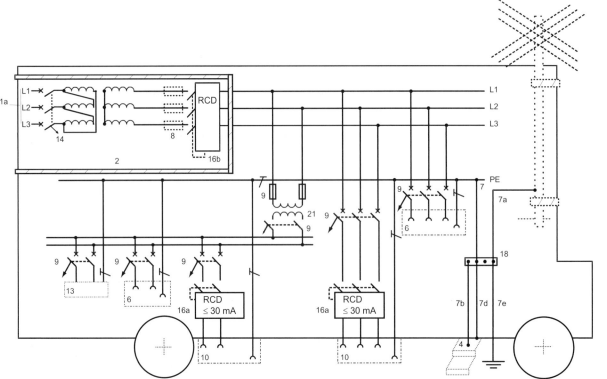

Fig 717.5 – Reserved for future use

Fig 717.6 – Example of connection to a fixed electrical installation with any type of earthing system using a simple separation transformer and an internal TN system, with or without an earth electrode

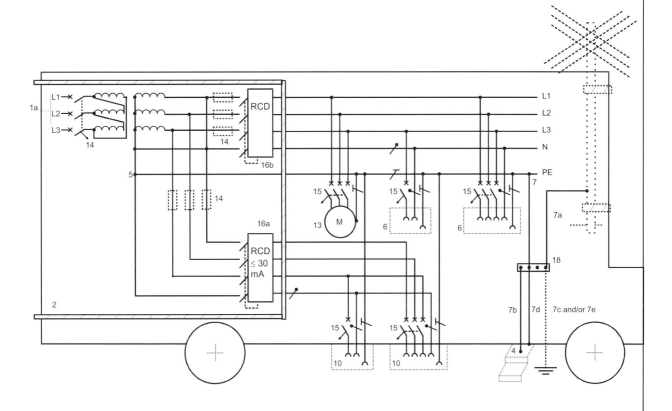

Fig 717.7 – Example of connection to a fixed electrical installation with any type of earthing system by using an IT system without automatic disconnection in the event of first fault

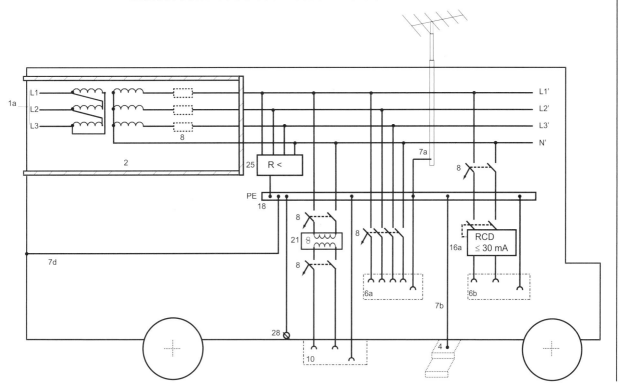

Key to Figures 717.1 to 7

1a   Connection of the unit to a supply through a transformer with simple separation

1b   Connection of the unit to a supply in which the protective measures are effective

1c   Connection to an LV generator set in accordance with Section 551

2   Class II or equivalent enclosure up to the first protective device providing automatic disconnection of supply

4   Conductive external staircase, if any

5   Connection of the neutral point (or, if not available, a line conductor) to the conductive structure of the unit

6   Socket-outlets for use exclusively within the unit

6a   Socket-outlets for use exclusively within the unit for reasons of continuity of supply in the event of first fault

6b   Socket-outlets for general use if explicitly required (operation of the RCD in the event of first fault cannot be excluded)

7   Main equipotential bonding in accordance with Regulation 717.411.3.1.2

7a   to an antenna pole, if any

7b   to the conductive external stairs, if any, in contact with the ground

7c   to a functional earth electrode, if required

7d   to the conductive structure of the unit

7e   to an earth electrode for protective purposes, if required

8   Protective devices, if required, for overcurrent and/or for protection by disconnection of supply in case of a second fault

9   Protective devices for overcurrent and for automatic disconnection of supply in case of a second fault

10   Socket-outlets for current-using equipment for use outside the unit

13   Current-using equipment for use exclusively within the unit

14   Overcurrent protective device, if required

15   Overcurrent protective device

16a   RCD having the characteristics specified in Regulation 415.1.1 for protection by automatic disconnection of supply for circuits of equipment for use outside the unit

16b   RCD for protection by automatic disconnection of supply for circuits of equipment for use inside the unit

18   Main earthing terminal or bar

21   Transformer with at least simple separation e.g. 230 V current-using equipment outside the unit

25   Insulation monitoring device or insulation fault location system including monitoring of the N conductor if distributed (disconnection only in the event of second fault)

28   Possible connection point to an existing lightning connection system in the vicinity (if any) for protection against lightning electromagnetic pulse (if any)

SECTION 721

ELECTRICAL INSTALLATIONS IN CARAVANS AND MOTOR CARAVANS

NOTE: In order not to mix regulations on different subjects, such as those for electrical installation of caravan parks with those for electrical installation inside caravans, two sections have been created:
Section 708, which concerns electrical installations in caravan parks, camping parks and similar locations and
Section 721, which concerns electrical installations in caravans and motor caravans.

721.1   Scope

The particular requirements of this section apply to the electrical installations of caravans and motor caravans at nominal voltages not exceeding 230/400 V a.c or 48 V d.c.

They do apply to those electrical circuits and equipment intended for the use of the caravan for habitation purposes.

They do not apply to those electrical circuits and equipment for automotive purposes, nor to installations covered by BS EN 1648-1 and BS EN 1648-2.

They do not apply to the electrical installations of mobile homes, residential park homes and transportable units.

NOTE 1: For mobile homes and residential park homes the general requirements apply.

NOTE 2: For transportable units see Section 717.

For the purposes of this section, caravans and motor caravans are referred to as "caravans".

The particular requirements of some other sections of Part 7 may also apply to such installations in caravans, e.g. Section 701.

721.31   Purpose, supplies and structure

721.313   Supplies

721.313.1.2   The nominal supply system voltage shall be chosen from IEC 60038.

The nominal a.c. supply voltage of the installation of the caravan shall not exceed 230 V single-phase, or 400 V three-phase.

The nominal d.c. supply voltage of the installation of the caravan shall not exceed 48 V.

721.41   Protection against electric shock

721.410.3   General requirements

721.410.3.3.1   Any portion of a caravan installation operating at extra-low voltage shall comply with the requirements of Section 414.

For extra-low voltage d.c. power sources, the following nominal voltages are generally applicable: 12 V, 24 V and 48 V.

In exceptional cases, where a.c. extra-low voltage is required, the following nominal voltages (rms) are generally applicable: 12 V, 24 V, 42 V and 48 V.

721.410.3.3.2   Except for shaver socket-outlets the protective measure electrical separation (Section 413) is not permitted.

721.410.3.5   The protective measures of obstacles and placing out of reach (Section 417) are not permitted.

721.410.3.6   The protective measures of non-conducting location (Regulation 418.1) and earth-free local equipotential bonding (Regulation 418.2) are not permitted.

721.411 Protective measure: Automatic disconnection of supply

721.411.1   Where protection by automatic disconnection of supply is used (Section 411), an RCD complying with BS EN 61008-1 or BS EN 61009-1 interrupting all live conductors shall be provided having the characteristics specified in Regulation 415.1.1, and the wiring system shall include a circuit protective conductor which shall be connected to:

  (i)   the protective contact of the inlet, and

  (ii)   the exposed-conductive-parts of the electrical equipment, and

  (iii)   the protective contacts of the socket-outlets.

721.411.3.1 Protective earthing and protective equipotential bonding

721.411.3.1.2   Protective equipotential bonding

Structural metallic parts which are accessible from within the caravan shall be connected through main protective bonding conductors to the main earthing terminal within the caravan.

721.43   Protection against overcurrent

721.43.1   Final circuits

Each final circuit shall be protected by an overcurrent protective device which disconnects all live conductors of that circuit.

721.5   Selection and erection of equipment

721.51   Common rules

721.510   Introduction

721.510.3   General

Where there is more than one electrically independent installation, each independent installation shall be supplied by a separate connecting device and shall be segregated in accordance with the relevant requirements of the Regulations.

721.514   Identification and notices

721.514.1   General

Instructions for use shall be provided with the caravan so that the caravan can be used safely.

The instructions shall comprise:

  (i)   a description of the installation

  (ii)   a description of the function of the RCD(s) and the use of the test button(s)

  (iii)   a description of the function of the main isolating switch

  (iv)   the text of the instructions of Figure 721.

If it is necessary to take precautions during user maintenance, appropriate details shall be given.

Fig 721 – Instructions for electricity supply

INSTRUCTIONS FOR ELECTRICITY SUPPLY

TO CONNECT

1. Before connecting the caravan installation to the mains supply, check that:

   (a) the supply available at the caravan pitch supply point is suitable for the caravan electrical installation and appliances, and

   (b) the voltage and frequency and current ratings are suitable, and

   (c) the caravan main switch is in the OFF position.

   Also, prior to use, examine the supply flexible cable to ensure there is no visible damage or deterioration.

2. Open the cover to the appliance inlet provided at the caravan supply point, if any, and insert the connector of the supply flexible cable.

3. Raise the cover of the electricity outlet provided on the pitch supply point and insert the plug of the supply cable.

THE CARAVAN SUPPLY FLEXIBLE CABLE MUST BE FULLY UNCOILED TO AVOID DAMAGE BY OVERHEATING

4. Switch on at the caravan main isolating switch.

5. Check the operation of residual current devices (RCDs) fitted in the caravan by depressing the test button(s) and reset.

IN CASE OF DOUBT OR, IF AFTER CARRYING OUT THE ABOVE PROCEDURE THE SUPPLY DOES NOT BECOME AVAILABLE, OR IF THE SUPPLY FAILS, CONSULT THE CARAVAN PARK OPERATOR OR THE OPERATOR'S AGENT OR A QUALIFIED ELECTRICIAN.

TO DISCONNECT

6. Switch off at the caravan main isolating switch, unplug the cable first from the caravan pitch supply point and then from the caravan inlet connector.

PERIODIC INSPECTION

Preferably not less than once every three years and annually if the caravan is used frequently, the caravan electrical installation and supply cable should be inspected and tested and a report on their condition obtained as prescribed in BS 7671 Requirements for Electrical Installations published by the Institution of Engineering and Technology and BSI.

721.521   Types of wiring system

721.521.2   The wiring systems shall be installed using one or more of the following:

   (i)   Insulated single-core cables, with flexible class 5 conductors, in non-metallic conduit

   (ii)   Insulated single-core cables, with stranded class 2 conductors (minimum of 7 strands), in non-metallic conduit

   (iii) Sheathed flexible cables.

All cables shall, as a minimum, meet the requirements of BS EN 60332-1-2.

Non-metallic conduits shall comply with BS EN 61386-21.

Cable management systems shall comply with BS EN 61386.

721.522   Selection and erection of wiring systems in relation to external influences

721.522.7   Vibration (AH)

721.522.7.1   As the wiring will be subjected to vibration, all wiring shall be protected against mechanical damage either by location or by enhanced mechanical protection. Wiring passing through metalwork shall be protected by means of suitable bushes or grommets, securely fixed in position. Precautions shall be taken to avoid mechanical damage due to sharp edges or abrasive parts.

721.522.8 Other mechanical stresses (AJ)

721.522.8.1.3 All cables, unless enclosed in rigid conduit, and all flexible conduit shall be supported at intervals not exceeding 0.4 m for vertical runs and 0.25 m for horizontal runs.

721.524 Cross-sectional areas of conductors

721.524.1 The cross-sectional area of every conductor shall be not less than 1.5 mm^2.

721.528 Proximity of wiring systems to other services

721.528.1 Proximity to electrical services

Cables of low voltage systems shall be run separately from the cables of extra-low voltage systems, in such a way, so far as is reasonably practicable, that there is no risk of physical contact between the two wiring systems.

721.528.3 Proximity to non-electrical services

721.528.3.4 No electrical equipment, including wiring systems, except ELV equipment for gas supply control, shall be installed in any gas cylinder storage compartment.

ELV cables and electrical equipment may only be installed within the LPG cylinder compartment if the installation serves the operation of the gas cylinder (e.g. indication of empty gas cylinders) or is for use within the compartment. Such electrical installations and components shall be constructed and installed so that they are not a potential source of ignition.

Where cables have to run through such a compartment, they shall be protected against mechanical damage by installation within a conduit system complying with the appropriate part of the BS EN 61386 series or within a ducting system complying with the appropriate part of the BS EN 50085 series.

Where installed, this conduit or ducting system shall be able to withstand an impact equivalent to AG3 without visible physical damage.

721.53 Protection, isolation, switching, control and monitoring

721.537 Isolation and switching

721.537.2 Isolation

721.537.2.1.1 Each installation shall be provided with a main disconnector which shall disconnect all live conductors and which shall be suitably placed for ready operation within the caravan. In an installation consisting of only one final circuit, the isolating switch may be the overcurrent protective device fulfilling the requirements for isolation.

721.537.2.1.1.1 A notice of such durable material as to be likely to remain easily legible throughout the life of the installation, shall be permanently fixed near the main isolating switch inside the caravan, bearing the text shown in Figure 721 in the appropriate language(s) in indelible and easily legible characters.

721.543 Protective conductors

721.543.2 Types of protective conductor

721.543.2.3 All circuit protective conductors shall be incorporated in a multicore cable or in a conduit together with the live conductors.

721.55 Other equipment

721.55.1 Inlets

721.55.1.1 Any a.c. electrical inlet on the caravan shall be an appliance inlet complying with BS EN 60309-1. If interchangeability is required the inlet shall comply with BS EN 60309-2.

721.55.1.2 The inlet shall be installed:

(i) not more than 1.8 m above ground level, and

(ii) in a readily accessible position, and

(iii) such that it shall have a minimum degree of protection of IP44 with or without a connector engaged, and

(iv) such that it shall not protrude significantly beyond the body of the caravan.

721.55.2 Accessories

721.55.2.1 Every low voltage socket-outlet, other than those supplied by an individual winding of an isolating transformer, shall incorporate an earth contact.

721.55.2.2 Every socket-outlet supplied at extra-low voltage shall have its voltage visibly marked.

721.55.2.3 Where an accessory is located in a position in which it is exposed to the effects of moisture it shall be constructed or enclosed so as to provide a degree of protection not less than IP44.

721.55.2.4 Each luminaire in a caravan shall preferably be fixed directly to the structure or lining of the caravan. Where a pendant luminaire is installed in a caravan, provision shall be made for securing the luminaire to prevent damage when the caravan is in motion.

Accessories for the suspension of pendant luminaires shall be suitable for the mass suspended and the forces associated with vehicle movement.

721.55.2.5 A luminaire intended for dual voltage operation shall comply with the appropriate standard.

721.55.2.6 The means of connection to the caravan pitch socket-outlet shall be supplied with the caravan and shall comprise the following (see Figure 708):

(i) A plug complying with BS EN 60309-2, and

(ii) a flexible cable of 25 m (±2 m) length, harmonized code designation H05RN-F (BS 7919) or equivalent, incorporating a protective conductor, with conductors to be identified according to Table 51 and of a cross-sectional area in accordance with Table 721, and

(iii) a connector, if any, compatible with the appliance inlet installed under Regulation 721.55.1.

TABLE 721 – Minimum cross-sectional areas of flexible cables for caravan connection

Rated current A	Minimum cross-sectional area mm^2
16	2.5
25	4
32	6
63	16
100	35

Annex A721 (Informative)
Guidance for extra-low voltage d.c. installations

NOTE: In general, the requirements of Section 721 are also applicable to an extra-low voltage d.c. installation. The following requirements should be applied in addition.

A721.31   Purpose, supplies and structure

A721.313   Supplies

A721.313.4   Sources of supply

The supply should be obtained from one or more of the following sources:

(i)   The electrical installation of the towing vehicle

(ii)   An auxiliary battery mounted in the caravan

(iii)   A low voltage d.c. supply via a transformer/rectifier unit complying with BS EN 60335-1 and BS EN 61558-2-6

(iv)   A d.c. generator that is driven by any form of energy

(v)   Solar photovoltaic (PV) power supply systems.

A721.514   Identification and notices

A721.514.1   General

The following information should be provided in the instructions for use and should be in the official language/s of the country in which the caravan is to be sold:

(i)   A warning worded as follows: "Any replacement of an auxiliary battery should be of the same type and specification as that originally fitted."

(ii)   Instructions on the maintenance and recharging of an auxiliary battery where it is fitted. Where a battery charger is provided, instructions on its safe use should be included.

(iii)   Instructions on selecting and installing an auxiliary battery, in a compartment, if the caravan installation is designed for the installation of an auxiliary battery.

(iv)   Details of the warning notice specified in A721.55.3.7 and its importance for safety.

(v)   In order to ensure safe operation of the electrical installation a simplified diagram of the wiring of the ELV and LV installation, with details of the cable colours and/or marking and the nominal values of the overcurrent protective devices should be provided.

(vi)   Type of appliances that can be used and from what source of supply.

(vii)   Instructions for the correct operation and maintenance of fitted appliances, as supplied by the appliance manufacturer.

(viii)   A warning worded as follows: "Always disconnect the electrical connector between the towing vehicle and the caravan before connecting an LV supply to the caravan and before charging the caravan battery by any other means."

A721.515   Prevention of mutual detrimental influence

A721.515.2 The ELV installation should be so installed that the protective measures of the LV installation for basic protection or for fault protection are not impaired.

It should be ensured that the protective conductors of the LV installation are not loaded by the operating currents of the ELV installation.

A721.521   Types of wiring system

A721.521.2   Cables should be of stranded construction and should comply with BS 6004, BS 6500, BS 7211 or BS 7919.

A721.523 Current-carrying capacities of cables

A721.523.1 The cross-sectional areas of the fixed wiring should be such that the permissible voltage drop is not exceeded.

A721.525 Voltage drop in consumers' installations

Under normal service conditions the voltage at the terminals of any fixed current-using equipment should be greater than the lower limit corresponding to the British or Harmonized Standard relevant to the equipment. Where the equipment is not the subject of a British or Harmonized Standard, the voltage at the terminals should be such as not to impair the safe functioning of that equipment. In the absence of precise data a voltage drop of 0.8 V from the power supply to the equipment may be allowed.

The voltage drop between the plug of the connector to the towing vehicle or LV battery charger and the auxiliary battery should not exceed 0.3 V.

The charging current I_c (A) to determine the voltage drop is established by the following formula:

$$I_c = \frac{c \times 0.1}{t}$$

where:

I_c is the charging current in A

c   is the battery capacity in Ah

t   is the charging period one h.

NOTE: Some battery manufacturers now rate batteries in Watt/hours (Wh).

A721.528 Proximity of wiring systems to other services

A721.528.3 Proximity to non-electrical services

A721.528.3.4 Cable runs and LPG installations

Cables including those used for road lighting and signalling - see Regulation 721.528.3.4.

A721.53 Protection, isolation, switching, control and monitoring

A721.533 Devices for protection against overcurrent

A721.533.1 General requirements

A721.533.1.5   The overcurrent protective device for the power supply from the towing vehicle should be fitted as near as possible to the auxiliary battery, but in no case more than 1 000 mm away. The overcurrent protective device for the auxiliary battery should be fitted at the end of the battery cable and before the fixed installation. The ELV output of the transformer/ rectifier unit and of the d.c. generator should be provided before distribution with an overcurrent protective device unless this is already incorporated within the device.

A721.533.1.6   Overcurrent protective devices should be either fuse links according to ISO 8820 or suitable circuit-breakers complying with BS EN 60898-2.

A721.533.1.7   Fuses should be protected to prevent accidental damage.

A721.533.1.8   Overcurrent protective devices should not be fitted in a fuel storage compartment or fuel storage housing intended for the storage of liquefied petroleum gas cylinders or in the compartment for housing an auxiliary battery.

A721.55 Other equipment

A721.55.1 Inlets

The inlet, when the plug is disconnected, should be protected against the ingress of water, foreign bodies and accidental damage.

A721.55.2 Accessories

A721.55.2.6 The means of connection to the towing vehicle should be supplied with the caravan and comprise the following:

(i) A plug complying with BS AU 149a and BS AU 177a or BS EN ISO 11446, and

(ii) a flexible cable with the number of cores with the minimum cross-sectional area and the allocation according to Table A721 and a length not exceeding 5 m, and

(iii) a connector complying with BS AU 149a and BS AU 177a or BS EN ISO 11446.

TABLE A721 – Functional allocation and cross-sectional areas of cores for caravan connectors

Core No.	Function	Contact numbers		Minimum cross-sectional area mm²
		BS EN ISO 11446	**BS AU 149a**	
1	Left-hand direction – indicator light	1	1	1.5
2	Rear fog light	2	2	1.5
3	Common return for core Nos. 1,2 and 4 to 8	3*	3*	2.5
4	Right-hand direction – indicator light	4	4	1.5
5	Right-hand rear position and marker lights, and rear registration-plate illumination device	5	5	1.5
6	Stop lights	6	6	1.5
7	Left-hand rear position and marker lights, and rear registration-plate illumination device	7	7	1.5
			BS AU 177a	
8	Reversing light	8	1	1.5
9	Continuous power supply	9	4	2.5
10	Power supply controlled by ignition switch	10	6	2.5
11	Return for core No. 10	11*	7*	2.5
12	Coding for coupled trailer	12	2	-
13	Return for core No. 9	13*	3*	2.5
14	No allocation	-	5	1.5

* These return circuits should not be connected electrically in the trailer.

A721.55.3 Auxiliary batteries

A721.55.3.1 Type of battery

An auxiliary battery should be of the rechargeable type.

NOTE: Non-rechargeable batteries are not auxiliary batteries. They may be used in caravans, provided that they are used in circuits separated from other sources of electrical supply.

A721.55.3.2 Capacity

An auxiliary battery should have a minimum capacity of 40 Ah at 20 h discharge rate.

NOTE: It is recommended to use a battery designed to be discharged over long periods at a relatively low current.

A721.55.3.3 Terminals

Auxiliary battery terminals should be clearly and durably marked "+" and "-". Connections to auxiliary battery terminals should be securely clamped or bolted to ensure continuous contact and should be insulated unless the auxiliary battery is provided with an insulating device.

A721.55.3.4 Location

An auxiliary battery should be placed in a separate compartment, with easy access for maintenance or removal, and secured to prevent movement of the battery, e. g. when the caravan is in motion.

A721.55.3.5 Auxiliary battery compartment

A tray should be installed under an auxiliary battery if the electrolyte of this battery is liquid.

The interior of an auxiliary battery compartment should be ventilated and protected against the corrosive effect of acid-laden gases, either by:

 (i)   installing a sealed auxiliary battery that incorporates an external ventilating kit that is taken to the exterior of the caravan, or

 (ii) installing an auxiliary battery in an enclosed battery compartment that is protected internally against corrosion and is ventilated to the exterior of the caravan by means of a suitable tube with a minimum inside diameter of 10 mm at the top of the auxiliary battery compartment, in accordance with the battery manufacturer's instructions or as supplied by the manufacturer of the auxiliary battery, or

 (iii) ventilating the compartment at low level and high level to the exterior of the caravan and constructing the interior of the compartment, including the sides of the ventilator openings, of acid-resistant material or providing it with an anticorrosive finish. If the compartment opens into the interior of the caravan, the lid should provide an air seal. The minimum free area of ventilation should be not less than 80 mm^2 at low level and not less than 80 mm^2 at high level.

If an auxiliary battery is not provided, then the position and instructions for the installation of the battery and compartment, in accordance with (i), (ii) or (iii), should be included in the instructions for use and a notice should be fixed in or near the proposed location stating: "For instructions on auxiliary battery installation, see the instructions for use".

The requirements concerning the protection against corrosion and ventilation are not applicable if batteries with bound electrolytes are used.

Where the manufacturer makes no provision for the installation of an auxiliary battery, the following statement should be made in the instructions for use: "This caravan has not been designed to accommodate an auxiliary battery. Do not fit one."

A721.55.3.6 Auxiliary battery cables

Cables from an auxiliary battery should be protected by additional sheathing or taping from the battery terminal up to the overcurrent protective device.

A721.55.3.7 Warning notice

A warning notice should be fixed in a prominent position near the auxiliary battery or displayed on the lid of the auxiliary battery compartment. This warning should be in the official language(s) of the country in which the caravan is to be sold and should state: "Switch off all appliances and lamps before disconnecting the auxiliary battery."

The auxiliary battery compartment should be additionally marked "Smoking prohibited" in accordance with BS 5499 and in the language(s) of the country in which the caravan is to be sold.

A721.55.4   Other sources of supply

A721.55.4.1 Generators and transformer/rectifier unit

If a supply is obtained from a generator or from a low voltage supply via a transformer/rectifier unit, the extra-low voltage at the output terminals of the supply unit should be maintained between 11 V minimum and 14 V maximum with applied loads varying from 0.5 A minimum up to the maximum rated load of the supply unit. Over the same load range, alternating voltage ripple should not exceed 1.2 V peak-to-peak.

A721.55.4.2 Regenerative sources

Regenerative energy sources, such as wind energy, solar energy etc., should be installed only for charging batteries.

Regenerative energy sources should only be operated with a device which prevents overcharging of the battery(ies).

A721.55.5 Charging of auxiliary battery and operation of refrigerator

A721.55.5.1 The circuit to charge an auxiliary battery should be separate from a circuit to operate a refrigerator.

A721.55.5.2 The charging circuit for an auxiliary battery should be completed only when the ignition of the towing vehicle is switched on.

A721.55.5.3 The 12 V heating facility of a refrigerator should be completed only when the ignition of the towing vehicle is switched on. This may be performed by a device built into the refrigerator.

A721.55.6 Terminal block

If the connection between the connecting cable(s) and the caravan's fixed wiring is by means of a terminal block, it should have a protective cover. If the terminal block is positioned externally it should have a cover with a degree of protection of at least IP34 according to BS EN 60529.

A721.55.7 Appliances

A721.55.7.1 General

The caravan manufacturer's technical specification should state whether an ELV appliance is suitable for use with a supply obtained from a d.c. generator or a transformer/rectifier unit.

Appliances suitable for operation on both 12 V a.c. and 12 V d.c. systems are allowed provided that a.c. and d.c. systems are segregated and interconnection is prevented.

A721.55.7.2 Selection and connection of appliances

All appliances should be fitted and connected in accordance with the appliance manufacturer's instructions. Where polarity-sensitive appliances are fitted and connected, only those should be used that have terminals clearly marked "-" and "+", or that have two conductors, indicating polarity by colour or by identification tags or sleeves marked "-" or " +".

A721.55.8 Socket-outlets

ELV socket-outlets should be two-pole non-reversible and should be of a different type from those provided for any low voltage installation. The voltage and maximum power rating of the circuit should be stated on or adjacent to the socket-outlets.

A721.55.9 Battery charger

If a battery charger is connected to a low voltage a.c. supply, it should comply with the relevant clauses of BS EN 60335-2-29. The d.c. output should either be electronically regulated or the maximum d.c. output of the charger in amperes should be limited to 10 % of the capacity of the auxiliary battery in Ah at 20 h discharge rate.

A721.55.10 External lights

Lights, such as door lamps, fixed outside on a caravan should be constructed or enclosed to provide protection against the ingress of water with a degree of protection of at least IP34 according to BS EN 60529.

SECTION 729

OPERATING AND MAINTENANCE GANGWAYS

729.1 Scope

The particular requirements of this section apply to basic protection and other aspects relating to the operation or maintenance of switchgear and controlgear within areas including gangways, where access is restricted to skilled or instructed persons.

729.3 Assessment of general characteristics

For restricted access areas the following apply:

(i) They shall be clearly and visibly marked by appropriate signs

(ii) They shall not provide access to unauthorised persons

(iii) Doors provided for closed restricted access areas shall allow easy evacuation by opening without the use of a key, tool or any other device not being part of the opening mechanism.

729.513 Accessibility

729.513.2 Requirements for operating and maintenance gangways

The width of gangways and access areas shall be adequate for work, operational access, emergency access, emergency evacuation and for transport of equipment.

Gangways shall permit at least a 90 degree opening of equipment doors or hinged panels (see also Annex A729).

729.513.2.1 Restricted access areas where basic protection is provided by barriers or enclosures

Where basic protection is provided by barriers or enclosures in accordance with Chapter 41, the following minimum dimensions apply (see Figure 729.1):

(i) Gangway width including between: barriers or enclosures and switch handles or circuit-breakers in the most onerous position, and barriers or enclosures or switch handles or circuit-breakers in the most onerous position and the wall	700 mm
(ii) Gangway width between barriers or enclosures or other barriers or enclosures and the wall	700 mm
(iii) Height of gangway to barrier or enclosure above floor	2000 mm
(iv) Live parts placed out of reach, see Regulation 417.3	2500 mm

NOTE: Where additional workspace is needed e.g. for special switchgear and controlgear assemblies, larger dimensions may be required.

Fig 729.1 – Gangways in installations with protection by barriers or enclosures

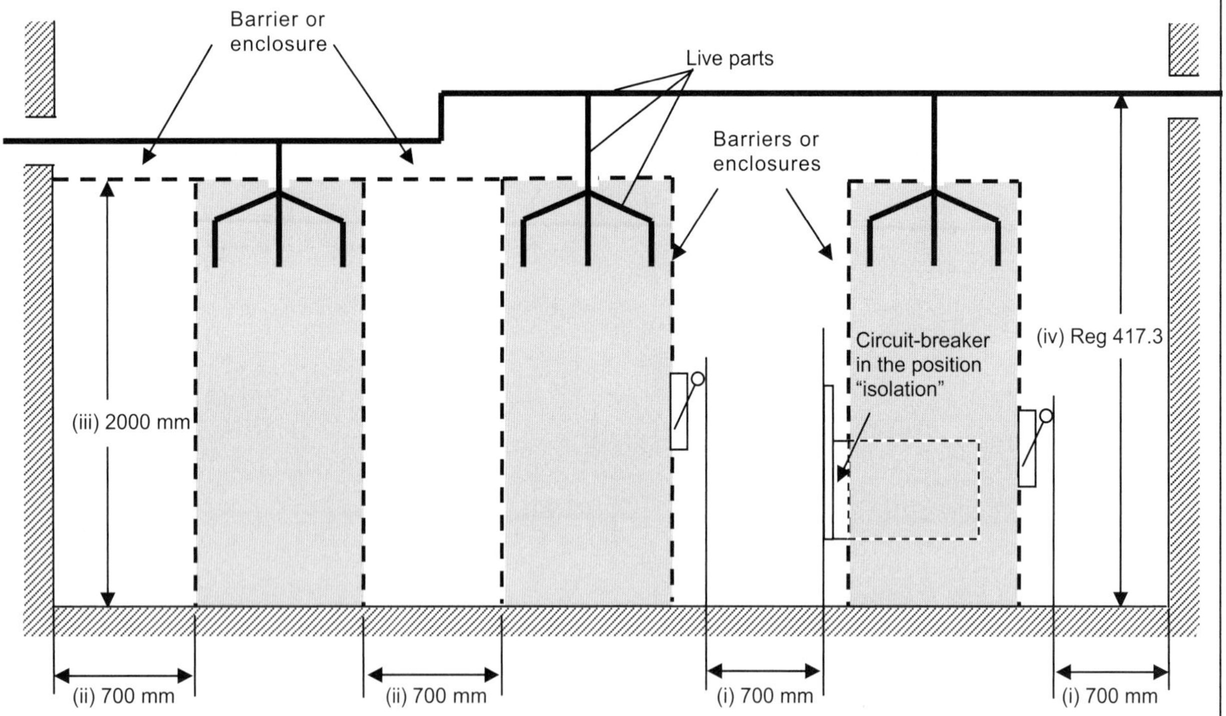

NOTE: The above dimensions apply after barriers and enclosures have been fixed and with circuit-breakers and switch handles in the most onerous position, including "isolation".

729.513.2.2 Restricted access areas where the protective measure of obstacles is applied

Where the protective measure of obstacles is used, the requirements of Section 417, Obstacles and placing out of reach, apply. The measure is for application in those parts of installations controlled or supervised by skilled persons.

The following minimum dimensions apply (see Figure 729.2):

(i) Gangway width including between: obstacles and switch handles or circuit-breakers in the most onerous position, and obstacles or switch handles or circuit-breakers in the most onerous position and the wall.	700 mm
(ii) Gangway width between obstacles or other obstacles and the wall	700 mm
(iii) Height of gangway to obstacles, barrier or enclosure above floor	2000 mm
(iv) Live parts placed out of reach, see Regulation 417.3	2500 mm

Fig 729.2 – Gangways in installations with protection by obstacles

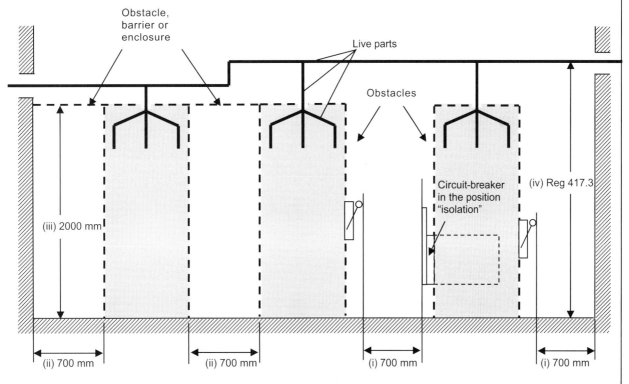

NOTE: The above dimensions apply after all obstacles, barriers and enclosures have been fixed and with circuit-breakers and switch handles in the most onerous position, including "isolation".

729.513.2.3 Access of gangways

Gangways longer than 10 m shall be accessible from both ends.

NOTE 1: This may be accomplished by placement of the equipment a minimum of 700 mm from all walls (see Figure 729.3) or by providing an access door, if needed, on the wall against which the equipment is positioned.

Closed restricted access areas with a length exceeding 20 m shall be accessible by doors from both ends.

NOTE 2: For closed restricted access areas with a length exceeding 6 m, accessibility from both ends is recommended.

Fig 729.3 – Examples of positioning of doors in closed restricted access areas

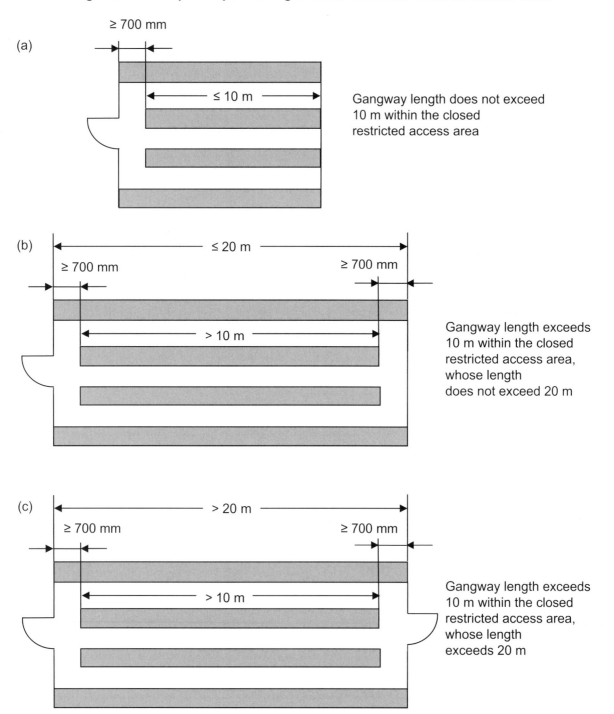

Doors giving access to gangways shall open outwards (see Figure 729.3) and they shall have the following minimum dimensions:

  (i) width   700 mm

  (ii) height 2000 mm.

Annex A729
(normative)
Additional requirements for closed restricted access areas

A729.1   Evacuation

For reason of easy evacuation the doors of any equipment inside the location shall close in the direction of the evacuation route (see Figure A729.1). Gangways shall permit at least a 90 degree opening of equipment doors or hinged panels (see Figure A729.2).

Fig A729.1 – Minimum passing width in case of evacuation – Case 1

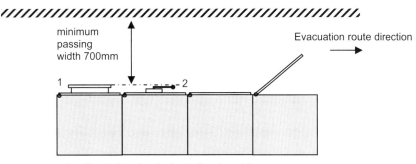

1   Circuit-breaker in the isolated position

2   Handles (e.g. for controls or equipment)

In the case of doors which can be fixed in the open position and circuit-breakers which are withdrawn fully for maintenance (completely extracted) a minimum distance of 500 mm shall be complied with between the door edge or circuit-breaker/ equipment edge and the opposite limitation of the gangway (see Figure A729.2).

Fig A729.2 – Minimum passing width in case of evacuation – Case 2

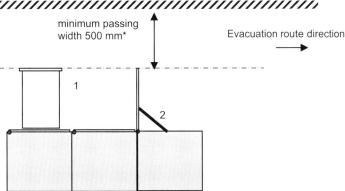

1   Circuit-breaker in the position "completely extracted"

2   Fixing device of a door

*   The minimum width gangway of 500 mm shall be taken into consideration between the wall and the circuit-breaker in the position "completely extracted" and equipment door in the 90 degree position.

SECTION 740

TEMPORARY ELECTRICAL INSTALLATIONS FOR STRUCTURES, AMUSEMENT DEVICES AND BOOTHS AT FAIRGROUNDS, AMUSEMENT PARKS AND CIRCUSES

740.1 Scope, object and fundamental principles

740.1.1 Scope

This section specifies the minimum electrical installation requirements to facilitate the safe design, installation and operation of temporarily erected mobile or transportable electrical machines and structures which incorporate electrical equipment. The machines and structures are intended to be installed repeatedly, without loss of safety, temporarily, at fairgrounds, amusement parks, circuses or similar places.

The object of this section is to define the electrical installation requirements for such structures and machines, being either integral parts or constituting the total amusement device.

This section does not apply to the internal electrical wiring of machines (see BS EN 60204-1).

NOTE 1: Guidance on temporary electrical systems for entertainment and related purposes is given in BS 7909.

NOTE 2: The permanent electrical installation is excluded from the scope.

740.3 Assessment of general characteristics

740.31 Purpose, supplies and structure

740.313 Supplies

740.313.1.1 Voltage

The nominal supply voltage of temporary electrical installations in booths, stands and amusement devices shall not exceed 230/400 V a.c. or 440 V d.c.

740.313.3 Supply from the public network

Irrespective of the number of sources of supply, the line and neutral conductors from different sources shall not be interconnected downstream of the origin of the temporary electrical installation. The instructions of the operator for the supply of the system to the public shall be followed.

740.4 Protection for safety

740.41 Protection against electric shock

740.410.3 General requirements

Automatic disconnection of supply to the temporary electrical installation shall be provided at the origin of the installation by one or more RCDs with a rated residual operating current not exceeding 300 mA. The RCD shall incorporate a time delay in accordance with BS EN 60947-2 or be of the S-type in accordance with BS EN 61008-1 or BS EN 61009-1 where necessary to provide discrimination with RCDs protecting final circuits.

740.410.3.5 The protective measure of obstacles (Regulation 417.2) is not permitted.

Placing out of arm's reach is acceptable for electric dodgems (see Regulation 740.55.9).

740.410.3.6 The protective measures of non-conducting location (Regulation 418.1) and earth-free local equipotential bonding (Regulation 418.2) are not permitted.

740.411 Protective measure: Automatic disconnection of supply

NOTE: For supplies to a.c. motors, RCDs, where used, should be of the time-delayed type in accordance with BS EN 60947-2 or be of the S-type in accordance with BS EN 61008-1 or BS EN 61009-1 where necessary to prevent unwanted tripping.

740.411.4 TN system

740.411.4.1 A PME earthing facility shall not be used as the means of earthing for an installation falling within the scope of this section.

NOTE: The ESQCR prohibit the use of a PME earthing facility as the means of earthing for the installation of a caravan or similar construction.

740.411.4.3 Where the type of system earthing is TN, a PEN conductor shall not be used downstream of the origin of the temporary electrical installation.

740.411.6 IT system

Where an alternative system is available, an IT system shall not be used. IT systems, however, may be used for d.c. applications where continuity of service is needed.

740.415 Additional protection

740.415.1 Additional protection: RCDs

All final circuits for:

(i) lighting,

(ii) socket-outlets rated up to 32 A, and

(iii) mobile equipment connected by means of a flexible cable with a current-carrying capacity up to 32 A

shall be protected by RCDs having the characteristics specified in Regulation 415.1.1.

NOTE: The requirement for additional protection relates to the increased risk of damage to cables.

The supply to a battery-operated emergency lighting circuit shall be connected to the same RCD protecting the lighting circuit.

This requirement does not apply to:

(iv) circuits protected by SELV or PELV, or

(v) circuits protected by electrical separation, or

(vi) lighting circuits placed out of arm's reach, provided they are not supplied by socket-outlets for household or similar purposes or socket-outlets according to BS EN 60309-1.

740.415.2 Additional protection: Supplementary equipotential bonding

740.415.2.1 In locations intended for livestock, supplementary bonding shall connect all exposed-conductive-parts and extraneous-conductive-parts that can be touched by livestock. Where a metal grid is laid in the floor, it shall be included within the supplementary bonding of the location (see Figure 705).

Extraneous-conductive-parts in, or on, the floor, e.g. concrete reinforcement in general or reinforcement of cellars for liquid manure, shall be connected to the supplementary equipotential bonding.

It is recommended that spaced floors made of prefabricated concrete elements be part of the equipotential bonding (see Figure 705). The supplementary equipotential bonding and the metal grid, if any, shall be erected so that it is durably protected against mechanical stresses and corrosion.

740.42 Protection against thermal effects

740.422.3 Nature of processed or stored materials

740.422.3.7 A motor which is automatically or remotely controlled and which is not continuously supervised shall be fitted with a manually reset protective device against excess temperature.

740.5 Selection and erection of equipment

740.51 Common rules

Switchgear and controlgear shall be placed in cabinets which can be opened only by the use of a key or a tool, except for those parts designed and intended to be operated by ordinary persons (BA1) as defined in Part 2 (see also Appendix 5).

740.512 Operational conditions and external influences

740.512.2 External influences

Electrical equipment shall have a degree of protection of at least IP44.

740.52 Wiring systems

740.521 Types of wiring system

740.521.1 Cables and cable management systems

Conduit systems shall comply with the relevant part of the BS EN 61386 series, cable trunking systems and cable ducting systems shall comply with the relevant part 2 of BS EN 50085, tray and ladder systems shall comply with BS EN 61537.

All cables shall meet the requirements of BS EN 60332-1-2.

Cables shall have a minimum rated voltage of 450 / 750 V, except that, within amusement devices, cables having a minimum rated voltage of 300 / 500 V may be used.

The routes of cables buried in the ground shall be marked at suitable intervals. Buried cables shall be protected against mechanical damage.

NOTE 1: Conduit classified as 450 N regarding protection against compression and classified as normal regarding protection against impact, according to BS EN 50086-2-4, is considered to fulfil the above requirement.

Armoured cables or cables protected against mechanical damage shall be used wherever there is a risk of mechanical damage due to external influence, e.g. > AG2. Mechanical protection shall be used in public areas and in areas where wiring systems are crossing roads or walkways.

The following methods are considered to meet the above requirements:

   (i)   conduit systems shall complying with BS EN 61386-21 with a classification of heavy regarding protection against compression, a classification of heavy regarding protection against impact, and, for metallic and composite conduit systems, class 3 protection against corrosion (i.e. medium protection inside and high protection outside)

   (ii)   cable trunking systems and cable ducting systems complying with BS EN 50085 series with a classification 5 J regarding protection against impact.

Where subjected to movement, wiring systems shall be of flexible construction. Where flexible conduit systems are provided they shall comply with BS EN 61386-23.

NOTE 2: Cables of type H07RNF or H07BN4-F (BS 7919) together with conduit complying with BS EN 61386-23 are deemed to satisfy this requirement.

740.526 Electrical connections

Joints shall not be made in cables except where necessary as a connection into a circuit. Where joints are made, these shall either use connectors in accordance with the relevant British or Harmonized Standard or the connection shall be made in an enclosure with a degree of protection of at least IPXXD or IP4X.

Where strain can be transmitted to terminals the connection shall incorporate cable anchorage(s).

740.53 Switchgear and controlgear

740.537 Isolation

740.537.1 General

Every electrical installation of a booth, stand or amusement device shall have its own means of isolation, switching and overcurrent protection, which shall be readily accessible.

740.537.2.1.1 Every separate temporary electrical installation for amusement devices and each distribution circuit supplying outdoor installations shall be provided with its own readily accessible and properly identified means of isolation.

740.537.2.2 Devices for isolation

A device for isolation shall disconnect all live conductors (line and neutral conductors).

740.55 Other equipment

740.55.1 Lighting installation

740.55.1.1 Luminaires

Every luminaire and decorative lighting chain shall have a suitable IP rating, be installed so as not to impair its ingress protection, and be securely attached to the structure or support intended to carry it. Its weight shall not be carried by the supply cable, unless it has been selected and erected for this purpose.

Luminaires and decorative lighting chains mounted less than 2.5 m (arm's reach) above floor level or otherwise accessible to accidental contact, shall be firmly fixed and so sited or guarded as to prevent risk of injury to persons or ignition of materials. Access to the fixed light source shall only be possible after removing a barrier or an enclosure which shall require the use of a tool.

Lighting chains shall use H05RN-F (BS 7919) cable or equivalent.

NOTE: Lighting chains may be used in any length provided the overcurrent protective device in the circuit is properly rated.

740.55.1.2 Lampholders

Insulation-piercing lampholders shall not be used unless the cables and lampholders are compatible and the lampholders are non-removable once fitted to the cable.

740.55.1.3 Lamps in shooting galleries

All lamps in shooting galleries and other sideshows where projectiles are used shall be suitably protected against accidental damage.

740.55.1.4 Floodlights

Where transportable floodlights are used, they shall be mounted so that the luminaire is inaccessible. Supply cables shall be flexible and have adequate protection against mechanical damage.

740.55.1.5 Fire risks from luminaires and floodlights

Luminaires and floodlights shall be so fixed and protected that a focusing or concentration of heat is not likely to cause ignition of any material.

740.55.3 Electric discharge lamp installations

Installations of any luminous tube, sign or lamp on a booth, stand or amusement device with an operating voltage higher than 230 V / 400 V a.c. shall comply with the Regulations 740.55.3.1 and 740.55.3.2.

740.55.3.1 Location

The luminous tube, sign or lamp shall be installed out of arm's reach or shall be adequately protected to reduce the risk of injury to persons.

740.55.3.2 Emergency switching device

A separate circuit shall be used to supply luminous tubes, signs or lamps, which shall be controlled by an emergency switch. The switch shall be easily visible, accessible and marked in accordance with the requirements of the local authority.

740.55.5 Safety isolating transformers and electronic convertors

Safety isolating transformers shall comply with BS EN 61558-2-6 or provide an equivalent degree of safety.

A manually reset protective device shall protect the secondary circuit of each transformer or electronic convertor.

Safety isolating transformers shall be mounted out of arm's reach or be mounted in a location that provides equal protection, e.g. in a panel or room that can only be accessed by a skilled or instructed person, and shall have adequate ventilation. Access by competent persons for testing or by a skilled person competent in such work for protective device maintenance shall be provided.

Electronic convertors shall conform to BS EN 61347-2-2.

Enclosures containing rectifiers and transformers shall be adequately ventilated and the vents shall not be obstructed when in use.

740.55.7 Socket-outlets and plugs

An adequate number of socket-outlets shall be installed to allow the user requirements to be met safely.

NOTE: In booths, stands and for fixed installations, one socket-outlet for each square metre or linear metre of wall is generally considered adequate.

Socket-outlets dedicated to lighting circuits placed out of arm's reach (in accordance with Regulation 740.415.1) shall be encoded or marked according to their purpose.

When used outdoor, plugs, socket-outlets and couplers shall comply with:
  (i)   BS EN 60309-2, or
  (ii) where interchangeability is not required, BS EN 60309-1.

However, socket-outlets according to the relevant National Standard may also be installed if they have suitable mechanical protection (equivalent to the requirements of BS EN 60309-1) and a rated current not exceeding 16 A.

NOTE: Suitable mechanical protection may be provided by the socket-outlet or by an enclosure.

740.55.8 Electrical supply

At each amusement device, there shall be a connection point readily accessible and permanently marked to indicate the following essential characteristics:
  (i)   Rated voltage
  (ii)   Rated current
  (iii) Rated frequency.

740.55.9 Electric dodgems

Electric dodgems shall only be operated at voltages not exceeding 50 V a.c. or 120 V d.c. The circuit shall be electrically separated from the supply mains by means of a transformer in accordance with BS EN 61558-2-4 or a motor-generator set.

740.551 Low voltage generating sets

740.551.8 Generators

All generators shall be so located or protected as to prevent danger and injury to people through inadvertent contact with hot surfaces and dangerous parts.

Electrical equipment associated with the generator shall be mounted securely and, if necessary, on anti-vibration mountings.

Where a generator supplies a temporary installation, forming part of a TN, TT or IT system, care shall be taken to ensure that the earthing arrangements shall be in accordance with Regulation 542.1 and, where earth electrodes are used, with Regulation 542.2.

The neutral conductor of the star-point of the generator shall, except for an IT system, be connected to the exposed-conductive-parts of the generator.

740.6 Inspection and testing

The electrical installation between its origin and any electrical equipment shall be inspected and tested after each assembly on site.

NOTE 1: Internal electrical wiring of roller coasters, electric dodgems and similar equipment are not considered as part of the verification.

NOTE 2: In special cases the number of the tests may be modified according to the type of temporary electrical installation.

SECTION 753

FLOOR AND CEILING HEATING SYSTEMS

753.1 Scope

This section applies to the installation of electric floor and ceiling heating systems which are erected as either thermal storage heating systems or direct heating systems. It does not apply to the installation of wall heating systems. Heating systems for use outdoors are not considered.

NOTE: A ceiling located under the roof of a building down to a vertical height of 1.50 m measured from the finished floor surface is also regarded as a ceiling within the meaning of these regulations.

753.4 Protection for safety

753.41 Protection against electric shock

753.410.3 General requirements

753.410.3.5 The protective measures of obstacles and placing out of reach (Section 417) are not permitted.

753.410.3.6 The protective measures of non-conducting location (Regulation 418.1) and earth-free local equipotential bonding (Regulation 418.2) are not permitted.

753 410.3.10 The protective measure of electrical separation (Section 413) is not permitted.

753.411 Protective measure: Automatic disconnection of supply

753.411.3.2 RCDs having the characteristics specified in Regulation 415.1.1 shall be used as disconnecting devices. In the case of heating units which are delivered from the manufacturer without exposed-conductive-parts, a suitable conductive covering, for example, a grid with a spacing of not more than 30 mm, shall be provided on site as an exposed-conductive-part above the floor heating elements or under the ceiling heating elements, and connected to the protective conductor of the electrical installation.

NOTE: Limitation of the rated heating power to 7.5 kW/230 V or 13 kW/400 V downstream of a 30 mA RCD may avoid unwanted tripping due to leakage capacitance. Values of leakage capacitance may be obtained from the manufacturer of the heating system.

753.415.1 Additional protection: RCDs

A circuit supplying heating equipment of Class II construction or equivalent insulation shall be provided with additional protection by the use of an RCD having the characteristics specified in Regulation 415.1.1.

753.42 Protection against thermal effects

753.423 Protection against burns

In floor areas where contact with skin or footwear is possible, the surface temperature of the floor shall be limited (for example, 35 °C).

753.424 Protection against overheating

753.424.3 Heating units

753.424.3.1 To avoid the overheating of floor or ceiling heating systems in buildings, one or more of the following measures shall be applied within the zone where heating units are installed to limit the temperature to a maximum of 80 °C:

(i) Appropriate design of the heating system

(ii) Appropriate installation of the heating system in accordance with the manufacturer's instructions

(iii) Use of protective devices.

Heating units shall be connected to the electrical installation via cold tails or suitable terminals.

Heating units shall be inseparably connected to cold tails, for example, by a crimped connection.

753.424.3.2 As the heating unit may cause higher temperatures or arcs under fault conditions, special measures to meet the requirements of Chapter 42 should be taken when the heating unit is installed close to easily ignitable building structures, such as placing on a metal sheet, in metal conduit or at a distance of at least 10 mm in air from the ignitable structure.

753.5 Selection and erection of equipment

753.51 Common rules

753.511 Compliance with standards

Flexible sheet heating elements shall comply with the requirements of BS EN 60335-2-96 or IEC 60800.

753.512 Operational conditions and external influences

753.512.1 Operational conditions

753.512.1.6 Precautions shall be taken not to stress the heating unit mechanically; for example, the material by which it is to be protected in the finished installation shall cover the heating unit as soon as possible.

753.512.2 External influences

753.512.2.5 Heating units for installation in ceilings shall have a degree of protection of not less than IPX1. Heating units for installation in a floor of concrete or similar material shall have a degree of ingress protection not less than IPX7 and shall have the appropriate mechanical properties.

753.514 Identification and notices

The designer of the installation/heating system or installer shall provide a plan for each heating system, containing the following details:

- (i) Manufacturer and type of heating units
- (ii) Number of heating units installed
- (iii) Length/area of heating units
- (iv) Rated power
- (v) Surface power density
- (vi) Layout of the heating units in the form of a sketch, a drawing, or a picture
- (vii) Position/depth of heating units
- (viii) Position of junction boxes
- (ix) Conductors, shields and the like
- (x) Heated area
- (xi) Rated voltage
- (xii) Rated resistance (cold) of heating units
- (xiii) Rated current of overcurrent protective device
- (xiv) Rated residual operating current of RCD
- (xv) The insulation resistance of the heating installation and the test voltage used
- (xvi) The leakage capacitance.

This plan shall be fixed to, or adjacent to, the distribution board of the heating system.

Furthermore, the requirements of Figure 753 apply to floor and ceiling heating systems to inform the owner and the user of the installation.

753.515 Prevention of mutual detrimental influence

753.515.4 Heating units shall not cross expansion joints of the building or structure.

753.515.5 The manufacturer's instructions concerning erection shall be followed during installation.

753.52 Wiring systems

753.520 Introduction

753.520.4 Heating-free areas

For the necessary attachment of room fittings, heating-free areas shall be provided in such a way that the heat emission is not prevented by such fittings.

753.522 Selection and erection in relation to external influence

753.522.1 Ambient temperature (AA)

753.522.1.3 For cold tails (circuit wiring) and control leads installed in the zone of heated surfaces, the increase of ambient temperature shall be taken into account.

753.522.4 Presence of solid foreign bodies (AE)

753.522.4.3 Where heating units are installed there shall be heating-free areas where drilling and fixing by screws, nails and the like are permitted. The installer shall inform other contractors that no penetrating means, such as screws for door stops, shall be used in the area where floor or ceiling heating units are installed.

Fig 753 – Information for the user of the installation

A description of the heating system shall be provided by the installer of the heating system to the owner of the building or his/her agent upon completion of the installation.

The description shall contain at least the following information:

a) Description of the construction of the heating system, which must include the installation depth of the heating units;

b) Location diagram with information concerning
   – the distribution of the heating circuits and their rated power;
   – the position of the heating units in each room;
   – conditions which have been taken into account when installing the heating units, for example, heating-free areas, complementary heating zones, unheated areas for fixing means penetrating into the floor covering;

c) Data on the control equipment used, with relevant circuit diagrams and the dimensioned position of floor temperature and weather conditions sensors, if any;

d) Data on the type of heating units and their maximum operating temperature.

The installer shall provide the owner with a description of the heating system including all necessary information, for example, to permit repair work. In addition, the installer shall provide instructions for use of the heating installation.

The designer/installer of the heating system shall hand over an appropriate number of instructions for use to the owner or his/her agent upon completion. One copy of the instructions for use shall be permanently fixed in or near each relevant distribution board.

The instructions for use shall include at least the following information:

a) Description of the heating system and its function;

b) Operation of the heating installation in the first heating period in the case of a new building, for example, regarding drying out;

c) Operation of the control equipment for the heating system in the dwelling area and the complementary heating zones as well, if any;

d) Information on restrictions on placing of furniture or similar. Information provided to the owner shall cover the restrictions, if any, including:
   • whether additional floor coverings are permitted, for example, carpets with a thickness of >10 mm may lead to higher floor temperatures which can adversely affect the performance of the heating system
   • where pieces of furniture solidly covering the floor and/or built-in cupboards may be placed on heating-free areas
   • where furniture, such as carpets, seating and rest furniture with pelmets, which in part do not solidly cover the floor, may not be placed in complementary heating zones, if any;

e) Information on restrictions on placing of furniture or similar;

f) In the case of ceiling heating systems, restrictions regarding the height of furniture. Cupboards of room height may be placed only below the area of ceiling where no heating elements are installed;

g) Dimensioned position of complementary heating zones and placing areas;

h) Statement that, in the case of thermal floor and ceiling heating systems, no fixing shall be made into the floor and ceiling respectively. Excluded from this requirement are unheated areas. Alternatives shall be given, where applicable.

APPENDICES

NOTE: Appendix 1 is normative, and is thus a requirement.
All other appendices are informative, and are provided as guidance.

APPENDIX 1 (Normative)

BRITISH STANDARDS TO WHICH REFERENCE IS MADE IN THE REGULATIONS

NOTE: Certain British Standards have been withdrawn since the issue of the 16th Edition in 2001. From the date of withdrawal, certificates and marks already awarded may continue to apply to production until a date specified in the superseding standard. During the period between these dates, the withdrawn standard may be specified in contracts. However, it should be noted that this appendix may not list such standards, as only current British Standards are listed with some references to superseded standards. Where standards are not dated they are a multiple standard.

BS or EN Number	Title	References
BS 67:1987 (1999)	Specification for ceiling roses	416.2.4 note 559.6.1.1
BS 88	The term "BS 88 series", when used in these Regulations, means BS 88-1, -2 and -3.	432.4 433.1.103 533.1 Table 53.4 Appx 4 sec 4 Appx 8 sec 4
BS 88-1:2007 BS 88-2:2010	Low-voltage fuses - Part 1: General requirements Low-voltage fuses - Part 2: Supplementary requirements for fuses for use by authorized persons (fuses mainly for industrial application). Examples of standardized systems of fuses A to J. BS 88-2:2010 has replaced BS 88-2.2:1988, and BS 88-6:1988 which have been withdrawn	Part 2 gG Gm Table 41.2 Table 41.4 Table 41.6 433.1.100 Appx 3 Fig 3A3(a) Fig 3A3(b)
BS 88-2.2:1988	Specification for fuses for use by authorized persons (mainly for industrial application). Additional requirements for fuses with fuse-links for bolted connections. Replaced by BS 88-2:2010 and withdrawn 1/3/2010	
BS 88-3:2010	Low-voltage fuses - Part 3: Supplementary requirements for fuses for use by unskilled persons (fuses mainly for household and similar applications) Examples of standardized systems of fuses A to F BS 88-3:2010 has replaced BS 1361:1971 which has been withdrawn.	Table 41.2 Table 41.4 433.1.100 533.1.1.2 Appx 3 Fig 3A1
BS 88-6:1988	Specification of supplementary requirements for fuses of compact dimensions for use in 240/415 V a.c. industrial and commercial electrical installations Replaced by BS 88-2:2010 and withdrawn 1/3/2010	
BS 196:1961	Specification for protected-type non-reversible plugs, socket-outlets cable-couplers and appliance-couplers with earthing contacts for single phase a.c. circuits up to 250 volts This standard has been withdrawn as the products have almost been replaced by those manufactured to BS EN 60309-1:1999 and and BS EN 60309-2:1999. The 'sliding earth' contact associated with BS 196 products has been found to be less reliable that the pin and socket tube design.	
BS 476 BS 476-4:1970 BS 476-12:1991	Fire tests on building materials and structures. Non-combustible test for materials Method of test for ignitability of products by direct flame impingement.	526.5(iii) 554.4.1
BS 546:1950 (1988)	Specification. Two-pole and earthing-pin plugs, socket-outlets and socket-outlet adaptors	Table 55.1 553.1.5 559.6.1.1(v) 705.512.2 705.553.1(iii)
BS 559:1998 (2005)	Specification for design, construction and installation of signs	110.1.3(i) 559.1 note 2
BS 646:1958 (1991)	Specification. Cartridge fuse-links (rated up to 5 amperes) for a.c. and d.c. service BS 646 remains current but the requirements for type B fuse-links have been replaced by BS 2950:1958	533.1 Table 55.1 553.1.5(i)
BS 951:1999	Electrical earthing. Clamps for earthing and bonding. Specification	514.13.1
BS 1361:1971 (1986)	Specification for cartridge fuses for a.c. circuits in domestic and similar premises Replaced by BS 88-3:2010 and withdrawn 1/3/2010	

BS or EN Number	Title	References
BS 1362:1973 (1992)	Specification for general purpose fuse links for domestic and similar purposes (primarily for use in plugs)	Table 41.2 Table 41.4 533.1 533.1.1.2 Table 53.4 Table 55.1 553.1.5(i)
BS 1363	13 A plugs, socket-outlets, connection units and adaptors.	433.1.103 Table 55.1 553.1.100 553.1.5 705.553.1(iii) Appx 15
BS 1363-1:1995 BS 1363-2:1995	Specification for rewirable and non-rewirable 13 A fused plugs Specification for 13 A switched and unswitched socket-outlets	Table 53.4 Table 53.4 559.6.1.1(v) 705.512.2
BS 1363-3:1995 BS 1363-4:1995	Specification for adaptors Specification for 13A fused connection units switched and unswitched	Table 53.4 559.6.1.1(vii) Appx 15 Fig 15B

BS 3036:1958 (1992)	Specification. Semi-enclosed electric fuses (ratings up to 100 amperes and 240 volts to earth)	Part 2 C$_f$
		Table 41.2
		Table 41.4
		432.4
		433.1.1 note 1
		433.1.101
		433.1.103
		533.1
		533.1.1.2
		Appx 3 Fig 3A2(a)
		   3A2(b)
		Appx 4 sec 3
		   sec 4
		   sec 5.1
		   sec 5.1.1(iii)
		   sec 6.1

BS 3535	Replaced by BS EN 61558-2-5	Appx 6 Cond. Report item 6.3
BS 3676	Switches for household and similar fixed electrical installations. Specification for general requirements. Now replaced by BS EN 60669-1 2000, but remains current. BS EN 60669-1: 2000 is dual numbered BS 3676: 2000	
BS 3858:1992 (2004)	Specification for binding and identification sleeves for use on electric cables and wires	514.3.2
BS 4177:1992	Specification for Cooker control units	Table 53.4
BS 4444:1989 (1995)	Guide to electrical earth monitoring and protective conductor proving	543.3.4 543.7.1.102(iii) 543.7.1.103(iv)
BS 4573:1970 (1979)	Specification for 2-pin reversible plugs and shaver socket-outlets	553.1.5(ii)
BS 4662:2006	Boxes for flush mounting of electrical accessories. Requirements and test methods and dimensions	530.4.2 559.6.1.1(viii)
BS 4727	Glossary of Electrotechnical power, telecommunications, electronics, lighting and colour terms	Part 2 first para
BS 5266	Emergency lighting	110.1.3(ii) 528.1 note 2 560.8.1 note 2 560.9 710.560.9.1(i)
BS 5266-1	Emergency lighting. Code of practice for the emergency lighting of premises	560.8.1 note 1(i)
BS 5467:1997	Electric cables. Thermosetting insulated, armoured cables for voltages of 600/1000 V and 1900/3300 V	522.6.100(ii) 522.6.101(i) 522.6.103(i) Appx 4 Table 4A3 Appx 7 Table 7C
BS 5499	Graphical symbols and signs. Safety signs, including fire safety signs. Current, but proposed for Withdrawal	A721.55.3.7

BS or EN Number	Title	References
BS 5655 BS 5655-1:1986 BS 5655-2:1988 BS 5655-11:2005 BS 5655-12:2005	Lifts and service lifts. Safety rules for the construction and installation of electric lifts (Applicable only to the modernization of existing lift installations) Safety rules for the construction and installation of hydraulic lifts (Applicable only to the modernization of existing lift installations) Code of practice for the undertaking of modifications to existing electric lifts (Applicable only to the modernization of existing lift installations) Code of practice for the undertaking of modifications to existing hydraulic lifts (Applicable only to the modernization of existing lift installations)	110.2(x)
BS 5733:1995	Specification for general requirements for electrical accessories	526.3(vi) 530.4.2 Table 53.4 559.6.1.1(vi)
BS 5803-5:1985	Thermal insulation for use in pitched roof spaces in dwellings. Specification for installation of man-made mineral fibre and cellulose fibre insulation	Appx 4 Table 4A2 items 100 to 103 Table 4D5
BS 5839 BS 5839-1:2002 (2008)	Fire detection and fire alarm systems for buildings Code of practice for system design, installation, commissioning and maintenance	528.1 note 2 560.8.1 note 2 560.10 710.560.11(vi) 560.8.1 note (ii) 612.3.2 note
BS 6004:2000 (2006)	Electric cables. PVC insulated, non-armoured cables for voltages up to and including 450/750 V, for electric power, lighting and internal wiring	717.411.3.1.2 note 717.52.2(i) 717.52.2(ii) A721.521.2 Appx 4 Table 4A3 Appx 7 Table 7B Appx 15 Fig 15B note
BS 6007:2006	Electric cables. Single core unsheathed heat resisting cables for voltages up to and including 450/750 V, for internal wiring	559.6.2.2(iii)
BS 6217	See note at end of this table	Appx 1 note
BS 6220:1983 (1999)	*Deleted by BS 7671:2008, Corrigendum (July 2008)*	
BS 6231:2006	Electric cables. Single core PVC insulated flexible cables of rated voltage 600/1000 V for switchgear and controlgear wiring	Appx 4 Table 4A3
BS 6346:1997 (2005)	Electric cables. PVC insulated, armoured cables for voltages of 600/1000 V and 1900/3300 V **NOTE:** This is not obsolete, it is current with a proposal for withdrawal (2005)	522.6.103(i) Appx 4 Table 4A3 Appx 7 Table 7C
BS 6351	Electric surface heating. This series has been withdrawn	
BS 6500:2000 (2005)	Electric cables. Flexible cords rated up to 300/500 V, for use with appliances and equipment intended for domestic, office and similar environments	A721.521.2 Appx 4 Table 4A3 Appx 7 Table 7D
BS 6701:2010	Telecommunications equipment and telecommunications cabling. Specification for installation, operation and maintenance	444.1(iii) 528.2 note 2
BS 6724:1997 (2008)	Electric cables. Thermosetting insulated, armoured cables for voltages of 600/1000 V and 1900/3300 V, having low emission of smoke and corrosive gases when affected by fire	522.6.100(ii) 522.6.101(i) 522.6.103(i) Appx 4 Table 4A3 Appx 7 Table 7C
BS 6891:2005 (2008)	Installation of low pressure gas pipework of up to 35 mm (R1 1/4) in domestic premises (2nd family gas).	528.3.4 note
BS 6907	Electrical installations for open-cast mines and quarries	110.1.3(viii)
BS 6972:1988	Specification for general requirements for luminaire supporting couplers for domestic, light industrial and commercial use	Table 53.4 559.6.1.1(ii)
BS 6991:1990	Specification for 6/10 A, two-pole weather-resistant couplers for household, commercial and light industrial equipment	553.2.1
BS 7001:1988	Specification for interchangeability and safety of a standardized luminaire supporting coupler	559.6.1.1(ii)
BS 7211:1998 (2005)	Electric cables. Thermosetting insulated, non-armoured cables for voltages up to and including 450/750 V, for electric power, lighting and internal wiring, and having low emission of smoke and corrosive gases when affected by fire	717.52.2(i) 717.52.2(ii) A721.521.2 Appx 4 Table 4A3
BS 7361-1:1991	Cathodic protection. Code of practice for land and marine applications. (Current but partially replaced by BS EN 15112:2006 and BS EN 13636:2004)	712.312.2 note

BS or EN Number	Title	References
BS 7375:2010	Code of practice for distribution of electricity on construction and building sites	704.0 704.411.3.1 note
BS 7430:1998	Code of practice for earthing	442.2 542.2.3 note 542.3.1 note
BS 7454:1991 (2008)	Method for calculation of thermally permissible short-circuit currents, taking into account non-adiabatic heating effects	Table 43.1 note 543.1.3
BS 7540	Electric cables. Guide to use for cables with a rated voltage not exceeding 450/750 V.	521.9.1 note
BS 7629-1:2008	Electric cables. Specification for 300/500 V fire resistant screened cables having low emission of smoke and corrosive gases when affected by fire. Multicore and multipair cables	560.8.1 note 2 Appx 4 Table 4A3
BS 7697:1993 (2004)	Nominal voltages for low voltage public electricity supply systems	Appx 2 sec 14
BS 7698-12:1998	Reciprocating internal combustion engine driven alternating current generating sets. Emergency power supply to safety devices.	560.6.13

BS 7769:2008	Electric cables. Calculation of the current rating.
	(Some parts of the BS 7769 series are now numbered BS IEC 60287 series, eventually all parts will be renumbered.)
		523.3
		Appx 4 sec 1
		  sec 2.1
		  sec 2.2
		  Table 4B3 note 2
		  Table 4C2 note 1 & 2
		  Table 4C3 note 1 & 2
		Appx 10 sec 1 note
		  sec 2 para 7

BS 7769-1.1:1997 BS 7769-1.2:1994 (2005) BS 7769-2.2:1997 (2005) BS 7769-2-2.1:1997 (2006) BS 7769-3.1:1997 (2005)	Has been superseded/withdrawn and replaced by BS IEC 60287-1-1:2006 Current rating equations (100% load factor) and calculation of losses. Sheath eddy current loss factors for two circuits in flat formation Thermal resistance. A method for calculating reduction factors for groups of cables in free air, protected from solar radiation Thermal resistance. Calculation of thermal resistance. Section 2.1: Calculation of thermal resistance Sections on operating conditions. Reference operating conditions and selection of cable type	
BS 7846:2009	Electric cables. 600/1000 V armoured fire-resistant cables having thermosetting insulation and low emission of smoke and corrosive gases when affected by fire	522.6.100(ii) 522.6.101(i) 522.6.103(i) 560.8.1 note 2 Appx 4 Table 4A3
BS 7889:1997	Electric cables. Thermosetting insulated, unarmoured cables for a voltage of 600/1000 V	Appx 4 Table 4A3
BS 7909	Code of practice for temporary electrical systems for entertainment and related purposes	110.1.3(ix) 706.1 711.1 717.1 740.1.1
BS 7919:2001 (2006)	Electric cables. Flexible cables rated up to 450/750V, for use with appliances and equipment intended for industrial and similar environments	702.522.23 704.522.8.11 705.422.8 note Fig 708 note 717.52.1 717.52.2(ii) 721.55.2.6(ii) A721.521.2 740.521.1 note 2 740.55.1.1 Appx 4 Table 4A3
BS 8434	Methods of test for assessment of the fire integrity of electric cables.	560.8.1(iii)
BS 8436:2004	Electric cables. 300/500 V screened electric cables having low emission of smoke and corrosive gases when affected by fire, for use in walls, partitions and building voids. Multicore cables	522.6.100(ii) 522.6.101(i) 522.6.103(i) Appx 4 Table 4A3
BS 8450:2006	Code of practice for installation of electrical and electronic equipment in ships	110.2(iv)
BS 8488:2009(2010)	Prefabricated wiring systems intended for permanent connection in fixed installations	521.100
BS 8491:2008	Method for assessment of fire integrity of large diameter power cables for use as components for smoke and heat control systems and certain other active fire safety systems	560.8.1(iii)

BS or EN Number	Title	References
BS 8519	Code of practice for the selection and installation of fire–resistant cables and systems for life safety and fire fighting applications	110.1.3(x) 560.8.1 note 1(iii) 560.8.1 note 2
BS AU 149a:1980 (1987)	Specification for electrical connections between towing vehicles and trailers with 6 V or 12 V electrical equipment: type 12 N (normal)	A721.55.2.6(i) A721.55.2.6(iii) Table A721
BS AU 177a:1980 (1987)	Specification for electrical connections between towing vehicles and trailers with 6 V or 12 V electrical equipment: type 12 S (supplementary)	A721.55.2.6(i) A721.55.2.6(iii) Table A721
BS EN 81 BS EN 81-1:1998(2009)	Safety rules for the construction and installation of lifts. Electric lifts (also known as BS 5655-1:1986 Lifts and service lifts … etc)	110.2(x) 528.3.5
BS EN 1648 BS EN 1648-1:2004 BS EN 1648-2:2005	Leisure accommodation vehicles. 12 V direct current extra low voltage electrical installations. Caravans 12 V direct current extra low voltage electrical installations. Motor caravans	708.1 note 1 721.1 708.1 note 1 721.1
BS EN 6100-1	*Deleted by BS 7671:2008, Corrigendum (July 2008)*	
BS EN 50085 BS EN 50085-1:2005 BS EN 50085-2-1:2006 BS EN 50085-2-3:2001 BS EN 50085-2-4:2001	Cable trunking and cable ducting systems for electrical installations. General requirements. BS EN 50085-1:1999 remains current Cable trunking systems and cable ducting systems intended for mounting on walls and ceilings Particular requirements for slotted cable trunking systems intended for installation in cabinets. Section 3: Slotted in cabinets. Particular requirements for service poles and service posts.	412.2.4.1(ii)(b) 422.2.1(iii) 422.3.4 521.6 527.1.6 527.1.5 711.521(ii) 717.52.2 717.528.3.4 721.528.3.4 740.521.1 740.521.1(ii) 422.4.103 522.6.100(iv) 522.6.101(iii) 522.6.103(iii) 705.522.16(ii)
BS EN 50086 BS EN 50086-1:1994 BS EN 50086-2-1:1996 BS EN 50086-2-2:1996 BS EN 50086-2-3:1996 BS EN 50086-2-4:1994	Specification for conduit systems for cable management. General requirements Replaced by BS EN 61386-1:2004 but remains current. Particular requirements. Rigid conduit systems Replaced by BS EN 61386-21:2004 but remains current., Particular requirements. Pliable conduit systems. pliable conduit systems Replaced by BS EN 61386-22:2004 but remains current. Particular requirements. Flexible conduit systems Replaced by BS EN 61386-23:2004 but remains current. Particular requirements. Conduit systems buried underground	527.1.5 527.1.6 740.521.1 note 1 Appx 4 Table 4B3
BS EN 50107 BS EN 50107-1:2002 BS EN 50107-2:2005	Signs and luminous-discharge-tube installations operating from a no-load rated output voltage exceeding 1 kV but not exceeding 10 kV. General requirements Requirements for earth-leakage and open-circuit protective devices	110.1.3(i) 559.1(v) 559.1 note 2
BS EN 50171:2001	Central power supply systems replaced by BS EN 60623 and BS EN 60896	

BS or EN Number	Title	References
BS EN 50174	Information technology – Cabling installation	443.1.1 444.1(v) 444.4.1 note A444.4 A444.5 528.2 note 2
BS EN 50174-1:2009 BS EN 50174-2:2009	Installation specification and quality assurance Installation planning and practices inside buildings	444.4.10(i) 444.4.10(ii) A444.4
BS EN 50174-3:2008	Installation technology. Cabling installation. Installation planning and practices outside buildings	A444.4
BS EN 50177-4-1	Coaxial cables – Sectional specification for cables for BCT cabling in accordance with EN 50173 – Indoor drop cables for systems operating at 5 MHz – 3 000 MHz	Table A444.1(v)
BS EN 50200:2006	Method of test for resistance to fire of unprotected small cables for use in emergency circuits	560.8.1(iii)
BS EN 50266	Common test methods for cables under fire conditions. Test for vertical flame spread of vertically-mounted bunched wires or cables. Standard withdrawn replaced by BS EN 60332-3-23.	
BS EN 50266-1:2001 (2006)	Apparatus Standard withdrawn replaced by BS EN 60332-3-10.	
BS EN 50266-2-1:2001 (2006)	Procedures. Category A F/R Standard withdrawn replaced by BS EN 60332-3-21.	
BS EN 50266-2-2:2001 (2006)	Procedures. Category A Standard withdrawn replaced by BS EN 60332-3-22.	
BS EN 50266-2-3:2001 (2006)	Procedures. Category B Standard withdrawn replaced by BS EN 60332-3-23.	
BS EN 50266-2-4:2001 (2006)	Procedures. Category C Standard withdrawn replaced by BS EN 60332-3-24.	
BS EN 50266-2-5:2001 (2006)	Procedures. Small cables. Category D Standard withdrawn replaced by BS EN 60332-3-25.	
BS EN 50281 BS EN 50281-1-1:1999	Electrical apparatus for use in the presence of combustible dust. Electrical apparatus protected by enclosures. Construction and testing Replaced by BS EN 60241-0:2006 and BS EN 61241-1:2004 but remains current.	110.1.3(iv)
BS EN 50281-1-2:1999	Electrical apparatus protected by enclosures. Selection, installation and maintenance Partially replaced by BS EN 61241-14:2004 and BS EN 61241-17: 2005	
BS EN 50281-2-1:1999	Test methods. Methods of determining minimum ignition temperatures	
BS EN 50288	Multi-element metallic cables used in analogue and digital communication and control.	Table A444.1(iv)
BS EN 50310:2010	Application of equipotential bonding and earthing in buildings with information technology equipment	444.1(iv) 444.4.10(iii) A444.1 note A444.1.3
BS EN 50362:2003	Method of test for resistance to fire of larger unprotected power and control cables for use in emergency circuits	
BS EN 50428:2005(2009)	Switches for household and similar fixed installations. Collateral standard. Switches and related accessories for use in home and building electronic systems (HBES)	Table 53.4
BS EN 50438:2007	Requirements for the connection of micro-cogenerators in parallel with public low-voltage distribution networks This document currently at DPC stage (Expired 2004/11/30)	551.1 note 551.7.4 551.7.5 note 551.7.6
BS EN 60068-2-11:1999	Environmental testing. Test methods. Tests. Test KA. Salt mist	Appx 5 AF2
BS EN 60073:2002	Basic and safety principles for man-machine interface, marking and identification. Coding principles for indicators and actuators.	514.1.1
BS EN 60079	Electrical apparatus for explosive gas atmospheres.	110.1.3(iii) App 5 BE3
BS EN 60079-10:2009 BS EN 60079-14:2008	Classification of hazardous areas Electrical installations in hazardous areas (other than mines)	532.1 note 2 422.3 532.1 note 2 560.1
BS EN 60079-17:2003	Inspection and maintenance of electrical installations in hazardous areas (other than mines)	612.1

BS or EN Number	Title	References
BS EN 60092-507:2000	Electrical installations in ships – Pleasure craft	709.1 note 2
BS EN 60146-2:2000	Semiconductor convertors. General requirements and line commutated convertors. Self-commutated semiconductor converters including direct d.c. converters	414.2 note 2
BS EN 60204 BS EN 60204-1:2006	Safety of machinery. Electrical equipment of machines. General requirements	110.2(xi) 537.3.1 537.4.1 537.5.4 711.55.4.1 717.1.(v) 740.1.1
BS EN 60228:2005	Conductors of insulated cables	Appx 4 sec 1
BS EN 60238:2004	Edison screw lampholders. BS EN 60238:1999 remains current.	416.2.4 note 559.6.1.8 612.6(ii)
BS EN 60255-22-1:2008	Electrical relays. Electrical disturbance tests for measuring relays and protection equipment. 1 MHz burst immunity tests	Appx 5 AM-24-2
BS EN 60269 BS EN 60269-1:2007 BS EN 60269-1:2007 +A1:2009 BS EN 60269-2:1995 BS HD 60269-2:2010 BS EN 60269-3:1995 BS HD 60269-3: 2010	Low-voltage fuses. General requirements withdrawn 1/3/2010 General requirements Supplementary requirements for fuses for use by authorized persons (fuses mainly for industrial application) Replaced by BS 88-2:2007 and BS EN 60269-1:2007 and withdrawn 1/3/2010 Low-voltage fuses - Part 2: Supplementary requirements for fuses for use by authorized persons (fuses mainly for industrial application) Examples of standardized systems of fuses A to J' (also numbered BS 88-2:2010) Supplementary requirements for fuses for use by unskilled persons (fuses mainly for household and similar applications) Replaced by BS 88-3:2007 and BS EN 60269-1:2007 and withdrawn 1/3/2010 Low-voltage fuses - Part 3: Supplementary requirements for fuses for use by unskilled persons (fuses mainly for household and similar applications) Examples of standardized systems of fuses A to F (also numbered BS 88-3:2010)	
BS EN 60309 BS EN 60309-1:1999 BS EN 60309-2:1999	Plugs, socket-outlets and couplers for industrial purposes. General requirements Dimensional interchangeability requirements for pin and contact-tube accessories	Table 53.4 551.4.4 note 705.553.1 709.553.1.8 721.55.1.1 740.415.1(vi) 740.55.7 543.7.1.101 543.7.1.102(ii) Table 55.1 553.1.5 553.2.1 559.6.1.1(v) 704.511.1 705.553.1(ii) 708.553.1.8 Fig 708 note 709.553.1.8 Fig 709.3 717.55.1 721.55.1.1 721.55.2.6 740.55.7(i)
BS EN 60320-1:2001	Appliance couplers for household and similar general purposes. General requirements	553.2.1

BS or EN Number	Title	References
BS EN 60332-1-2:2004	Tests on electric and optical fibre cables under fire conditions. Test for vertical flame propagation for a single insulated wire or cable. Procedure for 1 kW pre-mixed flame	422.3.4 422.4.102 527.1.3 527.1.4 560.8.1(i) (ii) & (iii) 711.521 721.521.2 740.521.1
BS EN 60332-3	Tests on electric and optical fibre cables under fire conditions. Test for vertical flame spread of vertically-mounted bunched wires or cables.	422.2.1(i) 422.3.4 527.1.3 711.521
BS EN 60335-1:2002 BS EN 60335-2-29:2004 BS EN 60335-2-41:2003 BS EN 60335-2-53:2003 BS EN 60335-2-71:2003 BS EN 60335-2-76:2005 BS EN 60335-2-96:2002	Household and similar electrical appliances. Safety. General requirements Particular requirements for battery chargers Particular requirements for pumps Particular requirements for sauna heating appliances Particular requirements for electrical heating appliances for breeding and rearing animals Particular requirements for electric fence energizers Particular requirements for flexible sheet heating elements for room heating	A721.313.4(iii) A721.55.9 702.55.3 703.55 705.422.6 110.2(xii) 705.1 note 110.1.3(vii) 753.511
BS EN 60417	See Note at end of this Table	412.2.1.1 note
BS EN 60439 BS EN 60439-1:1999 BS EN 60439-2:2000 BS EN 60439-3:1991 BS EN 60439-4:2004	Low-voltage switchgear and controlgear assemblies. Type-tested and partially type-tested assemblies. Replaced by BS EN 61439-1 and BS EN 61439-2 but remains current. Particular requirements for busbar trunking systems (busways) Particular requirements for low-voltage switchgear and controlgear assemblies intended to be installed in places where unskilled persons have access to their use. Distribution boards Particular requirements for assemblies for construction sites (ACS)	710.51.1 Part 2 LV switchge… 412.2.1.1(ii) 543.2.2(ii) 712.511.1 434.5.3 521.4 527.1.6 527.1.5 Appx 8 sec 1 sec 2 note 1 Appx 10 sec 2 note 530.3.4 704.511.1
BS EN 60445:2007	Basic and safety principles for man-machine interface, marking and identification. Identification of equipment terminals and of terminations of certain designated conductors, including general rules for an alphanumeric system	Appx 7 sec 1
BS EN 60446:2000	Basic and safety principles for man-machine interface, marking and identification. Identification of conductors by colours or numerals	Appx 7 sec 1
BS EN 60447:2007	Basic and safety principles for man-machine interface, marking and identification. Actuating principles	514.1.1
BS EN 60529:1992 (2004)	Specification for degrees of protection provided by enclosures (IP code)	527.2.3(i) & (ii) 702.512.2 708.512.2(i) & (ii) 708.553.1.8 A721.55.6 A721.55.10
BS EN 60570:2003	Electrical supply track systems for luminaires Replaces BS EN 60570:1997 and BS EN 60570-2-1:1995 which remain current	527.1.5 559.4.4

BS or EN Number	Title	References
BS EN 60598	Luminaires	559.6.1.1(iii) 559.6.1.1(iv) 559.6.2.2(i) & (ii) 702.55.2
BS EN 60598-1:2004	Luminaires. General requirements and tests. Replaced by BS EN 60598-1:2008 but remains current.	Table 55.2 and note
BS EN 60598-1:2008	Luminaires. General requirements and tests.	Table 55.2 and note
BS EN 60598-2-18:1994	Particular requirements. Luminaires for swimming pools and similar applications	702.55.2 702.55.3
BS EN 60598-2-23:1997	Particular requirements. Extra-low voltage lighting systems for filament lamps	559.11.4.1(ii) 711.559.4.2
BS EN 60598-2-24:1999	Particular requirements. Luminaires with limited surface temperatures	422.3.2 note 422.3.8(iii) Table 55.2
BS EN 60601	Medical electrical equipment. General requirements for basic safety and essential performance.	710.1 note 3 710.512.2.1 Fig 710.1
BS EN 60617	See Note at end of this Table	Table 53.4 note 1 Appx 1 note
BS EN 60623:2001	Secondary cells and batteries containing alkaline or other non-acid electrolytes. Vented nickel-cadmium prismatic rechargeable single cells	560.6.10 560.6.11
BS EN 60664-1:2007	Insulation coordination for equipment within low-voltage systems. Principles, requirements and tests	442.2.2 note 1 534.1 537.2.2.1
BS EN 60669 BS EN 60669-1:1999 +A2:2008	Switches for household and similar fixed electrical installations. General requirements. This replaces BS 3676.	416.2.4 note Table 53.4
BS EN 60669-2-1:2004	Particular requirements. Electronic switches	Table 53.4
BS EN 60669-2-2:2006	Particular requirements. Electromagnetic remote-control switches (RCS)	Table 53.4
BS EN 60669-2-3:2006	Particular requirements. Time delay switches (TDS)	Table 53.4
BS EN 60669-2-4:2005	Particular requirements. Isolating switches	Table 53.4
BS EN 60670 BS EN 60670-1:2005	Boxes and enclosures for electrical accessories for household and similar fixed electrical installations. General requirements	530.4.2 559.6.1.1(viii)
BS EN 60670-22:2006	Particular requirements for connecting boxes and enclosures	521.8.3 Appx 15 Fig 15A
BS EN 60684	Flexible insulating sleeving.	543.3.100
BS EN 60702-1:2002	Mineral insulated cables and their terminations with a rated voltage not exceeding 750 V. Cables	422.6(i) 433.1.103 522.6.100(ii) 522.6.101(i) 522.6.103(i) Table 52.1 543.4.8 560.8.1(i) 560.8.1 note 2 Appx 4 Table 4A3
BS EN 60702-2:2002	Mineral insulated cables and their terminations with a rated voltage not exceeding 750 V	560.8.1(i)
BS EN 60721 BS EN 60721-3-3:1995 (2005)	Classification of environmental conditions. Classification of groups of environmental parameters and their severities. Stationary use at weather protected locations	Appx 5 A …
BS EN 60721-3-4:1995 (2005)	Classification of groups of environmental parameters and their severities. Stationary use at non-weather protected locations	Appx 5 A …
BS EN 60896	Stationary lead-acid batteries. General requirements and methods of test. Vented types. General requirements and methods of test	560.6.10 560.6.11
BS EN 60898:1991	Specification for circuit-breakers for overcurrent protection for household and similar installations. Replaced by BS EN 60898-1:2003 but remains current	Table 41.3 Table 41.6 432.4 433.1.100 433.1.103 Table 53.4 Appx 3 Fig 3A4 Fig 3A5 Fig 3A6 Appx 4 sec 4 Appx 8 sec 4

BS or EN Number	Title	References
BS EN 60898-1:2003	Circuit breakers for a.c. operation	434.5.2 533.1 Appx 4 sec 5.5.2
BS EN 60898-2:2001	Circuit-breakers for a.c. and d.c. operation BS EN 60898-2:2001 remains current. (It was withdrawn in error and has been reinstated.)	434.5.2 533.1 A721.533.1.6
BS EN 60904-3:2008	Photovoltaic devices. Measurement principles for terrestrial photovoltaic (PV) solar devices with reference spectral irradiance data	Part 2 STC
BS EN 60947 BS EN 60947-2:2006	Low-voltage switchgear and control gear Circuit-breakers	432.4 433.1.100 433.1.103 533.1 Table 53 4 711.410.3.4 740.410.3 740.411 note Appx 4 sec 4 Appx 8 sec 4
BS EN 60947-3:2009	Switches, disconnectors, switch-disconnectors and fuse-combination units	533.1 Table 53 4
BS EN 60947-4-1:2001	Contactors and motor starters – Electromechanical contactor and motor starters	435.2 533.1 536.5.2 Table 53 4
BS EN 60947-5-1:2004	Control circuit devices and switching elements – Electromechanical control circuit devices	Table 53 4
BS EN 60947-6-1:2005	Multiple function equipment – Transfer switching equipment	533.1 Table 53 4 710.537.1
BS EN 60947-6-2:2003	Multiple function equipment – Control and protective switching devices (or equipment) (CPS)	533.1 Table 53 4
BS EN 60947-7 BS EN 60947-7-1:2002 BS EN 60947-7-2:2002	Specification for low-voltage switchgear and controlgear Ancillary equipment – Terminal blocks for copper conductors Ancillary equipment – Protective conductor terminal blocks for copper conductors	526.2 note 1
BS EN 60950-1	Information technology equipment. Safety. General requirements	444.4.9
BS EN 60998 BS EN 60998-2-1:2004 BS EN 60998-2-2:2004	Connecting devices for low-voltage circuits for household and similar purposes. Particular requirements for connecting devices as separate entities with screw-type clamping units Particular requirements for connecting devices as separate entities with screwless-type clamping units	526.2 note 1 559.11.6 559.11.6
BS EN 61000 BS EN 61000-2 BS EN 61000-4 BS EN 61000-6	Electromagnetic compatibility (EMC) BS EN 61000 is a multiple part standard Electromagnetic compatibility (EMC). Environment. Electromagnetic compatibility (EMC). Testing and measurement techniques. Electromagnetic compatibility (EMC). Generic standards.	444.1(vi) 515.2 Appx 4 sec 5.5.1 Appx 5 AM Appx 5 AM Table A444.1
BS EN 61008-1:2004 +12:2009	Residual current operated circuit-breakers without integral overcurrent protection for household and similar uses (RCCBs). General rules.	Introduction Chap 41 411.4.9 Table 41.5 Table 53.4 710.411.3.2.1 711.410.3.4 721.411.1 740.410.3 740.411 note Appx 3 Table 3A

BS or EN Number	Title	References

BS EN 61009-1:2004	Electrical accessories. Residual current operated circuit-breakers with integral overcurrent protection for household and similar uses (RCBOs). General rules.	Introduction Chap 41
		Table 41.3
		411.4.9
		Table 41.5
		Table 41.6
		432.4
		433.1.100
		433.1.103
		434.5.2
		Table 53.4
		533.1
		710.411.3.2.1
		711.410.3.4
		721.411.1
		740.410.3
		740.411 note
		Appx 3 Table 3A
		  Fig 3A4
		  Fig 3A5
		  Fig 3A6
		Appx 4 sec 4

BS EN 61034-2:2005	Measurement of smoke density of cables burning under defined conditions. Test procedure and requirements	422.2.1(i) para 6 422.2.1 711.521(i)
BS EN 61048:2006	Auxiliaries for lamps. Capacitors for use in tubular fluorescent and other discharge lamp circuits. General and safety requirements	559.8
BS EN 61095:2009	Specification for electromechanical contactors for household and similar purposes	Table 53.4

BS EN 61140:2002	Protection against electric shock. Common aspects for installation and equipment	Part 2 Class I ...
		  Class II ...
		  Class III ...
		410
		412.2.4.1 note 1
		Appx 5 BC

BS EN 61184:2008	Bayonet lampholders	416.2.4 note 559.6.1.7
BS EN 61215:2005	Crystalline silicon terrestrial photovoltaic (PV) modules. Design qualification and type approval	712.511.1
BS EN 61241 BS EN 61241-10:2004 BS EN 61241-14:2004 BS EN 61241-17:2005	Electrical apparatus for use in the presence of combustible dust. Replaced by BS EN 60079-10-2:2009 Selection and installation Inspection and maintenance of electrical installations in hazardous areas (other than mines). Replaced by BS EN 60079-17:2007 but remains current.	110.1.3(iv) 532.1 note 2 422.3 532.1 note 2 612.1
BS EN 61347 BS EN 61347-1:2001 BS EN 61347-2-2:2001	Lamp controlgear. General and safety requirements. Replaced by BS EN 601347-1:2008 but remains current. Particular requirements for d.c. or a.c. supplied electronic step-down convertors for filament lamps	711.55.6 Table 55.2 559.11.3.2 740.55.5
BS EN 61386 BS EN 61386-1:2004	Conduit systems for cable management. General requirements. Replaced by BS EN 61386-17:2008 but remains current.	412.2.4.1(ii)(b) 422.2.1(ii) A444.1.4 521.6 527.1.5 527.1.6 711.521(ii) 717.52.2 717.528.3.4 721.521.2 727.528.3.4 740.521.1 422.3.4 422.4.103

287

BS or EN Number	Title	References
BS EN 61386-21:2004	Particular requirements. Rigid conduit systems	522.6.100(iii) 522.6.101(ii) 522.6.103(ii) 705.522.16 705.522.16(i) 721.521.2 740.521.1(i)
BS EN 61386-22:2004	Particular requirements. Pliable conduit systems	
BS EN 61386-23:2004	Particular requirements. Flexible conduit systems	740.521.1 740.521.1 note 2
BS EN 61386-24:2010	Particular requirements. Conduit systems buried underground	522.8.10 note
BS EN 61534	Powertrack systems	422.2.1(v) 422.3.4 434.5.3 521.4 527.1.5 527.1.6 Appx 10 sec 2 note
BS EN 61534-1:2003	General requirements	Appx 8 sec 1 sec 2 note 2
BS EN 61534-21:2006	Particular requirements for powertrack systems intended for wall and ceiling mounting	
BS EN 61535:2009	Installation couplers intended for permanent connection to fixed installations	Part 2: Prefabricated wiring system 521.100 526.2 note 1 553.2.1
BS EN 61537:2007	Cable tray systems and cable ladder systems for cable management.	422.3.4 422.2.1(iv) 521.6 527.1.5 527.1.6 740.521.1
BS EN 61557	Electrical safety in low voltage distribution systems up to 1000 V a.c. and 1500 V d.c. Equipment for testing, measuring or monitoring of protective measures. General requirements	612.1 621.3
BS EN 61557-2:2007	Insulation resistance	Appx 13 Sect 1(2)
BS EN 61557-6:1998	Residual current devices (RCD) in TT, TN and IT systems. Replaced by BS EN 61557-6:2007 but remains current.	612.8.1 a) 2) 612.8.1 b) 2) 612.10
BS EN 61557-8:2007	Insulation monitoring devices for IT systems	538.1.1 710.411.6.3.1
BS EN 61557-9:2009	Equipment for insulation fault location in IT systems	538.2 710.411.6.3.3
BS EN 61558-1:1998	Safety of power transformers, power supply units and similar products. General requirements and tests Replaced by BS EN 61558-1:2005 +A1:2009 but remains current.	411.8.4.1(i) 717.411.6.2(ii)
BS EN 61558-2-1:2007	Particular requirements and tests for separating transformers and power supplies incorporating separating transformers for general applications	444.4.9
BS EN 61558-2-4:2009	Particular requirements for isolating transformers for general use	444.4.9 740.55.9
BS EN 61558-2-5:1998	Particular requirements for shaver transformers and shaver supply units. Replaced by BS EN 61558-2-5 but remains current.	Introduction note 2 Sec 701 553.1.5(ii) 701.512.2 701.512.3(ii) 701.512.3 Appx 6 Cond. Report Item 6.3
BS EN 61558-2-6:1998	Particular requirements for safety isolating transformers for general use	414.3(i) 444.4.9 559.11.3.1 Table 55.2 A721.313.4(iii) 740.55.5
BS EN 61558-2-15:2001	Particular requirements for isolating transformers for the supply of medical locations	444.4.9 710.512.1.1
BS EN 61558-2-23:2001	Particular requirements for transformers for construction sites. Replaced by BS EN 61558-2-23:2010 but remains current.	411.8.4.1(i)

BS or EN Number	Title	References
BS EN 61643	Low-voltage surge protective devices.	443.1.1 note 5 534.1
BS EN 61643-11:2002	Surge protective devices connected to low-voltage power systems. Requirements and tests	534.2.3.3 534.2.3.4.1 & 2 Appx 16 Table 16A
BS EN 61643-21:2001 (2009)	Surge protective devices connected to telecommunications and signalling networks. Performance requirements and testing methods	534.2.3.4.1
BS EN 61995-1	Devices for the connection of luminaires for household and similar purposes. General requirements	Table 53.4
BS EN 62020:1999	Electrical accessories. Residual current monitors for household and similar uses (RCMs)	538.4
BS EN 62040-1 BS EN 62040-3:2001	Uninterruptible power systems (UPS) Uninterruptible power systems (UPS). Method of specifying the performance and test requirements	560.6.12(iv) 560.6.12(iv)
BS EN 62208:2003	Empty enclosures for low-voltage switchgear and controlgear assemblies. General requirements	530.4.2
BS EN 62262:2002	Degrees of protection provided by enclosures for electrical equipment against external mechanical impacts (IK code)	708.512.2(iii) 709.512.2.1.4(iii)
BS EN 62305	Protection against lightning.	110.2(ix) 131.6.2 note 411.3.1.2 421.1.1 note 443.1.1 443.2.2 note 444.5.2(iv) 528.1 note 541.3 Appx 16 Table 16A
BS EN 62305-1:2006 BS EN 62305-2:2006 BS EN 62305-3:2006 BS EN 62305-4:2006	General requirements Risk management Physical damage to structures and life hazard Electrical and electronic systems within structures	Appx 5 AQ3 534.2.3.1.1 534.1 534.2.1 534.2.3.4.2 534.2.3.4.3 534.2.3.4.4 Appx 16 Table 16A Fig 16A5
BS EN 623951-1:2006	Electrical resistance trace heating systems for industrial and commercial applications. General and testing requirements.	
BS EN ISO 11446:2004	Road vehicles. Connectors for the electrical connection of towing and towed vehicles. 13-pole connectors for vehicles with 12 V nominal supply voltage	A721.55.2.6(i) & (iii) Table A721

NOTE on graphical symbols – IEC 60617 – is the central standards database for electrotechnical symbols. They used to be covered by BS 6217, BS EN 60417 and BS EN 60617. In 2002 IEC launched an 'on-line' database format for the symbol library, available on subscription from the IEC website. Following this decision, in 2004 CENELEC decided to cease publication of EN 60417 in 'paper' form, to withdraw the then-existing standards and formally to adopt the IEC database without any changes for use in Europe. Consequently, the British Standard versions have now also been withdrawn.

Other standards to which reference is made in the Regulations

BS IEC 60287	Electric cables. Calculation of the current rating. (Some parts of the BS 7769 series are now numbered BS IEC 60287 series, eventually all parts will be renumbered.)	523.3
		Appx 4 sec 1
		  sec 2.1
		  sec 2.2
		  Table 4B3 note 2
		  Table 4C2 note 1 & 2
		  Table 4C3 note 1 & 2
		Appx 10 sec 1 note
		  sec 2 para 9

BS IEC 60287-1-1:2006	Cable rating equations (100% load factor) and calculation of losses (general)	Appx 4 sec 5.6
BS IEC 61995-1	Devices for the connection of luminaires for household and similar purposes. General requirements	Table 53.4

HD 308:2001	Identification of cores in cables and flexible cords Please note, Table 51 basically implements the requirements of HD 308.	Preface Appx 7 Sec 1
HD 384.5.514	Now withdrawn	Appx 7 sec 1
HD 384.5.551:1997	Low voltage generating sets	Preface
HD 384.7.711 SI:2003	Exhibitions, shows and stands	Preface
HD 472 S1	BS 7697:1993 (2004) - Nominal voltages for low voltage public electricity supply systems	708.313.1.2 Appx 2 sec 14
HD 60364	Low-voltage electrical installations	Notes on the plan ...
HD 60364-1:2008	Fundamental principles, assessment of general ..., definitions	Preface
HD 60364-4-41:2007	Protection against electric shock	Preface Appx 5 BC
HD 60364-4-42:2001	Protection for safety - Protection against thermal effects	Preface Appx 5 BE2 Appx 5 CA2 Appx 5 CB2
HD 60364-4-43:2008	Protection against overcurrent	Preface
HD 60364-4-442:1997	Protection of low voltage installations against temporary overvoltages	Preface
HD 60364-4-443:2006	Protection against overvoltage	443.1.1 note 2
HD 60364-5-51:2006	Selection and erection of electrical equipment – Common rules	Preface Appx 5 Appx 5 BE2
HD 60364-5-52	Selection and erection of electrical equipment – Wiring systems	Appx 5 CB2 Appx 5 CB3 Appx 5 CB4
HD 60364-5-534:2008	Devices for protection against overvoltage	Preface
HD 60364-5-54:2007	Earthing arrangements	Preface
HD 60364-5-559	Outdoor lighting installations	Preface
HD 60364-6: 2007	Initial verification	Preface
HD 60364-7-701:2007	Locations containing a bath or shower	Preface
HD 60364-7-703:2005	Sauna heaters	Preface
HD 60364-7-704:2007	Construction and demolition site installations	Preface
HD 60364-7-705:2007	Agricultural and horticultural premises	Preface
HD 60364-7-706:2007	Locations with restricted movement	Preface
HD 60364-7-708:2009	Caravan parks, camping parks and similar locations	Preface
HD 60364-7-709:2009	Marinas and similar locations	Preface
HD 60364-7-712:2005	Solar photovoltaic (PV) power supply systems	Preface
HD 60364-7-721:2009	Electrical installations in caravans and motor caravans	Preface
HD 60364-7-729:2009	Operating and maintenance gangways	Preface
HD 60364-7-740:2006	Temporary electrical installations for structures, amusement devices	Preface
FprHD 60364-4-444:200X	Measures against electromagnetic disturbances	Preface
FprHD 60364-7-702:2009	Swimming pools and other basins	Preface
FprHD 60364-7-710:2010	Medical locations	Preface
FprHD 60364-7-717:2009	Mobile or transportable units	Preface
IEC 60038-Ed 7.0	IEC standard voltages	721.313.1.2

IEC 60331	Tests for electric cables under fire conditions - Circuit integrity. Test method for fire with shock at a temperature of at least 830 °C for cables of rated voltage up to and including 0,6/1,0 kV,	
IEC 60331-1	and with an overall diameter exceeding 20 mm	560.8.1(ii)
IEC 60331-2	and with an overall diameter not exceeding 20 mm	560.8.1(ii)
IEC 60331-3	tested in a metal enclosure	560.8.1(ii)
IEC 60364-4-44	Protection against voltage disturbances and electromagnetic disturbances	Preface
IEC 60449-am 1 Ed 1	Voltage bands for electrical installations of buildings	414.1.1
IEC 60502-1 Ed 2	Power cables with extruded insulation and their accessories for rated voltages from 1 kV (Um = 1,2 kV) up to 30 kV (Um = 36 kV) - Part 1: Cables for rated voltages of 1 kV (Um = 1,2 kV) and 3 kV (Um = 3,6 kV)	Table 52.1 Appx 4 sec 1
IEC 60617	Central standards database for electrotechnical symbols	514.9.2
IEC 60621-2 Ed 2	Electrical installations for outdoor sites under heavy conditions (including open-cast mines and quarries). Part 2: General protection requirements Withdrawn – no replacement.	704.1.1(vi)
IEC 60755-Ed 2	General requirements for residual current operated protective devices Withdrawn, see IEC 62423	
IEC 60800	Heating cables with a rated voltage of 300/500 V for comfort heating and prevention of ice formation	753.511
IEC 60884 Ed 3.1	Plugs and socket-outlets for household and similar purposes. Part 1. General requirements	Table 53.4
IEC 60906	IEC system of plugs and socket-outlets for household and similar purposes	Table 53.4
IEC 60974-9 Ed 1.0	Arc welding equipment. Installation and use	
IEC 61201 Ed 2	Extra-low voltage (ELV). Limit values. Also known as PD 6536	414.2 note 3
IEC 61662 TR2 Ed 1	Assessment of the risk of damage due to lightning	443.2.4 Note 3
IEC 61936-1 Ed 1	Power installations exceeding 1 kV a.c. - Part 1: Common rules	Fig 44.2 note 442.2.3
IEC 61995-1 Ed 1	Devices for the connection of luminaires for household and similar purposes - Part 1: General requirements	559.6.1.1(ix)
IEC 62305	Protection against lightning	443.1.1 note 2
IEC 62423 Ed 2 2009	Type F and type B residual current operated circuit breakers with and without integral overcurrent protection for household and similar use	710.411.3.2.1 712.411.3.2.1.2
IEC/TS 62081 Ed 1	Arc welding equipment. Installation and use. Withdrawn and replaced by IEC 60974-9	706.1
ISO 8820	Road vehicles. Fuse-links	A721.533.1.6
PD 6536:1992	Extra-low voltage (ELV). Limit values. Also known as IEC 61201	414.2 note 3

APPENDIX 2 (Informative)

STATUTORY REGULATIONS AND ASSOCIATED MEMORANDA

1. In the United Kingdom the following classes of electrical installations are required to comply with the Statutory Regulations indicated below. The regulations listed represent the principal legal requirements. Information concerning these regulations may be obtained from the appropriate authority also indicated below.

Provisions relating to electrical installations are also to be found in other legislation relating to particular activities.

(i)	Distributors' installations generally, subject to certain exemptions	Electricity Safety, Quality and Continuity Regulations 2002 as amended SI 2002 No 2665 SI 2006 No 1521	Secretary of State for Trade and Industry Secretary of State for Scotland
(ii)	Buildings generally subject to certain exemptions	The Building Regulations 2000 (as amended) (for England and Wales) SI 2000 No 2531	The Department for Communities and Local Government
		The Building (Scotland) Regulations 2004 Scottish SI 2004 No 406 (as amended)	The Scottish Executive
		Building Regulations (Northern Ireland) 2000 Statutory Rule 2000 No 389	The Department of Finance and Personnel
(iii)	Work activity Places of work Non-domestic installations	The Electricity at Work Regulations 1989 as amended SI 1989 No 635 SI 1996 No 192 SI 1997 No 1993 SI 1999 No 2024	Health and Safety Executive
		The Electricity at Work Regulations (Northern Ireland) 1991 Statutory Rule 1991 No. 13	Health and Safety Executive for Northern Ireland
(iv)	Cinematograph installations	Cinematograph (Safety) Regulations 1955, as amended made under the Cinematograph Act, 1909, and/or Cinematograph Act, 1952 SI 1982 No 1856	The Secretary of State for the Home Department, and The Scottish Executive
(v)	Machinery	The Supply of Machinery (Safety) Regulations 1992 as amended SI 1992 No 3073 SI 1994 No 2063	Department for Business, Innovation and Skills
(vi)	Theatres and other places licensed for public entertainment, music, dancing, etc.	Conditions of licence under: (a) in England and Wales, The Local Government Licensing Act 2003 (b) in Scotland, The Civic Government (Scotland) Act 1982	(a) Department for Culture, Media and Sport (b) The Scottish Executive
(vii)	High voltage luminous tube signs	As (a) and (b) above	As (a) and (b) above

2. Failure to comply in a consumer's installation in the United Kingdom with the requirements of Chapter 13 of BS 7671:2008, Requirements for Electrical Installations (the IET {formerly IEE} Wiring Regulations) places the distributor in the position of not being compelled to commence or, in certain circumstances, to continue to give, a supply of energy to that installation.

Under Regulation 26 of the Electricity Safety, Quality and Continuity Regulations 2002, any dispute which may arise between a consumer and the distributor having reference to the consumer's installation shall be determined by a person nominated by the Secretary of State (or the Scottish Executive in relation to disputes arising in Scotland) on the application of the consumer or consumer's authorized agent or the distributor.

Regulation 28 of the Electricity Safety, Quality and Continuity Regulations 2002 requires distributors to provide the following information to relevant persons free of charge:

The maximum prospective short-circuit current at the supply terminals

The maximum earth loop impedance of the earth fault path outside the installation (Z_e)

The type and rating of the distributor's protective device or devices nearest to the supply terminals

The type of earthing system applicable to the connection

The number of phases of the supply

The frequency of the supply and the extent of the permitted variations

The voltage of the supply and the extent of the permitted variations.

3. Where it is intended to use protective multiple earthing the distributor and the consumer must comply with the Electricity Safety, Quality and Continuity Regulations 2002.

4. For further guidance on the application of the Electricity at Work Regulations, reference may be made to the following publication:

(i) Memorandum of guidance on the Electricity at Work Regulations 1989 (HSR25).

5. For installations in potentially explosive atmospheres reference should be made to:

(i) the Electricity at Work Regulations 1989 (SI 1989 No 635)

(ii) the Dangerous Substances and Explosive Atmospheres Regulations (DSEAR) 2002 (SI 2002 No 2776)

(iii) the Petroleum (Consolidation) Act 1928

(iv) the Equipment and Protective Systems Intended for Use in Potentially Explosive Atmospheres Regulations 1996 (SI 1996 No 192)

(v) relevant British or Harmonized Standards.

Under the Petroleum (Consolidation) Act 1928 local authorities are empowered to grant licences in respect of premises where petroleum spirit is stored and as the authorities may attach such conditions as they think fit, the requirements may vary from one local authority to another. Guidance may be obtained from the Energy Institute (APEA/IP) publication Design, Construction, Modification, Maintenance and Decommissioning of Filling Stations.

6. For installations in theatres and other places of public entertainment, and on caravan parks, the requirements of the licensing authority should be ascertained. Model Standards were issued by the Department of the Environment in 1977 under the Caravan Sites and Control of Development Act 1960 as guidance for local authorities.

7. The Electrical Equipment (Safety) Regulations 1994 (SI 1994 No 3260), administered by the Department for Business, Innovation and Skills (BIS), contain requirements for safety of equipment designed or suitable for general use. Information on the application of the Regulations is given in guidance issued by BIS.

8. The Plugs and Sockets etc. (Safety) Regulations 1994 (SI 1994 No 1768) made under the Consumer Safety Act 1978, administered by the Department for Business, Innovation and Skills, contains requirements for the safety of plugs, sockets, adaptors and fuse links etc. designed for use at a voltage of not less than 200 volts.

9. The Health and Safety (Safety Signs and Signals) Regulations 1996 (SI 1996 No 341) require employers to ensure that safety signs are provided. Guidance from the Health and Safety Executive L64, Safety Signs and Signals, specifies signs including emergency escape, first aid and fire safety signs.

10. The Management of Health and Safety at Work Regulations 1999 (SI 1999 No 3242) require employers and self-employed persons to assess risks to workers and others who may be affected by their work or business. This is intended to enable them to identify measures they need to take to comply with the law. The Health and Safety

Commission has published an Approved Code of Practice L21, Management of health and safety at work, which gives advice that has special legal status.

11. The Provision and Use of Work Equipment Regulations 1998 (SI 1998 No 2306) require employers to ensure that all work equipment is suitable for the purpose for which it is used, is properly maintained and that appropriate training is given. The Health and Safety Commission has published an Approved Code of Practice L22, Safe use of work equipment, which gives advice that has special legal status.

12. The Electromagnetic Compatibility Regulations 2005 (SI 2005 No 281) provide requirements for electrical and electronic products for electromagnetic compatibility.

13. Other Regulations relevant to electrical installation include:

> The Personal Protective Equipment at Work Regulations 2002
> (European Directive 89/656/EEC, HSE Publication L25)
>
> The Workplace (Health, Safety and Welfare) Regulations 1992
> (European Directive 89/654/EEC, HSE Publication L24)
>
> The Manual Handling Operations Regulations 1992
> (European Directive 90/269/EEC, HSE Publication L23)
>
> The Work at Heights Regulations 2005
>
> The Construction (Design and Management) Regulations 2007
> (European Directive 92/57/EEC, HSE Publication L144), SI 2007 No 320.

14. In November 1988 the European electrical standards body CENELEC agreed on harmonization of low voltage electricity supplies within Europe (CENELEC document HD 472 S1), implemented by BS 7697 Nominal voltages for low voltage public electricity supply systems. The measure is intended to harmonize mains electricity supplies at 230 V within Europe. CENELEC has proposed three stages of harmonization. Two stages of harmonization have taken place, these being shown below.

Effective date	Nominal voltage	Permitted tolerance	Permitted voltage range
Pre-1995	240 V	+6 % / -6 %	225.6 - 254.4 V
1 January 1995	230 V	+10 % / -6 %	216.2 - 253.0 V

APPENDIX 3 (Informative)
TIME/CURRENT CHARACTERISTICS OF
OVERCURRENT PROTECTIVE DEVICES AND RCDs

FUSES:

This appendix gives the time/current characteristics of the following overcurrent protective devices:

Figure 3A1   Fuses to BS 88-3 fuse system C

Figures 3A2(a) & 3A2(b) Semi-enclosed fuses to BS 3036

Figures 3A3(a) & 3A3(b) Fuses to BS 88-2 – fuse systems E (bolted) and G (clip-in)

CIRCUIT-BREAKERS:

Figure 3A4   Type B to BS EN 60898 and the overcurrent characteristics of RCBOs to BS EN 61009-1

Figure 3A5   Type C to BS EN 60898 and the overcurrent characteristics of RCBOs to BS EN 61009-1

Figure 3A6   Type D to BS EN 60898 and the overcurrent characteristics of RCBOs to BS EN 61009-1

In all of these cases time/current characteristics are based on the slowest operating times for compliance with the Regulations and have been used as the basis for determining the limiting values of earth fault loop impedance prescribed in Chapter 41.

Maximum earth fault loop impedance

Regulation 411.3.2 specifies maximum disconnection times for circuits. Regulations 411.4.6 to 9 provide maximum earth fault loop impedances (Z_S) that will result in protective devices operating within the required disconnection times.

The maximum earth fault loop impedance for a protective device is given by:

$$Z_S = \frac{U_0}{I_a}$$

where:

U_0 is the nominal a.c. rms line voltage to Earth.

I_a is the current causing operation of the protective device within the specified time.

The tabulated values are applicable for supplies from distribution network operators. For other supplies the designer will need to determine the nominal voltage and calculate Z_S accordingly.

RCDs:

Table 3A gives the time/current performance criteria for RCDs to BS EN 61008-1 and BS EN 61009-1.

TABLE 3A – Time/current performance criteria for RCDs to BS EN 61008-1 and BS EN 61009-1

RCD type	Rated residual operating current $I_{\Delta n}$ mA	Residual current mA	Trip time ms	Residual current mA	Trip time ms	Residual current mA	Trip time ms
General Non-delay	10	10	300 max.	20	150 max.	50	40 max.
	30	30		60		150	
	100	100		200		500	
	300	300		600		1500	
	500	500		1000		2500	
Delay 'S'	100	100	130 min. 500 max.	200	60 min. 200 max.	500	40 min. 150 max.
	300	300		600		1500	
	500	500		1000		2500	

Fig 3A1 – Fuses to BS 88-3 fuse system C

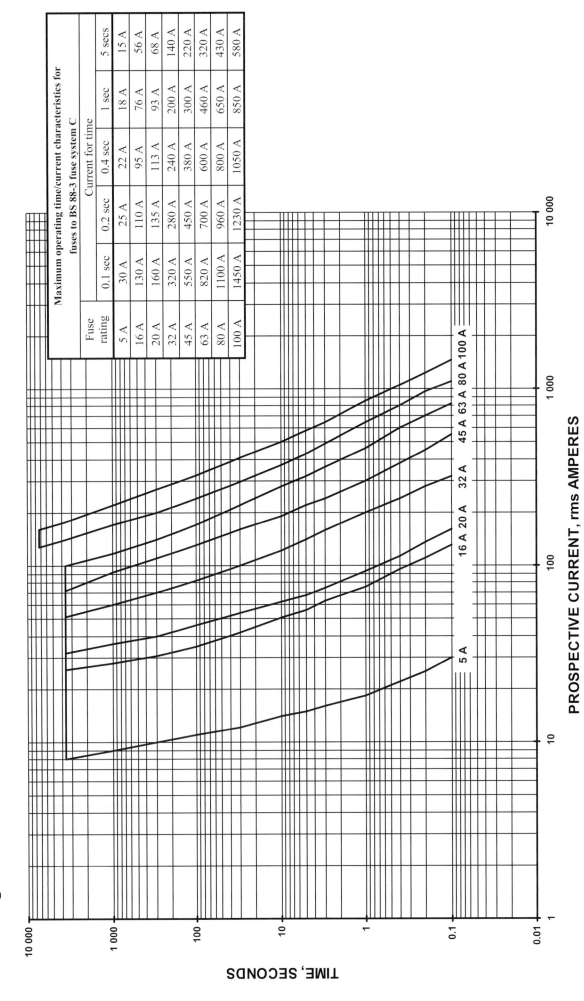

Maximum operating time/current characteristics for fuses to BS 88-3 fuse system C

Fuse rating	Current for time					
	0.1 sec	0.2 sec	0.4 sec	1 sec	5 secs
5 A	30 A	25 A	22 A	18 A	15 A
16 A	130 A	110 A	95 A	76 A	56 A
20 A	160 A	135 A	113 A	93 A	68 A
32 A	320 A	280 A	240 A	200 A	140 A
45 A	550 A	450 A	380 A	300 A	220 A
63 A	820 A	700 A	600 A	460 A	320 A
80 A	1100 A	960 A	800 A	650 A	430 A
100 A	1450 A	1230 A	1050 A	850 A	580 A

PROSPECTIVE CURRENT, rms AMPERES

TIME, SECONDS

5 A   16 A   20 A   32 A   45 A   63 A   80 A   100 A

Fig 3A2(a) – Semi-enclosed fuses to BS 3036

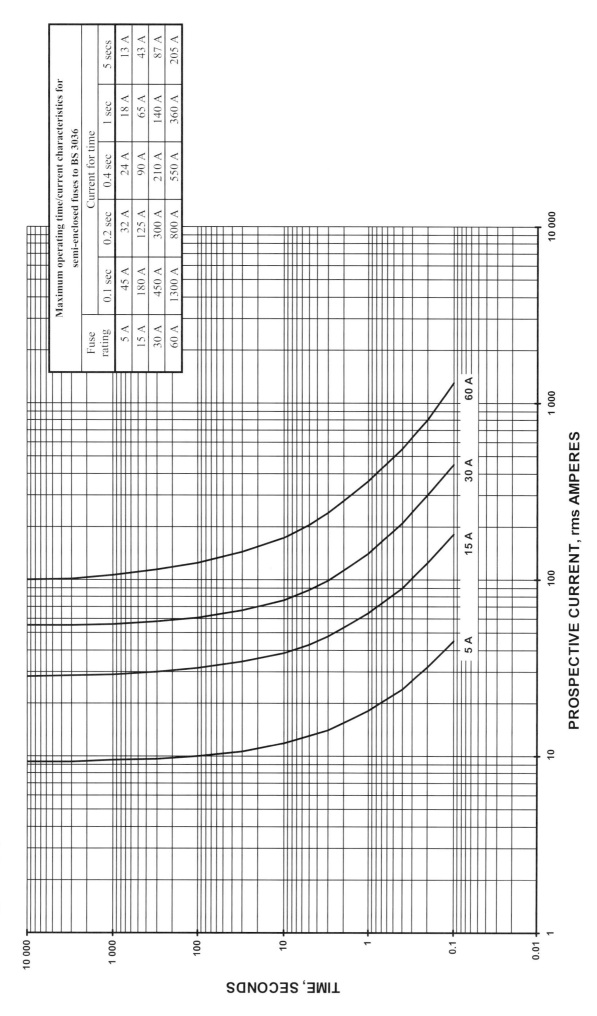

Fuse rating	Maximum operating time/current characteristics for semi-enclosed fuses to BS 3036					
	Current for time					
	0.1 sec	0.2 sec	0.4 sec	1 sec	5 secs	
5 A	45 A	32 A	24 A	18 A	13 A	
15 A	180 A	125 A	90 A	65 A	43 A	
30 A	450 A	300 A	210 A	140 A	87 A	
60 A	1300 A	800 A	550 A	360 A	205 A	

TIME, SECONDS

PROSPECTIVE CURRENT, rms AMPERES

Fig 3A2(b) – Semi-enclosed fuses to BS 3036

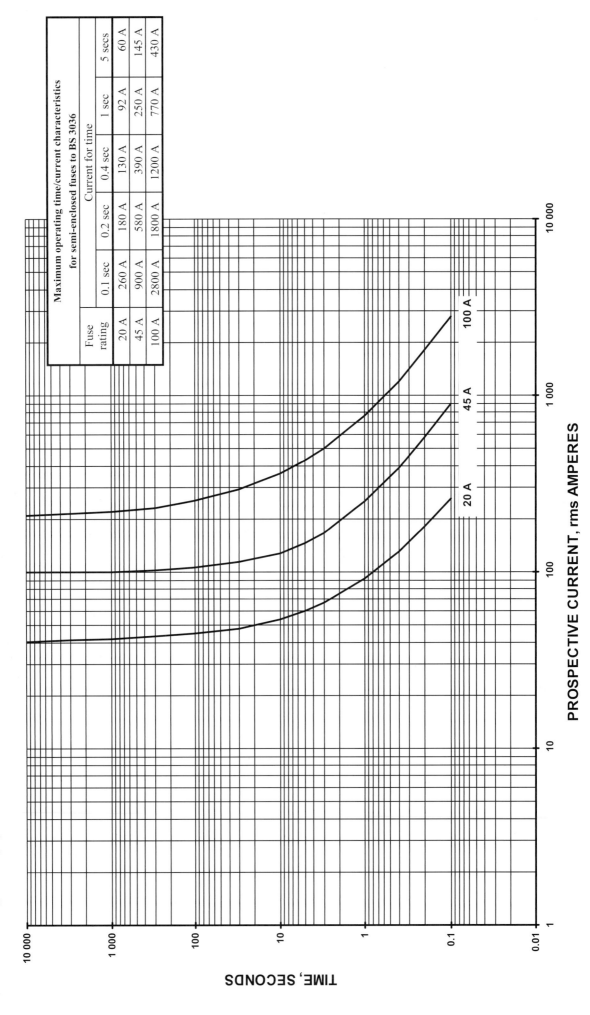

Maximum operating time/current characteristics for semi-enclosed fuses to BS 3036						
Fuse rating	Current for time					
	0.1 sec	0.2 sec	0.4 sec	1 sec	5 secs	
20 A	260 A	180 A	130 A	92 A	60 A	
45 A	900 A	580 A	390 A	250 A	145 A	
100 A	2800 A	1800 A	1200 A	770 A	430 A	

TIME, SECONDS

PROSPECTIVE CURRENT, rms AMPERES

Fig 3A3(a) – Fuses to BS 88-2 fuse systems E and G

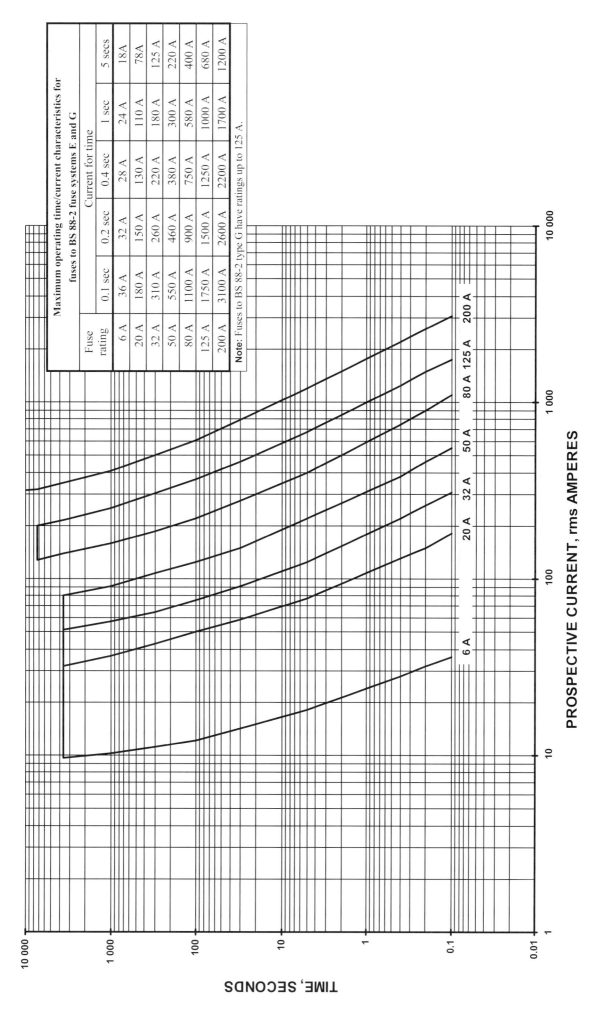

Fuse rating	Maximum operating time/current characteristics for fuses to BS 88-2 fuse systems E and G Current for time					
	0.1 sec	0.2 sec	0.4 sec	1 sec	5 secs
6 A	36 A	32 A	28 A	24 A	18A
20 A	180 A	150 A	130 A	110 A	78A
32 A	310 A	260 A	220 A	180 A	125 A
50 A	550 A	460 A	380 A	300 A	220 A
80 A	1100 A	900 A	750 A	580 A	400 A
125 A	1750 A	1500 A	1250 A	1000 A	680 A
200 A	3100 A	2600 A	2200 A	1700 A	1200 A

Note: Fuses to BS 88-2 type G have ratings up to 125 A.

PROSPECTIVE CURRENT, rms AMPERES

TIME, SECONDS

Fig 3A3(b) – Fuses to BS 88-2 fuse systems E and G

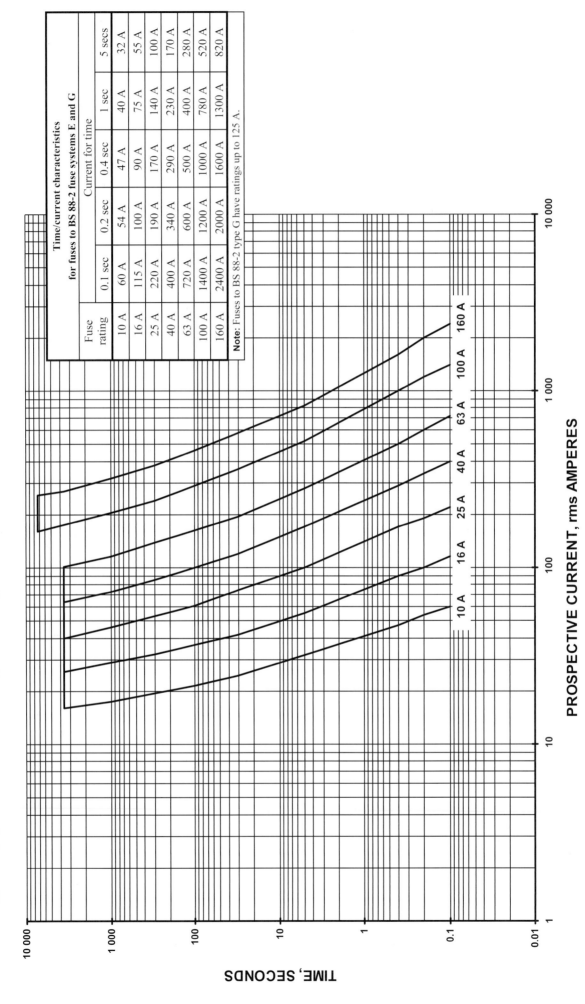

Time/current characteristics for fuses to BS 88-2 fuse systems E and G

Fuse rating	Current for time					
	0.1 sec	0.2 sec	0.4 sec	1 sec	5 secs	
10 A	60 A	54 A	47 A	40 A	32 A	
16 A	115 A	100 A	90 A	75 A	55 A	
25 A	220 A	190 A	170 A	140 A	100 A	
40 A	400 A	340 A	290 A	230 A	170 A	
63 A	720 A	600 A	500 A	400 A	280 A	
100 A	1400 A	1200 A	1000 A	780 A	520 A	
160 A	2400 A	2000 A	1600 A	1300 A	820 A	

Note: Fuses to BS 88-2 type G have ratings up to 125 A.

PROSPECTIVE CURRENT, rms AMPERES

TIME, SECONDS

Fig 3A4 – Type B circuit-breakers to BS EN 60898 and RCBOs to BS EN 61009-1

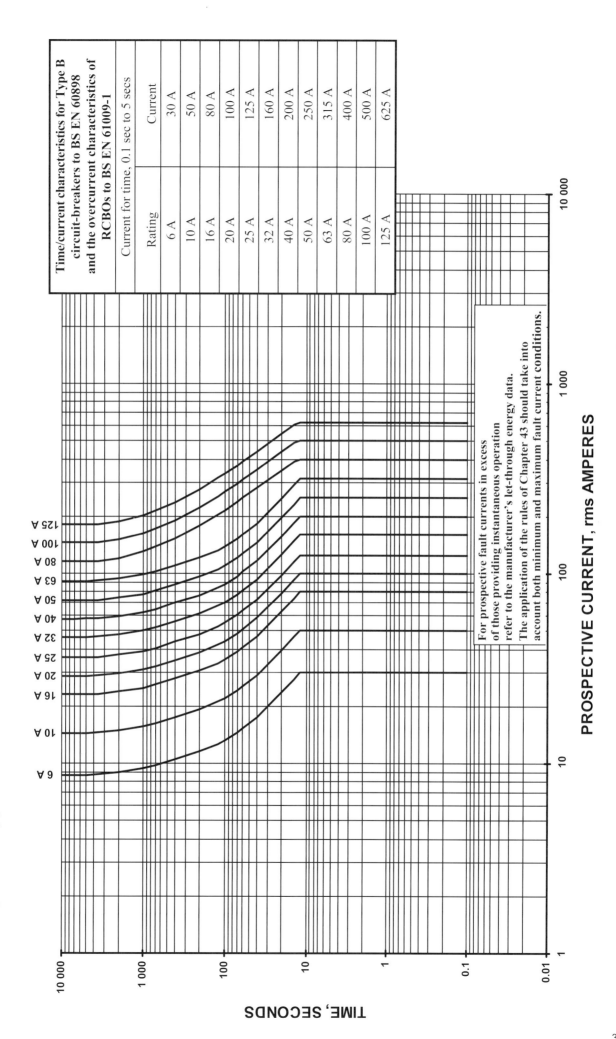

Time/current characteristics for Type B circuit-breakers to BS EN 60898 and the overcurrent characteristics of RCBOs to BS EN 61009-1

Current for time, 0.1 sec to 5 secs

Rating	Current
6 A	30 A
10 A	50 A
16 A	80 A
20 A	100 A
25 A	125 A
32 A	160 A
40 A	200 A
50 A	250 A
63 A	315 A
80 A	400 A
100 A	500 A
125 A	625 A

For prospective fault currents in excess of those providing instantaneous operation refer to the manufacturer's let-through energy data.

The application of the rules of Chapter 43 should take into account both minimum and maximum fault current conditions.

PROSPECTIVE CURRENT, rms AMPERES

TIME, SECONDS

Fig 3A5 – Type C circuit-breakers to BS EN 60898 and RCBOs to BS EN 61009-1

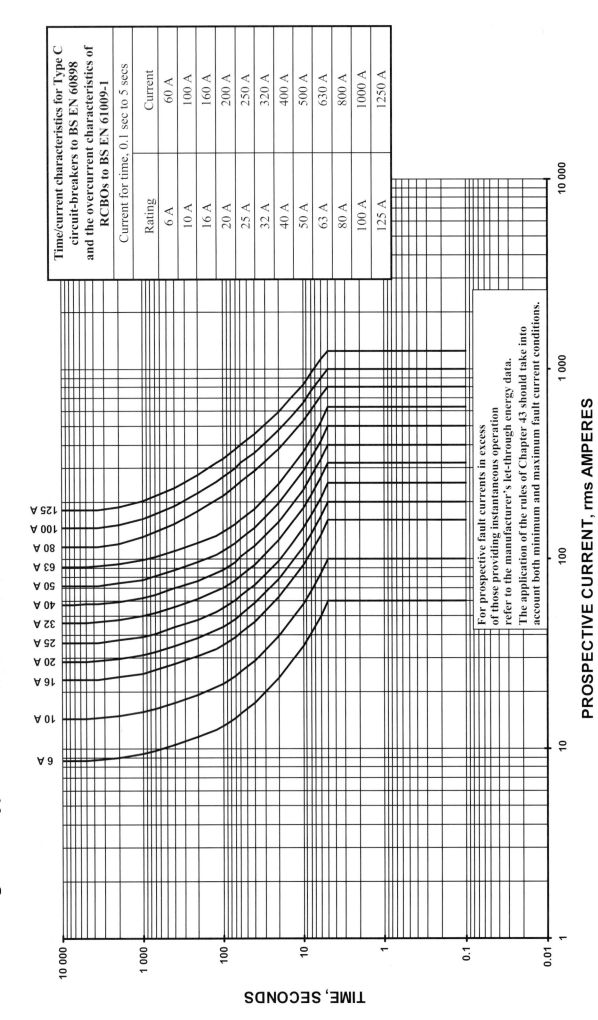

Time/current characteristics for Type C circuit-breakers to BS EN 60898 and the overcurrent characteristics of RCBOs to BS EN 61009-1

Current for time, 0.1 sec to 5 secs

Rating	Current
6 A	60 A
10 A	100 A
16 A	160 A
20 A	200 A
25 A	250 A
32 A	320 A
40 A	400 A
50 A	500 A
63 A	630 A
80 A	800 A
100 A	1000 A
125 A	1250 A

For prospective fault currents in excess of those providing instantaneous operation refer to the manufacturer's let-through energy data.

The application of the rules of Chapter 43 should take into account both minimum and maximum fault current conditions.

TIME, SECONDS

PROSPECTIVE CURRENT, rms AMPERES

Fig 3A6 – Type D circuit-breakers to BS EN 60898 and RCBOs to BS EN 61009-1

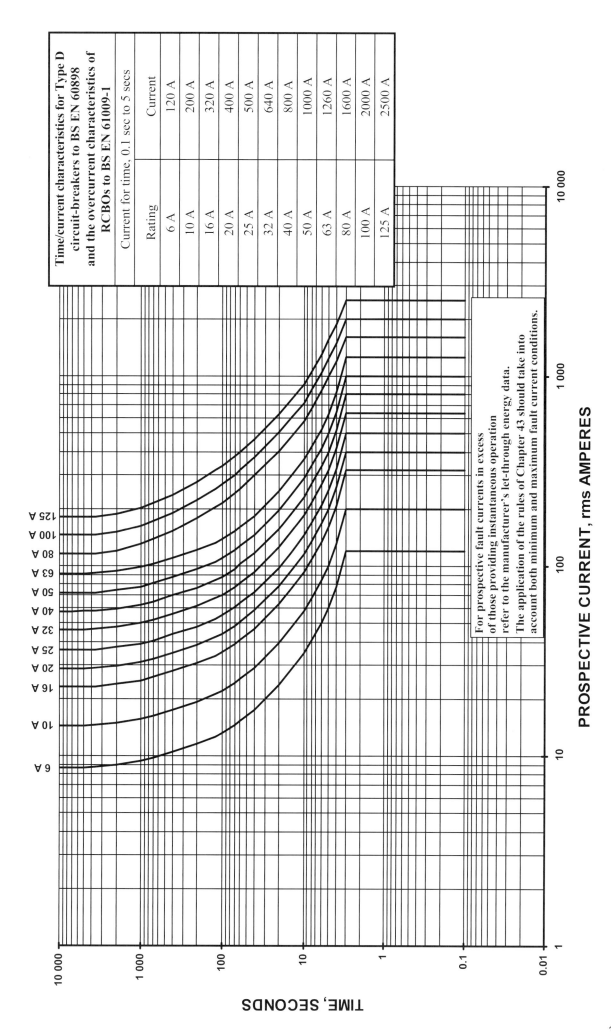

Time/current characteristics for Type D circuit-breakers to BS EN 60898 and the overcurrent characteristics of RCBOs to BS EN 61009-1

Current for time, 0.1 sec to 5 secs

Rating	Current
6 A	120 A
10 A	200 A
16 A	320 A
20 A	400 A
25 A	500 A
32 A	640 A
40 A	800 A
50 A	1000 A
63 A	1260 A
80 A	1600 A
100 A	2000 A
125 A	2500 A

For prospective fault currents in excess of those providing instantaneous operation refer to the manufacturer's let-through energy data.

The application of the rules of Chapter 43 should take into account both minimum and maximum fault current conditions.

PROSPECTIVE CURRENT, rms AMPERES

TIME, SECONDS

APPENDIX 4 (Informative)

CURRENT-CARRYING CAPACITY AND VOLTAGE DROP

FOR CABLES

CONTENTS

Tables:

APPENDIX 4 (Informative)

CURRENT-CARRYING CAPACITY AND VOLTAGE DROP FOR CABLES

1 INTRODUCTION

The recommendations of this appendix are intended to provide for a satisfactory life of conductors and insulation subjected to the thermal effects of carrying current for prolonged periods of time in normal service. Other considerations affect the choice of cross-sectional area of conductors, such as the requirements for protection against electric shock (Chapter 41), protection against thermal effects (Chapter 42), overcurrent protection (Chapter 43), voltage drop (Section 525), and limiting temperatures for terminals of equipment to which the conductors are connected (Section 526).

This appendix applies to non-sheathed and sheathed cables having a nominal voltage rating not exceeding 1 kV a.c. or 1.5 kV d.c.

The values in Tables 4D1A to 4J4A have been derived in accordance with the methods given in BS 7769 (BS IEC 60287) using such dimensions as specified in the international standard IEC 60502-1 and conductor resistances given in BS EN 60228. Known practical variations in cable construction (e.g. form of conductor) and manufacturing tolerances result in a spread of possible dimensions and hence current-carrying capacities for each conductor size. Tabulated current-carrying capacities have been selected in such a way as to take account of this spread of values with safety and to lie on a smooth curve when plotted against conductor cross-sectional area.

For multicore cables having conductors with a cross-sectional area of 25 mm^2 or larger, either circular or shaped conductors are permissible. Tabulated values have been derived from dimensions appropriate to shaped conductors.

All the current-carrying capacities given are based on the ambient temperature and conductor/sheath operating temperature stated in Tables 4D1A to 4F2A and 4G1A to 4J4A.

2 CIRCUIT PARAMETERS

2.1 Ambient Temperature

The current-carrying capacities in this appendix are based upon the following reference ambient temperatures:

  (i) For non-sheathed and sheathed cables in air, irrespective of the Installation Method: 30°C

  (ii) For buried cables, either directly in the soil or in ducts in the ground: 20 °C.

Where the ambient temperature in the intended location of the non-sheathed or sheathed cables differs from the reference ambient temperature, the appropriate rating factors given in Tables 4B1 and 4B2 are to be applied to the values of current-carrying capacity set out in Tables 4D1A to 4J4A. For buried cables, further correction is not needed if the soil temperature exceeds the selected ambient temperature by an amount up to 5 °C for only a few weeks a year.

The rating factors in Tables 4B1 and 4B2 do not take account of the increase, if any, due to solar or other infrared radiation. Where non-sheathed or sheathed cables are subject to such radiation, the current-carrying capacity may be derived by the methods specified in BS 7769 (BS IEC 60287).

2.2 Soil Thermal Resistivity

The current-carrying capacities tabulated in this appendix for cables in the ground are based upon a soil thermal resistivity of 2.5 K.m/W and are intended to be applied to cables laid in and around buildings. For other installations, where investigations establish more accurate values of soil thermal resistivity appropriate for the load to be carried, the values of current-carrying capacity may be derived by the methods of calculation given in BS 7769 (BS IEC 60287) or obtained from the cable manufacturer.

In locations where the effective soil thermal resistivity is higher than 2.5 K.m/W, an appropriate reduction in current-carrying capacity should be made or the soil immediately around the cables should be replaced by a more suitable material. Such cases can usually be recognized by very dry ground conditions. Rating factors for soil thermal resistivities other than 2.5 K.m/W are given in Table 4B3.

2.3 Groups of cables containing more than one circuit

2.3.1 Methods of Installation A to D in Table 4A2

Current-carrying capacities given in Tables 4D1A to 4J4A apply to single circuits consisting of:

(i) two non-sheathed cables or two single-core cables, or one two-core cable

(ii) three non-sheathed cables or three single-core cables, or one three-core cable.

Where more non-sheathed cables, other than bare mineral insulated cables not exposed to touch, are installed in the same group, the group rating factors specified in Tables 4C1 to 4C3 need to be applied.

NOTE: The group rating factors have been calculated on the basis of prolonged steady-state operation at a 100% load factor for all live conductors. Where the loading is less than 100% as a result of the conditions of operation of the installation, the group rating factors may be higher.

2.3.2 Methods of Installation E and F in Table 4A2

The current-carrying capacities of Tables 4D1A to 4J4A apply to these Reference Methods.

For installations on perforated trays, cleats and similar, current-carrying capacities for both single circuits and groups are obtained by multiplying the capacities given for the relevant arrangements of non-sheathed or sheathed cables in free air, as indicated in Tables 4D1A to 4J4A, by the applicable group rating factors given in Tables 4C4 and 4C5. No group rating factors are required for bare mineral insulated cables not exposed to touch, Tables 4G1A and 4G2A refer.

NOTE 1: Group rating factors have been calculated as averages for the range of conductor sizes, cable types and installation condition considered. Attention is drawn to the notes under each table. In some instances, a more precise calculation may be required.

NOTE 2: Group rating factors have been calculated on the basis that the group consists of similar, equally loaded non-sheathed or sheathed cables. Where a group contains various sizes of non-sheathed or sheathed cables, caution should be exercised over the current loading of the smaller cables (see 2.3.3 below).

NOTE 3: A group of similar cables is taken to be a group where the current-carrying capacity of all the cables is based on the same maximum permissible conductor temperature and where the range of conductor sizes in the group spans not more than three adjacent standard sizes.

2.3.3 Groups of cables containing different sizes

Tabulated group rating factors are applicable to groups consisting of similar equally loaded cables. The calculation of rating factors for groups containing different sizes of equally loaded sheathed or non-sheathed cables is dependent on the total number in the group and the mix of sizes. Such factors cannot be tabulated but must be calculated for each group. The method of calculation of such factors is outside the scope of this appendix. Two specific examples of where such calculations may be advisable are given below.

2.3.3.1 Groups in conduit systems, cable trunking systems or cable ducting systems

For a group containing different sizes of non-sheathed or sheathed cables in conduit systems, cable trunking systems or cable ducting systems, a simple formula for calculation of the group rating factor is:

$$F = \frac{1}{\sqrt{n}}$$

where

F   is the group rating factor

n   is the number of circuits in the group.

The group rating factor obtained by this equation will reduce the danger of overloading the smaller sizes but may lead to under-utilization of the larger sizes. Such under-utilization can be avoided if large and small sizes of non-sheathed or sheathed cable are not mixed in the same group.

The use of a method of calculation specifically intended for groups containing different sizes of non-sheathed or sheathed cable in conduit will produce a more precise group rating factor.

2.3.3.2 Groups of cables on trays

Where a group contains different sizes of non-sheathed or sheathed cable, caution must be exercised over the current loading of the smaller sizes. It is preferable to use a method of calculation specifically intended for groups containing different sizes of non-sheathed or sheathed cables.

The group rating factor obtained in accordance with the formula in 2.3.3.1 will provide a value which may be safely applied, but which may result in under-utilisation.

2.4 Conductors

The current-carrying capacities and voltage drops tabulated in this appendix are based on cables having solid conductors (Class 1), or stranded conductors (Class 2), except for Tables 4F1A to 4F3B. To obtain the correct current-carrying capacity or voltage drop for cable types similar to those covered by Tables 4D1, 4D2, 4E1 and 4E2 but with flexible conductors (Class 5), the tabulated values are multiplied by the following factors:

Cable size	Current-carrying capacity	Voltage drop
≤ 16 mm²	0.95	1.10
≥ 25 mm²	0.97	1.06

3 RELATIONSHIP OF CURRENT-CARRYING CAPACITY TO OTHER CIRCUIT PARAMETERS

The relevant symbols used in the Regulations are as follows:

I_z the current-carrying capacity of a cable for continuous service, under the particular installation conditions concerned.

I_t the value of current tabulated in this appendix for the type of cable and installation method concerned, for a single circuit in the ambient temperature stated in the current-carrying capacity tables.

I_b the design current of the circuit, i.e. the current intended to be carried by the circuit in normal service.

I_n the rated current or current setting of the protective device.

I_2 the operating current (i.e. the fusing current or tripping current for the conventional operating time) of the device protecting the circuit against overload.

C   a rating factor to be applied where the installation conditions differ from those for which values of current-carrying capacity are tabulated in this appendix. The various rating factors are identified as follows:

    C_a for ambient temperature

    C_c for circuits buried in the ground

    C_d for depth of burial

    C_f for semi-enclosed fuse to BS 3036

    C_g for grouping

    C_i for thermal insulation

    C_s for thermal resistivity of soil.

The rated current or current setting of the protective device (I_n) must not be less than the design current (I_b) of the circuit, and the rated current or current setting of the protective device (I_n) must not exceed the lowest of the current-carrying capacities (I_z) of any of the conductors of the circuit.

Where the overcurrent device is intended to afford protection against overload, I_2 must not exceed 1.45 I_z and I_n must not exceed I_z (see paragraph 4 below).

Where the overcurrent device is intended to afford fault current protection only, I_n can be greater than I_z and I_2 can be greater than 1.45 I_z. The protective device must be selected for compliance with Regulation 434.5.2.

4 OVERLOAD PROTECTION

Where overload protection is required, the type of protection does not affect the current-carrying capacity of a cable for continuous service (I_z) but it may affect the choice of conductor size. The operating conditions of a cable are influenced not only by the limiting conductor temperature for continuous service, but also by the conductor temperature which might be attained during the conventional operating time of the overload protective device, in the event of an overload.

This means that the operating current of the protective device must not exceed 1.45 I_z. Where the protective device is a fuse to BS 88 series, a circuit-breaker to BS EN 60898 or BS EN 60947-2 or a residual current circuit-breaker with integral overcurrent protection to BS EN 61009-1 (RCBO), this requirement is satisfied by selecting a value of I_z not less than I_n.

In practice, because of the standard steps in ratings of fuses and circuit-breakers, it is often necessary to select a value of I_n exceeding I_b. In that case, because it is also necessary for I_z in turn to be not less than the selected value of I_n, the choice of conductor cross-sectional area may be dictated by the overload conditions and the current-carrying capacity (I_z) of the conductors will not always be fully utilised.

The size needed for a conductor protected against overload by a BS 3036 semi-enclosed fuse can be obtained by the use of a rating factor, $1.45/2 = 0.725$, which results in the same degree of protection as that afforded by other overload protective devices. This factor is to be applied to the nominal rating of the fuse as a divisor, thus indicating the minimum value of I_t required of the conductor to be protected. In this case also, the choice of conductor size is dictated by the overload conditions and the current-carrying capacity (I_z) of the conductors cannot be fully utilised.

The tabulated current-carrying capacities for cables direct in ground or in ducts in the ground, given in this appendix, are based an ambient temperature of 20 °C. The factor of 1.45 that is applied in Regulation 433.1.1 when considering overload protection assumes that the tabulated current-carrying capacities are based on an ambient temperature of 30 °C. To achieve the same degree of overload protection where a cable is "in a duct in the ground" or "buried direct" as compared with other installation methods a rating factor of 0.9 is applied as a multiplier to the tabulated current-carrying capacity.

5 DETERMINATION OF THE SIZE OF CABLE TO BE USED

Having established the design current (I_b) of the circuit under consideration, the appropriate procedure described in paragraphs 5.1 and 5.2 below will enable the designer to determine the size of the cable it will be necessary to use.

As a preliminary step it is useful to identify the length of the cable run and the permissible voltage drop for the equipment being supplied, as this may be an overriding consideration (see Section 525 and paragraph 6 of this appendix). The permissible voltage drop in mV, divided by I_b and by the length of run, will give the value of voltage drop in mV/A/m which can be tolerated. A voltage drop not exceeding that value is identified in the appropriate table and the corresponding cross-sectional area of conductor needed on this account can be read off directly before any other calculations are made.

The conductor size necessary from consideration of the conditions of normal load and overload is then determined. All rating factors affecting I_z (i.e. for factors for ambient temperature, grouping and thermal insulation) can, if desired, be applied to the values of I_t as multipliers. This involves a process of trial and error until a cross-sectional area is reached which ensures that I_z is not less than I_b and not less than I_n of any protective device it is intended to select. In any event, if a rating factor for protection by a semi-enclosed fuse is necessary, this has to be applied to I_n as a divisor. It is therefore more convenient to apply all the rating factors to I_n as divisors.

This method is used in items 5.1 and 5.2 and produces a value of current and that value (or the next larger value) can be readily located in the appropriate table of current-carrying capacity and the corresponding cross-sectional area of conductor can be identified directly. It should be noted that the value of I_t appearing against the chosen cross-sectional area is not I_z. It is not necessary to know I_z where the size of conductor is chosen by this method.

5.1 Where overload protection is afforded by a device listed in Regulation 433.1.100 or a semi-enclosed fuse to BS 3036

5.1.1 For single circuits

(i) Divide the rated current of the protective device (I_n) by any applicable rating factors for ambient temperature (C_a) , soil thermal resistivity (C_s) and depth of burial (C_d) given in Tables 4B1 to 4B4.

For cables installed above ground C_s and $C_d = 1$.

(ii) Then further divide by any applicable rating factor for thermal insulation (C_i).

(iii) Then further divide by the applicable rating factor for the type of protective device or installation condition (C_f, C_c):

$$I_t \geq \frac{I_n}{C_a\,C_s\,C_d\,C_i\,C_f\,C_c} \qquad \text{Equation 1}$$

    a) Where the protective device is a semi-enclosed fuse to BS 3036, $C_f = 0.725$. Otherwise $C_f = 1$

    b) Where the cable installation method is 'in a duct in the ground' or 'buried direct', $C_c = 0.9$. For cables installed above ground $C_c = 1$.

The size of cable to be used is to be such that its tabulated current-carrying capacity (I_t) is not less than the value of rated current of the protective device adjusted as above.

5.1.2 For groups

(i) In addition to the factors given in 5.1.1, divide the rated current of the protective device (I_n) by the applicable rating factor for grouping (C_g) given in Tables 4C1 to 4C6:

$$I_t \geq \frac{I_n}{C_g\, C_a\, C_s\, C_d\, C_i\, C_f\, C_c} \qquad \text{Equation 2}$$

Alternatively, I_t may be obtained from the following formulae, provided that the circuits of the group are not liable to simultaneous overload:

$$I_t \geq \frac{I_b}{C_g\, C_a\, C_s\, C_d\, C_i\, C_f\, C_c} \qquad \text{Equation 3}$$

$$I_t \geq \frac{1}{C_a C_s C_d C_i} \sqrt{\left(\frac{I_n}{C_f C_c}\right)^2 + 0.48\, I_b^{\,2}\left(\frac{1-C_g^{\,2}}{C_g^{\,2}}\right)} \qquad \text{Equation 4}$$

The size of cable to be used is to be such that its tabulated single-circuit current-carrying capacity (I_t) is not less than the value of I_t calculated in accordance with equation 2 above or, where equations 3 and 4 are used, not less than the larger of the resulting two values of I_t.

5.2 Where overload protection is not required

Where Regulation 433.3.1 applies, and the cable under consideration is not required to be protected against overload, the design current of the circuit (I_b) is to be divided by any applicable rating factors, and the size of the cable to be used is to be such that its tabulated current-carrying capacity (I_t) for the installation method concerned is not less than the value of I_b adjusted as above, i.e.:

$$I_t \geq \frac{I_b}{C_g\, C_a\, C_s\, C_d\, C_i\, C_c} \qquad \text{Equation 5}$$

5.3 Other frequencies

Current ratings stated in the tables are for d.c. and 50/60 Hz a.c. The current-carrying capacity of cables carrying, for example, balanced 400 Hz a.c. compared with the current-carrying capacity at 50 Hz, may be no more than 50 %. For small cables (e.g. as may be used to supply individual loads), the difference in the 50 Hz and the 400 Hz current-carrying capacities may be negligible. Current rating and voltage drop vary with frequency. Suitable ratings should be obtained from the manufacturer.

5.4 Effective current-carrying capacity

The current-carrying capacity of a cable corresponds to the maximum current that can be carried in specified conditions without the conductors exceeding the permissible limit of steady-state temperature for the type of insulation concerned.

The values of current tabulated represent the effective current-carrying capacity only where no rating factor is applicable. Otherwise, the current-carrying capacity corresponds to the tabulated value multiplied by the appropriate factor or factors for ambient temperature, grouping and thermal insulation as well as depth of burial and soil thermal resistivity, for buried cables, as applicable. Where harmonic currents are present further factors may need to be applied. See section 5.5 of this appendix.

Irrespective of the type of overcurrent protective device associated with the conductors concerned, the ambient temperature rating factors to be used when calculating current-carrying capacity (as opposed to those used when selecting cable sizes) are those given in Tables 4B1 and 4B2.

5.5 Rating factors for triple harmonic currents in four-core and five-core cables with four cores carrying current

5.5.1 Rating factors

Regulation 523.6.3 states that, where the neutral conductor carries current without a corresponding reduction in load of the line conductors, the current in the neutral conductor shall be taken into account in ascertaining the current-carrying capacity of the circuit.

This section is intended to cover the situation where there is current flowing in the neutral of a balanced three-phase system. Such neutral currents are due to the line currents having a harmonic content which does not cancel in the neutral. The most significant harmonic which does not cancel in the neutral is usually the third harmonic. The magnitude of the neutral current due to the third harmonic may exceed the magnitude of the power frequency line current. In such a case the neutral current will have a significant effect on the current-carrying capacity of the cables of the circuit.

The rating factors given in this appendix apply to balanced three-phase circuits; it is recognized that the situation is more onerous if only two of the three phases are loaded. In this situation, the neutral conductor will carry the harmonic currents in addition to the unbalanced current. Such a situation can lead to overloading of the neutral conductor.

Equipment likely to cause significant harmonic currents includes, for example, variable-speed motor drives, fluorescent lighting banks and d.c. power supplies such as those found in computers. Further information on harmonic disturbances can be found in BS EN 61000.

The rating factors given in the following table only apply to cables where the neutral conductor is within a four-core or five-core cable and is of the same material and cross-sectional area as the line conductors. These rating factors have been calculated on the basis of third harmonic currents measured with respect to the fundamental frequency of the line current. Where the total harmonic distortion is more than 15%, due to the third harmonic or multiples thereof, e.g. 9th, 15th, etc. then lower rating factors are applicable. Where there is an imbalance between phases of more than 50 % then lower rating factors may be applicable.

The tabulated rating factors, when applied to the current-carrying capacity of a cable with three loaded conductors, will give the current-carrying capacity of a cable with four loaded conductors where the current in the fourth conductor is due to harmonics. The rating factors also take the heating effect of the harmonic current in the line conductors into account.

Where the neutral current is expected to be higher than the line current then the cable size should be selected on the basis of the neutral current.

Where the cable size selection is based on a neutral current which is not significantly higher than the line current it is necessary to reduce the tabulated current-carrying capacity for three loaded conductors.

If the neutral current is more than 135 % of the line current and the cable size is selected on the basis of the neutral current then the three line conductors will not be fully loaded. The reduction in heat generated by the line conductors offsets the heat generated by the neutral conductor to the extent that it is not necessary to apply any rating factor to the current-carrying capacity for three loaded conductors, to take account of the effect of four loaded conductors.

TABLE 4Aa – Rating factors for triple harmonic currents in four-core and five-core cables

Third harmonic content of line current* %	Rating factor	
	Size selection is based on line current	Size selection is based on neutral current
0 – 15	1.0	–
>15 – 33	0.86	–
>33 – 45	–	0.86
> 45	–	1.0

*NOTE: The third harmonic content expressed as total harmonic distortion.

5.5.2 Example of the application of rating factor for third harmonic currents

Consider a three-phase circuit with a design load (fundamental current) of 58 A is to be installed using a four-core $90^{\circ}C$ thermosetting insulated cable. The cable will be installed in a group with 3 other circuits on a perforated cable tray (method E or F) in an expected maximum ambient temperature of $35^{\circ}C$. The cable will be protected at its origin using a circuit-breaker to BS EN 60898-1.

Case 1: Load does not produce third harmonic currents

The design current, I_b, of the three-phase load is 58 A

To satisfy Regulation 433.1.1, $I_n \geq I_b$, so the rated current of the circuit-breaker, I_n, is selected to be 63 A. The required tabulated current-carrying capacity, I_t, under the above operational conditions is to satisfy:

$$I_t \geq \frac{I_n}{C_a C_g}$$ (where circuits of the group are assumed to be liable to simultaneous overload)

From Table 4B1, $C_a = 0.96$ and from Table 4C1, $C_g = 0.77$

$$I_t = \frac{63}{0.96 \times 0.77} = 85.2 \text{ A}$$

From Table 4E4A, a 16 mm^2 cable with copper conductors and steel wire armour has a tabulated current-carrying capacity of 99 A and hence it is suitable if third harmonic currents are not present in the circuit.

Case 2: Load produces an additional third harmonic content – THD-i = 20%

For the second case it is assumed that the above load is expected to produce third harmonic distortion of 20% in addition to the fundamental line current.

The fundamental line current of the above load is 58 A. Since the third harmonic content is between 15-33%, the cable sizing is based upon the line current. Because the load has 20% third harmonic the design current used for the selection of the protective device is given by:

$$I_{bh} = 58 \times \sqrt{1^2 + 0.2^2} = 59.1 \text{ A}$$

where: I_{bh} = design current including the effect of third harmonic currents

To satisfy Regulation 433.1.1, $I_n \geq I_b$, so the rated current of the circuit-breaker, I_n, is selected to be 63 A. In addition, to comply with the Regulation 431.2.3, overcurrent detection shall be provided for the neutral conductor. Therefore, a 4-pole protective device with overcurrent protection of the neutral should be provided. The required tabulated current-carrying capacity, I_t, under the above operational conditions is to satisfy:

$$I_t \geq \frac{I_n}{C_a C_g \times 0.86} \quad \text{(where circuits of the group are assumed to be liable to simultaneous overload)}$$

The factor of 0.86 is taken from the above table.

Applying the above grouping and temperature rating factors, the required tabulated current-carrying capacity is found to be 99.1 A.

From Table 4E4A, a 16 mm^2 cable has a tabulated current-carrying capacity of 99 A, thus a rule-based system may select a 25 mm^2 cable whereas a designer may exercise judgement and select a 16 mm^2 cable.

Case 3: Load produces third harmonic content – THD-i =42%

The load is expected to produce third harmonic distortion of 42% of the fundamental line current.

The fundamental line current of the above load is 58 A. Since the third harmonic content is between 33-45%, the cable sizing is based upon the neutral current with a rating factor of 0.86 applied to the current-carrying capacity of the cable. In addition, to comply with Regulation 431.2.3, overcurrent detection must be provided for the neutral conductor. Therefore, a 4-pole protective device with overcurrent protection of the neutral should be provided.

The neutral current arising from third harmonics is given by

$$I_{bn} = \frac{3h}{100} I_{nL}$$

where: I_{bn} = neutral current due to third harmonic currents

I_{nL} = fundamental line current

h = third harmonic as a percentage of the fundamental line current.

Hence, the neutral current of the circuit is $I_{bn} = 3 \times 0.42 \times 58 = 73$ A.

Therefore, the design current of the circuit due to third harmonics is 73 A.

To satisfy Regulation 433.1.1, $I_n \geq I_b$, so the rated current of the circuit-breaker, I_n, is selected to be 80 A. The required tabulated current-carrying capacity, I_t, under the above operational conditions is to satisfy:

$$I_t \geq \frac{I_n}{C_a C_g \times 0.86} \quad \text{(where circuits of the group are assumed to be liable to simultaneous overload)}$$

Applying the above grouping and temperature rating factors, the required tabulated current-carrying capacity is found to be 125.8 A.

From Table 4E4A, a 25 mm^2 cable is necessary to compensate for the additional thermal effect due to third harmonic current.

All the above cable selections are based on the current-carrying capacity of the cable; voltage drop and other aspects of design have not been considered.

5.6 Harmonic currents in line conductors

Section 5.5 covers the effect of additive harmonic currents flowing in the neutral conductor. The rating factors given in section 5.5 take account of the heating effect of the third harmonic in the neutral as well as the heating effect of the third harmonic in each of the line conductors.

Where other harmonics are present, e.g. 5th, 7th etc, the heating effect of these harmonics in the line conductors has to be taken into account. For smaller sizes, less than 50 mm^2, the effect of harmonic currents can be taken into account by applying the following factor, C_h, to the fundamental design current.

$$C_h = \sqrt{\frac{I_f^2 + I_{h5}^2 + ... + I_{hn}^2}{I_f^2}}$$

where: I_f = 50 Hz current

I_{h5} = 5th harmonic current

I_{hn} = nth harmonic current

For larger conductor sizes the increase in conductor resistance, due to skin and proximity effects, at higher frequencies has to be taken into account. The resistance at harmonic frequencies can be calculated using the equations given in BS IEC 60287-1-1.

6 TABLES OF VOLTAGE DROP

In the tables, values of voltage drop are given for a current of one ampere for a metre run, i.e. for a distance of 1 m along the route taken by the cables, and represent the result of the voltage drops in all the circuit conductors. The values of voltage drop assume that the conductors are at their maximum permitted normal operating temperature.

The values in the tables, for a.c. operation, apply to frequencies in the range 49 to 61 Hz and for single-core armoured cables the tabulated values apply where the armour is bonded to earth at both ends. The values of voltage drop for cables operating at higher frequencies may be substantially greater.

For a given run, to calculate the voltage drop (in mV) the tabulated value of voltage drop per ampere per metre for the cable concerned has to be multiplied by the length of the run in metres and by the current the cable is intended to carry, namely, the design current of the circuit (I_b) in amperes. For three-phase circuits the tabulated mV/A/m values relate to the line voltage and balanced conditions have been assumed.

For cables having conductors of 16 mm^2 or less cross-sectional area, their inductances can be ignored and (mV/A/m)$_r$ values only are tabulated. For cables having conductors greater than 16 mm^2 cross-sectional area the impedance values are given as (mV/A/m)$_z$, together with the resistive component (mV/A/m)$_r$ and the reactive component (mV/A/m)$_x$.

The direct use of the tabulated (mV/A/m)$_r$ or (mV/A/m)$_z$ values, as appropriate, may lead to pessimistically high calculated values of voltage drop or, in other words, to unnecessarily low values of permitted circuit lengths. For example, where the design current of a circuit is significantly less than the effective current-carrying capacity of the chosen cable, the actual voltage drop would be less than the calculated value because the conductor temperature (and hence their resistance) will be less than that on which the tabulated mV/A/m had been based.

As regards power factor in a.c. circuits, the use of the tabulated mV/A/m values (for the larger cable sizes, the tabulated (mV/A/m)$_z$ values) leads to a calculated value of the voltage drop higher than the actual value. In some cases it may be advantageous to take account of the load power factor when calculating voltage drop.

Where a more accurate assessment of the voltage drop is desirable the following methods may be used.

6.1　　Correction for operating temperature

For cables having conductors of cross-sectional area 16 mm^2 or less, the design value of mV/A/m is obtained by multiplying the tabulated value by a factor C_t, given by:

$$C_t = \frac{230 + t_p - \left(C_a^2 \, C_g^2 \, C_s^2 \, C_d^2 - \dfrac{I_b^2}{I_t^2}\right)(t_p - 30)}{230 + t_p} \qquad \text{Equation 6}$$

where t_p is the maximum permitted normal operating temperature (°C).

This equation applies only where the overcurrent protective device is other than a BS 3036 fuse and where the actual ambient temperature is equal to or greater than 30 °C.

NOTE:　For convenience, the above equation is based on the approximate resistance-temperature coefficient of 0.004 per °C at 20 °C for both copper and aluminium conductors.

For cables having conductors of cross-sectional area greater than 16 mm^2, only the resistive component of the voltage drop is affected by the temperature and the factor C_t is therefore applied only to the tabulated value of $(mV/A/m)_r$ and the design value of $(mV/A/m)_z$ is given by the vector sum of $C_t \, (mV/A/m)_r$ and $(mV/A/m)_x$.

For very large conductor sizes, where the resistive component of voltage drop is much less than the corresponding reactive part (i.e. when $x/r \geq 3$), this rating factor need not be considered.

6.2　　Correction for load power factor

For cables having conductors of cross-sectional area 16 mm^2 or less, the design value of mV/A/m is obtained approximately by multiplying the tabulated value by the power factor of the load, cos Ø.

For cables having conductors of cross-sectional area greater than 16 mm^2, the design value of mV/A/m is given approximately by:

cos Ø (tabulated $(mV/A/m)_r$) + sin Ø (tabulated $(mV/A/m)_x$)

For single-core cables in flat formation the tabulated values apply to the outer cables and may underestimate for the voltage drop between an outer cable and the centre cable for cross-sectional areas above 240 mm^2, and power factors greater than 0.8.

6.3　　Correction for both operating temperature and load power factor

For paragraphs 6.1 and 6.2 above, where it is considered appropriate to correct the tabulated mV/A/m values for both operating temperature and load power factor, the design figure for mV/A/m is given by:

(i)　for cables having conductors of cross-sectional area 16 mm^2 or less

C_t cos Ø (tabulated mV/A/m)

(ii)　for cables having conductors of cross-sectional area greater than 16 mm^2

C_t cos Ø (tabulated $(mV/A/m)_r$) + sin Ø (tabulated $(mV/A/m)_x$).

6.4 Voltage drop in consumers' installations

The voltage drop between the origin of an installation and any load point should not be greater than the values in the table below expressed with respect to the value of the nominal voltage of the installation.

The calculated voltage drop should include any effects due to harmonic currents.

TABLE 4Ab – Voltage drop

	Lighting	Other uses
(i) Low voltage installations supplied directly from a public low voltage distribution system	3%	5%
(ii) Low voltage installation supplied from private LV supply (*)	6%	8%

(*) The voltage drop within each final circuit should not exceed the values given in (i).

Where the wiring systems of the installation are longer than 100 m, the voltage drops indicated above may be increased by 0.005% per metre of the wiring system beyond 100 m, without this increase being greater than 0.5%.

The voltage drop is determined from the demand of the current-using equipment, applying diversity factors where applicable, or from the value of the design current of the circuit.

NOTE 1: A greater voltage drop may be acceptable for a motor circuit during starting and for other equipment with a high inrush current, provided that in both cases it is ensured that the voltage variations remain within the limits specified in the relevant equipment standard.

NOTE 2: The following temporary conditions are excluded:
- voltage transients
- voltage variations due to abnormal operation

7 METHODS OF INSTALLATION

Table 4A2 lists the methods of installation for which this appendix provides guidance for the selection of the appropriate cable size. Table 4A3 lists the appropriate tables for selection of current ratings for specific cable constructions. The Reference Methods are those methods of installation for which the current-carrying capacities given in Tables 4D1A to 4J4A have been determined (see 7.1 below).

The use of other methods is not precluded and in that case the evaluation of current-carrying capacity may need to be based on experimental work.

7.1 Reference Methods

The Reference Methods are those methods of installation for which the current-carrying capacity has been determined by test or calculation.

NOTE: It is impractical to calculate and publish current ratings for every installation method, since many would result in the same current rating. Therefore a suitable (limited) number of current ratings have been calculated which cover all of the installation methods stated in Table 4A2 and have been called Reference Methods.

Reference Method A, for example, Installation Methods 1 and 2 of Table 4A2 (non-sheathed cables and multicore cables in conduit in a thermally insulated wall).

The wall consists of an outer weatherproof skin, thermal insulation and an inner skin of wood or wood-like material having a thermal conductance of at least 10 W/m^2.K. The conduit is fixed such that it is close to, but not necessarily touching, the inner skin. Heat from the cables is assumed to escape through the inner skin only. The conduit can be metal or plastic.

Reference Method B, for example, Installation Method 4 of Table 4A2 (non-sheathed cables in conduit mounted on a wooden or masonry wall) and Installation Method 5 of Table 4A2 (multicore cable in conduit on a wooden or masonry wall).

The conduit is mounted on a wooden wall such that the gap between the conduit and the surface is less than 0.3 times the conduit diameter. The conduit can be metal or plastic. Where the conduit is fixed to a masonry wall the current-carrying capacity of the non-sheathed or sheathed cable may be higher.

Reference Method C (clipped direct), for example, Installation Method 20 of Table 4A2 (single-core or multicore cables on a wooden or masonry wall)

Cable mounted on a wooden wall so that the gap between the cable and the surface is less than 0.3 times the cable diameter. Where the cable is fixed to or embedded in a masonry wall the current-carrying capacity may be higher.

NOTE: The term 'masonry' is taken to include brickwork, concrete, plaster and similar (but excluding thermally insulating materials).

Reference Method D, for example, Installation Method 70 of Table 4A2 (multicore armoured cable in conduit or in cable ducting in the ground).

The cable is drawn into a 100 mm diameter plastic, earthenware or metallic duct laid in direct contact with soil having a thermal resistivity of 2.5 K.m/W and at a depth of 0.7 m. The values given for this method are those stated in this appendix and are based on conservative installation parameters. If the specific installation parameters are known (thermal resistance of the ground, ground ambient temperature, cable depth), reference can be made to the cable manufacturer or the ERA 69-30 series of publications, which may result in a smaller cable size being selected.

NOTE: The current-carrying capacity for cables laid in direct contact with soil having a thermal resistivity of 2.5 K.m/W and at a depth of 0.7 m is approximately 10 % higher than the values tabulated for Reference Method D.

Reference Methods E, F and G, for example, Installation Methods 31 to 35 of Table 4A (single-core or multicore cables in free air).

The cable is supported such that the total heat dissipation is not impeded. Heating due to solar radiation and other sources is to be taken into account. Care is to be taken that natural air convection is not impeded. In practice, a clearance between a cable and any adjacent surface of at least 0.3 times the cable external diameter for multicore cables or 1.0 times the cable diameter for single-core cables is sufficient to permit the use of current-carrying capacities appropriate to free air conditions.

7.2   Other Methods

Cable on a floor: Reference Method C applies for current rating purposes.

Cable under a ceiling: This installation may appear similar to Reference Method C but because of the reduction in natural air convection, Reference Method B is to be used for the current rating.

Cable tray systems: A perforated cable tray has a regular pattern of holes that occupy at least 30% of the area of the base of the tray. The current-carrying capacity for cables attached to perforated cable trays should be taken as Reference Method E or F. The current-carrying capacity for cables attached to unperforated cable trays (no holes or holes that occupy less than 30% of the area of the base of the tray) is to be taken as Reference Method C.

Cable ladder system: This is a construction which offers a minimum of impedance to the air flow around the cables, i.e. supporting metalwork under the cables occupies less than 10% of the plan area. The current-carrying capacity for cables on ladder systems should be taken as Reference Method E or F.

Cable cleats, cable ties and cable hangers: Cable supports hold the cable at intervals along its length and permit substantially complete free air flow around the cable. The current-carrying capacity for cable cleats, cable ties and cable hangers should be taken as Reference Method E or F.

Cable installed in a ceiling: This is similar to Reference Method A. It may be necessary to apply the rating factors due to higher ambient temperatures that may arise in junction boxes and similar mounted in the ceiling.

NOTE: Where a junction box in the ceiling is used for the supply to a luminaire, the heat dissipation from the luminaire may provide higher ambient temperatures than permitted in Tables 4D1A to 4J4A (see also Regulation 522.2.1). The temperature may be between 40 °C and 50 °C, and a rating factor according to Table 4B1 must be applied.

General notes to all tables in this appendix

NOTE 1: Current-carrying capacities are tabulated for methods of installation which are commonly used for fixed electrical installations. The tabulated capacities are for continuous steady-state operation (100 % load factor) for d.c. or a.c. of nominal frequency 50 Hz and take no account of harmonic content.

NOTE 2: Table 4A2 itemises the reference methods of installation to which the tabulated current-carrying capacities refer.

TABLE 4A1 – Schedule of Installation Methods in relation to conductors and cables

Conductors and cables		Installation Method							
		Without fixings	Clipped direct	Conduit systems	Cable trunking systems*	Cable ducting systems	Cable ladder, cable tray, cable brackets	On insulators	Support wire
Bare conductors		np	np	np	np	np	np	P	np
Non-sheathed cable		np	np	P¹	P¹ ²	P¹	np¹	P	np
Sheathed cables (including armoured and mineral insulated)	Multicore	P	P	P	P	P	P	N/A	P
	Single-core	N/A	P	P	P	P	P	N/A	P

P   Permitted.

np   Not permitted.

N/A   Not applicable, or not normally used in practice.

* including skirting trunking and flush floor trunking

[1]   Non-sheathed cables which are used as protective conductors or protective bonding conductors need not be laid in conduits or ducts

[2]   Non-sheathed cables are acceptable if the trunking system provides at least the degree of protection IPXXD or IP4X and if the cover can only be removed by means of a tool or a deliberate action.

TABLE 4A2 – Schedule of Installation Methods of cables (including Reference Methods) for determining current-carrying capacity

NOTE 1: The illustrations are not intended to depict actual product or installation practices but are indicative of the method described.

NOTE 2: The installation and reference methods stated are in line with IEC. However, not all methods have a corresponding rating for all cable types.

Installation Method			Reference Method to be used to determine current-carrying capacity
Number	Examples	Description	
1	Room	Non-sheathed cables in conduit in a thermally insulated wall with an inner skin having a thermal conductance of not less than 10 W/m²K [c]	A
2	Room	Multicore cable in conduit in a thermally insulated wall with an inner skin having a thermal conductance of not less than 10 W/m²K [c]	A
3	Room	Multicore cable direct in a thermally insulated wall with an inner skin having a thermal conductance of not less than 10 W/m²K [c]	A
4		Non-sheathed cables in conduit on a wooden or masonry wall or spaced less than 0.3 × conduit diameter from it [c]	B
5		Multicore cable in conduit on a wooden or masonry wall or spaced less than 0.3 × conduit diameter from it [c]	B
6 7	6 7	Non-sheathed cables in cable trunking on a wooden or masonry wall 6 - run horizontally [b] 7 - run vertically [b, c]	B
8 9	8 9	Multicore cable in cable trunking on a wooden or masonry wall 8 - run horizontally [b] 9 - run vertically [b, c]	B*
10		Non-sheathed cables in suspended cable trunking [b]	B
11	10 11	Multicore cable in suspended cable trunking [b]	B
12		Non-sheathed cables run in mouldings [c,e]	A

b   Values given for Installation Method B in Appendix 4 are for a single circuit. Where there is more than one circuit in the trunking the group rating factor given in Table 4C1 is applicable, irrespective of the presence of an internal barrier or partition.

c   Care is needed where the cable runs vertically and ventilation is restricted. The ambient temperature at the top of the vertical section can be much higher.

e   The thermal resistivity of the enclosure is assumed to be poor because of the material of construction and possible air spaces. Where the construction is thermally equivalent to Installation Methods 6 or 7, Reference Method B may be used.

*   Still under consideration in IEC.

Installation Method			Reference Method to be used to determine current-carrying capacity
Number	**Examples**	**Description**	
13 *14*		*Deleted by BS 7671:2008 Amendment No 1*	
15		Non-sheathed cables in conduit or single-core or multicore cable in architrave [c, f]	A
16		Non-sheathed cables in conduit or single-core or multicore cable in window frames [c, f]	A
20		Single-core or multicore cables: - fixed on (clipped direct), or spaced less than 0.3 × cable diameter from a wooden or masonry wall [c]	C
21		Single-core or multicore cables: - fixed directly under a wooden or masonry ceiling	C (Higher than standard ambient temperatures may occur with this installation method)
22		Single-core or multicore cables: - spaced from a ceiling	E, F or G* (Higher than standard ambient temperatures may occur with this installation method)
23		*Not used.*	
30		Single-core or multicore cables: - on unperforated tray run horizontally or vertically [c, h]	C with item 2 of Table 4C1
31		Single-core or multicore cables: - on perforated tray run horizontally or vertically [c, h]	E or F
32		Single-core or multicore cables: - on brackets or on a wire mesh tray run horizontally or vertically [c, h]	E or F

c Care is needed where the cable runs vertically and ventilation is restricted. The ambient temperature at the top of the vertical section can be much higher.

f The thermal resistivity of the enclosure is assumed to be poor because of the material of construction and possible air spaces. Where the construction is thermally equivalent to Installation Methods 6, 7, 8, or 9, Reference Method B may be used.

h D_e = the external diameter of a multicore cable:
- 2.2 x the cable diameter when three single-core cables are bound in trefoil, or
- 3 x the cable diameter when three single-core cables are laid in flat formation.

* Still under consideration in IEC.

Installation Method			Reference Method to be used to determine current-carrying capacity
Number	Examples	Description	
33		Single-core or multicore cables: - spaced more than 0.3 times the cable diameter from a wall	**E, F or G**[g]
34		Single-core or multicore cables: - on a ladder [c]	**E or F**
35		Single-core or multicore cable suspended from or incorporating a support wire or harness	**E or F**
36		Bare or non-sheathed cables on insulators	**G**
40	D_e V	Single-core or multicore cable in a building void [c, h, i]	Where $1.5\ D_e \leq V < 20\ D_e$ use **B**
41	D_e V	Non-sheathed cables in conduit in a building void in masonry having a thermal resistivity not greater than 2 K.m/W [c, i, j]	Where $1.5\ D_e \leq V$ use **B**
42	D_e V	Single-core or multicore cable in conduit in a building void in masonry having a thermal resistivity not greater than 2 K.m/W [c, j]	Where $1.5\ D_e \leq V$ use **B**
43	D_e V	Non-sheathed cables in cable ducting in a building void in masonry having a thermal resistivity not greater than 2 K.m/W [c, i, j]	Where $1.5\ D_e \leq V$ use **B**
44	D_e V	Single-core or multicore cable in cable ducting in a building void in masonry having a thermal resistivity not greater than 2 K.m/W [c, i, j]	Where $1.5\ D_e \leq V$ use **B**
45	D_e V	Non-sheathed cables in cable ducting in masonry having a thermal resistivity not greater than 2 K.m/W [c, h, i]	Where $1.5\ D_e \leq V < 50\ D_e$ use **B**
46	D_e V	Single-core or multicore cable in cable ducting in masonry having a thermal resistivity not greater than 2 K.m/W [c, h, i]	Where $1.5\ D_e \leq V < 50\ D_e$ use **B**
47	D_e V	Single-core or multicore cable: - in a ceiling void - in a suspended floor [h, i]	Where $1.5\ D_e \leq V < 50\ D_e$ use **B**

c Care is needed where the cable runs vertically and ventilation is restricted. The ambient temperature at the top of the vertical section can be much higher.

g The factors in Table 4C1 may also be used.

h D_e = the external diameter of a multicore cable:
- 2.2 x the cable diameter when three single-core cables are bound in trefoil, or
- 3 x the cable diameter when three single-core cables are laid in flat formation.

i V = the smaller dimension or diameter of a masonry duct or void,
or the vertical depth of a rectangular duct, floor or ceiling void or channel.

j D_e = external diameter of conduit or vertical depth of cable ducting.

TABLE 4A2 *(continued)*

	Installation Method		Reference Method to be used to determine current-carrying capacity
Number	Examples	Description	
50		Non-sheathed cables in flush cable trunking in the floor	B
51		Multicore cable in flush cable trunking in the floor	B
52		Non-sheathed cables in flush trunking [c]	B
53		Multicore cable in flush trunking [c]	B
54		Non-sheathed cables or single-core cables in conduit in an unventilated cable channel run horizontally or vertically [c, i, k, m]	Where $1.5 D_e \leq V$ use **B**
55		Non-sheathed cables in conduit in an open or ventilated cable channel in the floor [l, m]	B
56		Sheathed single-core or multicore cable in an open or ventilated cable channel run horizontally or vertically [m]	B
57		Single-core or multicore cable direct in masonry having a thermal resistivity not greater than 2 K.m/W - without added mechanical protection [n, o]	C
58		Single-core or multicore cable direct in masonry having a thermal resistivity not greater than 2 K.m/W - with added mechanical protection [n, o] (e.g. capping)	C
59		Non-sheathed cables or single-core cables in conduit in masonry having a thermal resistivity not greater than 2 K.m/W [o]	B
60		Multicore cables in conduit in masonry having a thermal resistivity not greater than 2 K.m/W [o]	B

c   Care is needed where the cable runs vertically and ventilation is restricted. The ambient temperature at the top of the vertical section can be much higher.

k D_e = external diameter of conduit.

i   V = the smaller dimension or diameter of a masonry duct or void, or the vertical depth of a rectangular duct, floor or ceiling void or channel.
The depth of the channel is more important than the width.

l   For multicore cable installed as Method 55, use current-carrying capacity for Reference Method B.

m   It is recommended that these Installation Methods are used only in areas where access is restricted to authorized persons so that the reduction in current-carrying capacity and the fire hazard due to the accumulation of debris can be prevented.

n   For cables having conductors not greater than 16 mm^2, the current-carrying capacity may be higher.

o   Thermal resistivity of masonry is not greater than 2 K.m/W. The term masonry is taken to include brickwork, concrete, plaster and the like (excludes thermally insulating materials).

TABLE 4A2 *(continued)*

Number	Installation Method — Examples	Installation Method — Description	Reference Method to be used to determine current-carrying capacity
70		Multicore armoured cable in conduit or in cable ducting in the ground	**D** **For multicore armoured cable only**
71		*Deleted by BS 7671:2008 Amendment No 1*	
72		Sheathed, armoured or multicore cables direct in the ground: - without added mechanical protection (see note)	D
73		Sheathed, armoured or multicore cables direct in the ground: with added mechanical protection (e.g. cable covers) (see note)	D

NOTE: The inclusion of directly buried cables is satisfactory where the soil thermal resistivity is of the order of 2.5 K.m/W. For lower soil resistivities, the current-carrying capacity for directly buried cables is appreciably higher than for cables in ducts.

TABLE 4A2 *(continued –*
Installation methods specifically for flat twin and earth cables in thermal insulation)

Number	Installation Method — Examples	Installation Method — Description	Reference Method to be used to determine current-carrying capacity
100		Installation methods for flat twin and earth cable clipped direct to a wooden joist, or touching the plasterboard ceiling surface, above a plasterboard ceiling with thermal insulation not exceeding 100 mm in thickness having a minimum U value of 0.1 W/m²K	Table 4D5
101		Installation methods for flat twin and earth cable clipped direct to a wooden joist, or touching the plasterboard ceiling surface, above a plasterboard ceiling with thermal insulation exceeding 100 mm in thickness having a minimum U value of 0.1 W/m²K	Table 4D5
102		Installation methods for flat twin and earth cable in a stud wall with thermal insulation with a minimum U value of 0.1 W/m²K with the cable touching the inner wall surface, or touching the plasterboard ceiling surface, and the inner skin having a minimum U value of 10 W/m²K	Table 4D5
103		Installation methods for flat twin and earth cable in a stud wall with thermal insulation with a minimum U value of 0.1 W/m²K with the cable not touching the inner wall surface	Table 4D5

Wherever practicable, a cable is to be fixed in a position such that it will not be covered with thermal insulation.

Regulation 523.9, BS 5803-5: Appendix C: Avoidance of overheating of electric cables, Building Regulations Approved document B and Thermal insulation: avoiding risks, BR 262, BRE, 2001 refer.

TABLE 4A2 *(continued –*
Installation methods for cables enclosed in infloor concrete troughs)

	Installation Method		Reference Method to be used to determine current-carrying capacity
Number	Examples	Description	
117		Cables supported on the wall of an open or ventilated infloor concrete trough with spacing as follows: - Sheathed single-core cables in free air (any supporting metalwork under the cables occupying less than 10% of plan area). - Two or three cables vertically one above the other, minimum distance between cable surfaces equal to the overall cable diameters, distance from the wall not less than ½ the cable diameter. - Two or three cables horizontally with spacing as above	E or F
118		Cables in enclosed trench 450 mm wide by 300 mm deep (minimum dimensions) including 100 mm cover - Two to six single–core cables with surfaces separated by a minimum of one cable diameter - One or two groups of three single-core cables in trefoil formation - One to four 2-core cables or one to three cables of 3 or 4 cores with all cables separated by a minimum of 50 mm	E or F using rating factors in Table 4C6
119		Cables enclosed in an infloor concrete trough 450 mm wide by 600 mm deep (minimum dimensions) including 100 mm cover. Six to twelve single-core cables arranged in flat groups of two or three on the vertical trench wall with cables separated by one cable diameter and a minimum of 50 mm between groups or two to four groups of three single-core cables in trefoil formation with a minimum of 50 mm between trefoil formations or four to eight 2-core cables or three to six cables of 3 or 4 cores with cables separated by a minimum of 75 mm. All cables spaced at least 25 mm from trench wall.	E or F using rating factors in Table 4C6

322

Installation Method			Reference Method to be used to determine current-carrying capacity
Number	Examples	Description	
120		Cables enclosed in an infloor concrete trough 600mm wide by 760 mm deep (minimum dimensions) including 100 mm cover. Twelve to twenty four single-core cables arranged in either flat formation of two or three cables in a group with cables separated by one cable diameter and each cable group separated by a minimum of 50 mm either horizontally or vertically or single-core cables in trefoil formation with each group or trefoil formation separated by a minimum of 50 mm either horizontally or vertically or eight to sixteen 2-core cables or six to twelve cables of 3 or 4 cores with cables separated by a minimum of 75 mm either horizontally or vertically. All cables spaced at least 25 mm from trench wall.	**E or F using rating factors in Table 4C6**

TABLE 4A3 – Schedule of cable specifications and current rating tables

Specification number	Specification title	Applicable current rating Tables	Conductor operating temperature
BS 5467	Electric cables – Thermosetting insulated armoured cables for voltages of 600/1000 V and 1900/3300 V.	4E3, 4E4, 4J3, 4J4	90 °C
BS 6004	Electric cables – PVC insulated, non-armoured cables for voltages up to and including 450/750 V, for electric power, lighting and internal wiring.	4D1, 4D2	70 °C
BS 6004	Thermoplastic insulated and sheathed flat cable with protective conductor to Table 8.	4D5	70 °C
BS 6231	Electric cables – single-core PVC insulated flexible cables of rated voltage 600/1000 V for switchgear and controlgear wiring.	4D1	70 °C*
BS 6346 (obsolete)	Electric cables – PVC insulated, armoured cables for voltages of 600/1000 V and 1900/3300 V.	4D3, 4D4, 4H3, 4H4, 4J3, 4J4	70 °C
BS 6500	Electric cables – Flexible cords rated up to 300/500 V, for use with appliances and equipment intended for domestic, office and similar environments.	4F3	60 °C, 90 °C
BS 6724	Electric cables – Thermosetting insulated, armoured cables for voltages of 600/1000 V and 1900/3300 V, having low emission of smoke and corrosive gases when affected by fire.	4E3, 4E4, 4J3, 4J4	90 °C
BS 7211	Electric cables – Thermosetting insulated, non-armoured cables for voltages up to and including 450/750 V, for electric power, lighting and internal wiring, and having low emission of smoke and corrosive gases when affected by fire.	4E1, 4E2	90 °C
BS 7629-1	Specification for 300/500 V fire-resistant electric cables having low emission of smoke and corrosive gases when affected by fire - Part 1: Multicore cables.	4D2	70 °C
BS 7846	Electric cables – 600/1000 V armoured fire-resistant cables having thermosetting insulation and low emission of smoke and corrosive gases when affected by fire.	4E3, 4E4, 4J3, 4J4	90 °C
BS 7889	Electric cables – Thermosetting insulated, unarmoured cables for a voltage of 600/1000 V.	4E1	90 °C
BS 7919	Electric cables – Flexible cables rated up to 450/750 V, for use with appliances and equipment intended for industrial and similar environments.	4F1, 4F2, 4F3	60 °C, 90 °C 180 °C
BS 8436	Electric cables – 300/500 V screened electric cables having low emission of smoke and corrosive gases when affected by fire, for use in walls, partitions and building voids - multicore cables.	4D2	70 °C
BS EN 60702-1	Mineral insulated cables and their terminations with a rated voltage not exceeding 750 V – cables	4G1, 4G2	70 °C**, 105 °C**

*   Cables to BS 6231 when installed in conduit or trunking are rated to 70 °C.

**   Sheath operating temperature.

TABLE 4B1 – Rating factors (C_a) for ambient air temperatures other than 30 °C

Ambient temperature [a] °C	Insulation				
	60 °C thermosetting	70 °C thermoplastic	90 °C thermosetting	Mineral [a]	
				Thermoplastic covered or bare and exposed to touch 70 °C	Bare and not exposed to touch 105 °C
25	1.04	1.03	1.02	1.07	1.04
30	1.00	1.00	1.00	1.00	1.00
35	0.91	0.94	0.96	0.93	0.96
40	0.82	0.87	0.91	0.85	0.92
45	0.71	0.79	0.87	0.78	0.88
50	0.58	0.71	0.82	0.67	0.84
55	0.41	0.61	0.76	0.57	0.80
60		0.50	0.71	0.45	0.75
65		–	0.65	–	0.70
70		–	0.58	–	0.65
75		–	0.50	–	0.60
80		–	0.41	–	0.54
85		–	–	–	0.47
90		–	–	–	0.40
95		–	–	–	0.32

a   For higher ambient temperatures, consult manufacturer.

TABLE 4B2 – Rating factors (C_a) for ambient ground temperatures other than 20 °C

Ground temperature °C	Insulation	
	70 °C thermoplastic	90 °C thermosetting
10	1.10	1.07
15	1.05	1.04
20	1.00	1.00
25	0.95	0.96
30	0.89	0.93
35	0.84	0.89
40	0.77	0.85
45	0.71	0,80
50	0.63	0.76
55	0.55	0.71
60	0.45	0.65
65	–	0.60
70	–	0.53
75	–	0.46
80	–	0.38

TABLE 4B3 – Rating factors (C_s) for cables buried direct in the ground or in an underground conduit system to BS EN 50086-2-4 for soil thermal resistivities other than 2.5 K.m/W to be applied to the current-carrying capacities for Reference Method D

Thermal resistivity, K.m/W	0.5	0.8	1	1.2	1.5	2	2.5	3
Rating factor for cables in buried ducts	1.28	1.20	1.18	1.13	1.1	1.05	1	0.96
Rating factor for direct buried cables	1.88	1.62	1.5	1.40	1.28	1.12	1	0.90

NOTE 1: The rating factors given have been averaged over the range of conductor sizes and types of installation included in the relevant tables in this appendix. The overall accuracy of rating factors is within ± 5%.

NOTE 2: Where more precise values are required they may be calculated by methods given in BS 7769 (BS IEC 60287).

NOTE 3: The rating factors are applicable to ducts buried at depths of up to 0.8 m.

TABLE 4B4 – Rating factors (C_d) for depths of laying other than 0.7 m for direct buried cables and cables in buried ducts

Depth of laying, m	Buried direct	In buried ducts
0.5	1.03	1.02
0.7	1.00	1.00
1	0.97	0.98
1.25	0.95	0.96
1.5	0.94	0.95
1.75	0.93	0.94
2	0.92	0.93
2.5	0.90	0.92
3	0.89	0.91

TABLE 4B5 – Rating factors for cables having more than 4 loaded cores

Number of loaded cores	5	6	7	10	12	14	19
Rating factor	0.72	0.67	0.63	0.56	0.53	0.51	0.45
Number of loaded cores	24	27	30	37	44	46	48
Rating factor	0.42	0.40	0.39	0.36	0.34	0.33	0.33

NOTE 1: The current-carrying capacity for a cable in the size range 1.5 to 4 mm^2, having more than 4 loaded cores, is obtained by multiplying the current-carrying capacity of a 2-core, having the same insulation type, by the factor selected from this table. The current–carrying capacity for the 2-core cable is that for the installation condition to be used for the multicore cable.

NOTE 2: If, due to known operating conditions, a core is expected to carry not more than 30% of its current-carrying capacity in the multicore cable it may be ignored for the purpose of determining the number of cores in the cable.

NOTE 3: If, due to known operating conditions, a core is expected to carry not more than 30% of its rating, after applying the rating factor for the total number of current-carrying cores, it may be ignored for the purpose of obtaining the rating factor for the number of loaded cores.

For example, the current-carrying capacity of a cable having N loaded cores would normally be obtained by multiplying the current-carrying capacity of a 2-core, having the same insulation type, by the factor selected from this table for N cores. That is $I_{zlc} = I_{t2c} \times C_{gN}$

where:

I_{zlc} is the current-carrying capacity of the multicore cable after applying the rating factor for the total number of current-carrying cores

I_{t2c} is the tabulated current-carrying capacity of a 2-core cable, having the same insulation type as the multi-core cable

C_{gN} is the rating factor from Table 4B5 for the total number of current-carrying cores

However, if M cores in the cable carry loads which are not greater than $0.3 \times I_{t2c} \times C_{gN}$, the current-carrying capacity can be obtained by using the rating factor corresponding to (N-M) cores.

The "not greater than $0.3 \times I_{t2c} \times C_{gN}$" calculation should be applied before the adjacent multicore cable grouping factor, if applicable, from Table 4C1. The 30% rule should not be further applied to any adjacent cable grouping factor calculations.

I_{zlc} should be greater than or equal to I_n or I_b as appropriate, divided by the relevant rating factor(s) C, that is
$I_{zlc} \geq I_n$ or I_b / C

TABLE 4C1 – Rating factors for one circuit or one multicore cable or for a group of circuits, or a group of multicore cables, to be used with current-carrying capacities of Tables 4D1A to 4J4A

Item	Arrangement (cables touching)	Number of circuits or multicore cables												To be used with current-carrying capacities, Reference Method
		1	2	3	4	5	6	7	8	9	12	16	20	
1.	Bunched in air, on a surface, embedded or enclosed	1.00	0.80	0.70	0.65	0.60	0.57	0.54	0.52	0.50	0.45	0.41	0.38	A to F
2.	Single layer on wall or floor	1.00	0.85	0.79	0.75	0.73	0.72	0.72	0.71	0.70	0.70	0.70	0.70	C
3.	Single layer multicore on a perforated horizontal or vertical cable tray system	1.00	0.88	0.82	0.77	0.75	0.73	0.73	0.72	0.72	0.72	0.72	0.72	E
4.	Single layer multicore on cable ladder system or cleats etc.,	1.00	0.87	0.82	0.80	0.80	0.79	0.79	0.78	0.78	0.78	0.78	0.78	

NOTE 1: These factors are applicable to uniform groups of cables, equally loaded.

NOTE 2: Where horizontal clearances between adjacent cables exceed twice their overall diameter, no rating factor need be applied.

NOTE 3: The same factors are applied to:
- groups of two or three single-core cables;
- multicore cables.

NOTE 4: If a group consists of both two- and three-core cables, the total number of cables is taken as the number of circuits, and the corresponding factor is applied to the tables for two loaded conductors for the two-core cables, and to the Tables for three loaded conductors for the three-core cables.

NOTE 5: If a group consists of n single-core cables it may either be considered as $n/2$ circuits of two loaded conductors or $n/3$ circuits of three loaded conductors.

NOTE 6: The rating factors given have been averaged over the range of conductor sizes and types of installation included in Tables 4D1A to 4J4A and the overall accuracy of tabulated values is within 5%.

NOTE 7: For some installations and for other methods not provided for in the above table, it may be appropriate to use factors calculated for specific cases, see for example Tables 4C4 and 4C5.

NOTE 8: Where cables having differing conductor operating temperature are grouped together, the current rating is to be based upon the lowest operating temperature of any cable in the group.

NOTE 9: If, due to known operating conditions, a cable is expected to carry not more than 30 % of its *grouped* rating, it may be ignored for the purpose of obtaining the rating factor for the rest of the group.

For example, a group of N loaded cables would normally require a group rating factor of C_g applied to the tabulated I_t. However, if M cables in the group carry loads which are not greater than 0.3 $C_g I_t$ amperes the other cables can be sized by using the group rating factor corresponding to (N-M) cables.

TABLE 4C2 – Rating factors for more than one circuit, cables buried directly in the ground – Reference Method D in Tables 4D4A to 4J4A
multicore cables

Number of circuits	Cable-to-cable clearance (α)				
	Nil (cables touching)	One cable diameter	0.125 m	0.25 m	0.5 m
2	0.75	0.80	0.85	0.90	0.90
3	0.65	0.70	0.75	0.80	0.85
4	0.60	0.60	0.70	0.75	0.80
5	0.55	0.55	0.65	0.70	0.80
6	0.50	0.55	0.60	0.70	0.80

Multicore cables

α $\quad\quad\quad\quad$ α

NOTE 1: Values given apply to an installation depth of 0.7 m and a soil thermal resistivity of 2.5 K.m/W. These are average values for the range of cable sizes and types quoted for Tables 4D4A to 4J4A. The process of averaging, together with rounding off, can result in some cases in errors of up to ±10%. (Where more precise values are required they may be calculated by methods given in BS 7769 (BS IEC 60287)).

NOTE 2: In case of a thermal resistivity lower than 2.5 K.m/W the rating factors can, in general, be increased and can be calculated by the methods given in BS 7769 (BS IEC 60287).

TABLE 4C3 – Rating factors for more than one circuit, cables in ducts buried in the ground – Reference Method D in Tables 4D4A to 4J4A (Multicore cables in single-way ducts)

Number of cables	Duct-to-duct clearance (α)			
	Nil (ducts touching)	0.25 m	0.5 m	1.0 m
2	0.85	0.90	0.95	0.95
3	0.75	0.85	0.90	0.95
4	0.70	0.80	0.85	0.90
5	0.65	0.80	0.85	0.90
6	0.60	0.80	0.80	0.90

Multicore cables

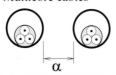

α

NOTE 1: Values given apply to an installation depth of 0.7 m and a soil thermal resistivity of 2.5 K.m/W. They are average values for the range of cable sizes and types quoted for Tables 4D4A to 4J4A. The process of averaging, together with rounding off, can result in some cases in errors of up to ±10%. (Where more precise values are required they may be calculated by methods given in BS 7769 (BS IEC 60287).)

NOTE 2: In case of a thermal resistivity lower than 2.5 K.m/W the rating factors can, in general, be increased and can be calculated by the methods given in BS 7769 (BS IEC 60287).

TABLE 4C4 – Rating factors for groups of more than one multicore cable, to be applied to reference current-carrying capacities for multicore cables in free air – Reference Method E in Tables 4D2A to 4J4A

Installation Method in Table 4A2			Number of trays or ladders	Number of cables per tray or ladder					
				1	2	3	4	6	9
Perforated cable tray systems (Note 3)	31	Touching ≥ 20 mm ≥ 300 mm	1	See item 3 of Table 4C1					
			2	1.00	0.87	0.80	0.77	0.73	0.68
			3	1.00	0.86	0.79	0.76	0.71	0.66
			6	1.00	0.84	0.77	0.73	0.68	0.64
		Spaced De ≥ 20 mm	1	1.00	1.00	0.98	0.95	0.91	–
			2	1.00	0.99	0.96	0.92	0.87	–
			3	1.00	0.98	0.95	0.91	0.85	–
Vertical perforated cable tray systems (Note 4)	31	Touching ≥ 225 mm	1	See item 3 of Table 4C1					
			2	1.00	0.88	0.81	0.76	0.71	0.70
		Spaced ≥ 225 mm De	1	1.00	0.91	0.89	0.88	0.87	–
			2	1.00	0.91	0.88	0.87	0.85	–
Unperforated cable tray systems	30	Touching ≥ 20 mm ≥ 300 mm	1	0.97	0.84	0.78	0.75	0.71	0.68
			2	0.97	0.83	0.76	0.72	0.68	0.63
			3	0.97	0.82	0.75	0.71	0.66	0.61
			6	0.97	0.81	0.73	0.69	0.63	0.58
Cable ladder systems, cleats, wire mesh tray, etc. (Note 3)	32 33 34	Touching ≥ 20 mm ≥ 300 mm	1	See item 4 of Table 4C1					
			2	1.00	0.86	0.80	0.78	0.76	0.73
			3	1.00	0.85	0.79	0.76	0.73	0.70
			6	1.00	0.84	0.77	0.73	0.68	0.64
		Spaced De ≥ 20 mm	1	1.00	1.00	1.00	1.00	1.00	–
			2	1.00	0.99	0.98	0.97	0.96	–
			3	1.00	0.98	0.97	0.96	0.93	–

NOTE 1: Values given are averages for the cable types and range of conductor sizes considered in Tables 4D2A to 4J4A. The spread of values is generally less than 5%.

NOTE 2: Factors apply to single layer groups of cables as shown above and do not apply when cables are installed in more than one layer touching each other. Values for such installations may be significantly lower and must be determined by an appropriate method.

NOTE 3: Values are given for vertical spacing between cable trays of 300 mm and at least 20 mm between cable trays and wall. For closer spacing the factors should be reduced.

NOTE 4: Values are given for horizontal spacing between cable trays of 225 mm with cable trays mounted back to back. For closer spacing the factors should be reduced.

TABLE 4C5 – Rating factors for groups of one or more circuits of single-core cables to be applied to reference current-carrying capacity for one circuit of single-core cables in free air – Reference Method F in Tables 4D1A to 4J3A

Installation Method in Table 4A2			Number of trays or ladders	Number of three-phase circuits per tray or ladder			Use as a multiplier to rating for
				1	2	3	
Perforated cable tray systems (Note 3)	31	Touching ≥ 300 mm, ≥ 20 mm	1 2 3	0.98 0.96 0.95	0.91 0.87 0.85	0.87 0.81 0.78	Three cables in horizontal formation
Vertical perforated cable tray systems (Note 4)	31	Touching ≥ 225 mm	1 2	0.96 0.95	0.86 0.84	– –	Three cables in vertical formation
Cable ladder systems, cleats, wire mesh tray, etc. (Note 3)	32 33 34	Touching ≥ 300 mm, ≥ 20 mm	1 2 3	1.00 0.98 0.97	0.97 0.93 0.90	0.96 0.89 0.86	Three cables in horizontal formation
Perforated cable tray systems (Note 3)	31	≥ 2D$_e$, D$_e$ ≥ 300 mm, ≥ 20 mm	1 2 3	1.00 0.97 0.96	0.98 0.93 0.92	0.96 0.89 0.86	Three cables in trefoil formation
Vertical perforated cable tray systems (Note 4)	31	Spaced ≥ 225 mm, ≥ 2D$_e$, D$_e$	1 2	1.00 1.00	0.91 0.90	0.89 0.86	
Cable ladder systems, cleats, wire mesh tray, etc. (Note 3)	32 33 34	≥ 2D$_e$, D$_e$ ≥ 300 mm, ≥ 20 mm	1 2 3	1.00 0.97 0.96	1.00 0.95 0.94	1.00 0.93 0.90	

NOTE 1: Values given are averages for the cable types and range of conductor sizes considered in Tables 4D1A to 4J3A. The spread of values is generally less than 5%.

NOTE 2: Factors apply to single layer groups of cables (or trefoil groups) as shown above and do not apply when cables are installed in more than one layer touching each other. Values for such installations may be significantly lower and must be determined by an appropriate method.

NOTE 3: Values are given for vertical spacing between cable trays of 300 mm and at least 20 mm between cable trays and wall. For closer spacing the factors should be reduced.

NOTE 4: Values are given for horizontal spacing between cable trays of 225 mm with cable trays mounted back to back. For closer spacing the factors should be reduced.

NOTE 5: For circuits having more than one cable in parallel per phase, each three-phase set of conductors is to be considered as a circuit for the purpose of this table.

TABLE 4C6 – Rating factors for cables enclosed in infloor concrete troughs (Installation Methods 118 to 120 of Table 4A2)

The rating factors tabulated below relate to the disposition of cables illustrated in items 118 to 120 of Table 4A2 and are applicable to the current-carrying capacities for Reference Methods E and F as given in the relevant tables of this appendix.

	Rating factor									
Conductor cross-sectional area	Installation method 118				Installation method 119			Installation method 120		
	2 single-core cables, or 1 three- or four-core cable	3 single-core cables, or 2 two-core cables	4 single-core cables, or 2 three- or four-core cables	6 single-core cables, 4 two-core cables, or 3 three- or four-core cables	6 single-core cables, 4 two-core cables, or 3 three- or four-core cables	8 single-core cables, or 4 three- or four-core cables	12 single-core cables, 8 two-core cables, or 6 three- or four-core cables	12 single-core cables, 8 two-core cables, or 6 three- or four-core cables	18 single-core cables, 12 two-core cables, or 9 three- or four-core cables	24 single-core cables, 16 two-core cables, or 12 three- or four-core cables
1	2	3	4	5	6	7	8	9	10	11
(mm^2)										
4	0.93	0.90	0.87	0.82	0.86	0.83	0.76	0.81	0.74	0.69
6	0.92	0.89	0.86	0.81	0.86	0.82	0.75	0.80	0.73	0.68
10	0.91	0.88	0.85	0.80	0.85	0.80	0.74	0.78	0.72	0.66
16	0.91	0.87	0.84	0.78	0.83	0.78	0.71	0.76	0.70	0.64
25	0.90	0.86	0.82	0.76	0.81	0.76	0.69	0.74	0.67	0.62
35	0.89	0.85	0.81	0.75	0.80	0.74	0.68	0.72	0.66	0.60
50	0.88	0.84	0.79	0.74	0.78	0.73	0.66	0.71	0.64	0.59
70	0.87	0.82	0.78	0.72	0.77	0.72	0.64	0.70	0.62	0.57
95	0.86	0.81	0.76	0.70	0.75	0.70	0.63	0.68	0.60	0.55
120	0.85	0.80	0.75	0.69	0.73	0.68	0.61	0.66	0.58	0.53
150	0.84	0.78	0.74	0.67	0.72	0.67	0.59	0.64	0.57	0.51
185	0.83	0.77	0.73	0.65	0.70	0.65	0.58	0.63	0.55	0.49
240	0.82	0.76	0.71	0.63	0.69	0.63	0.56	0.61	0.53	0.48
300	0.81	0.74	0.69	0.62	0.68	0.62	0.54	0.59	0.52	0.46
400	0.80	0.73	0.67	0.59	0.66	0.60	0.52	0.57	0.50	0.44
500	0.78	0.72	0.66	0.58	0.64	0.58	0.51	0.56	0.48	0.43
630	0.77	0.71	0.65	0.56	0.63	0.57	0.49	0.54	0.47	0.41

NOTES:

1. The factors in Table 4C6 are applicable to groups of cables all of one size. The value of current derived from application of the appropriate factors is the maximum current to be carried by any of the cables in the group.

2. If, due to known operating conditions, a cable is expected to carry not more than 30% of its *grouped* rating, it may be ignored for the purpose of obtaining the rating factor for the rest of the group.

3. Where cables having different conductor operating temperatures are grouped together the current rating should be based on the lowest operating temperature of any cable in the group.

4. When the number of cables used differs from those stated in the table, the rating factor for the next higher stated number of cables should be used.

TABLE 4D1A – Single-core 70 °C thermoplastic insulated cables, non-armoured, with or without sheath
(COPPER CONDUCTORS)

Ambient temperature: 30 °C
Conductor operating temperature: 70 °C

CURRENT-CARRYING CAPACITY (amperes):

Conductor cross-sectional area	Reference Method A (enclosed in conduit in thermally insulating wall etc.)		Reference Method B (enclosed in conduit on a wall or in trunking etc.)		Reference Method C (clipped direct)		Reference Method F (in free air or on a perforated cable tray horizontal or vertical)					
							Touching			Spaced by one diameter	
	2 cables, single-phase a.c. or d.c.	3 or 4 cables, three-phase a.c.	2 cables, single-phase a.c. or d.c.	3 or 4 cables, three-phase a.c.	2 cables, single-phase a.c. or d.c. flat and touching	3 or 4 cables, three-phase a.c. flat and touching or trefoil	2 cables, single-phase a.c. or d.c. flat	3 cables, three-phase a.c flat	3 cables, three-phase a.c trefoil	2 cables, single-phase a.c. or d.c. or 3 cables three-phase a.c. flat	
										Horizontal	Vertical
1	2	3	4	5	6	7	8	9	10	11	12
(mm²)	(A)	(A)	(A)	(A)	(A)	(A)	(A)	(A)	(A)	(A)	(A)
1	11	10.5	13.5	12	15.5	14	-	-	-	-	-
1.5	14.5	13.5	17.5	15.5	20	18	-	-	-	-	-
2.5	20	18	24	21	27	25	-	-	-	-	-
4	26	24	32	28	37	33	-	-	-	-	-
6	34	31	41	36	47	43	-	-	-	-	-
10	46	42	57	50	65	59	-	-	-	-	-
16	61	56	76	68	87	79	-	-	-	-	-
25	80	73	101	89	114	104	131	114	110	146	130
35	99	89	125	110	141	129	162	143	137	181	162
50	119	108	151	134	182	167	196	174	167	219	197
70	151	136	192	171	234	214	251	225	216	281	254
95	182	164	232	207	284	261	304	275	264	341	311
120	210	188	269	239	330	303	352	321	308	396	362
150	240	216	300	262	381	349	406	372	356	456	419
185	273	245	341	296	436	400	463	427	409	521	480
240	321	286	400	346	515	472	546	507	485	615	569
300	367	328	458	394	594	545	629	587	561	709	659
400	-	-	546	467	694	634	754	689	656	852	795
500	-	-	626	533	792	723	868	789	749	982	920
630	-	-	720	611	904	826	1005	905	855	1138	1070
800	-	-	-	-	1030	943	1086	1020	971	1265	1188
1000	-	-	-	-	1154	1058	1216	1149	1079	1420	1337

NOTE:
For cables having flexible conductors, see section 2.4 of this Appendix for adjustment factors for current-carrying capacity and voltage drop.

TABLE 4D1B

VOLTAGE DROP (per ampere per metre):

Conductor operating temperature: 70 °C

Descriptive column groupings (column numbers as printed in the table):
- **1** — Conductor cross-sectional area (mm²)
- **2** — 2 cables, d.c. (mV/A/m)
- **2 cables, single-phase a.c.** — **3** Reference Methods A & B (enclosed in conduit or trunking) (mV/A/m); **4** Reference Methods C & F (clipped direct, on tray or in free air), Cables touching (mV/A/m); **5** Cables spaced* (mV/A/m)
- **3 or 4 cables, three-phase a.c.** — **6** Reference Methods A & B (enclosed in conduit or trunking) (mV/A/m); **7** Reference Methods C & F (clipped direct, on tray or in free air), Cables touching, Trefoil (mV/A/m); **8** Cables touching, Flat (mV/A/m); **9** Cables spaced*, Flat (mV/A/m)

1 (mm²)	2 dc	3 r	3 x	3 z	4 r	4 x	4 z	5 r	5 x	5 z	6 r	6 x	6 z	7 r	7 x	7 z	8 r	8 x	8 z	9 r	9 x	9 z
1	44		44			44			44			38			38			38			38	
1.5	29		29			29			29			25			25			25			25	
2.5	18		18			18			18			15			15			15			15	
4	11		11			11			11			9.5			9.5			9.5			9.5	
6	7.3		7.3			7.3			7.3			6.4			6.4			6.4			6.4	
10	4.4		4.4			4.4			4.4			3.8			3.8			3.8			3.8	
16	2.8		2.8			2.8			2.8			2.4			2.4			2.4			2.4	
25	1.75	1.80	0.33	1.80	1.75	0.20	1.75	1.75	0.29	1.80	1.50	0.29	1.55	1.50	0.175	1.50	1.50	0.25	1.55	1.50	0.32	1.55
35	1.25	1.30	0.31	1.30	1.25	0.195	1.25	1.25	0.28	1.30	1.10	0.27	1.10	1.10	0.170	1.10	1.10	0.24	1.10	1.10	0.32	1.15
50	0.93	0.95	0.30	1.00	0.93	0.190	0.95	0.93	0.28	0.97	0.81	0.26	0.85	0.80	0.165	0.82	0.80	0.24	0.84	0.80	0.32	0.86
70	0.63	0.65	0.29	0.72	0.63	0.185	0.66	0.63	0.27	0.69	0.56	0.25	0.61	0.55	0.160	0.57	0.55	0.24	0.60	0.55	0.31	0.63
95	0.46	0.49	0.28	0.56	0.47	0.180	0.50	0.47	0.27	0.54	0.42	0.24	0.48	0.41	0.155	0.43	0.41	0.23	0.47	0.40	0.31	0.51
120	0.36	0.39	0.27	0.47	0.37	0.175	0.41	0.37	0.26	0.45	0.33	0.23	0.41	0.32	0.150	0.36	0.32	0.23	0.40	0.32	0.30	0.44
150	0.29	0.31	0.27	0.41	0.30	0.175	0.34	0.29	0.26	0.39	0.27	0.23	0.36	0.26	0.150	0.30	0.26	0.23	0.34	0.26	0.30	0.40
185	0.23	0.25	0.27	0.37	0.24	0.170	0.29	0.24	0.26	0.35	0.22	0.23	0.32	0.21	0.145	0.26	0.21	0.22	0.31	0.21	0.30	0.36
240	0.180	0.195	0.26	0.33	0.185	0.165	0.25	0.185	0.25	0.31	0.17	0.23	0.29	0.160	0.145	0.22	0.160	0.22	0.27	0.160	0.29	0.34
300	0.145	0.160	0.26	0.31	0.150	0.165	0.22	0.150	0.25	0.29	0.14	0.23	0.27	0.130	0.140	0.190	0.130	0.22	0.25	0.130	0.29	0.32
400	0.105	0.130	0.26	0.29	0.120	0.160	0.20	0.115	0.25	0.27	0.12	0.22	0.25	0.105	0.140	0.175	0.105	0.21	0.24	0.100	0.29	0.31
500	0.086	0.110	0.26	0.28	0.098	0.155	0.185	0.093	0.24	0.26	0.10	0.22	0.25	0.086	0.135	0.160	0.086	0.21	0.23	0.081	0.29	0.30
630	0.068	0.094	0.25	0.27	0.081	0.155	0.175	0.076	0.24	0.25	0.08	0.22	0.24	0.072	0.135	0.150	0.072	0.21	0.22	0.066	0.28	0.29
800	0.053	-	-		0.068	0.150	0.165	0.061	0.24	0.25				0.060	0.130	0.145	0.060	0.21	0.22	0.053	0.28	0.29
1000	0.042	-	-		0.059	0.150	0.160	0.050	0.24	0.24				0.052	0.130	0.140	0.052	0.20	0.21	0.044	0.28	0.28

NOTE: * Spacings larger than one cable diameter will result in a larger voltage drop.

TABLE 4D2A – Multicore 70 °C thermoplastic insulated and thermoplastic sheathed cables, non-armoured (COPPER CONDUCTORS)

Ambient temperature: 30 °C
Conductor operating temperature: 70 °C

CURRENT-CARRYING CAPACITY (amperes):

Conductor cross-sectional area	Reference Method A (enclosed in conduit in thermally insulating wall etc.)		Reference Method B (enclosed in conduit on a wall or in trunking etc.)		Reference Method C (clipped direct)		Reference Method E (in free air or on a perforated cable tray etc, horizontal or vertical)	
	1 two-core cable*, single-phase a.c. or d.c.	1 three-core cable* or 1 four-core cable, three-phase a.c.	1 two-core cable*, single-phase a.c. or d.c.	1 three-core cable* or 1 four-core cable, three-phase a.c.	1 two-core cable*, single-phase a.c. or d.c.	1 three-core cable* or 1 four-core cable, three-phase a.c.	1 two-core cable*, single-phase a.c. or d.c.	1 three-core cable* or 1 four-core cable, three-phase a.c.
1	2	3	4	5	6	7	8	9
(mm²)	(A)	(A)	(A)	(A)	(A)	(A)	(A)	(A)
1	11	10	13	11.5	15	13.5	17	14.5
1.5	14	13	16.5	15	19.5	17.5	22	18.5
2.5	18.5	17.5	23	20	27	24	30	25
4	25	23	30	27	36	32	40	34
6	32	29	38	34	46	41	51	43
10	43	39	52	46	63	57	70	60
16	57	52	69	62	85	76	94	80
25	75	68	90	80	112	96	119	101
35	92	83	111	99	138	119	148	126
50	110	99	133	118	168	144	180	153
70	139	125	168	149	213	184	232	196
95	167	150	201	179	258	223	282	238
120	192	172	232	206	299	259	328	276
150	219	196	258	225	344	299	379	319
185	248	223	294	255	392	341	434	364
240	291	261	344	297	461	403	514	430
300	334	298	394	339	530	464	593	497
400	-	-	470	402	634	557	715	597

* with or without a protective conductor

NOTE:
For cables having flexible conductors, see section 2.4 of this Appendix for adjustment factors for current-carrying capacity and voltage drop.

TABLE 4D2B

VOLTAGE DROP (per ampere per metre): Conductor operating temperature: 70 °C

Conductor cross-sectional area	Two-core cable, d.c.	Two-core cable, single-phase a.c.			Three- or four-core cable, three-phase a.c.		
1	2	3			4		
(mm²)	(mV/A/m)	(mV/A/m)			(mV/A/m)		
1	44	44			38		
1.5	29	29			25		
2.5	18	18			15		
4	11	11			9.5		
6	7.3	7.3			6.4		
10	4.4	4.4			3.8		
16	2.8	2.8			2.4		
		r	x	z	r	x	z
25	1.75	1.75	0.170	1.75	1.50	0.145	1.50
35	1.25	1.25	0.165	1.25	1.10	0.145	1.10
50	0.93	0.93	0.165	0.94	0.80	0.140	0.81
70	0.63	0.63	0.160	0.65	0.55	0.140	0.57
95	0.46	0.47	0.155	0.50	0.41	0.135	0.43
120	0.36	0.38	0.155	0.41	0.33	0.135	0.35
150	0.29	0.30	0.155	0.34	0.26	0.130	0.29
185	0.23	0.25	0.150	0.29	0.21	0.130	0.25
240	0.180	0.190	0.150	0.24	0.165	0.130	0.21
300	0.145	0.155	0.145	0.21	0.135	0.130	0.185
400	0.105	0.115	0.145	0.185	0.100	0.125	0.160

COPPER CONDUCTORS

TABLE 4D3A – Single-core armoured 70 °C thermoplastic insulated cables
(non-magnetic armour)
(COPPER CONDUCTORS)

Ambient temperature: 30 °C
Conductor operating temperature: 70 °C

CURRENT-CARRYING CAPACITY (amperes):

	Reference Method C (clipped direct) Touching		Reference Method F (in free air or on a perforated cable tray, horizontal or vertical) Touching								
						Spaced by one cable diameter					
Conductor cross-sectional area	2 cables, single-phase a.c. or d.c. flat	3 or 4 cables, three-phase a.c. flat	2 cables, single-phase a.c. or d.c. flat	3 cables, three-phase a.c. flat	3 cables, three-phase a.c. trefoil	2 cables, d.c.		2 cables, single-phase a.c.		3 or 4 cables, three-phase a.c.	
						Horizontal	Vertical	Horizontal	Vertical	Horizontal	Vertical
1	2	3	4	5	6	7	8	9	10	11	12
(mm²)	(A)	(A)	(A)	(A)	(A)	(A)	(A)	(A)	(A)	(A)	(A)
50	193	179	205	189	181	229	216	229	217	230	212
70	245	225	259	238	231	294	279	287	272	286	263
95	296	269	313	285	280	357	340	349	332	338	313
120	342	309	360	327	324	415	396	401	383	385	357
150	393	352	413	373	373	479	458	449	429	436	405
185	447	399	469	422	425	548	525	511	489	490	456
240	525	465	550	492	501	648	622	593	568	566	528
300	594	515	624	547	567	748	719	668	640	616	578
400	687	575	723	618	657	885	851	737	707	674	632
500	763	622	805	673	731	1035	997	810	777	721	676
630	843	669	891	728	809	1218	1174	893	856	771	723
800	919	710	976	777	886	1441	1390	943	905	824	772
1000	975	737	1041	808	945	1685	1627	1008	967	872	816

TABLE 4D3B

VOLTAGE DROP (per ampere per metre): Conductor operating temperature: 70 °C

Reference Methods C & F (clipped direct, on tray or free air)

Conductor cross-sectional area	2 cables, d.c.	2 cables, single-phase a.c.						3 or 4 cables, three-phase a.c.								
		touching			spaced*			trefoil and touching			flat and touching			flat and spaced*		
		3			4			5			6			7		
	2															
(mm²)	(mV/A/m)	(mV/A/m)			(mV/A/m)			(mV/A/m)			(mV/A/m)			(mV/A/m)		
1		r	x	z	r	x	z	r	x	z	r	x	z	r	x	z
50	0.93	0.93	0.22	0.95	0.92	0.30	0.97	0.80	0.190	0.82	0.79	0.26	0.84	0.79	0.34	0.86
70	0.63	0.64	0.21	0.68	0.66	0.29	0.72	0.56	0.180	0.58	0.57	0.25	0.62	0.59	0.32	0.68
95	0.46	0.48	0.20	0.52	0.51	0.28	0.58	0.42	0.175	0.45	0.44	0.25	0.50	0.47	0.31	0.57
120	0.36	0.39	0.195	0.43	0.42	0.28	0.50	0.33	0.170	0.37	0.36	0.24	0.43	0.40	0.30	0.50
150	0.29	0.31	0.190	0.37	0.34	0.27	0.44	0.27	0.165	0.32	0.30	0.24	0.38	0.34	0.30	0.45
185	0.23	0.26	0.190	0.32	0.29	0.27	0.39	0.22	0.160	0.27	0.25	0.23	0.34	0.29	0.29	0.41
240	0.180	0.20	0.180	0.27	0.23	0.26	0.35	0.175	0.160	0.23	0.20	0.23	0.30	0.24	0.28	0.37
300	0.145	0.160	0.180	0.24	0.190	0.26	0.32	0.140	0.155	0.21	0.165	0.22	0.28	0.20	0.28	0.34
400	0.105	0.140	0.175	0.22	0.180	0.24	0.30	0.120	0.130	0.195	0.160	0.21	0.26	0.21	0.25	0.32
500	0.086	0.120	0.170	0.21	0.165	0.23	0.29	0.105	0.145	0.180	0.145	0.20	0.25	0.190	0.24	0.30
630	0.068	0.105	0.165	0.195	0.150	0.22	0.27	0.091	0.145	0.170	0.135	0.195	0.23	0.175	0.22	0.28
800	0.053	0.095	0.160	0.185	0.145	0.21	0.25	0.082	0.140	0.160	0.125	0.180	0.22	0.170	0.195	0.26
1000	0.042	0.091	0.155	0.180	0.140	0.190	0.24	0.079	0.135	0.155	0.125	0.165	0.21	0.165	0.170	0.24

NOTE: * Spacings larger than one cable diameter will result in a larger voltage drop.

TABLE 4D4A – Multicore armoured 70 °C thermoplastic insulated cables (COPPER CONDUCTORS)

Air ambient temperature: 30 °C
Ground ambient temperature: 20 °C
Conductor operating temperature: 70 °C

CURRENT-CARRYING CAPACITY (amperes):

Conductor cross-sectional area	Reference Method C (clipped direct)		Reference Method E (in free air or on a perforated cable tray etc, horizontal or vertical)		Reference Method D (direct in ground or in ducting in ground, in or around buildings)	
	1 two-core cable, single-phase a.c. or d.c.	1 three- or four-core cable, three-phase a.c.	1 two-core cable, single-phase a.c. or d.c.	1 three- or four-core cable, three-phase a.c.	1 two-core cable, single-phase a.c. or d.c.	1 three- or four-core cable, three-phase a.c.
1	2	3	4	5	6	7
(mm²)	(A)	(A)	(A)	(A)	(A)	(A)
1.5	21	18	22	19	22	18
2.5	28	25	31	26	29	24
4	38	33	41	35	37	30
6	49	42	53	45	46	38
10	67	58	72	62	60	50
16	89	77	97	83	78	64
25	118	102	128	110	99	82
35	145	125	157	135	119	98
50	175	151	190	163	140	116
70	222	192	241	207	173	143
95	269	231	291	251	204	169
120	310	267	336	290	231	192
150	356	306	386	332	261	217
185	405	348	439	378	292	243
240	476	409	516	445	336	280
300	547	469	592	510	379	316
400	621	540	683	590	–	–

TABLE 4D4B

VOLTAGE DROP (per ampere per metre): Conductor operating temperature: 70 °C

Conductor cross-sectional area	Two-core cable, d.c.	Two-core cable, single-phase a.c.			Three- or four-core cable, three-phase a.c.		
1	2	3			4		
(mm²)	(mV/A/m)	(mV/A/m)			(mV/A/m)		
1.5	29	29			25		
2.5	18	18			15		
4	11	11			9.5		
6	7.3	7.3			6.4		
10	4.4	4.4			3.8		
16	2.8	2.8			2.4		
		r	x	z	r	x	z
25	1.75	1.75	0.170	1.75	1.50	0.145	1.50
35	1.25	1.25	0.165	1.25	1.10	0.145	1.10
50	0.93	0.93	0.165	0.94	0.80	0.140	0.81
70	0.63	0.63	0.160	0.65	0.55	0.140	0.57
95	0.46	0.47	0.155	0.50	0.41	0.135	0.43
120	0.36	0.38	0.155	0.41	0.33	0.135	0.35
150	0.29	0.30	0.155	0.34	0.26	0.130	0.29
185	0.23	0.25	0.150	0.29	0.21	0.130	0.25
240	0.180	0.190	0.150	0.24	0.165	0.130	0.21
300	0.145	0.155	0.145	0.21	0.135	0.130	0.185
400	0.105	0.115	0.145	0.185	0.100	0.125	0.160

TABLE 4D5 – 70 °C thermoplastic insulated and sheathed flat cable with protective conductor (COPPER CONDUCTORS)

CURRENT-CARRYING CAPACITY (amperes) and VOLTAGE DROP (per ampere per metre):

Ambient temperature: 30 °C
Conductor operating temperature: 70 °C

Conductor cross-sectional area	Method 100# (above a plasterboard ceiling covered by thermal insulation not exceeding 100 mm in thickness)	Method 101# (above a plasterboard ceiling covered by thermal insulation exceeding 100 mm in thickness)	Method 102# (in a stud wall with thermal insulation with cable touching the inner wall surface)	Method 103# (in a stud wall with thermal insulation with cable not touching the inner wall surface)	Reference Method C* (clipped direct)	Reference Method A* (enclosed in conduit in an insulated wall)	Voltage drop (per ampere per metre)
1	2	3	4	5	6	7	8
(mm²)	(A)	(A)	(A)	(A)	(A)	(A)	(mV/A/m)
1	13	10.5	13	8	16	11.5	44
1.5	16	13	16	10	20	14.5	29
2.5	21	17	21	13.5	27	20	18
4	27	22	27	17.5	37	26	11
6	34	27	35	23.5	47	32	7.3
10	45	36	47	32	64	44	4.4
16	57	46	63	42.5	85	57	2.8

A*   For full installation method refer to Table 4A2 Installation Method 2 but for flat twin and earth cable

C*   For full installation method refer to Table 4A2 Installation Method 20 but for flat twin and earth cable

100#   For full installation method refer to Table 4A2 Installation Method 100

101#   For full installation method refer to Table 4A2 Installation Method 101

102#   For full installation method refer to Table 4A2 Installation Method 102

103#   For full installation method refer to Table 4A2 Installation Method 103

Wherever practicable, a cable is to be fixed in a position such that it will not be covered with thermal insulation.
Regulation 523.9, BS 5803-5: Appendix C: Avoidance of overheating of electric cables,
Building Regulations Approved document B and Thermal insulation: avoiding risks, BR 262, BRE, 2001 refer.

THIS PAGE HAS BEEN INTENTIONALLY LEFT BLANK

TABLE 4E1A – Single-core 90 °C thermosetting insulated cables, non-armoured, with or without sheath
(COPPER CONDUCTORS)

Ambient temperature: 30 °C
Conductor operating temperature: 90 °C

CURRENT-CARRYING CAPACITY (amperes):

Conductor cross-sectional area	Reference Method A (enclosed in conduit in thermally insulating wall etc.)		Reference Method B (enclosed in conduit on a wall or in trunking etc.)		Reference Method C (clipped direct)		Reference Method F (in free air or on a perforated cable tray etc horizontal or vertical etc) Touching			Reference Method G (in free air) Spaced by one cable diameter	
	2 cables, single-phase a.c. or d.c.	3 or 4 cables, three-phase a.c.	2 cables, single-phase a.c. or d.c.	3 or 4 cables, three-phase a.c.	2 cables, single-phase a.c. flat and touching	3 or 4 cables, three-phase a.c. flat and touching or trefoil	2 cables, single-phase a.c. or d.c. flat	3 cables, three-phase a.c. flat	3 cables, three-phase a.c trefoil	2 cables, single-phase a.c. or d.c. or 3 cables three-phase a.c. flat Horizontal	Vertical
1	2	3	4	5	6	7	8	9	10	11	12
(mm²)	(A)	(A)	(A)	(A)	(A)	(A)	(A)	(A)	(A)	(A)	(A)
1	14	13	17	15	19	17.5	-	-	-	-	-
1.5	19	17	23	20	25	23	-	-	-	-	-
2.5	26	23	31	28	34	31	-	-	-	-	-
4	35	31	42	37	46	41	-	-	-	-	-
6	45	40	54	48	59	54	-	-	-	-	-
10	61	54	75	66	81	74	-	-	-	-	-
16	81	73	100	88	109	99	-	-	-	-	-
25	106	95	133	117	143	130	161	141	135	182	161
35	131	117	164	144	176	161	200	176	169	226	201
50	158	141	198	175	228	209	242	216	207	275	246
70	200	179	253	222	293	268	310	279	268	353	318
95	241	216	306	269	355	326	377	342	328	430	389
120	278	249	354	312	413	379	437	400	383	500	454
150	318	285	393	342	476	436	504	464	444	577	527
185	362	324	449	384	545	500	575	533	510	661	605
240	424	380	528	450	644	590	679	634	607	781	719
300	486	435	603	514	743	681	783	736	703	902	833
400	-	-	683	584	868	793	940	868	823	1085	1008
500	-	-	783	666	990	904	1083	998	946	1253	1169
630	-	-	900	764	1130	1033	1254	1151	1088	1454	1362
800	-	-	-	-	1288	1179	1358	1275	1214	1581	1485
1000	-	-	-	-	1443	1323	1520	1436	1349	1775	1671

NOTES:

1. Where it is intended to connect the cables in this table to equipment or accessories designed to operate at a temperature lower than the maximum operating temperature of the cable, the cables should be rated at the maximum operating temperature of the equipment or accessory (see Regulation 512.1.5).

2. Where it is intended to group a cable in this table with other cables, the cable should be rated at the lowest of the maximum operating temperatures of any of the cables in the group (see Regulation 512.1.5).

3. For cables having flexible conductors see section 2.4 of this appendix for adjustment factors for current-carrying capacity and voltage drop.

TABLE 4E1B

VOLTAGE DROP (per ampere per metre): Conductor operating temperature: 90 °C

Column groups:
- [1] Conductor cross-sectional area (mm²)
- [2] 2 cables, d.c. (mV/A/m)
- 2 cables, single-phase a.c.:
  - [3] Reference Methods A & B (enclosed in conduit or trunking) (mV/A/m)
  - Reference Methods C, F & G (clipped direct, on tray or in free air): [4] Cables touching (mV/A/m); [5] Cables spaced* (mV/A/m)
- 3 or 4 cables, three-phase a.c.:
  - [6] Reference Methods A & B (enclosed in conduit or trunking) (mV/A/m)
  - Reference Methods C, F & G (clipped direct, on tray or in free air): [7] Cables touching, Trefoil (mV/A/m); [8] Cables touching, Flat (mV/A/m); [9] Cables spaced*, Flat (mV/A/m)

[1] Area (mm²)	[2] d.c.	[3] r	[3] x	[3] z	[4] r	[4] x	[4] z	[5] r	[5] x	[5] z	[6] r	[6] x	[6] z	[7] r	[7] x	[7] z	[8] r	[8] x	[8] z	[9] r	[9] x	[9] z
1	46	46			46			46			40			40			40			40		
1.5	31	31			31			31			27			27			27			27		
2.5	19	19			19			19			16			16			16			16		
4	12	12			12			12			10			10			10			10		
6	7.9	7.9			7.9			7.9			6.8			6.8			6.8			6.8		
10	4.7	4.7			4.7			4.7			4.0			4.0			4.0			4.0		
16	2.9	2.9			2.9			2.9			2.5			2.5			2.5			2.5		
25	1.85	1.85	0.31	1.90	1.85	0.190	1.85	1.85	0.28	1.85	1.60	0.27	1.65	1.60	0.165	1.60	1.60	0.190	1.60	1.60	0.27	1.65
35	1.35	1.35	0.29	1.35	1.35	0.180	1.35	1.35	0.27	1.35	1.15	0.25	1.15	1.15	0.155	1.15	1.15	0.180	1.15	1.15	0.26	1.20
50	0.99	1.00	0.29	1.05	0.99	0.180	1.00	0.99	0.27	1.00	0.87	0.25	0.90	0.86	0.155	0.87	0.86	0.180	0.87	0.86	0.26	0.89
70	0.68	0.70	0.28	0.75	0.68	0.175	0.71	0.68	0.26	0.73	0.60	0.24	0.65	0.59	0.150	0.61	0.59	0.175	0.62	0.59	0.25	0.65
95	0.49	0.51	0.27	0.58	0.49	0.170	0.52	0.49	0.26	0.56	0.44	0.23	0.50	0.43	0.145	0.45	0.43	0.170	0.46	0.43	0.25	0.49
120	0.39	0.41	0.26	0.48	0.39	0.165	0.43	0.39	0.25	0.47	0.35	0.23	0.42	0.34	0.140	0.37	0.34	0.165	0.38	0.34	0.24	0.42
150	0.32	0.33	0.26	0.43	0.32	0.165	0.36	0.32	0.25	0.41	0.29	0.23	0.37	0.28	0.140	0.31	0.28	0.165	0.32	0.28	0.24	0.37
185	0.25	0.27	0.26	0.37	0.26	0.165	0.30	0.25	0.25	0.36	0.23	0.23	0.32	0.22	0.140	0.26	0.22	0.165	0.28	0.22	0.24	0.33
240	0.190	0.21	0.26	0.33	0.20	0.160	0.25	0.195	0.25	0.31	0.185	0.22	0.29	0.170	0.140	0.22	0.170	0.165	0.24	0.170	0.24	0.29
300	0.155	0.175	0.25	0.31	0.160	0.160	0.22	0.155	0.25	0.29	0.150	0.22	0.27	0.140	0.140	0.195	0.135	0.160	0.21	0.135	0.24	0.27
400	0.120	0.140	0.25	0.29	0.130	0.155	0.20	0.125	0.24	0.27	0.125	0.22	0.25	0.110	0.135	0.175	0.110	0.160	0.195	0.110	0.24	0.26
500	0.093	0.120	0.25	0.28	0.105	0.155	0.185	0.098	0.24	0.26	0.100	0.22	0.24	0.090	0.135	0.160	0.088	0.160	0.180	0.085	0.24	0.25
630	0.072	0.100	0.25	0.27	0.086	0.155	0.175	0.078	0.24	0.25	0.088	0.21	0.23	0.074	0.135	0.150	0.071	0.160	0.170	0.068	0.23	0.24
800	0.056	0.072	–	–	0.072	0.150	0.170	0.064	0.24	0.25		–		0.062	0.130	0.145	0.059	0.155	0.165	0.055	0.23	0.29
1000	0.045	0.063	–	–	0.063	0.150	0.165	0.054	0.24	0.24		–		0.055	0.130	0.140	0.050	0.155	0.165	0.047	0.23	0.27

NOTE: * Spacings larger than one cable diameter will result in a larger voltage drop.

343

TABLE 4E2A – Multicore 90 °C thermosetting insulated and thermoplastic sheathed cables, non-armoured
(COPPER CONDUCTORS)

CURRENT-CARRYING CAPACITY (amperes):

Ambient temperature: 30 °C
Conductor operating temperature: 90 °C

Conductor cross-sectional area	Reference Method A (enclosed in conduit in thermally insulating wall etc.)		Reference Method B (enclosed in conduit on a wall or in trunking etc.)		Reference Method C (clipped direct)		Reference Method E (free air or on a perforated cable tray etc., horizontal or vertical)	
	1 two-core cable*, single-phase a.c. or d.c.	1 three- or four-core cable*, three-phase a.c.	1 two-core cable*, single-phase a.c. or d.c.	1 three- or four-core cable*, three-phase a.c.	1 two-core cable*, single-phase a.c. or d.c.	1 three- or four-core cable*, three-phase a.c.	1 two-core cable*, single-phase a.c. or d.c.	1 three- or four-core cable*, three-phase a.c.
1	2	3	4	5	6	7	8	9
(mm²)	(A)	(A)	(A)	(A)	(A)	(A)	(A)	(A)
1	14.5	13	17	15	19	17	21	18
1.5	18.5	16.5	22	19.5	24	22	26	23
2.5	25	22	30	26	33	30	36	32
4	33	30	40	35	45	40	49	42
6	42	38	51	44	58	52	63	54
10	57	51	69	60	80	71	86	75
16	76	68	91	80	107	96	115	100
25	99	89	119	105	138	119	149	127
35	121	109	146	128	171	147	185	158
50	145	130	175	154	209	179	225	192
70	183	164	221	194	269	229	289	246
95	220	197	265	233	328	278	352	298
120	253	227	305	268	382	322	410	346
150	290	259	334	300	441	371	473	399
185	329	295	384	340	506	424	542	456
240	386	346	459	398	599	500	641	538
300	442	396	532	455	693	576	741	621
400	-	-	625	536	803	667	865	741

NOTES:

1. Where it is intended to connect the cables in this table to equipment or accessories designed to operate at a temperature lower than the maximum operating temperature of the cable, the cables should be rated at the maximum operating temperature of the equipment or accessory (see Regulation 512.1.5).

2. Where it is intended to group a cable in this table with other cables, the cable should be rated at the lowest of the maximum operating temperatures of any of the cables in the group (see Regulation 512.1.5).

3. For cables having flexible conductors see section 2.4 of this appendix for adjustment factors for current-carrying capacity and voltage drop.

* with or without a protective conductor

TABLE 4E2B

VOLTAGE DROP (per ampere per metre):

Conductor operating temperature: 90 °C

Conductor cross-sectional area	Two-core cable, d.c.	Two-core cable, single-phase a.c.			Three- or four-core cable, three-phase a.c.		
1	2	3			4		
(mm²)	(mV/A/m)	(mV/A/m)			(mV/A/m)		
1	46	46			40		
1.5	31	31			27		
2.5	19	19			16		
4	12	12			10		
6	7.9	7.9			6.8		
10	4.7	4.7			4.0		
16	2.9	2.9			2.5		
		r	x	z	r	x	z
25	1.85	1.85	0.160	1.90	1.60	0.140	1.65
35	1.35	1.35	0.155	1.35	1.15	0.135	1.15
50	0.98	0.99	0.155	1.00	0.86	0.135	0.87
70	0.67	0.67	0.150	0.69	0.59	0.130	0.60
95	0.49	0.50	0.150	0.52	0.43	0.130	0.45
120	0.39	0.40	0.145	0.42	0.34	0.130	0.37
150	0.31	0.32	0.145	0.35	0.28	0.125	0.30
185	0.25	0.26	0.145	0.29	0.22	0.125	0.26
240	0.195	0.200	0.140	0.24	0.175	0.125	0.21
300	0.155	0.160	0.140	0.21	0.140	0.120	0.185
400	0.120	0.130	0.140	0.190	0.115	0.120	0.165

TABLE 4E3A – Single-core armoured 90 °C thermosetting insulated cables (non-magnetic armour) (COPPER CONDUCTORS)

CURRENT-CARRYING CAPACITY (amperes):

Ambient temperature: 30 °C

Conductor operating temperature: 90 °C

Conductor cross-sectional area	Reference Method C (clipped direct) Touching		Reference Method F (in free air or on a perforated cable tray, horizontal or vertical)									
	2 cables, single-phase a.c. or d.c. flat	3 or 4 cables, three-phase a.c. flat	Touching			Spaced by one cable diameter					
			2 cables, single-phase a.c. or d.c. flat	3 cables, three-phase a.c. flat	3 cables, three-phase a.c. trefoil	2 cables, d.c.		2 cables, single-phase a.c.		3 or 4 cables, three-phase a.c.	
						Horizontal	Vertical	Horizontal	Vertical	Horizontal	Vertical
1	2	3	4	5	6	7	8	9	10	11	12
(mm²)	(A)	(A)	(A)	(A)	(A)	(A)	(A)	(A)	(A)	(A)	(A)
50	237	220	253	232	222	284	270	282	266	288	266
70	303	277	322	293	285	356	349	357	337	358	331
95	367	333	389	352	346	446	426	436	412	425	393
120	425	383	449	405	402	519	497	504	477	485	449
150	488	437	516	462	463	600	575	566	539	549	510
185	557	496	587	524	529	688	660	643	614	618	574
240	656	579	689	612	625	815	782	749	714	715	666
300	755	662	792	700	720	943	906	842	805	810	755
400	853	717	899	767	815	1137	1094	929	889	848	797
500	962	791	1016	851	918	1314	1266	1032	989	923	871
630	1082	861	1146	935	1027	1528	1474	1139	1092	992	940
800	1170	904	1246	987	1119	1809	1744	1204	1155	1042	978
1000	1261	961	1345	1055	1214	2100	2026	1289	1238	1110	1041

NOTES:

1. Where it is intended to connect the cables in this table to equipment or accessories designed to operate at a temperature lower than the maximum operating temperature of the cable, the cables should be rated at the maximum operating temperature of the equipment or accessory (see Regulation 512.1.5).

2. Where it is intended to group a cable in this table with other cables, the cable should be rated at the lowest of the maximum operating temperatures of any of the cables in the group (see Regulation 512.1.5).

VOLTAGE DROP (per ampere per metre): Conductor operating temperature: 90 °C

TABLE 4E3B

Conductor cross-sectional area	2 cables, d.c.	Reference Methods C & F (clipped direct, on tray or in free air)																
		2 cables, single-phase a.c.					3 or 4 cables, three-phase a.c.											
		touching		spaced*			trefoil and touching			flat and touching			flat and spaced*					
1	2	3		4			5			6			7					
(mm²)	(mV/A/m)	(mV/A/m)		(mV/A/m)			(mV/A/m)			(mV/A/m)			(mV/A/m)					
		r	x	z	r	x	z	r	x	z	r	x	z	r	x	z		
50	0.98	0.99	0.21	1.00	0.98	0.29	1.00	0.86	0.180	0.87	0.84	0.25	0.88	0.84	0.33	0.90		
70	0.67	0.68	0.200	0.71	0.69	0.29	0.75	0.59	0.170	0.62	0.60	0.25	0.65	0.62	0.32	0.70		
95	0.49	0.51	0.195	0.55	0.53	0.28	0.60	0.44	0.170	0.47	0.46	0.24	0.52	0.49	0.31	0.58		
120	0.39	0.41	0.190	0.45	0.43	0.27	0.51	0.35	0.165	0.39	0.38	0.24	0.44	0.41	0.30	0.51		
150	0.31	0.33	0.185	0.38	0.36	0.27	0.45	0.29	0.160	0.33	0.31	0.23	0.39	0.34	0.29	0.45		
185	0.25	0.27	0.185	0.33	0.30	0.26	0.40	0.23	0.160	0.28	0.26	0.23	0.34	0.29	0.29	0.41		
240	0.195	0.21	0.180	0.28	0.24	0.26	0.35	0.180	0.155	0.24	0.21	0.22	0.30	0.24	0.28	0.37		
300	0.155	0.170	0.175	0.25	0.195	0.25	0.32	0.145	0.150	0.21	0.170	0.22	0.28	0.20	0.27	0.34		
400	0.115	0.145	0.170	0.22	0.180	0.24	0.30	0.125	0.150	0.195	0.160	0.21	0.27	0.20	0.27	0.33		
500	0.093	0.125	0.170	0.21	0.165	0.24	0.29	0.105	0.145	0.180	0.145	0.20	0.25	0.190	0.24	0.31		
630	0.073	0.105	0.165	0.195	0.150	0.23	0.27	0.092	0.145	0.170	0.135	0.195	0.24	0.175	0.23	0.29		
800	0.056	0.090	0.160	0.190	0.145	0.23	0.27	0.086	0.140	0.165	0.130	0.180	0.23	0.175	0.195	0.26		
1000	0.045	0.092	0.155	0.180	0.140	0.21	0.25	0.080	0.135	0.155	0.125	0.170	0.21	0.165	0.180	0.24		

NOTE: * Spacings larger than one cable diameter will result in a larger voltage drop.

TABLE 4E4A – Multicore armoured 90 °C thermosetting insulated cables (COPPER CONDUCTORS)

Air ambient temperature: 30 °C
Ground ambient temperature: 20 °C
Conductor operating temperature: 90 °C

CURRENT-CARRYING CAPACITY (amperes):

Conductor cross-sectional area	Reference Method C (clipped direct)		Reference Method E (in free air or on a perforated cable tray etc, horizontal or vertical)		Reference Method D (direct in ground or in ducting in ground, in or around buildings)	
	1 two-core cable, single-phase a.c. or d.c.	1 three- or 1 four-core cable, three-phase a.c.	1 two-core cable, single-phase a.c. or d.c.	1 three- or 1 four-core cable, three-phase a.c.	1 two-core cable, single-phase a.c. or d.c.	1 three- or 1 four-core cable, three-phase a.c.
1	2	3	4	5	6	7
(mm²)	(A)	(A)	(A)	(A)	(A)	(A)
1.5	27	23	29	25	25	21
2.5	36	31	39	33	33	28
4	49	42	52	44	43	36
6	62	53	66	56	53	44
10	85	73	90	78	71	58
16	110	94	115	99	91	75
25	146	124	152	131	116	96
35	180	154	188	162	139	115
50	219	187	228	197	164	135
70	279	238	291	251	203	167
95	338	289	354	304	239	197
120	392	335	410	353	271	223
150	451	386	472	406	306	251
185	515	441	539	463	343	281
240	607	520	636	546	395	324
300	698	599	732	628	446	365
400	787	673	847	728	-	-

NOTES:
1. Where it is intended to connect the cables in this table to equipment or accessories designed to operate at a temperature lower than the maximum operating temperature of the cable, the cables should be rated at the maximum operating temperature of the equipment or accessory (see Regulation 512.1.5).
2. Where it is intended to group a cable in this table with other cables, the cable should be rated at the lowest of the maximum operating temperatures of any of the cables in the group (see Regulation 512.1.5).

TABLE 4E4B

VOLTAGE DROP (per ampere per metre): Conductor operating temperature: 90 °C

Conductor cross-sectional area	Two-core cable, d.c.	Two-core cable, single-phase a.c.			Three- or four-core cable, three-phase a.c.		
1	2	3			4		
(mm²)	(mV/A/m)	(mV/A/m)			(mV/A/m)		
1.5	31	31			27		
2.5	19	19			16		
4	12	12			10		
6	7.9	7.9			6.8		
10	4.7	4.7			4.0		
16	2.9	2.9			2.5		
		r	x	z	r	x	z
25	1.85	1.85	0.160	1.90	1.60	0.140	1.65
35	1.35	1.35	0.155	1.35	1.15	0.135	1.15
50	0.98	0.99	0.155	1.00	0.86	0.135	0.87
70	0.67	0.67	0.150	0.69	0.59	0.130	0.60
95	0.49	0.50	0.150	0.52	0.43	0.130	0.45
120	0.39	0.40	0.145	0.42	0.34	0.130	0.37
150	0.31	0.32	0.145	0.35	0.28	0.125	0.30
185	0.25	0.26	0.145	0.29	0.22	0.125	0.26
240	0.195	0.20	0.140	0.24	0.175	0.125	0.21
300	0.155	0.16	0.140	0.21	0.140	0.120	0.185
400	0.120	0.13	0.140	0.190	0.115	0.120	0.165

TABLE 4F1A – 60 °C thermosetting insulated flexible cables with sheath, non-armoured
(COPPER CONDUCTORS)

CURRENT-CARRYING CAPACITY (amperes): Ambient temperature: 30 °C
Conductor operating temperature: 60 °C

Conductor cross-sectional area	Single-phase a.c. or d.c. 1 two-core cable, with or without protective conductor	Three-phase a.c. 1 three-core, four-core or five-core cable	Single-phase a.c. or d.c. 2 single-core cables, touching
1	2	3	4
(mm²)	(A)	(A)	(A)
4	30	26	-
6	39	34	-
10	51	47	-
16	73	63	-
25	97	83	-
35	-	102	140
50	-	124	175
70	-	158	216
95	-	192	258
120	-	222	302
150	-	255	347
185	-	291	394
240	-	343	471
300	-	394	541
400	-	-	644
500	-	-	738
630	-	-	861

NOTES:

1. The current ratings tabulated are for cables in free air but may also be used for cables resting on a surface. If the cable is to be wound on a drum on load the ratings should be reduced in accordance with NOTE 2 below and for cables which may be covered, NOTE 3 below.

2. *Flexible cables wound on reeling drums*

   The current ratings of cables used on reeling drums are to be reduced by the following factors:

   a) Radial type drum
      ventilated: 85 %
      unventilated: 75 %

   b) Ventilated cylindrical type drum
      1 layer of cable: 85 %
      2 layers of cable: 65 %
      3 layers of cable: 45 %
      4 layers of cable: 35 %

   A radial type drum is one where spiral layers of cable are accommodated between closely spaced flanges; if fitted with solid flanges the ratings given above should be reduced and the drum is described as non-ventilated. If the flanges have suitable apertures the drum is described as ventilated.

   A ventilated cylindrical cable drum is one where layers of cable are accommodated between widely spaced flanges and the drum and end flanges have suitable ventilating apertures.

3. Where cable may be covered over or coiled up whilst on load, or the air movement over the cable restricted, the current rating should be reduced.

   It is not possible to specify the amount of reduction but the table of rating factors for reeling drums can be used as a guide.

TABLE 4F1B

VOLTAGE DROP (per ampere per metre):

Conductor operating temperature: 60 °C

Conductor cross-sectional area	Two-core cable, d.c.	Two-core cable, single-phase a.c.			1 three-core, four-core or five-core cable, three-phase a.c.			2 single-core cables, touching			
								d.c.	Single-phase a.c.*		
1	2	3			4			5	6		
(mm²)	(mV/A/m)	(mV/A/m)			(mV/A/m)			(mV/A/m)	(mV/A/m)		
		r	x	z	r	x	z		r	x	z
4	12	12			10			-	-	-	-
6	7.8	7.8			6.7			-	-	-	-
10	4.6	4.6			4.0			-	-	-	-
16	2.9	2.9			2.5			-	-	-	-
25	1.80	1.80	0.175	1.85	1.55	0.150	1.55	-	-	-	-
35	-	-	-	-	1.10	0.150	1.15	1.31	1.31	0.21	1.32
50	-	-	-	-	0.83	0.145	0.84	0.91	0.91	0.21	0.93
70	-	-	-	-	0.57	0.140	0.58	0.64	0.64	0.20	0.67
95	-	-	-	-	0.42	0.135	0.44	0.49	0.49	0.195	0.53
120	-	-	-	-	0.33	0.135	0.36	0.38	0.38	0.190	0.43
150	-	-	-	-	0.27	0.130	0.30	0.31	0.31	0.190	0.36
185	-	-	-	-	0.22	0.130	0.26	0.25	0.25	0.190	0.32
240	-	-	-	-	0.170	0.130	0.21	0.190	0.195	0.185	0.27
300	-	-	-	-	0.135	0.125	0.185	0.150	0.155	0.180	0.24
400	-	-	-	-	-	-	-	0.115	0.120	0.175	0.21
500	-	-	-	-	-	-	-	0.090	0.099	0.170	0.20
630	-	-	-	-	-	-	-	0.068	0.079	0.170	0.185

NOTE: * A larger voltage drop will result if the cables are spaced.

COPPER CONDUCTORS

TABLE 4F2A – 90 °C and 180 °C thermosetting insulated flexible cables with sheath, non-armoured
(COPPER CONDUCTORS)

CURRENT-CARRYING CAPACITY (amperes): Ambient temperature: 30 °C
Conductor operating temperature: 90 °C

Conductor cross-sectional area	Single-phase a.c. or d.c. — 1 two-core cable, with or without protective conductor	Three-phase a.c. — 1 three-core, four-core or five-core cable	Single-phase a.c. or d.c. — 2 single-core cables, touching
1	2	3	4
(mm²)	(A)	(A)	(A)
4	42	37	–
6	55	49	–
10	76	66	–
16	103	89	–
25	136	119	200
35	–	146	250
50	–	177	310
70	–	225	369
95	–	273	432
120	–	316	497
150	–	363	564
185	–	414	673
240	–	487	773
300	–	560	–
400	–	–	924
500	–	–	1062
630	–	–	1242

NOTES:

1. The current ratings tabulated are for cables in free air but may also be used for cables resting on a surface. If the cable is to be wound on a drum on load the ratings should be reduced in accordance with NOTE 2 below and for cables which may be covered, NOTE 3 below.

2. *Flexible cables wound on reeling drums*
The current ratings of cables used on reeling drums are to be reduced by the following factors:

a) Radial type drum		b) Ventilated cylindrical type drum	
ventilated:	85 %	1 layer of cable:	85 %
unventilated:	75 %	2 layers of cable:	65 %
		3 layers of cable:	45 %
		4 layers of cable:	35 %

A radial type drum is one where spiral layers of cable are accommodated between closely spaced flanges; if fitted with solid flanges the ratings given above should be reduced and the drum is described as non-ventilated. If the flanges have suitable apertures the drum is described as ventilated.

A ventilated cylindrical cable drum is one where layers of cable are accommodated between widely spaced flanges and the drum and end flanges have suitable ventilating apertures.

3. Where cable may be covered over or coiled up whilst on load, or the air movement over the cable restricted, the current rating should be reduced.
It is not possible to specify the amount of reduction but the table of rating factors for reeling drums can be used as a guide.

4. For 180 °C cables, the rating factors for ambient temperature allow a conductor operating temperature up to 150 °C. Consult the cable manufacturer for further information.

5. Where it is intended to connect the cables in this table to equipment or accessories designed to operate at a temperature lower than the maximum operating temperature of the cable, the cables should be rated at the maximum operating temperature of the equipment or accessory (see Regulation 512.1.5).

6. Where it is intended to group a cable in this table with other cables, the cable should be rated at the lowest of the maximum operating temperatures of any of the cables in the group (see Regulation 512.1.5).

RATING FACTOR FOR AMBIENT TEMPERATURE

90 °C thermosetting insulated cables:

Ambient temperature	35 °C	40 °C	45 °C	50 °C	55 °C	60 °C	65 °C	70 °C	75 °C	80 °C	85 °C
Rating factor	0.95	0.91	0.86	0.82	0.76	0.70	0.64	0.57	0.50	0.40	0.28

180 °C thermosetting insulated cables:

Ambient temperature	35 to 90 °C	95 °C	100 °C	105 °C	110 °C	115 °C	120 °C	125 °C	130 °C	135 °C	140 °C	145 °C
Rating factor	1.0	0.96	0.91	0.86	0.81	0.76	0.70	0.64	0.57	0.50	0.40	0.28

TABLE 4F2B

VOLTAGE DROP (per ampere per metre):

Conductor operating temperature: 90 °C

Conductor cross-sectional area	1 two-core or 2 single-core cables, d.c.	Two-core cable, single-phase a.c.			1 three-core, four-core or five-core cable, three-phase a.c.			2 single-core cables touching Single-phase a.c.*		
1	2	3			4			5		
(mm²)	(mV/A/m)	(mV/A/m)			(mV/A/m)			(mV/A/m)		
		r	x	z	r	x	z	r	x	z
4	13.2	13.2			11.1			–		
6	8.5	8.5			7.4			–		
10	5.1	5.1			4.4			–		
16	3.2	3.2			2.7			–		
25	2.03	2.03	0.175	2.04	1.73	0.15	1.73	–	–	–
35	1.42	–	–	–	1.22	0.15	1.23	1.44	0.21	1.46
50	1.00	–	–	–	0.91	0.145	0.93	1.00	0.21	1.02
70	0.71	–	–	–	0.62	0.14	0.64	0.71	0.20	0.73
95	0.54	–	–	–	0.47	0.135	0.49	0.54	0.195	0.57
120	0.42	–	–	–	0.37	0.135	0.39	0.42	0.190	0.46
150	0.34	–	–	–	0.29	0.130	0.32	0.34	0.190	0.39
185	0.27	–	–	–	0.24	0.130	0.27	0.27	0.190	0.33
240	0.21	–	–	–	0.188	0.130	0.23	0.21	0.185	0.28
300	0.167	–	–	–	0.147	0.125	0.195	0.173	0.180	0.25
400	0.127	–	–	–	–	–	–	0.132	0.175	0.22
500	0.100	–	–	–	–	–	–	0.107	0.170	0.20
630	0.074	–	–	–	–	–	–	0.085	0.170	0.190

NOTES:

1. The voltage drop figures given above are based on a conductor operating temperature of 90 °C and are therefore not accurate when the operating temperature is in excess of 90 °C. In the case of the 180 °C cables with a conductor temperature of 150 °C the above resistive values should be increased by a factor of 1.2.

2. * A larger voltage drop will result if the cables are spaced.

TABLE 4F3A – Flexible cables, non-armoured (COPPER CONDUCTORS)

CURRENT-CARRYING CAPACITY (amperes): and MASS SUPPORTABLE (kg):

Conductor cross-sectional area	Current-carrying capacity		Maximum mass supportable by twin flexible cable (see Regulations 522.7.2 and 559.6.1.5)
	Single-phase a.c.	Three-phase a.c.	
1	2	3	4
(mm²)	(A)	(A)	(kg)
0.5	3	3	2
0.75	6	6	3
1	10	10	5
1.25	13	–	5
1.5	16	16	5
2.5	25	20	5
4	32	25	5

Where cable is on a reel see the notes to Table 4F1A.

RATING FACTOR FOR AMBIENT TEMPERATURE

60 °C thermoplastic or thermosetting insulated cable:

Ambient temperature	35 °C	40 °C	45 °C	50 °C	55 °C
Rating factor	0.91	0.82	0.71	0.58	0.41

90 °C thermoplastic or thermosetting insulated cable:

Ambient temperature	35 to 50 °C	55 °C	60 °C	65 °C	70 °C
Rating factor	1.0	0.96	0.83	0.67	0.47

150 °C flexible cable:

Ambient temperature	35 to 120 °C	125 °C	130 °C	135 °C	140 °C	145 °C
Rating factor	1.0	0.96	0.85	0.74	0.60	0.42

Glass fibre flexible cable:

Ambient temperature	35 to 150 °C	155 °C	160 °C	165 °C	170 °C	175 °C
Rating factor	1.0	0.92	0.82	0.71	0.57	0.40

TABLE 4F3B

VOLTAGE DROP (per ampere per metre):

Conductor operating temperature: 60 °C*

Conductor cross-sectional area	d.c. or single-phase a.c.	Three-phase a.c.
1	2	3
(mm²)	(mV/A/m)	(mV/A/m)
0.5	93	80
0.75	62	54
1	46	40
1.25	37	–
1.5	32	27
2.5	19	16
4	12	10

NOTE: * The tabulated values above are for 60 °C thermoplastic or thermosetting insulated flexible cables and for other types of flexible cable they are to be multiplied by the following factors:

For 90 °C thermoplastic or thermosetting insulated   1.09
150 °C   1.31
185 °C glass fibre   1.43

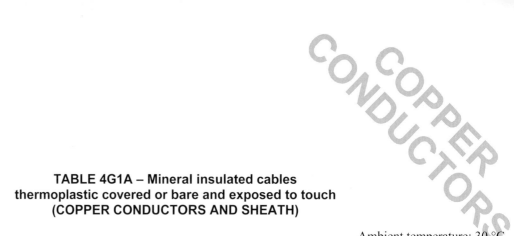

TABLE 4G1A – Mineral insulated cables
thermoplastic covered or bare and exposed to touch
(COPPER CONDUCTORS AND SHEATH)

Ambient temperature: 30 °C

CURRENT-CARRYING CAPACITY (amperes):

Sheath operating temperature: 70 °C

Conductor cross-sectional area	Reference Method C (clipped direct)			Reference Methods E, F and G (in free air or on a perforated cable tray etc, horizontal or vertical)				
	Single-phase a.c. or d.c.	Three-phase a.c.		Single-phase a.c. or d.c.	Three-phase a.c.			
	2 single-core cables touching or 1 two-core cable	3 single-core cables in trefoil or 1 three-core or four-core cable	3 single-core cables flat and touching, horizontal or vertical	2 single-core cables touching or 1 two-core cable	3 single-core cables in trefoil or 1 three-core or four-core cable	3 single-core cables flat and touching	3 single-core cables flat and spaced by one cable diameter	
							vertical	horizontal
1	2	3	4	5	6	7	8	9
(mm²)	(A)	(A)	(A)	(A)	(A)	(A)	(A)	(A)
Light duty 500 V								
1	18.5	15	17	19.5	16.5	18	20	23
1.5	23	19	21	25	21	23	26	29
2.5	31	26	29	33	28	31	34	39
4	40	35	38	44	37	41	45	51
Heavy duty 750 V								
1	19.5	16	18	21	17.5	20	22	25
1.5	25	21	23	26	22	26	28	32
2.5	34	28	31	36	30	34	37	43
4	45	37	41	47	40	45	49	56
6	57	48	52	60	51	57	62	71
10	77	65	70	82	69	77	84	95
16	102	86	92	109	92	102	110	125
25	133	112	120	142	120	132	142	162
35	163	137	147	174	147	161	173	197
50	202	169	181	215	182	198	213	242
70	247	207	221	264	223	241	259	294
95	296	249	264	317	267	289	309	351
120	340	286	303	364	308	331	353	402
150	388	327	346	416	352	377	400	454
185	440	371	392	472	399	426	446	507
240	514	434	457	552	466	496	497	565

NOTES:

1. For single-core cables, the sheaths of the circuit are assumed to be connected together at both ends.
2. For bare cables exposed to touch, the tabulated values should be multiplied by 0.9.

TABLE 4G1B

VOLTAGE DROP (per ampere per metre): Sheath operating temperature: 70 °C

Conductor cross-sectional area	Single-phase a.c. or d.c.						Three-phase a.c.														
	2 single-core cables touching			1 two-core cable			1 three- or four-core cable			3 single-core cables in trefoil formation			3 single-core cables flat and touching			3 single-core cables flat and spaced by one cable diameter*		
1	2			3			4			5			6			7		
(mm²)	(mV/A/m)			(mV/A/m)			(mV/A/m)			(mV/A/m)			(mV/A/m)			(mV/A/m)		
1	42			42			36			36			36			36		
1.5	28			28			24			24			24			24		
2.5	17			17			14			14			14			14		
4	10			10			9.1			9.1			9.1			9.1		
6	7			7			6.0			6.0			6.0			6.0		
10	4.2			4.2			3.6			3.6			3.6			3.6		
16	2.6			2.6			2.3			2.3			2.3			2.3		
	r	x	z	r	x	z	r	x	z	r	x	z	r	x	z	r	x	z
25	1.65	0.200	1.65	1.65	0.145	1.65	1.45	0.125	1.45	1.45	0.170	1.45	1.45	0.25	1.45	1.45	0.32	1.50
35	1.20	0.195	1.20	-	-	-	-	-	-	1.05	0.165	1.05	1.05	0.24	1.10	1.05	0.31	1.10
50	0.89	0.185	0.91	-	-	-	-	-	-	0.78	0.160	0.80	0.79	0.24	0.83	0.82	0.31	0.87
70	0.62	0.180	0.64	-	-	-	-	-	-	0.54	0.155	0.56	0.55	0.23	0.60	0.58	0.30	0.65
95	0.46	0.175	0.49	-	-	-	-	-	-	0.40	0.150	0.43	0.41	0.22	0.47	0.44	0.29	0.53
120	0.37	0.170	0.41	-	-	-	-	-	-	0.32	0.150	0.36	0.33	0.22	0.40	0.36	0.28	0.46
150	0.30	0.170	0.34	-	-	-	-	-	-	0.26	0.145	0.30	0.29	0.21	0.36	0.32	0.27	0.42
185	0.25	0.165	0.29	-	-	-	-	-	-	0.21	0.140	0.26	0.25	0.21	0.32	0.28	0.26	0.39
240	0.190	0.160	0.25	-	-	-	-	-	-	0.165	0.140	0.22	0.21	0.20	0.29	0.26	0.25	0.36

NOTE: * Spacings larger than one cable diameter will result in a larger voltage drop.

TABLE 4G2A – Mineral insulated cables
bare and neither exposed to touch nor in contact with combustible materials
(COPPER CONDUCTORS AND SHEATH)

CURRENT-CARRYING CAPACITY (amperes):

Ambient temperature: 30 °C
Sheath operating temperature: 105 °C

Conductor cross-sectional area	Reference Method C (clipped direct)			Reference Methods E, F and G (in free air or on a perforated cable tray etc, horizontal or vertical)				
	Single-phase a.c. or d.c.	Three-phase a.c.		Single-phase a.c. or d.c.	Three-phase a.c.			
	2 single-core cables touching or 1 two-core cable	3 single-core cables in trefoil or 1 three-core or four-core cable	3 single-core cables flat and touching, horizontal or vertical	2 single-core cables touching or 1 two-core cable	3 single-core cables in trefoil or 1 three-core or four-core cable	3 single-core cables flat and touching	3 single-core cables flat and spaced by one cable diameter	
							vertical	horizontal
1	2	3	4	5	6	7	8	9
(mm²)	(A)	(A)	(A)	(A)	(A)	(A)	(A)	(A)
Light duty 500 V								
1	22	19	21	24	21	23	26	29
1.5	28	24	27	31	26	29	33	37
2.5	38	33	36	41	35	39	43	49
4	51	44	47	54	46	51	56	64
Heavy duty 750 V								
1	24	20	24	26	22	25	28	32
1.5	31	26	30	33	28	32	35	40
2.5	42	35	41	45	38	43	47	54
4	55	47	53	60	50	56	61	70
6	70	59	67	76	64	71	78	89
10	96	81	91	104	87	96	105	120
16	127	107	119	137	115	127	137	157
25	166	140	154	179	150	164	178	204
35	203	171	187	220	184	200	216	248
50	251	212	230	272	228	247	266	304
70	307	260	280	333	279	300	323	370
95	369	312	334	400	335	359	385	441
120	424	359	383	460	385	411	441	505
150	485	410	435	526	441	469	498	565
185	550	465	492	596	500	530	557	629
240	643	544	572	697	584	617	624	704

NOTES:

1. For single-core cables, the sheaths of the circuit are assumed to be connected together at both ends.
2. No rating factor for grouping need be applied.
3. Where a conductor operates at a temperature exceeding 70 °C it should be ascertained that the equipment connected to the conductor is suitable for the conductor operating temperature (see Regulation 512.1.5).

TABLE 4G2B

VOLTAGE DROP (per ampere per metre):

Sheath operating temperature: 105 °C

Conductor cross-sectional area	Single-phase a.c. or d.c.						Three-phase a.c.												
	2 single-core cables touching			1 two-core cable			1 three- or four-core cable			3 single-core cables in trefoil formation			3 single-core cables flat and touching			3 single-core cables flat and spaced by one cable diameter*			
1	2			3			4			5			6			7			
(mm²)	(mV/A/m)			(mV/A/m)			(mV/A/m)			(mV/A/m)			(mV/A/m)			(mV/A/m)			
1	47			47			40			40			40			40			
1.5	31			31			27			27			27			27			
2.5	19			19			16			16			16			16			
4	12			12			10			10			10			10			
6	7.8			7.8			6.8			6.8			6.8			6.8			
10	4.7			4.7			4.1			4.1			4.1			4.1			
16	3.0			3.0			2.6			2.6			2.6			2.6			
	r	x	z	r	x	z	r	x	z	r	x	z	r	x	z	r	x	z	
25	1.85	0.180	1.85	1.85	0.145	1.85	1.60	0.125	1.60	1.60	0.160	1.65	1.60	0.23	1.65	1.60	0.31	1.65	
35	1.35	0.175	1.35	-	-	-	-	-	-	1.15	0.155	1.20	1.15	0.23	1.20	1.20	0.30	1.25	
50	1.00	0.170	1.00	-	-	-	-	-	-	0.87	0.150	0.88	0.88	0.22	0.91	0.90	0.29	0.95	
70	0.69	0.165	0.71	-	-	-	-	-	-	0.60	0.145	0.62	0.61	0.22	0.65	0.63	0.29	0.70	
95	0.51	0.160	0.54	-	-	-	-	-	-	0.45	0.140	0.47	0.46	0.21	0.50	0.48	0.28	0.56	
120	0.41	0.160	0.44	-	-	-	-	-	-	0.36	0.135	0.38	0.37	0.21	0.42	0.39	0.28	0.48	
150	0.33	0.155	0.36	-	-	-	-	-	-	0.29	0.135	0.32	0.31	0.20	0.37	0.34	0.27	0.43	
185	0.27	0.150	0.31	-	-	-	-	-	-	0.23	0.130	0.27	0.26	0.20	0.33	0.29	0.26	0.39	
240	0.21	0.150	0.26	-	-	-	-	-	-	0.180	0.130	0.22	0.22	0.195	0.29	0.26	0.25	0.36	

NOTE: * Spacings larger than one cable diameter will result in a larger voltage drop.

ALUMINIUM CONDUCTORS

TABLE 4H1A – Single-core 70 °C thermoplastic insulated cables, non-armoured, with or without sheath (ALUMINIUM CONDUCTORS)

CURRENT-CARRYING CAPACITY (amperes):

Ambient temperature: 30 °C
Conductor operating temperature: 70 °C

Conductor cross-sectional area	Reference Method A (enclosed in conduit in thermally insulating wall etc.)		Reference Method B (enclosed in conduit on a wall or in trunking etc.)		Reference Method C (clipped direct)		Reference Method F (in free air on a perforated cable tray, horizontal or vertical)				
							Touching			Spaced by one diameter	
										2 cables, single-phase a.c. or d.c. or 3 cables three-phase a.c. flat	
	2 cables, single-phase a.c. or d.c.	3 or 4 cables, three-phase a.c.	2 cables, single-phase a.c. or d.c.	3 or 4 cables, three-phase a.c.	2 cables, single-phase a.c. or d.c. flat and touching	3 or 4 cables, three-phase a.c. flat and touching or trefoil	2 cables, single-phase a.c. or d.c. flat	3 cables, three-phase a.c. flat	3 cables, three-phase a.c trefoil	Horizontal	Vertical
1	2	3	4	5	6	7	8	9	10	11	12
(mm²)	(A)	(A)	(A)	(A)	(A)	(A)	(A)	(A)	(A)	(A)	(A)
50	93	84	118	104	125	110	149	133	128	169	152
70	118	107	150	133	160	140	192	173	166	217	196
95	142	129	181	161	195	170	235	212	203	265	241
120	164	149	210	186	226	197	273	247	237	308	282
150	189	170	234	204	261	227	316	287	274	356	327
185	215	194	266	230	298	259	363	330	316	407	376
240	252	227	312	269	352	305	430	392	375	482	447
300	289	261	358	306	406	351	497	455	434	557	519
380	-	-	413	352	511	472	543	502	507	625	584
480	-	-	477	405	591	546	629	582	590	726	680
600	-	-	545	462	679	626	722	669	680	837	787
740	-	-	-	-	771	709	820	761	776	956	902
960	-	-	-	-	900	823	953	886	907	1125	1066
1200	-	-	-	-	1022	926	1073	999	1026	1293	1229

TABLE 4H1B

VOLTAGE DROP (per ampere per metre):

Conductor operating temperature: 70 °C

Conductor cross-sectional area	2 cables, d.c.	2 cables, single-phase a.c.										3 or 4 cables, three-phase a.c.											
		Reference Methods A & B (enclosed in conduit or trunking)			Reference Methods C & F (clipped direct, on tray or in free air)								Reference Methods A & B (enclosed in conduit or trunking)			Reference Methods C & F (clipped direct, on tray or in free air)								
					Cables touching				Cables spaced*							Cables touching, Trefoil			Cables touching, Flat			Cables spaced*, Flat		
1	2	3			4				5				6			7			8			9		
(mm²)	(mV/A/m)	(mV/A/m)			(mV/A/m)				(mV/A/m)				(mV/A/m)			(mV/A/m)			(mV/A/m)			(mV/A/m)		
		r	x	z	r	x	z		r	x	z		r	x	z	r	x	z	r	x	z	r	x	z
50	1.55	1.60	0.30	1.60	1.55	0.190	1.55		1.55	0.28	1.55		1.35	0.26	1.40	1.35	0.165	1.35	1.35	0.24	1.35	1.35	0.32	1.40
70	1.05	1.10	0.30	1.15	1.05	0.185	1.05		1.05	0.27	1.10		0.94	0.26	0.97	0.91	0.160	0.92	0.91	0.24	0.94	0.91	0.31	0.96
95	0.77	0.81	0.29	0.86	0.77	0.185	0.79		0.77	0.27	0.82		0.70	0.25	0.74	0.67	0.160	0.69	0.67	0.23	0.71	0.67	0.31	0.74
120	0.61	0.64	0.29	0.70	0.61	0.180	0.64		0.61	0.27	0.67		0.55	0.25	0.61	0.53	0.155	0.55	0.53	0.23	0.58	0.53	0.31	0.61
150	0.49	0.51	0.28	0.59	0.49	0.175	0.52		0.49	0.26	0.55		0.45	0.24	0.51	0.42	0.155	0.45	0.42	0.23	0.48	0.42	0.30	0.52
185	0.39	0.42	0.28	0.50	0.40	0.175	0.43		0.39	0.26	0.47		0.36	0.24	0.44	0.34	0.150	0.37	0.34	0.23	0.41	0.34	0.30	0.46
240	0.30	0.32	0.27	0.42	0.30	0.170	0.35		0.30	0.26	0.40		0.28	0.24	0.37	0.26	0.150	0.30	0.26	0.22	0.35	0.26	0.30	0.40
300	0.24	0.26	0.27	0.37	0.24	0.170	0.30		0.24	0.26	0.35		0.23	0.23	0.32	0.21	0.145	0.26	0.21	0.22	0.31	0.21	0.30	0.36
380	0.190	0.22	0.27	0.35	0.195	0.165	0.26		0.195	0.25	0.32		0.190	0.23	0.30	0.170	0.145	0.22	0.170	0.22	0.28	0.170	0.29	0.34
480	0.150	0.18	0.26	0.32	0.155	0.165	0.23		0.155	0.25	0.29		0.155	0.23	0.27	0.140	0.140	0.195	0.140	0.22	0.26	0.135	0.29	0.32
600	0.120	0.150	0.26	0.30	0.130	0.160	0.21		0.125	0.25	0.28		0.125	0.22	0.26	0.110	0.140	0.180	0.110	0.22	0.24	0.110	0.29	0.31
740	0.099	-	-	-	0.105	0.160	0.190		0.100	0.25	0.27		-	-	-	0.094	0.135	0.165	0.094	0.21	0.23	0.089	0.29	0.30
960	0.075	-	-	-	0.086	0.155	0.180		0.082	0.24	0.26		-	-	-	0.077	0.135	0.155	0.077	0.21	0.22	0.071	0.29	0.29
1200	0.060	-	-	-	0.074	0.155	0.170		0.068	0.24	0.25		-	-	-	0.066	0.135	0.150	0.066	0.21	0.22	0.059	0.28	0.29

NOTE: * Spacings larger than one cable diameter will result in a larger voltage drop.

TABLE 4H2A – Multicore 70 °C thermoplastic insulated and thermoplastic sheathed cables, non-armoured
(ALUMINIUM CONDUCTORS)

Ambient temperature: 30 °C
Conductor operating temperature: 70 °C

CURRENT-CARRYING CAPACITY (amperes):

Conductor cross-sectional area	Reference Method A (enclosed in conduit in thermally insulating wall etc.)		Reference Method B (enclosed in conduit on a wall or in trunking etc.)		Reference Method C (clipped direct)		Reference Method E (in free air or on a perforated cable tray etc, horizontal or vertical)	
	1 two-core cable, single-phase a.c. or d.c.	1 three- or four-core cable, three-phase a.c.	1 two-core cable, single-phase a.c. or d.c.	1 three- or four-core cable, three-phase a.c.	1 two-core cable, single-phase a.c. or d.c.	1 three- or four-core cable, three-phase a.c.	1 two-core cable, single-phase a.c. or d.c.	1 three- or four-core cable, three-phase a.c.
1	2	3	4	5	6	7	8	9
(mm^2)	(A)	(A)	(A)	(A)	(A)	(A)	(A)	(A)
16	44	41	54	48	66	59	73	61
25	58	53	71	62	83	73	89	78
35	71	65	86	77	103	90	111	96
50	86	78	104	92	125	110	135	117
70	108	98	131	116	160	140	173	150
95	130	118	157	139	195	170	210	183
120	-	135	-	160	-	197	-	212
150	-	155	-	176	-	227	-	245
185	-	176	-	199	-	259	-	280
240	-	207	-	232	-	305	-	330
300	-	237	-	265	-	351	-	381

TABLE 4H2B

VOLTAGE DROP (per ampere per metre):

Conductor operating temperature: 70 °C

Conductor cross-sectional area	Two-core cable, d.c.	Two-core cable, single-phase a.c.			Three- or four-core cable, three-phase a.c.		
1	2	3			4		
(mm²)	(mV/A/m)	(mV/A/m)			(mV/A/m)		
16	4.5	4.5			3.9		
		r	x	z	r	x	z
25	2.9	2.9	0.175	2.9	2.5	0.150	2.5
35	2.1	2.1	0.170	2.1	1.80	0.150	1.80
50	1.55	1.55	0.170	1.55	1.35	0.145	1.35
70	1.05	1.05	0.165	1.05	0.90	0.140	0.92
95	0.77	0.77	0.160	0.79	0.67	0.140	0.68
120	-	-	-	-	0.53	0.135	0.55
150	-	-	-	-	0.42	0.135	0.44
185	-	-	-	-	0.34	0.135	0.37
240	-	-	-	-	0.26	0.130	0.30
300	-	-	-	-	0.21	0.130	0.25

ALUMINIUM CONDUCTORS

TABLE 4H3A – Single-core armoured 70 °C thermoplastic insulated cables (non-magnetic armour) (ALUMINIUM CONDUCTORS)

CURRENT-CARRYING CAPACITY (amperes):

Ambient temperature: 30 °C
Conductor operating temperature: 70 °C

Conductor cross-sectional area	Reference Method C (clipped direct)			Reference Method F (in free air or on a perforated cable tray, horizontal or vertical)									
	Touching			Touching			Spaced by one cable diameter					
	2 cables, single-phase a.c. or d.c. flat	3 or 4 cables, three-phase a.c. flat	2 cables, single-phase a.c. or d.c. flat	3 cables, three-phase a.c. flat	3 cables, three-phase a.c. trefoil	2 cables, d.c.		2 cables, single-phase a.c.		3 or 4 cables, three-phase a.c.	
						Horizontal	Vertical	Horizontal	Vertical	Horizontal	Vertical
1	2	3	4	5	6	7	8	9	10	11	12
(mm²)	(A)	(A)	(A)	(A)	(A)	(A)	(A)	(A)	(A)	(A)	(A)
50	143	133	152	141	131	167	157	168	159	169	155
70	183	168	194	178	168	214	202	212	200	213	196
95	221	202	234	214	205	261	247	259	245	255	236
120	255	233	270	246	238	303	288	299	285	293	272
150	294	267	310	282	275	349	333	340	323	335	312
185	334	303	352	319	315	400	382	389	371	379	354
240	393	354	413	374	372	472	452	457	437	443	415
300	452	405	474	427	430	545	523	520	498	505	475
380	518	452	543	479	497	638	613	583	559	551	518
480	586	501	616	534	568	742	715	655	629	604	568
600	658	550	692	589	642	859	828	724	696	656	618
740	728	596	769	642	715	986	952	802	770	707	666
960	819	651	868	706	808	1171	1133	866	832	770	726
1200	893	692	952	756	880	1360	1317	938	902	822	774

TABLE 4H3B

VOLTAGE DROP (per ampere per metre): Conductor operating temperature: 70 °C

Conductor cross-sectional area	2 cables, d.c.	2 cables, single-phase a.c.						3 or 4 cables, three-phase a.c.								
1	2	3 touching			4 spaced*			5 trefoil and touching			6 flat and touching			7 flat and spaced*		
(mm²)	(mV/A/m)	(mV/A/m)			(mV/A/m)			(mV/A/m)			(mV/A/m)			(mV/A/m)		
		r	x	z	r	x	z	r	x	z	r	x	z	r	x	z
50	1.55	1.55	0.23	1.55	1.55	0.31	1.55	1.35	0.195	1.35	1.35	0.27	1.35	1.30	0.34	1.35
70	1.05	1.05	0.22	1.10	1.05	0.30	1.10	0.92	0.190	0.93	0.93	0.26	0.96	0.95	0.33	1.00
95	0.77	0.78	0.21	0.81	0.81	0.29	0.86	0.68	0.185	0.70	0.70	0.25	0.75	0.73	0.32	0.80
120	0.61	0.62	0.21	0.66	0.65	0.29	0.71	0.54	0.180	0.57	0.57	0.25	0.62	0.60	0.32	0.68
150	0.49	0.50	0.20	0.54	0.53	0.28	0.60	0.44	0.175	0.47	0.46	0.24	0.52	0.50	0.31	0.58
185	0.39	0.41	0.195	0.45	0.44	0.28	0.52	0.35	0.170	0.39	0.38	0.24	0.45	0.42	0.30	0.51
240	0.30	0.32	0.190	0.37	0.34	0.27	0.44	0.28	0.165	0.32	0.30	0.23	0.38	0.33	0.29	0.44
300	0.24	0.26	0.185	0.32	0.28	0.26	0.39	0.22	0.160	0.27	0.24	0.23	0.34	0.28	0.29	0.40
380	0.190	0.22	0.185	0.28	0.26	0.25	0.36	0.185	0.155	0.24	0.22	0.22	0.32	0.27	0.26	0.38
480	0.150	0.180	0.180	0.25	0.22	0.25	0.33	0.155	0.155	0.22	0.195	0.22	0.29	0.24	0.25	0.35
600	0.120	0.150	0.175	0.23	0.195	0.24	0.31	0.130	0.150	0.200	0.170	0.21	0.27	0.21	0.24	0.32
740	0.097	0.135	0.170	0.22	0.180	0.23	0.29	0.115	0.145	0.185	0.160	0.20	0.26	0.200	0.22	0.30
960	0.075	0.115	0.160	0.200	0.165	0.21	0.27	0.100	0.140	0.175	0.150	0.185	0.24	0.190	0.195	0.27
1200	0.060	0.110	0.155	0.190	0.160	0.180	0.24	0.094	0.140	0.170	0.145	0.160	0.22	0.185	0.165	0.25

Reference Methods C & F (clipped direct, on tray or in free air)

NOTE: * Spacings larger than one cable diameter will result in a larger voltage drop.

TABLE 4H4A – Multicore armoured 70 °C thermoplastic insulated cables (ALUMINIUM CONDUCTORS)

Air Ambient temperature: 30 °C
Ground Ambient temperature: 20 °C
Conductor operating temperature: 70 °C

CURRENT-CARRYING CAPACITY (amperes):

Conductor cross-sectional area	Reference Method C (clipped direct)		Reference Method E (in free air or on a perforated cable tray etc, horizontal or vertical)		Reference Method D (direct in ground or in ducting in ground, in or around buildings)	
	1 two-core cable, single-phase a.c. or d.c.	1 three- or 1 four-core cable, three-phase a.c.	1 two-core cable, single-phase a.c. or d.c.	1 three- or 1 four-core cable, three-phase a.c.	1 two-core cable, single-phase a.c. or d.c.	1 three- or 1 four-core cable, three-phase a.c.
1	2	3	4	5	6	7
(mm²)	(A)	(A)	(A)	(A)	(A)	(A)
16	68	58	71	61	77	64
25	89	76	94	80	93	77
35	109	94	115	99	109	91
50	131	113	139	119	135	112
70	165	143	175	151	159	132
95	199	174	211	186	–	–
120	–	202	–	216	–	150
150	–	232	–	250	–	169
185	–	265	–	287	–	190
240	–	312	–	342	–	218
300	–	360	–	399	–	247

TABLE 4H4B

VOLTAGE DROP (per ampere per metre):   Conductor operating temperature: 70 °C

Conductor cross-sectional area	Two-core cable, d.c.	Two-core cable, single-phase a.c.			Three- or four-core cable, three-phase a.c.		
1	2	3			4		
(mm²)	(mV/A/m)	(mV/A/m)			(mV/A/m)		
16	4.5	4.5			3.9		
		r	x	z	r	x	z
25	2.9	2.9	0.175	2.9	2.5	0.150	2.5
35	2.1	2.1	0.170	2.1	1.80	0.150	1.80
50	1.55	1.55	0.170	1.55	1.35	0.145	1.35
70	1.05	1.05	0.165	1.05	0.90	0.140	0.92
95	0.77	0.77	0.160	0.79	0.67	0.140	0.68
120	-	-	-	-	0.53	0.135	0.55
150	-	-	-	-	0.42	0.135	0.44
185	-	-	-	-	0.34	0.135	0.37
240	-	-	-	-	0.26	0.130	0.30
300	-	-	-	-	0.21	0.130	0.25

TABLE 4J1A – Single-core 90 °C thermosetting insulated cables, non-armoured, with or without sheath (ALUMINIUM CONDUCTORS)

CURRENT-CARRYING CAPACITY (amperes):

Ambient temperature: 30 °C
Conductor operating temperature: 90 °C

Conductor cross-sectional area	Reference Method A (enclosed in conduit in thermally insulating wall etc.)		Reference Method B (enclosed in conduit on a wall or in trunking etc.)		Reference Method C (clipped direct)		Reference Method F (in free air or on a perforated cable tray horizontal or vertical etc) Touching			Reference Method G (in free air) Spaced by one cable diameter — 2 cables, single-phase a.c. or d.c. or 3 cables three-phase a.c. flat	
	2 cables, single-phase a.c. or d.c.	3 or 4 cables, three-phase a.c.	2 cables, single-phase a.c. or d.c.	3 or 4 cables, three-phase a.c.	2 cables, single-phase a.c. or d.c. flat and touching	3 or 4 cables, three-phase a.c. flat and touching or trefoil	2 cables, single-phase a.c. or d.c. flat	3 cables, three-phase a.c. flat	3 cables, three-phase a.c trefoil	Horizontal	Vertical
1	2	3	4	5	6	7	8	9	10	11	12
(mm²)	(A)	(A)	(A)	(A)	(A)	(A)	(A)	(A)	(A)	(A)	(A)
50	125	113	157	140	154	136	184	165	159	210	188
70	158	142	200	179	198	174	237	215	206	271	244
95	191	171	242	217	241	211	289	264	253	332	300
120	220	197	281	251	280	245	337	308	296	387	351
150	253	226	307	267	324	283	389	358	343	448	408
185	288	256	351	300	371	323	447	413	395	515	470
240	338	300	412	351	439	382	530	492	471	611	561
300	387	344	471	402	508	440	613	571	544	708	652
380	-	-	-	-	658	594	679	628	638	798	742
480	-	-	-	-	765	692	786	728	743	927	865
600	-	-	-	-	871	791	903	836	849	1058	990
740	-	-	-	-	1001	911	1025	951	979	1218	1143
960	-	-	-	-	1176	1072	1191	1108	1151	1440	1355
1200	-	-	-	-	1333	1217	1341	1249	1307	1643	1550

NOTES:

1. Where it is intended to connect the cables in this table to equipment or accessories designed to operate at a temperature lower than the maximum operating temperature of the cable, the cables should be rated at the maximum operating temperature of the equipment or accessory (see Regulation 512.1.5).

2. Where it is intended to group a cable in this table with other cables, the cable should be rated at the lowest of the maximum operating temperatures of any of the cables in the group (see Regulation 512.1.5).

VOLTAGE DROP (per ampere per metre): Conductor operating temperature: 90 °C

TABLE 4J1B

Conductor cross-sectional area	2 cables, d.c.	2 cables, single-phase a.c.									3 or 4 cables, three-phase a.c.												
		Reference Methods A & B (enclosed in conduit or trunking)			Reference Methods C, F & G (clipped direct, on tray or in free air)						Reference Methods A & B (enclosed in conduit or trunking)			Reference Methods C, F & G (clipped direct, on tray or in free air)								
					Cables touching			Cables spaced*						Cables touching, Trefoil			Cables touching, Flat			Cables spaced*, Flat		
	2	3			4			5			6			7			8			9		
(mm²)	(mV/A/m)	(mV/A/m)			(mV/A/m)			(mV/A/m)			(mV/A/m)			(mV/A/m)			(mV/A/m)			(mV/A/m)		
		r	x	z	r	x	z	r	x	z	r	x	z	r	x	z	r	x	z	r	x	z
50	1.65	1.70	0.30	1.72	1.65	0.190	1.66	1.65	0.28	1.68	1.44	0.26	1.46	1.44	0.165	1.45	1.44	0.24	1.46	1.44	0.32	1.48
70	1.13	1.17	0.30	1.21	1.12	0.185	1.14	1.12	0.27	1.15	1.00	0.26	1.04	0.97	0.160	0.98	0.97	0.24	1.00	0.97	0.31	1.02
95	0.82	0.86	0.29	0.91	0.82	0.185	0.84	0.82	0.27	0.94	0.75	0.25	0.79	0.71	0.160	0.73	0.71	0.23	0.75	0.71	0.31	0.78
120	0.65	0.68	0.29	0.74	0.65	0.180	0.67	0.65	0.27	0.70	0.59	0.25	0.64	0.57	0.155	0.59	0.57	0.23	0.61	0.57	0.31	0.64
150	0.53	0.54	0.28	0.61	0.52	0.175	0.55	0.52	0.26	0.58	0.48	0.24	0.54	0.45	0.155	0.47	0.45	0.23	0.50	0.45	0.30	0.54
185	0.42	0.45	0.28	0.53	0.43	0.175	0.46	0.42	0.26	0.49	0.38	0.24	0.45	0.36	0.150	0.39	0.36	0.23	0.43	0.36	0.30	0.47
240	0.32	0.34	0.27	0.43	0.32	0.170	0.36	0.32	0.26	0.41	0.30	0.24	0.38	0.28	0.150	0.32	0.28	0.22	0.35	0.28	0.30	0.41
300	0.26	0.28	0.27	0.38	0.26	0.170	0.31	0.26	0.26	0.36	0.25	0.23	0.34	0.22	0.145	0.27	0.22	0.22	0.31	0.22	0.30	0.37
380	0.20	-	-	-	0.21	0.165	0.27	0.21	0.25	0.33	0.20	0.23	0.31	0.180	0.145	0.23	0.180	0.22	0.28	0.180	0.29	0.34
480	0.160	-	-	-	0.170	0.165	0.23	0.165	0.25	0.30	0.165	0.23	0.28	0.150	0.140	0.20	0.150	0.22	0.27	0.145	0.29	0.32
600	0.130	-	-	-	0.140	0.160	0.21	0.135	0.25	0.28	0.135	0.22	0.26	0.120	0.140	0.185	0.120	0.22	0.25	0.120	0.29	0.31
740	0.105	-	-	-	0.115	0.160	0.19	0.110	0.25	0.27	-	-	-	0.100	0.135	0.170	0.100	0.21	0.23	0.095	0.29	0.30
960	0.080	-	-	-	0.092	0.155	0.18	0.087	0.24	0.26	-	-	-	0.082	0.135	0.160	0.082	0.21	0.23	0.076	0.29	0.30
1200	0.064	-	-	-	0.079	0.155	0.17	0.073	0.24	0.25	-	-	-	0.070	0.135	0.150	0.070	0.21	0.22	0.063	0.28	0.29

NOTE: * Spacings larger than one cable diameter will result in a larger voltage drop.

TABLE 4J2A – Multicore 90 °C thermosetting insulated and thermoplastic sheathed cables, non-armoured (ALUMINIUM CONDUCTORS)

CURRENT-CARRYING CAPACITY (amperes):

Ambient temperature: 30 °C
Conductor operating temperature: 90 °C

Conductor cross-sectional area	Reference Method A (enclosed in conduit in thermally insulating wall etc.)		Reference Method B (enclosed in conduit on a wall or in trunking etc.)		Reference Method C (clipped direct)		Reference Method E (in free air or on a perforated cable tray etc, horizontal or vertical)	
	1 two-core cable, single-phase a.c. or d.c.	1 three- or four-core cable, three-phase a.c.	1 two-core cable, single-phase a.c. or d.c.	1 three- or four-core cable, three-phase a.c.	1 two-core cable, single-phase a.c. or d.c.	1 three- or four-core cable, three-phase a.c.	1 two-core cable, single-phase a.c. or d.c.	1 three- or four-core cable, three-phase a.c.
1	2	3	4	5	6	7	8	9
(mm²)	(A)	(A)	(A)	(A)	(A)	(A)	(A)	(A)
16	60	55	72	64	84	76	91	77
25	78	71	94	84	101	90	108	97
35	96	87	115	103	126	112	135	120
50	115	104	138	124	154	136	164	146
70	145	131	175	156	198	174	211	187
95	175	157	210	188	241	211	257	227
120	-	180	-	216	-	245	-	263
150	-	206	-	240	-	283	-	304
185	-	233	-	272	-	323	-	347
240	-	273	-	318	-	382	-	409
300	-	313	-	364	-	440	-	471

NOTES:
1. Where it is intended to connect the cables in this table to equipment or accessories designed to operate at a temperature lower than the maximum operating temperature of the cable, the cables should be rated at the maximum operating temperature of the equipment or accessory (see Regulation 512.1.5).

2. Where it is intended to group a cable in this table with other cables, the cable should be rated at the lowest of the maximum operating temperatures of any of the cables in the group (see Regulation 512.1.5).

TABLE 4J2B

VOLTAGE DROP (per ampere per metre): Conductor operating temperature: 90 °C

Conductor cross-sectional area	Two-core cable, d.c.	Two-core cable, single-phase a.c.			Three- or four-core cable, three-phase a.c.		
1	2	3			4		
(mm²)	(mVA/m)	(mV/A/m)			(mV/A/m)		
16	4.8	4.8			4.2		
		r	x	z	r	x	z
25	3.1	3.1	0.165	3.1	2.7	0.140	2.7
35	2.2	2.2	0.160	2.2	1.90	0.140	1.95
50	1.60	1.65	0.160	1.65	1.40	0.135	1.45
70	1.10	1.10	0.155	1.15	0.96	0.135	0.97
95	0.82	0.82	0.150	0.84	0.71	0.130	0.72
120	-	-	-	-	0.56	0.130	0.58
150	-	-	-	-	0.45	0.130	0.47
185	-	-	-	-	0.37	0.130	0.39
240	-	-	-	-	0.28	0.125	0.31
300	-	-	-	-	0.23	0.125	0.26

TABLE 4J3A – Single-core armoured 90 °C thermosetting insulated cables (non-magnetic armour) (ALUMINIUM CONDUCTORS)

Ambient temperature: 30 °C
Conductor operating temperature: 90 °C

CURRENT-CARRYING CAPACITY (amperes):

Conductor cross-sectional area	Reference Method C (clipped direct) Touching		Reference Method F (in free air or on a perforated cable tray, horizontal or vertical)									
			Touching			Spaced by one cable diameter					
	2 cables, single-phase a.c. or d.c. flat	3 or 4 cables, three-phase a.c. flat	2 cables, single-phase a.c. or d.c. flat	3 cables, three-phase a.c. flat	3 cables, three-phase a.c. trefoil	2 cables, d.c. Horizontal	2 cables, d.c. Vertical	2 cables, single-phase a.c. Horizontal	2 cables, single-phase a.c. Vertical	3 or 4 cables, three-phase a.c. Horizontal	3 or 4 cables, three-phase a.c. Vertical
1	2	3	4	5	6	7	8	9	10	11	12
(mm²)	(A)	(A)	(A)	(A)	(A)	(A)	(A)	(A)	(A)	(A)	(A)
50	179	165	192	176	162	216	197	212	199	215	192
70	228	209	244	222	207	275	253	269	254	270	244
95	276	252	294	267	252	332	307	328	310	324	296
120	320	291	340	308	292	384	357	378	358	372	343
150	368	333	390	352	337	441	411	429	409	424	394
185	419	378	444	400	391	511	480	490	467	477	447
240	494	443	521	468	465	605	572	576	549	554	523
300	568	508	597	536	540	701	666	654	624	626	595
380	655	573	688	608	625	812	780	735	704	693	649
480	747	642	786	685	714	942	906	825	790	765	717
600	836	706	880	757	801	1076	1036	909	872	832	780
740	934	764	988	824	897	1250	1205	989	950	890	835
960	1056	838	1121	911	1014	1488	1435	1094	1052	970	911
1200	1163	903	1236	990	1118	1715	1658	1187	1141	1043	980

NOTES:

1. Where it is intended to connect the cables in this table to equipment or accessories designed to operate at a temperature lower than the maximum operating temperature of the cable, the cables should be rated at the maximum operating temperature of the equipment or accessory (see Regulation 512.1.5).

2. Where it is intended to group a cable in this table with other cables, the cable should be rated at the lowest of the maximum operating temperatures of any of the cables in the group (see Regulation 512.1.5).

VOLTAGE DROP (per ampere per metre):   Conductor operating temperature: 90 °C

TABLE 4J3B

Conductor cross-sectional area	2 cables, d.c.	Reference Methods C & F (clipped direct, on tray or in free air)															
		2 cables, single-phase a.c.						3 or 4 cables, three-phase a.c.									
		touching			spaced*			trefoil and touching			flat and touching			flat and spaced*		
		3			4			5			6			7		
(mm²)	(mV/A/m)	(mV/A/m)			(mV/A/m)			(mV/A/m)			(mV/A/m)			(mV/A/m)		
1	2	r	x	z	r	X	z	r	x	z	r	x	z	r	x	z
50	1.60	1.60	0.22	1.60	1.60	0.30	1.60	1.40	0.185	1.40	1.40	0.26	1.40	1.35	0.34	1.40
70	1.10	1.10	0.21	1.15	1.10	0.29	1.15	0.96	0.180	0.98	0.97	0.25	1.00	0.99	0.33	1.05
95	0.82	0.83	0.20	0.85	0.85	0.29	0.90	0.71	0.175	0.74	0.74	0.25	0.78	0.76	0.32	0.83
120	0.66	0.66	0.20	0.69	0.69	0.28	0.74	0.57	0.170	0.60	0.60	0.24	0.64	0.63	0.31	0.70
150	0.52	0.53	0.195	0.57	0.56	0.28	0.62	0.46	0.170	0.49	0.49	0.24	0.54	0.52	0.30	0.60
185	0.42	0.43	0.190	0.47	0.46	0.27	0.54	0.38	0.165	0.41	0.40	0.24	0.47	0.44	0.30	0.53
240	0.32	0.34	0.185	0.39	0.37	0.27	0.45	0.29	0.160	0.34	0.32	0.23	0.39	0.35	0.29	0.46
300	0.26	0.27	0.185	0.33	0.30	0.26	0.40	0.24	0.160	0.29	0.26	0.23	0.34	0.29	0.29	0.41
380	0.21	0.23	0.180	0.29	0.26	0.25	0.36	0.195	0.155	0.25	0.23	0.22	0.32	0.27	0.27	0.38
480	0.160	0.185	0.175	0.25	0.23	0.25	0.34	0.160	0.155	0.22	0.20	0.21	0.29	0.24	0.26	0.35
600	0.130	0.160	0.175	0.24	0.20	0.24	0.31	0.135	0.150	0.20	0.175	0.21	0.27	0.22	0.25	0.33
740	0.105	0.140	0.170	0.22	0.190	0.22	0.29	0.120	0.145	0.190	0.165	0.195	0.26	0.21	0.22	0.30
960	0.080	0.120	0.160	0.20	0.170	0.21	0.27	0.105	0.140	0.175	0.150	0.180	0.24	0.195	0.195	0.28
1200	0.064	0.105	0.160	0.190	0.155	0.20	0.25	0.093	0.135	0.165	0.140	0.175	0.22	0.180	0.185	0.26

NOTE: * Spacings larger than one cable diameter will result in a larger voltage drop.

TABLE 4J4A – Multicore armoured 90 °C thermosetting insulated cables (ALUMINIUM CONDUCTORS)

CURRENT-CARRYING CAPACITY (amperes):

Air Ambient temperature: 30 °C
Ground Ambient temperature: 20 °C
Conductor operating temperature: 90 °C

Conductor cross-sectional area	Reference Method C (clipped direct)		Reference Method E (in free air or on a perforated cable tray etc, horizontal or vertical)		Reference Method D (direct in ground or in ducting in ground, in or around buildings)	
	1 two-core cable, single-phase a.c. or d.c.	1 three- or 1 four-core cable, three-phase a.c.	1 two-core cable, single-phase a.c. or d.c.	1 three- or 1 four-core cable, three-phase a.c.	1 two-core cable, single-phase a.c. or d.c.	1 three- or 1 four-core cable, three-phase a.c.
1	2	3	4	5	6	7
(mm²)	(A)	(A)	(A)	(A)	(A)	(A)
16	82	71	85	74	71	59
25	108	92	112	98	90	75
35	132	113	138	120	108	90
50	159	137	166	145	128	106
70	201	174	211	185	158	130
95	242	214	254	224	186	154
120	–	249	–	264	–	174
150	–	284	–	305	–	197
185	–	328	–	350	–	220
240	–	386	–	418	–	253
300	–	441	–	488	–	286

NOTES:

1. Where it is intended to connect the cables in this table to equipment or accessories designed to operate at a temperature lower than the maximum operating temperature of the cable, the cables should be rated at the maximum operating temperature of the equipment or accessory (see Regulation 512.1.5).

2. Where it is intended to group a cable in this table with other cables, the cable should be rated at the lowest of the maximum operating temperatures of any of the cables in the group (see Regulation 512.1.5).

TABLE 4J4B

VOLTAGE DROP (per ampere per metre): Conductor operating temperature: 90 °C

Conductor cross-sectional area	Two-core cable, d.c.	Two-core cable, single-phase a.c.			Three- or four-core cable, three-phase a.c.		
1	2	3			4		
(mm²)	(mV/A/m)	(mV/A/m)			(mV/A/m)		
16	4.8	4.8			4.2		
		r	x	z	r	x	z
25	3.1	3.1	0.165	3.1	2.7	0.140	2.7
35	2.2	2.2	0.160	2.2	1.90	0.140	1.95
50	1.60	1.65	0.160	1.65	1.40	0.135	1.45
70	1.10	1.10	0.155	1.15	0.96	0.135	0.97
95	0.82	0.82	0.150	0.84	0.71	0.130	0.72
120	-	-	-	-	0.56	0.130	0.58
150	-	-	-	-	0.45	0.130	0.47
185	-	-	-	-	0.37	0.130	0.39
240	-	-	-	-	0.28	0.125	0.31
300	-	-	-	-	0.23	0.125	0.26

APPENDIX 5 (Informative)

CLASSIFICATION OF EXTERNAL INFLUENCES

This appendix gives the classification and codification of external influences.

NOTE: The appendix is an extract from HD 60364-5-51.

Each condition of external influence is designated by a code comprising a group of two capital letters and a number, as follows:

The first letter relates to the general category of external influence:

A   Environment

B   Utilisation

C   Construction of buildings

The second letter relates to the nature of the external influence:

...   **A**

...   **B**

...   **C**

The number relates to the class within each external influence:

...   ...   1

...   ...   2

...   ...   3

For example, the code **AA4** signifies:

A   =   Environment

AA   =   Environment - Ambient temperature

AA4   =   Environment - Ambient temperature in the range of -5 °C to +40 °C.

NOTE: The codification given in this appendix is not intended to be used for marking equipment.

CONCISE LIST OF EXTERNAL INFLUENCES

A — Environment

AA *Ambient (°C)*

AA1	-60 °C	+5 °C
AA2	-40 °C	+5 °C
AA3	-25 °C	+5 °C
AA4	-5 °C	+40 °C
AA5	+5 °C	+40 °C
AA6	+5 °C	+60 °C
AA7	-25 °C	+55 °C
AA8	-50 °C	+40 °C

AB *Temperature and humidity*

AC *Altitude (metres)*
- AC1 ≤ 2000 metres
- AC2 > 2000 metres

AD *Water*
- AD1 Negligible
- AD2 Drops
- AD3 Sprays
- AD4 Splashes
- AD5 Jets
- AD6 Waves
- AD7 Immersion
- AD8 Submersion

AE *Foreign bodies*
- AE1 Negligible
- AE2 Small
- AE3 Very small
- AE4 Light dust
- AE5 Moderate dust
- AE6 Heavy dust

AF *Corrosion*
- AF1 Negligible
- AF2 Atmospheric
- AF3 Intermittent
- AF4 Continuous

AG *Impact*
- AG1 Low
- AG2 Medium
- AG3 High

AH *Vibration*
- AH1 Low
- AH2 Medium
- AH3 High

AJ *Other mechanical stresses*

AK *Flora*
- AK1 No hazard
- AK2 Hazard

AL *Fauna*
- AL1 No hazard
- AL2 Hazard

AM *Electromagnetic …*
- AM1 Level
- AM2 Signalling voltages
- AM3 Voltage amplitude variations
- AM4 Voltage unbalance
- AM5 Power frequency variations
- AM6 Induced low-frequency voltage
- AM7 DC current in AC network
- AM8 Radiated magnetic fields
- AM9 Electric fields
- AM21 High-frequency etc…
- AM22 Conducted…nano…
- AM23 Conducted…micro…
- AM24 Conducted oscillatory…
- AM25 Radiated HF
- AM31 Electrostatic discharges
- AM41 Ionization

AN *Solar*
- AN1 Low
- AN2 Medium
- AN3 High

AP *Seismic*
- AP1 Negligible
- AP2 Low
- AP3 Medium
- AP4 High

AQ *Lightning*
- AQ1 Negligible
- AQ2 Indirect
- AQ3 Direct

AR *Movement of air*
- AR1 Low
- AR2 Medium
- AR3 High

AS *Wind*
- AS1 Low
- AS2 Medium
- AS3 High

B — Utilisation

BA *Capability*
- BA1 Ordinary
- BA2 Children
- BA3 Handicapped
- BA4 Instructed
- BA5 Skilled

BB *Resistance*

BC *Contact with Earth*
- BC1 None
- BC2 Low
- BC3 Frequent
- BC4 Continuous

BD *Evacuation*
- BD1 Normal
- BD2 Difficult
- BD3 Crowded
- BD4 Difficult and crowded

BE *Materials*
- BE1 No risk
- BE2 Fire risk
- BE3 Explosion risk
- BE4 Contamination risk

C — Buildings

CA *Materials*
- CA1 Non-combustible
- CA2 Combustible

CB *Structure*
- CB1 Negligible
- CB2 Fire propagation
- CB3 Structural movement
- CB4 Flexible

A ENVIRONMENT:

Code	External influences	Characteristics required for selection and erection of equipment	Reference for information only
A	*Environmental conditions*		
AA	*Ambient temperature* The ambient temperature is that of the ambient air where the equipment is to be installed It is assumed that the ambient temperature includes the effects of other equipment installed in the same location The ambient temperature to be considered for the equipment is the temperature at the place where the equipment is to be installed resulting from the influence of all other equipment in the same location, when operating, not taking into account the thermal contribution of the equipment to be installed Lower and upper limits of ranges of ambient temperature:		
AA1	−60 °C +5 °C	Specially designed equipment or appropriate arrangements[a]	Includes temperature range of BS EN 60721-3-3, class 3K8, with high air temperature restricted to +5 °C. Part of temperature range of BS EN 60721-3-4, class 4K4, with low air temperature restricted to −60 °C and high air temperature restricted to +5 °C
AA2	−40 °C +5 °C		Part of temperature range of BS EN 60721-3-3, class 3K7, with high air temperature restricted to +5 °C. Includes part of temperature range of BS EN 60721-3-4, class 4K3, with high air temperature restricted to +5 °C
AA3	−25 °C +5 °C		Part of temperature range of BS EN 60721-3-3, class 3K6, with high air temperature restricted to +5 °C. Includes temperature range of BS EN 60721-3-4, class 4K1, with high air temperature restricted to +5 °C
AA4	−5 °C +40 °C	Normal (in certain cases special precautions may be necessary)	Part of temperature range of BS EN 60721-3-3, class 3K5, with high air temperature restricted to +40 °C
AA5	+5°C +40 °C	Normal	Identical to temperature range of BS EN 60721-3-3, class 3K3

[a] May necessitate certain supplementary precautions (e.g. special lubrication).

[b] This means that ordinary equipment will operate safely under the described external influences.

[c] This means that special arrangements should be made, for example, between the designer of the installation and the equipment manufacturer, e.g. for specially designed equipment.

A ENVIRONMENT *(cont.)*:

Code	External influences			Characteristics required for selection and erection of equipment	Reference for information only
AA6	+5 °C +60 °C			Specially designed equipment or appropriate arrangements[a]	Part of temperature range of BS EN 60721-3-3, class 3K7, with low air temperature restricted to +5 °C and high air temperature restricted to +60 °C. Includes temperature range of BS EN 60721-3-4, class 4K4 with low air temperature restricted to +5 °C
AA7	−25 °C +55 °C			Specially designed equipment or appropriate arrangements[a]	– Identical to temperature range of BS EN 60721-3-3, class 3K6
AA8	−50 °C +40 °C				– Identical to temperature range of BS EN 60721-3-4, class 4K3
	Ambient temperature classes are applicable only where humidity has no influence The average temperature over a 24 h period must not exceed 5 °C below the upper limits Combination of two ranges to define some environments may be necessary. Installations subject to temperatures outside the ranges require special consideration				

Code	*Atmospheric humidity*				
AB	Air temperature °C a) low b) high	Relative humidity % c) low d) high	Absolute humidity g/m³ e) low f) high		
AB1	−60 +5	3 100	0.003 7	Indoor and outdoor locations with extremely low ambient temperatures Appropriate arrangements should be made[c]	Includes temperature range of BS EN 60721-3-3, class 3K8, with high air temperature restricted to +5 °C. Part of temperature range of BS EN 60721-3-4, class 4K4, with low air temperature restricted to −60 °C and high air temperature restricted to +5 °C
AB2	−40 +5	10 100	0.1 7	Indoor and outdoor locations with low ambient temperatures Appropriate arrangements should be made[c]	Part of temperature range of BS EN 60721-3-3, class 3K7, with high temperature restricted to +5 °C. Part of temperature range of BS EN 60721-3-4, class 4K4, with low air temperature restricted to −40 °C and high air temperature restricted to +5 °C

[a]   May necessitate certain supplementary precautions (e.g. special lubrication).

[b]   This means that ordinary equipment will operate safely under the described external influences.

[c]   This means that special arrangements should be made, for example, between the designer of the installation and the equipment manufacturer, e.g. for specially designed equipment.

A ENVIRONMENT *(cont.)*:

Code	External influences			Characteristics required for selection and erection of equipment	Reference for information only
	Air temperature °C a) low b) high	Relative humidity % c) low d) high	Absolute humidity g/m^3 e) low f) high		
AB3	−25　　+5	10　　100	0.5　　7	Indoor and outdoor locations with low ambient temperatures. Appropriate arrangements should be made[c]	Part of temperature range of BS EN 60721-3-3, class 3K6, with high air temperature restricted to +5 °C. Includes temperature range of BS EN 60721-3-4, class 4K1, with high air temperature range restricted to +5 °C
AB4	−5　　+40	5　　95	1　　29	Weather protected locations having neither temperature nor humidity control. Heating may be used to raise low ambient temperatures. Normal[b]	Identical with temperature range of BS EN 60721-3-3, class 3K5. The high air temperature restricted to +40 °C
AB5	+5　　+40	5　　85	1　　25	Weather protected locations with temperature control. Normal[b]	Identical with temperature range of BS EN 60721-3-3, class 3K3
AB6	+5　　+60	10　　100	1　　35	Indoor and outdoor locations with extremely high ambient temperatures, influence of cold ambient temperatures is prevented. Occurrence of solar and heat radiation. Appropriate arrangements should be made[c]	Part of temperature range of BS EN 60721-3-3, class 3K7, with low air temperature restricted to +5 °C and high air temperature restricted to +60 °C. Includes temperature range of BS EN 60721-3-4, class 4K4, with low air temperature restricted to +5 °C
AB7	−25　　+55	10　　100	0.5　　29	Indoor weather protected locations having neither temperature nor humidity control, the locations may have openings directly to the open air and be subjected to solar radiation. Appropriate arrangements must be made[c]	Identical to temperature range of BS EN 60721-3-3, class 3K6
AB8	−50　　+40	15　　100	0.04　　36	Outdoor and non-weather protected locations, with low and high temperatures. Appropriate arrangements should be made[c]	Identical to temperature range of BS EN 60721-3-4, class 4K3

[a]　May necessitate certain supplementary precautions (e.g. special lubrication).

[b]　This means that ordinary equipment will operate safely under the described external influences.

[c]　This means that special arrangements should be made, for example, between the designer of the installation and the equipment manufacturer, e.g. for specially designed equipment.

A ENVIRONMENT *(cont.)*:

Code	External influences	Characteristics required for selection and erection of equipment	Reference for information only
AC AC1 AC2	*Altitude* ≤2 000 m >2 000 m	Normal[b] May necessitate special precautions such as the application of derating factors **NOTE** For some equipment special arrangements may be necessary at altitudes of 1 000 m and above	
AD AD1	*Presence of water* Negligible	IPX0 Outdoor and non-weather protected locations, with low and high temperatures	BS EN 60721-3-4 class 4Z6
AD2	Free-falling drops	IPX1 or IPX2 Location in which water vapour occasionally condenses as drops or where steam may occasionally be present	BS EN 60721-3-3 class 3Z7
AD3	Sprays	IPX3 Locations in which sprayed water forms a continuous film on floors and/or walls	BS EN 60721-3-3 class 3Z8 BS EN 60721-3-4 class 4Z7
AD4	Splashes	IPX4 Locations where equipment may be subjected to splashed water; this applies, for example, to certain external luminaires, construction site equipment	BS EN 60721-3-3 class 3Z9 BS EN 60721-3-4 class 4Z7
AD5	Jets	IPX5 Locations where hose water is used regularly (yards, car-washing bays)	BS EN 60721-3-3 class 3Z10 BS EN 60721-3-4 class 4Z8
AD6	Waves	IPX6 Seashore locations such as piers, beaches, quays, etc	BS EN 60721-3-4 class 4Z9
AD7	Immersion	IPX7 Locations which may be flooded and/or where water may be at maximum 150 mm above the highest point of equipment, the lowest part of equipment being not more than 1 m below the water surface	
AD8	Submersion	IPX8 Locations such as swimming pools where electrical equipment is permanently and totally covered with water under a pressure greater than 0.1 bar	

[a] May necessitate certain supplementary precautions (e.g. special lubrication).

[b] This means that ordinary equipment will operate safely under the described external influences.

[c] This means that special arrangements should be made, for example, between the designer of the installation and the equipment manufacturer, e.g. for specially designed equipment.

A ENVIRONMENT *(cont.)*:

Code	External influences	Characteristics required for selection and erection of equipment	Reference for information only
AE	*Presence of foreign solid bodies*	IPXX see also Section 416	
AE1	Negligible	IP0X	BS EN 60721-3-3, class 3S1 BS EN 60721-3-4, class 4S1
AE2	Small objects (2.5 mm)	IP3X Tools and small objects are examples of foreign solid bodies of which the smallest dimension is at least 2.5 mm	BS EN 60721-3-3, class 3S2 BS EN 60721-3-4, class 4S2
AE3	Very small objects (1 mm)	IP4X Wires are examples of foreign solid bodies of which the smallest dimension is not less than 1 mm	BS EN 60721-3-3, class 3S3 BS EN 60721-3-4, class 4S3
AE4	Light dust	IP5X if dust penetration is not harmful to the functioning of the equipment IP6X if dust should not penetrate equipment	BS EN 60721-3-3, class 3S2 BS EN 60721-3-4, class 4S2 BS EN 60721-3-3, class 3S3 BS EN 60721-3-4, class 4S3
AE5	Moderate dust		
AE6	Heavy dust	IP6X	BS EN 60721-3-3, class 3S4 BS EN 60721-3-4, class 4S4
AF	*Presence of corrosive of polluting substances*		
AF1	Negligible	Normal[b]	BS EN 60721-3-3, class 3C1 BS EN 60721-3-4, class 4C1
AF2	Atmospheric	According to the nature of substances (for example, satisfaction of salt mist test according to BS EN 60068-2-11) Installations situated by the sea or near industrial zones producing serious atmospheric pollution, such as chemical works, cement works; this type of pollution arises especially in the production of abrasive, insulating or conductive dusts	BS EN 60721-3-3, class 3C2 BS EN 60721-3-4, class 4C2
AF3	Intermittent or accidental	Protection against corrosion according to equipment specification Locations where some chemical products are handled in small quantities and where these products may come only accidentally into contact with electrical equipment; such conditions are found in factory laboratories, other laboratories or in locations where hydrocarbons are used (boiler-rooms, garages, etc.)	BS EN 60721-3-3, class 3C3 BS EN 60721-3-4, class 4C3
AF4	Continuous	Equipment specially designed according to the nature of substances For example, chemical works	BS EN 60721-3-3, class 3C4 BS EN 60721-3-4, class 4C4
AG	*Mechanical stress* *Impact*		
AG1	Low severity	Normal, e.g. household and similar equipment	BS EN 60721-3-3, classes 3M1/3M2/3M3 BS EN 60721-3-4, classes 4M1/4M2/4M3
AG2	Medium severity	Standard industrial equipment, where applicable, or reinforced protection	BS EN 60721-3-3, classes 3M4/3M5/3M6 BS EN 60721-3-4, classes 4M4/4M5/4M6
AG3	High severity	Reinforced protection	BS EN 60721-3-3, classes 3M7/3M8 BS EN 60721-3-4, classes 4M7/4M8

[a] May necessitate certain supplementary precautions (e.g. special lubrication).

[b] This means that ordinary equipment will operate safely under the described external influences.

[c] This means that special arrangements should be made, for example, between the designer of the installation and the equipment manufacturer, e.g. for specially designed equipment.

A ENVIRONMENT *(cont.)*:

Code	External influences	Characteristics required for selection and erection of equipment	Reference for information only
AH AH1	*Vibration* Low severity	Normal[b] Household and similar conditions where the effects of vibration are generally negligible	BS EN 60721-3-3, classes 3M1/3M/3M3 BS EN 60721-3-4, classes 4M1/4M2/4M3
AH2	Medium severity	Usual industrial conditions Specially designed equipment or special arrangements	BS EN 60721-3-3, classes 3M4/3M5/3M6 BS EN 60721-3-4, classes 4M4/4M5/4M6
AH3	High severity	Industrial installations subject to severe conditions	BS EN 60721-3-3, classes 3M7/3M8 BS EN 60721-3-4, classes 4M7/4M8
AJ	*Other mechanical stresses*	Under consideration	
AK AK1	*Presence of flora and/or mould growth* No hazard	Normal[b]	BS EN 60721-3-3, class 3B1 BS EN 60721-3-4, class 4B1
AK2	Hazard	The hazard depends on local conditions and the nature of flora. Distinction should be made between harmful growth of vegetation or conditions for promotion of mould growth Special protection, such as: – increased degree of protection (see AE) – special materials or protective coating of enclosures – arrangements to exclude flora from location	BS EN 60721-3-3, class 3B2 BS EN 60721-3-4, class 4B2
AL AL1	*Presence of fauna* No hazard	Normal[b]	BS EN 60721-3-3, class 3B1 BS EN 60721-3-4, class 4B1
AL2	Hazard	The hazard depends on the nature of the fauna Distinction should be made between: – presence of insects in harmful quantity or of an aggressive nature – presence of small animals or birds in harmful quantity or of an aggressive nature Protection may include: – an appropriate degree of protection against penetration of foreign solid bodies (see AE) – sufficient mechanical resistance (see AG) – precautions to exclude fauna from the location (such as cleanliness, use of pesticides) – special equipment or protective coating of enclosures	BS EN 60721-3-3, class 3B2 BS EN 60721-3-4, class 4B2

[a] May necessitate certain supplementary precautions (e.g. special lubrication).

[b] This means that ordinary equipment will operate safely under the described external influences.

[c] This means that special arrangements should be made, for example, between the designer of the installation and the equipment manufacturer, e.g. for specially designed equipment.

A ENVIRONMENT *(cont.)*:

Code	External influences	Characteristics required for selection and erection of equipment	Reference for information only
AM	*Electromagnetic, electrostatic, or ionizing influences* *Low-frequency electromagnetic phenomena (conducted or radiated)* *Harmonics, interharmonics*		BS EN 61000-2 series and BS EN 61000-4 series
AM-1-1	Controlled level	Care should be taken that the controlled situation is not impaired	Lower than table 1 of BS EN 61000-2-2
AM-1-2	Normal level	Special measures in the design of the installation, e.g. filters	Complying with table 1 of BS EN 61000-2-2
AM-1-3	High level		Locally higher than table 1 of BS EN 61000-2-2
	Signalling voltages		
AM-2-1	Controlled level	Possibly: blocking circuits	Lower than specified below
AM-2-2	Medium level	No additional requirement	BS EN 61000-2-1 and BS EN 61000-2-2
AM-2-3	High level	Appropriate measures	
	Voltage amplitude variations		
AM-3-1	Controlled level		
AM-3-2	Normal level	Compliance with BS 7671 Chapter 44	
AM-4	*Voltage unbalance*		Compliance with BS EN 61000-2-2
AM-5	*Power frequency variations*		±1 Hz according to BS EN 61000-2-2
AM-6	*Induced low-frequency voltages* No classification	Refer to BS 7671 Chapter 44 High withstand of signal and control systems of switchgear and controlgear	ITU-T
AM-7	*Direct current in a.c. networks (321.10.1.7)* No classification	Measures to limit their presence in level and time in the current-using equipment or their vicinity	
	Radiated magnetic fields		
AM-8-1	Medium level	Normal[b]	Level 2 of BS EN 61000-4-8
AM-8-2	High level	Protection by appropriate measures e.g. screening and/or separation	Level 4 of BS EN 61000-4-8

[a]   May necessitate certain supplementary precautions (e.g. special lubrication).

[b]   This means that ordinary equipment will operate safely under the described external influences.

[c]   This means that special arrangements should be made, for example, between the designer of the installation and the equipment manufacturer, e.g. for specially designed equipment.

A ENVIRONMENT *(cont.)*:

Code	External influences	Characteristics required for selection and erection of equipment	Reference for information only
	Electric fields		
AM-9-1	Negligible level	Normal[b]	
AM-9-2	Medium level	Refer to BS EN 61000-2-5	BS EN 61000-2-5
AM-9-3	High level	Refer to BS EN 61000-2-5	
AM-9-4	Very high level	Refer to BS EN 61000-2-5	
	High-frequency electromagnetic phenomena conducted, induced or radiated (continuous or transient)		
	Induced oscillatory voltages or currents		
AM-21	No classification	Normal[b]	BS EN 61000-4-6
	Conducted unidirectional transients of the nanosecond time scale		BS EN 61000-4-4
AM-22-1	Negligible level	Protective measures are necessary	Level 1
AM-22-2	Medium level	Protective measures are necessary (see 321.10.2.2)	Level 2
AM-22-3	High level	Normal equipment	Level 3
AM-22-4	Very high level	High immunity equipment	Level 4
	Conducted unidirectional transients of microsecond to millisecond time scale		
AM-23-1	Controlled level	Impulse withstand of equipment and overvoltage protective means chosen taking into account the nominal supply voltage and the impulse withstand category according to BS 7671 Chapter 44	BS 7671 Chapter 44
AM-23-2	Medium level		
AM-23-3	High level		BS 7671 Chapter 44
	Conducted oscillatory transients		
AM-24-1	Medium level	Refer to BS EN 61000-4-12	BS EN 61000-4-12
AM-24-2	High level	Refer to BS EN 60255-22-1	BS EN 60255-22-1
	Radiated high-frequency phenomena		BS EN 61000-4-3
AM-25-1	Negligible level		Level 1
AM-25-2	Medium level	Normal[b]	Level 2
AM-25-3	High level	Reinforced level	Level 3
	Electrostatic discharges		BS EN 61000-4-2
AM-31-1	Small level	Normal[b]	Level 1
AM-31-2	Medium level	Normal[b]	Level 2
AM-31-3	High level	Normal[b]	Level 3
AM-31-4	Very high level	Reinforced	Level 4
AM-41-1	*Ionization* No classification	Special protection such as: – Spacings from source – Interposition of screens, enclosure by special materials	

[a] May necessitate certain supplementary precautions (e.g. special lubrication).

[b] This means that ordinary equipment will operate safely under the described external influences.

[c] This means that special arrangements should be made, for example, between the designer of the installation and the equipment manufacturer, e.g. for specially designed equipment.

A ENVIRONMENT *(cont.)*:

Code	External influences	Characteristics required for selection and erection of equipment	Reference for information only
AN	*Solar radiation*		
AN1	Low	Normal[b]	BS EN 60721-3-3
AN2	Medium	Appropriate arrangements must be made[c]	BS EN 60721-3-3
AN3	High	Appropriate arrangements must be made[c] Such arrangements could be: – material resistant to ultraviolet radiation – special colour coating – interposition of screens	BS EN 60721-3-4
AP	*Seismic effects*		
AP1	Negligible	Normal	
AP2	Low severity	Under consideration	
AP3	Medium severity		
AP4	High severity	Vibration which may cause the destruction of the building is outside the classification Frequency is not taken into account in the classification; however, if the seismic wave resonates with the building, seismic effects must be specially considered. In general, the frequency of seismic acceleration is between 0 Hz and 10 Hz	
AQ	*Lightning*		
AQ1	Negligible	Normal	
AQ2	Indirect exposure	In accordance with Section 443 Installations supplied by overhead lines	
AQ3	Direct exposure	If lightning protection is necessary it should be arranged according to BS EN 62305-1 Parts of installations located outside buildings The risks AQ2 and AQ3 relate to regions with a particularly high level of thunderstorm activity	
AR	*Movement of air*		
AR1	Low	Normal[b]	
AR2	Medium	Appropriate arrangements should be made[c]	
AR3	High	Appropriate arrangements should be made[c]	
AS	*Wind*		
AS1	Low	Normal[b]	
AS2	Medium	Appropriate arrangements should be made[c]	
AS3	High	Appropriate arrangements should be made[c]	

[a] May necessitate certain supplementary precautions (e.g. special lubrication).

[b] This means that ordinary equipment will operate safely under the described external influences.

[c] This means that special arrangements should be made, for example, between the designer of the installation and the equipment manufacturer, e.g. for specially designed equipment.

B UTILISATION:

Code	External influences	Characteristics required for selection and erection of equipment	Reference for information only
BA	*Capability of persons*		
BA1	Ordinary	Normal[b]	inaccessibility of electrical equipment. Limitation of temperature of accessible surfaces
BA2	Children	Equipment of degrees of protection higher than IP2X. Inaccessibility of equipment with external surface temperature exceeding 80 °C (60 °C for nurseries and the like)	
BA3	Handicapped	According to the nature of the handicap	
BA4	Instructed	Equipment not protected against direct contact admitted solely in locations which are accessible only to duly authorized persons	
BA5	Skilled		
BB	*Electrical resistance of the human body* — Under construction		
BC	*Contact of persons with Earth potential*		
		Class of equipment according to BS EN 61140	
		0-0I I II III	
BC1	None	A Y A A	
BC2	Low	A A A A	413.3 of HD 60364-4-41
BC3	Frequent	X A A A	
BC4	Continuous	Under construction	
		A = equipment permitted X = equipment prohibited Y = permitted if used as Class 0	
BD	*Conditions of evacuation in an emergency*		
BD1	Low density / easy exit	Normal[b]	
BD2	Low density / difficult exit	Equipment made of material retarding the spread of flame and evolution of smoke and toxic gases. Detailed requirements are under consideration	
BD3	High density / easy exit		
BD4	High density / difficult exit		
BE	*Nature of processed or stored materials*		
BE1	No significant risk	Normal[b]	
BE2	Fire risks	Equipment made of material retarding the spread of flame. Arrangements such that a significant temperature rise or a spark within electrical equipment cannot initiate an external fire	HD 60364-4-42 HD 60364-5-51
BE3	Explosion risks	Barns, woodworking shops, paper factories	
		Requirements for electrical apparatus for explosive atmospheres (see BS EN 60079),	
BE4	Contamination risks	Oil refineries, hydrocarbon stores	
		Appropriate arrangements, such as: protection against falling debris from broken lamps and other fragile objects screens against harmful radiation such as infrared or ultraviolet	
		Foodstuff industries, kitchens certain precautions may be necessary, in the event of fault, to prevent processed materials being contaminated by electrical equipment, e.g. by broken lamps	

[a]   May necessitate certain supplementary precautions (e.g. special lubrication).

[b]   This means that ordinary equipment will operate safely under the described external influences.

[c]   This means that special arrangements should be made, for example, between the designer of the installation and the equipment manufacturer, e.g. for specially designed equipment.

C CONSTRUCTION OF BUILDINGS:

Code	External influences	Characteristics required for selection and erection of equipment	Reference for information only
CA	*Construction materials*		
CA1	Non-combustible	Normal[b]	HD 60364-4-42
CA2	Combustible	Under construction Wooden buildings	
CB	*Building design*		
CB1	Negligible risks	Normal[b]	
CB2	Propagation of fire	Equipment made of material retarding the propagation of fire including fires not originating from the electrical installation. Fire barriers **NOTE:** Fire detectors may be provided High-rise buildings Forced ventilation systems	HD 60364-4-42 HD 60364-5-52
CB3	Movement	Contraction or expansion joints in electrical wiring Buildings of considerable length or erected on unstable ground	Contraction or expansion joints HD 60364-5-52
CB4	Flexible or unstable	Under consideration Tents, air-support structures, false ceilings, removable partitions. Installations to be structurally self-supporting	Flexible wiring HD 60364-5-52

[a]   May necessitate certain supplementary precautions (e.g. special lubrication).

[b]   This means that ordinary equipment will operate safely under the described external influences.

[c]   This means that special arrangements should be made, for example, between the designer of the installation and the equipment manufacturer, e.g. for specially designed equipment.

APPENDIX 6 (Informative)

MODEL FORMS FOR CERTIFICATION AND REPORTING

Introduction

(i)   The Electrical Installation Certificate required by Part 6 should be made out and signed or otherwise authenticated by a competent person or persons in respect of the design, construction, inspection and testing of the work.

(ii)   The Minor Works Certificate required by Part 6 should be made out and signed or otherwise authenticated by a competent person in respect of the design, construction, inspection and testing of the minor work.

(iii)   The Electrical Installation Condition Report required by Part 6 should be made out and signed or otherwise authenticated by a competent person in respect of the inspection and testing of an installation.

(iv)   Competent persons will, as appropriate to their function under (i) (ii) and (iii) above, have a sound knowledge and experience relevant to the nature of the work undertaken and to the technical standards set down in these Regulations, be fully versed in the inspection and testing procedures contained in these Regulations and employ adequate testing equipment.

(v)   Electrical Installation Certificates will indicate the responsibility for design, construction, inspection and testing, whether in relation to new work or further work on an existing installation.

Where design, construction, inspection and testing are the responsibility of one person a Certificate with a single-signature declaration in the form shown below may replace the multiple signatures section of the model form.

FOR DESIGN, CONSTRUCTION, INSPECTION & TESTING

I being the person responsible for the Design, Construction, Inspection & Testing of the electrical installation (as indicated by my signature below), particulars of which are described above, having exercised reasonable skill and care when carrying out the Design, Construction, Inspection & Testing, hereby CERTIFY that the said work for which I have been responsible is to the best of my knowledge and belief in accordance with BS 7671:2008, amended to ....(date) except for the departures, if any, detailed as follows.

(vi)   A Minor Works Certificate will indicate the responsibility for design, construction, inspection and testing of the work described on the certificate.

(vii)   An Electrical Installation Condition Report will indicate the responsibility for the inspection and testing of an existing installation within the extent and limitations specified on the report.

(viii) Schedules of inspection and schedules of test results as required by Part 6 should be issued with the associated Electrical Installation Certificate or Electrical Installation Condition Report.

(ix)   When making out and signing a form on behalf of a company or other business entity, individuals should state for whom they are acting.

(x)   Additional forms may be required as clarification, if needed by ordinary persons, or in expansion, for larger or more complex installations.

(xi)   The IET Guidance Note 3 provides further information on inspection and testing and for periodic inspection, testing and reporting.

ELECTRICAL INSTALLATION CERTIFICATE
(REQUIREMENTS FOR ELECTRICAL INSTALLATIONS - BS 7671 [IET WIRING REGULATIONS])

DETAILS OF THE CLIENT

....

INSTALLATION ADDRESS

....

....

DESCRIPTION AND EXTENT OF THE INSTALLATION Tick boxes as appropriate

Description of installation:	New installation ☐
Extent of installation covered by this Certificate:	Addition to an existing installation ☐
	Alteration to an existing installation ☐
(Use continuation sheet if necessary) see continuation sheet No:	

FOR DESIGN

I/We being the person(s) responsible for the design of the electrical installation (as indicated by my/our signatures below), particulars of which are described above, having exercised reasonable skill and care when carrying out the design hereby CERTIFY that the design work for which I/we have been responsible is to the best of my/our knowledge and belief in accordance with BS 7671:2008, amended to .... (date) except for the departures, if any, detailed as follows:

Details of departures from BS 7671 (Regulations 120.3 and 133.5):

The extent of liability of the signatory or the signatories is limited to the work described above as the subject of this Certificate.

For the DESIGN of the installation: **(Where there is mutual responsibility for the design)

Signature: ....   Date: ....   Name (IN BLOCK LETTERS): .... Designer No 1

Signature: ....   Date: ....   Name (IN BLOCK LETTERS): .... Designer No 2**

FOR CONSTRUCTION

I/We being the person(s) responsible for the construction of the electrical installation (as indicated by my/our signatures below), particulars of which are described above, having exercised reasonable skill and care when carrying out the construction hereby CERTIFY that the construction work for which I/we have been responsible is to the best of my/our knowledge and belief in accordance with BS 7671:2008, amended to ....(date) except for the departures, if any, detailed as follows:

Details of departures from BS 7671 (Regulations 120.3 and 133.5):

The extent of liability of the signatory is limited to the work described above as the subject of this Certificate.

For CONSTRUCTION of the installation:

Signature: ....   Date: ....   Name (IN BLOCK LETTERS): .... Constructor

FOR INSPECTION & TESTING

I/We being the person(s) responsible for the inspection & testing of the electrical installation (as indicated by my/our signatures below), particulars of which are described above, having exercised reasonable skill and care when carrying out the inspection & testing hereby CERTIFY that the work for which I/we have been responsible is to the best of my/our knowledge and belief in accordance with BS 7671:2008, amended to ....(date) except for the departures, if any, detailed as follows:

Details of departures from BS 7671 (Regulations 120.3 and 133.5):

The extent of liability of the signatory is limited to the work described above as the subject of this Certificate.

For INSPECTION AND TESTING of the installation:

Signature: ....   Date: ....   Name (IN BLOCK LETTERS): .... Inspector

NEXT INSPECTION

I/We the designer(s), recommend that this installation is further inspected and tested after an interval of not more than .... years/months.

PARTICULARS OF SIGNATORIES TO THE ELECTRICAL INSTALLATION CERTIFICATE

Designer (No 1)

Name: .... Company: ....

Address: ....

.... Postcode: .... Tel No: ....

Designer (No 2)
(if applicable)

Name: .... Company: ....

Address: ....

.... Postcode: .... Tel No: ....

Constructor

Name: .... Company: ....

Address: ....

.... Postcode: .... Tel No: ....

Inspector

Name: .... Company: ....

Address: ....

.... Postcode: .... Tel No: ....

SUPPLY CHARACTERISTICS AND EARTHING ARRANGEMENTS Tick boxes and enter details, as appropriate

Earthing arrangements	Number and Type of Live Conductors		Nature of Supply Parameters	Supply Protective Device Characteristics
TN-C ☐ TN-S ☐ TN-C-S ☐ TT ☐ IT ☐	a.c. ☐	d.c. ☐	Nominal voltage, $U/U_0^{(1)}$ V	Type: ....
	1-phase, 2-wire ☐	2-wire ☐	Nominal frequency, $f^{(1)}$ .... Hz	
	2-phase, 3-wire ☐	3-wire ☐	Prospective fault current, $I_{pf}^{(2)}$ .... kA	Rated current..............A
Other ☐ sources of supply (to be detailed on attached schedules)	3-phase, 3-wire ☐	other ☐	External loop impedance, $Z_e^{(2)}$ .... Ω	
	3-phase, 4-wire ☐		*(Note: (1) by enquiry, (2) by enquiry or by measurement)*	

PARTICULARS OF INSTALLATION REFERRED TO IN THE CERTIFICATE Tick boxes and enter details, as appropriate

Means of Earthing	Maximum Demand
Distributor's facility ☐	Maximum demand (load) .... kVA / Amps ^{Delete as appropriate}

Details of Installation Earth Electrode *(where applicable)*

Installation earth electrode ☐

Type (e.g. rod(s), tape etc)	Location	Electrode resistance to Earth
....	....	.... Ω

Main Protective Conductors

Earthing conductor: material .... csa ....mm^2 Continuity and connection verified ☐

Main protective bonding
conductors   material .... csa ....mm^2 Continuity and connection verified ☐

To incoming water and/or gas service ☐   To other elements: ....

Main Switch or Circuit-breaker

BS, Type and No. of poles ....   Current rating ....A   Voltage rating V

Location ....   Fuse rating or setting.................A

Rated residual operating current $I_{\Delta n}$ = ... mA, and operating time of .... ms (at $I_{\Delta n}$) ^(applicable only where an RCD is suitable and is used as a main circuit-breaker)

COMMENTS ON EXISTING INSTALLATION (in the case of an addition or alteration see Section 633):

....

....

....

....

SCHEDULES
The attached Schedules are part of this document and this Certificate is valid only when they are attached to it.
.... Schedules of Inspections and .... Schedules of Test Results are attached.
(Enter quantities of schedules attached).

ELECTRICAL INSTALLATION CERTIFICATE
NOTES:

1. The Electrical Installation Certificate is to be used only for the initial certification of a new installation or for an addition or alteration to an existing installation where new circuits have been introduced.

   It is not to be used for a Periodic Inspection, for which an Electrical Installation Condition Report form should be used. For an addition or alteration which does not extend to the introduction of new circuits, a Minor Electrical Installation Works Certificate may be used.

   The "original" Certificate is to be given to the person ordering the work (Regulation 632.1). A duplicate should be retained by the contractor.

2. This Certificate is only valid if accompanied by the Schedule of Inspections and the Schedule(s) of Test Results.

3. The signatures appended are those of the persons authorized by the companies executing the work of design, construction, inspection and testing respectively. A signatory authorized to certify more than one category of work should sign in each of the appropriate places.

4. The time interval recommended before the first periodic inspection must be inserted (see IET Guidance Note 3 for guidance).

5. The page numbers for each of the Schedules of Test Results should be indicated, together with the total number of sheets involved.

6. The maximum prospective value of fault current (I_{pf}) recorded should be the greater of either the prospective value of short-circuit current or the prospective value of earth fault current.

7. The proposed date for the next inspection should take into consideration the frequency and quality of maintenance that the installation can reasonably be expected to receive during its intended life, and the period should be agreed between the designer, installer and other relevant parties.

ELECTRICAL INSTALLATION CERTIFICATE
GUIDANCE FOR RECIPIENTS (to be appended to the Certificate)

This safety Certificate has been issued to confirm that the electrical installation work to which it relates has been designed, constructed, inspected and tested in accordance with British Standard 7671 (the IET Wiring Regulations).

You should have received an "original" Certificate and the contractor should have retained a duplicate. If you were the person ordering the work, but not the owner of the installation, you should pass this Certificate, or a full copy of it including the schedules, immediately to the owner.

The "original" Certificate should be retained in a safe place and be shown to any person inspecting or undertaking further work on the electrical installation in the future. If you later vacate the property, this Certificate will demonstrate to the new owner that the electrical installation complied with the requirements of British Standard 7671 at the time the Certificate was issued. The Construction (Design and Management) Regulations require that, for a project covered by those Regulations, a copy of this Certificate, together with schedules, is included in the project health and safety documentation.

For safety reasons, the electrical installation will need to be inspected at appropriate intervals by a competent person. The maximum time interval recommended before the next inspection is stated on Page 1 under "NEXT INSPECTION".

This Certificate is intended to be issued only for a new electrical installation or for new work associated with an addition or alteration to an existing installation. It should not have been issued for the inspection of an existing electrical installation. An "Electrical Installation Condition Report" should be issued for such an inspection.

MINOR ELECTRICAL INSTALLATION WORKS CERTIFICATE
(REQUIREMENTS FOR ELECTRICAL INSTALLATIONS - BS 7671 [IET WIRING REGULATIONS])
To be used only for minor electrical work which does not include the provision of a new circuit

---

PART 1:Description of minor works

1. Description of the minor works

2. Location/Address

3. Date minor works completed

4. Details of departures, if any, from BS 7671:2008

PART 2:Installation details

1. System earthing arrangement   TN-C-S ☐   TN-S ☐   TT ☐

2. Method of fault protection

3. Protective device for the modified circuit   Type .... Rating .... A

Comments on existing installation, including adequacy of earthing and bonding arrangements (see Regulation 132.16):

PART 3:Essential Tests
Earth continuity satisfactory   ☐

Insulation resistance:

Line/neutral .... MΩ

Line/earth .... MΩ

Neutral/earth.................................... MΩ

Earth fault loop impedance .... Ω

Polarity satisfactory   ☐

RCD operation (if applicable). Rated residual operating current $I_{\Delta n}$ ....mA and operating time of ....ms (at $I_{\Delta n}$)

PART 4:Declaration

I/We CERTIFY that the said works do not impair the safety of the existing installation, that the said works have been designed, constructed, inspected and tested in accordance with BS 7671:2008 (IET Wiring Regulations), amended to .... (date) and that the said works, to the best of my/our knowledge and belief, at the time of my/our inspection, complied with BS 7671 except as detailed in Part 1 above.

Name: ....

For and on behalf of: ....

Address: ....

....

Signature: ....

Position: ....

Date:..

MINOR ELECTRICAL INSTALLATION WORKS CERTIFICATE
NOTES:

The Minor Works Certificate is intended to be used for additions and alterations to an installation that do not extend to the provision of a new circuit. Examples include the addition of socket-outlets or lighting points to an existing circuit, the relocation of a light switch etc. This Certificate may also be used for the replacement of equipment such as accessories or luminaires, but not for the replacement of distribution boards or similar items. Appropriate inspection and testing, however, should always be carried out irrespective of the extent of the work undertaken.

MINOR ELECTRICAL INSTALLATION WORKS CERTIFICATE
GUIDANCE FOR RECIPIENTS (to be appended to the Certificate)

This Certificate has been issued to confirm that the electrical installation work to which it relates has been designed, constructed, inspected and tested in accordance with British Standard 7671 (the IET Wiring Regulations).

You should have received an "original" Certificate and the contractor should have retained a duplicate. If you were the person ordering the work, but not the owner of the installation, you should pass this Certificate, or a copy of it, to the owner. A separate Certificate should have been received for each existing circuit on which minor works have been carried out. This Certificate is not appropriate if you requested the contractor to undertake more extensive installation work, for which you should have received an Electrical Installation Certificate.

The Certificate should be retained in a safe place and be shown to any person inspecting or undertaking further work on the electrical installation in the future. If you later vacate the property, this Certificate will demonstrate to the new owner that the minor electrical installation work carried out complied with the requirements of British Standard 7671 at the time the Certificate was issued.

SCHEDULE OF INSPECTIONS (for new installation work only)

Methods of protection against electric shock

Both basic and fault protection:

- [ ] (i) SELV
- [ ] (ii) PELV
- [ ] (iii) Double insulation
- [ ] (iv) Reinforced insulation

Basic protection:

- [ ] (i) Insulation of live parts
- [ ] (ii) Barriers or enclosures
- [ ] (iii) Obstacles
- [ ] (iv) Placing out of reach

Fault protection:

(i) Automatic disconnection of supply:

- [ ] Presence of earthing conductor
- [ ] Presence of circuit protective conductors
- [ ] Presence of protective bonding conductors
- [ ] Presence of supplementary bonding conductors
- [ ] Presence of earthing arrangements for combined protective and functional purposes
- [ ] Presence of adequate arrangements for other sources, where applicable
- [ ] FELV
- [ ] Choice and setting of protective and monitoring devices (for fault and/or overcurrent protection)

(ii) Non-conducting location:

- [ ] Absence of protective conductors

(iii) Earth-free local equipotential bonding:

- [ ] Presence of earth-free local equipotential bonding

(iv) Electrical separation:

- [ ] Provided for **one item** of current-using equipment
- [ ] Provided for **more than one item** of current-using equipment

Additional protection:

- [ ] Presence of residual current devices(s)
- [ ] Presence of supplementary bonding conductors

Prevention of mutual detrimental influence

- [ ] (a) Proximity to non-electrical services and other influences
- [ ] (b) Segregation of Band I and Band II circuits or use of Band II insulation
- [ ] (c) Segregation of safety circuits

Identification

- [ ] (a) Presence of diagrams, instructions, circuit charts and similar information
- [ ] (b) Presence of danger notices and other warning notices
- [ ] (c) Labelling of protective devices, switches and terminals
- [ ] (d) Identification of conductors

Cables and conductors

- [ ] Selection of conductors for current-carrying capacity and voltage drop
- [ ] Erection methods
- [ ] Routing of cables in prescribed zones
- [ ] Cables incorporating earthed armour or sheath, or run within an earthed wiring system, or otherwise adequately protected against nails, screws and the like
- [ ] Additional protection provided by 30 mA RCD for cables concealed in walls (where required in premises not under the supervision of a skilled or instructed person)
- [ ] Connection of conductors
- [ ] Presence of fire barriers, suitable seals and protection against thermal effects

General

- [ ] Presence and correct location of appropriate devices for isolation and switching
- [ ] Adequacy of access to switchgear and other equipment
- [ ] Particular protective measures for special installations and locations
- [ ] Connection of single-pole devices for protection or switching in line conductors only
- [ ] Correct connection of accessories and equipment
- [ ] Presence of undervoltage protective devices
- [ ] Selection of equipment and protective measures appropriate to external influences
- [ ] Selection of appropriate functional switching devices

Inspected by ....

Date ....

NOTES:
✓ to indicate an inspection has been carried out and the result is satisfactory
N/A to indicate that the inspection is not applicable to a particular item
An entry must be made in every box.

ELECTRICAL INSTALLATION CONDITION REPORT

SECTION A. DETAILS OF THE CLIENT / PERSON ORDERING THE REPORT

Name ....

Address ....

....

SECTION B. REASON FOR PRODUCING THIS REPORT ....

....

Date(s) on which inspection and testing was carried out ....

SECTION C. DETAILS OF THE INSTALLATION WHICH IS THE SUBJECT OF THIS REPORT

Occupier ....

Address ....

....

Description of premises (tick as appropriate)

Domestic ☐   Commercial ☐   Industrial ☐   Other (include brief description) ☐   ....

Estimated age of wiring system ....years

Evidence of additions / alterations   Yes ☐   No ☐   Not apparent ☐   If yes, estimate age ....years

Installation records available? (Regulation 621.1)   Yes ☐   No ☐   Date of last inspection .... (date)

SECTION D. EXTENT AND LIMITATIONS OF INSPECTION AND TESTING

Extent of the electrical installation covered by this report

....

....

Agreed limitations including the reasons (see Regulation 634.2) ....

....

Agreed with: ....

Operational limitations including the reasons (see page no..............)   ....

....

The inspection and testing detailed in this report and accompanying schedules have been carried out in accordance with BS 7671: 2008 (IET Wiring Regulations) as amended to ....

It should be noted that cables concealed within trunking and conduits, under floors, in roof spaces, and generally within the fabric of the building or underground, have **not** been inspected unless specifically agreed between the client and inspector prior to the inspection.

SECTION E. SUMMARY OF THE CONDITION OF THE INSTALLATION

General condition of the installation (in terms of electrical safety) ....

....

....

Overall assessment of the installation in terms of its suitability for continued use

SATISFACTORY / UNSATISFACTORY* (Delete as appropriate)

*An unsatisfactory assessment indicates that dangerous (code C1) and/or potentially dangerous (code C2) conditions have been identified.

SECTION F. RECOMMENDATIONS

Where the overall assessment of the suitability of the installation for continued use above is stated as UNSATISFACTORY, I / we recommend that any observations classified as *'Danger present'* (code C1) or *'Potentially dangerous'* (code C2) are acted upon as a matter of urgency.

Investigation without delay is recommended for observations identified as *'further investigation required'*.

Observations classified as *'Improvement recommended'* (code C3) should be given due consideration.

Subject to the necessary remedial action being taken, I / we recommend that the installation is further inspected and tested by ....(date)

SECTION G. DECLARATION

I/We, being the person(s) responsible for the inspection and testing of the electrical installation (as indicated by my/our signatures below), particulars of which are described above, having exercised reasonable skill and care when carrying out the inspection and testing, hereby declare that the information in this report, including the observations and the attached schedules, provides an accurate assessment of the condition of the electrical installation taking into account the stated extent and limitations in section D of this report.

Inspected and tested by:	Report authorised for issue by:
Name (Capitals)	Name (Capitals) ....
Signature	Signature ....
For/on behalf of	For/on behalf of ....
Position	Position ....
Address	Address ....
Date	Date ....

SECTION H. SCHEDULE(S)

............schedule(s) of inspection and ....schedule(s) of test results are attached.

The attached schedule(s) are part of this document and this report is valid only when they are attached to it.

SECTION I. SUPPLY CHARACTERISTICS AND EARTHING ARRANGEMENTS

Earthing arrangements	Number and Type of Live Conductors		Nature of Supply Parameters	Supply Protective Device
TN-C ☐ TN-S ☐ TN-C-S ☐ TT ☐ IT ☐	a.c. ☐ 1-phase, 2-wire ☐ 2 phase, 3-wire ☐ 3 phase, 3-wire ☐ 3 phase, 4-wire ☐ Confirmation of supply polarity ☐	d.c. ☐ 2-wire ☐ 3-wire ☐	Nominal voltage, U / U_0[1] .... V Nominal frequency, f[1] .... Hz Prospective fault current, I_{pf}[2] .... kA External loop impedance, Z_e[2] ....Ω Note: (1) by enquiry (2) by enquiry or by measurement	BS (EN) .... Type .... Rated current ....A

Other sources of supply (as detailed on attached schedule) ☐

SECTION J. PARTICULARS OF INSTALLATION REFERRED TO IN THE REPORT

Means of Earthing   Details of Installation Earth Electrode *(where applicable)*

Means of Earthing	Details of Installation Earth Electrode *(where applicable)*
Distributor's facility ☐ Installation earth electrode ☐	Type .... Location .... Resistance to Earth .... Ω

Main Protective Conductors

Earthing conductor	Material	Csa ....mm^2	Connection / continuity verified ☐
Main protective bonding conductors	Material	Csa ....mm^2	Connection / continuity verified ☐

To incoming water service ☐	To incoming gas service ☐	To incoming oil service ☐	To structural steel ☐

To lightning protection ☐   To other incoming service(s) ☐   Specify ....

Main Switch / Switch-Fuse / Circuit-Breaker / RCD

Location .... BS(EN) .... No of poles	Current rating .... A Fuse / device rating or setting .... A Voltage rating V	**If RCD main switch** Rated residual operating current ($I_{\Delta n}$) ....mA Rated time delay ....ms Measured operating time(at $I_{\Delta n}$) ....ms

SECTION K. OBSERVATIONS

Referring to the attached schedules of inspection and test results, and subject to the limitations specified at the *Extent and limitations of inspection and testing* section

No remedial action is required ☐   The following observations are made ☐ (see below):

OBSERVATION(S)	CLASSIFICATION CODE	FURTHER INVESTIGATION REQUIRED (YES / NO)
....	....	....
....	....	....
....	....	....
....	....	....
....	....	....
....	....	....
....	....	....
....	....	....
....	....	....
....	....	....
....	....	....
....	....	....
....	....	....
....	....	....
....	....	....
....	....	....
....	....	....
....	....	....

One of the following codes, as appropriate, has been allocated to each of the observations made above to indicate to the person(s) responsible for the installation the degree of urgency for remedial action.

C1 – Danger present. Risk of injury. Immediate remedial action required

C2 – Potentially dangerous - urgent remedial action required

C3 – Improvement recommended

CONDITION REPORT
Notes for the person producing the Report:

1. This Report should only be used for reporting on the condition of an existing electrical installation. An installation which was designed to an earlier edition of the Regulations and which does not fully comply with the current edition is not necessarily unsafe for continued use, or requires upgrading. Only damage, deterioration, defects, dangerous conditions and non-compliance with the requirements of the Regulations, which may give rise to danger, should be recorded.

2. The Report, normally comprising at least six pages, should include schedules of both the inspection and the test results. Additional pages may be necessary for other than a simple installation. The number of each page should be indicated, together with the total number of pages involved.

3. The reason for producing this Report, such as change of occupancy or landlord's periodic maintenance, should be identified in Section B.

4. Those elements of the installation that are covered by the Report and those that are not should be identified in Section D (Extent and limitations). These aspects should have been agreed with the person ordering the report and other interested parties before the inspection and testing commenced. Any operational limitations, such as inability to gain access to parts of the installation or an item of equipment, should also be recorded in Section D.

5. The maximum prospective value of fault current (I_{pf}) recorded should be the greater of either the prospective value of short-circuit current or the prospective value of earth fault current.

6. Where an installation has an alternative source of supply a further schedule of supply characteristics and earthing arrangements based upon Section I of this Report should be provided.

7. A summary of the condition of the installation in terms of safety should be clearly stated in Section E. Observations, if any, should be categorised in Section K using the coding C1 to C3 as appropriate. Any observation given a code C1 or C2 classification should result in the overall condition of the installation being reported as unsatisfactory.

8. Wherever practicable, **items classified as 'Danger present' (C1) should be made safe on discovery**. Where this is not practical the owner or user should be given written notification as a matter of urgency.

9. Where an observation requires further investigation because the inspection has revealed an apparent deficiency which could not, owing to the extent or limitations of the inspection, be fully identified, this should be indicated in the column headed "Further investigation required" within Section K.

10. If the space available for observations in Section K is insufficient, additional pages should be provided as necessary.

11. The date by which the next Electrical Installation Condition Report is recommended should be given in Section F. The interval between inspections should take into account the type and usage of the installation and its overall condition.

CONDITION REPORT
GUIDANCE FOR RECIPIENTS
(to be appended to the Report)

This Report is an important and valuable document which should be retained for future reference.

1. The purpose of this Condition Report is to confirm, so far as reasonably practicable, whether or not the electrical installation is in a satisfactory condition for continued service (see Section E). The Report should identify any damage, deterioration, defects and/or conditions which may give rise to danger (see Section K).

2. The person ordering the Report should have received the "original" Report and the inspector should have retained a duplicate.

3. The "original" Report should be retained in a safe place and be made available to any person inspecting or undertaking work on the electrical installation in the future. If the property is vacated, this Report will provide the new owner /occupier with details of the condition of the electrical installation at the time the Report was issued.

4. Where the installation incorporates a residual current device (RCD) there should be a notice at or near the device stating that it should be tested quarterly. **For safety reasons it is important that this instruction is followed.**

5. Section D (Extent and Limitations) should identify fully the extent of the installation covered by this Report and any limitations on the inspection and testing. The inspector should have agreed these aspects with the person ordering the Report and with other interested parties (licensing authority, insurance company, mortgage provider and the like) before the inspection was carried out.

6. Some operational limitations such as inability to gain access to parts of the installation or an item of equipment may have been encountered during the inspection. The inspector should have noted these in Section D.

7. For items classified in Section K as C1 ("Danger present"), **the safety of those using the installation is at risk,** and it is recommended that a competent person undertakes the necessary remedial work immediately.

8. For items classified in Section K as C2 ("Potentially dangerous"), **the safety of those using the installation may be at risk** and it is recommended that a competent person undertakes the necessary remedial work as a matter of urgency.

9. Where it has been stated in Section K that an observation requires further investigation the inspection has revealed an apparent deficiency which could not, due to the extent or limitations of the inspection, be fully identified. Such observations should be investigated as soon as possible. A further examination of the installation will be necessary, to determine the nature and extent of the apparent deficiency (see Section F).

10 For safety reasons, the electrical installation should be re-inspected at appropriate intervals by a competent person. The recommended date by which the next inspection is due is stated in Section F of the Report under 'Recommendations' and on a label at or near to the consumer unit / distribution board.

CONDITION REPORT INSPECTION SCHEDULE
GUIDANCE FOR THE INSPECTOR

1. Section 1.0. Where inadequacies in the distributor's equipment are encountered the inspector should advise the person ordering the work to inform the appropriate authority.

2. Older installations designed prior to BS 7671:2008 may not have been provided with RCDs for additional protection. The absence of such protection should as a minimum be given a code C3 classification (item 5.12).

3. The schedule is not exhaustive.

4. Numbers in brackets are Regulation references to specified requirements.

CONDITION REPORT INSPECTION SCHEDULE FOR
DOMESTIC AND SIMILAR PREMISES WITH UP TO 100 A SUPPLY
Note: This form is suitable for many types of smaller installation not exclusively domestic.

OUTCOMES	Acceptable condition	✓	Unacceptable condition	State **C1** or **C2**	Improvement recommended	State **C3**	Not verified	**N/V**	Limitation	**LIM**	Not applicable	**N/A**

ITEM NO	DESCRIPTION	OUTCOME (Use codes above. Provide additional comment where appropriate. C1, C2 and C3 coded items to be recorded in Section K of the Condition Report)	Further investigation required? (**Y** or **N**)
1.0	**DISTRIBUTOR'S / SUPPLY INTAKE EQUIPMENT**		
1.1	Service cable condition		
1.2	Condition of service head		
1.3	Condition of tails - Distributor		
1.4	Condition of tails - Consumer		
1.5	Condition of metering equipment		
1.6	Condition of isolator (where present)		
2.0	**PRESENCE OF ADEQUATE ARRANGEMENTS FOR OTHER SOURCES SUCH AS MICROGENERATORS (551.6; 551.7)**		
3.0	**EARTHING / BONDING ARRANGEMENTS (411.3; Chap 54)**		
3.1	Presence and condition of distributor's earthing arrangement (542.1.2.1; 542.1.2.2)		
3.2	Presence and condition of earth electrode connection where applicable (542.1.2.3)		
3.3	Provision of earthing / bonding labels at all appropriate locations (514.11)		
3.4	Confirmation of earthing conductor size (542.3; 543.1.1)		
3.5	Accessibility and condition of earthing conductor at MET (543.3.2)		
3.6	Confirmation of main protective bonding conductor sizes (544.1)		
3.7	Condition and accessibility of main protective bonding conductor connections (543.3.2; 544.1.2)		
3.8	Accessibility and condition of all protective bonding connections (543.3.2)		
4.0	**CONSUMER UNIT(S) / DISTRIBUTION BOARD(S)**		
4.1	Adequacy of working space / accessibility to consumer unit / distribution board (132.12; 513.1)		
4.2	Security of fixing (134.1.1)		
4.3	Condition of enclosure(s) in terms of IP rating etc (416.2)		
4.4	Condition of enclosure(s) in terms of fire rating etc (526.5)		
4.5	Enclosure not damaged/deteriorated so as to impair safety (621.2(iii))		
4.6	Presence of main linked switch (as required by 537.1.4)		
4.7	Operation of main switch (functional check) (612.13.2)		
4.8	Manual operation of circuit-breakers and RCDs to prove disconnection (612.13.2)		
4.9	Correct identification of circuit details and protective devices (514.8.1; 514.9.1)		
4.10	Presence of RCD quarterly test notice at or near consumer unit / distribution board (514.12.2)		
4.11	Presence of non-standard (mixed) cable colour warning notice at or near consumer unit / distribution board (514.14)		
4.12	Presence of alternative supply warning notice at or near consumer unit / distribution board (514.15)		
4.13	Presence of other required labelling (please specify) (Section 514)		
4.14	Examination of protective device(s) and base(s); correct type and rating (no signs of unacceptable thermal damage, arcing or overheating) (421.1.3)		
4.15	Single-pole protective devices in line conductor only (132.14.1; 530.3.2)		
4.16	Protection against mechanical damage where cables enter consumer unit / distribution board (522.8.1; 522.8.11)		
4.17	Protection against electromagnetic effects where cables enter consumer unit / distribution board / enclosures (521.5.1)		
4.18	RCD(s) provided for fault protection – includes RCBOs (411.4.9; 411.5.2; 531.2)		
4.19	RCD(s) provided for additional protection - includes RCBOs (411.3.3; 415.1)		

OUTCOMES	Acceptable condition	✔	Unacceptable condition	State **C1** or **C2**	Improvement recommended	State **C3**	Not verified **N/V**	Limitation **LIM**	Not applicable **N/A**

ITEM NO	DESCRIPTION	OUTCOME (Use codes above. Provide additional comment where appropriate. C1, C2 and C3 coded items to be recorded in Section K of the Condition Report)	Further investigation required? (**Y** or **N**)
5.0	**FINAL CIRCUITS**		
5.1	Identification of conductors (514.3.1)		
5.2	Cables correctly supported throughout their run (522.8.5)		
5.3	Condition of insulation of live parts (416.1)		
5.4	Non-sheathed cables protected by enclosure in conduit, ducting or trunking (521.10.1)		
	▪ To include the integrity of conduit and trunking systems (metallic and plastic)		
5.5	Adequacy of cables for current-carrying capacity with regard for the type and nature of installation (Section 523)		
5.6	Coordination between conductors and overload protective devices (433.1; 533.2.1)		
5.7	Adequacy of protective devices: type and rated current for fault protection (411.3)		
5.8	Presence and adequacy of circuit protective conductors (411.3.1.1; 543.1)		
5.9	Wiring system(s) appropriate for the type and nature of the installation and external influences (Section 522)		
5.10	Concealed cables installed in prescribed zones (see Section D. *Extent and limitations*) (522.6.101)		
5.11	Concealed cables incorporating earthed armour or sheath, or run within earthed wiring system, or otherwise protected against mechanical damage from nails, screws and the like (see Section D. *Extent and limitations*) (522.6.101; 522.6.103)		
5.12	Provision of additional protection by RCD not exceeding 30 mA:		
	▪ for all socket-outlets of rating 20 A or less provided for use by ordinary persons unless an exception is permitted (411.3.3)		
	▪ for supply to mobile equipment not exceeding 32 A rating for use outdoors (411.3.3)		
	▪ for cables concealed in walls or partitions (522.6.102; 522.6.103)		
5.13	Provision of fire barriers, sealing arrangements and protection against thermal effects (Section 527)		
5.14	Band II cables segregated / separated from Band I cables (528.1)		
5.15	Cables segregated / separated from communications cabling (528.2)		
5.16	Cables segregated / separated from non-electrical services (528.3)		
5.17	Termination of cables at enclosures – indicate extent of sampling in Section D of the report (Section 526)		
	▪ Connections soundly made and under no undue strain (526.6)		
	▪ No basic insulation of a conductor visible outside enclosure (526.98)		
	▪ Connections of live conductors adequately enclosed (526.5)		
	▪ Adequately connected at point of entry to enclosure (glands, bushes etc.) (522.8.5)		
5.18	Condition of accessories including socket-outlets, switches and joint boxes (621.2 (iii))		
5.19	Suitability of accessories for external influences (512.2)		

ITEM NO	DESCRIPTION	OUTCOME	Further investigation
6.0	**LOCATION(S) CONTAINING A BATH OR SHOWER**		
6.1	Additional protection for all low voltage (LV) circuits by RCD not exceeding 30 mA (701.411.3.3)		
6.2	Where used as a protective measure, requirements for SELV or PELV met (701.414.4.5)		
6.3	Shaver sockets comply with BS EN 61558-2-5 formally BS 3535 (701.512.3)		
6.4	Presence of supplementary bonding conductors, unless not required by BS 7671:2008 (701.415.2)		
6.5	Low voltage (e.g. 230 volt) socket-outlets sited at least 3 m from zone 1 (701.512.3)		
6.6	Suitability of equipment for external influences for installed location in terms of IP rating (701.512.2)		
6.7	Suitability of equipment for installation in a particular zone (701.512.3)		
6.8	Suitability of current-using equipment for particular position within the location (701.55)		

ITEM NO	DESCRIPTION	OUTCOME	Further investigation
7.0	**OTHER PART 7 SPECIAL INSTALLATIONS OR LOCATIONS**		
7.1	List all other special installations or locations present, if any. (Record separately the results of particular inspections applied.)		

Inspected by:

Name (Capitals) .... Signature .... Date ....

GENERIC SCHEDULE OF TEST RESULTS

DB reference no ....

Details of circuits and/or installed equipment vulnerable to damage when testing ....

Location ....

Zs at DB (Ω) ....

I_{pf} at DB (kA) ....

Correct supply polarity confirmed ☐

Phase sequence confirmed (where appropriate) ☐

Details of test instruments used (state serial and/or asset numbers)

Continuity ....

Insulation resistance ....

Earth fault loop impedance ....

RCD ....

Earth electrode resistance ....

Tested by:

Name (Capitals) ....

Signature .... Date ....

Circuit details									Test results													
		Overcurrent device				Conductor details			Ring final circuit continuity (Ω)			Continuity (Ω) $(R_1 + R_2)$ or R_2		Insulation Resistance (MΩ)		Polarity	Z_s (Ω)	RCD			Remarks (continue on a separate sheet if necessary)	
Circuit Description		BS (EN)	type	rating (A)	breaking capacity (kA)	Reference Method	Live (mm²)	cpc (mm²)	r_1 (line)	r_n (neutral)	r_2 (cpc)	$(R_1 + R_2)$ *	R_2	Live-Live	Live-E			@ $I_{\Delta n}$	@ $5I_{\Delta n}$	Test button operation		
																		(ms)				
Circuit number																						
1	2	3	4	5	6	7	8	9	10	11	12	13	14	15	16	17	18	19	20	21	22	

* Where there are no spurs connected to a ring final circuit this value is also the $(R_1 + R_2)$ of the circuit.

EXAMPLES OF ITEMS REQUIRING INSPECTION
FOR AN ELECTRICAL INSTALLATION CONDITION REPORT

A visual inspection should firstly be made of the external condition of all electrical equipment which is not concealed.

Further detailed inspection, including partial dismantling of equipment as required, should be carried out as agreed with the person ordering the work. (621.2)

These examples are not exhaustive. Numbers in brackets are Regulation references.

ELECTRICAL INTAKE EQUIPMENT

- Service cable
- Service cut-out/fuse
- Meter tails – Distributor
- Meter tails – Consumer
- Metering equipment
- Isolator

Where inadequacies in distributor's equipment are encountered, it is recommended that the person ordering the report informs the appropriate authority.

PRESENCE OF ADEQUATE ARRANGEMENTS FOR PARALLEL
OR SWITCHED ALTERNATIVE SOURCES (551.6; 551.7)

AUTOMATIC DISCONNECTION OF SUPPLY

- Main earthing / bonding arrangements (411.3; Chap 54)
    1. Presence of distributor's earthing arrangement (542.1.2.1; 542.1.2.2), or presence of installation earth electrode arrangement (542.1.2.3)
    2. Adequacy of earthing conductor size (542.3; 543.1.1)
    3. Main protective earthing conductor connections (542.3.2)
    4. Accessibility of earthing conductor connections (543.3.2)
    5. Adequacy of main protective bonding conductor sizes (544.1)
    6. Main protective bonding conductor connections (543.3.2; 544.1.2)
    7. Accessibility of all protective bonding connections (543.3.2)
    8. Provision of earthing / bonding labels at all appropriate locations (514.11)
- FELV

OTHER METHODS OF PROTECTION
(Where any of the methods listed below are employed details should be provided on separate sheets)

- Non-conducting location (418.1)
- Earth-free local equipotential bonding (418.2)
- Electrical separation (Section 413; 418.3)
- Double insulation (Section 412)
- Reinforced insulation (Section 412)

DISTRIBUTION EQUIPMENT

- Adequacy of working space / accessibility to equipment (132.12; 513.1)
- Security of fixing (134.1.1)
- Condition of insulation of live parts (416.1)
- Adequacy / security of barriers (416.2)
- Condition of enclosure(s) in terms of IP rating etc (416.2)
- Condition of enclosure(s) in terms of fire rating etc (421.1.6; 526.5)
- Enclosure not damaged / deteriorated so as to impair safety (621.2(iii))
- Presence and effectiveness of obstacles (417.2)
- Placing out of reach (417.3)
- Presence of main switch(es), linked where required (537.1.2; 537.1.4)

- Operation of main switch(es) (functional check) (612.13.2)
- Manual operation of circuit-breakers and RCDs to prove disconnection (612.13..2)
- Confirmation that integral test button / switch causes RCD(s) to trip when operated (functional check) (612.13.1)
- RCD(s) provided for fault protection – includes RCBOs (414.4.9; 411.5.2; 531.2)
- RCD(s) provided for additional protection, where required - includes RCBOs (411.3.3; 415.1)
- Presence of RCD quarterly test notice at or near equipment, where required (514.12.2)
- Presence of diagrams, charts or schedules at or near equipment, where required (514.9.1)
- Presence of non-standard (mixed) cable colour warning notice at or near equipment, where required (514.14)
- Presence of alternative supply warning notice at or near equipment, where required (514.15)
- Presence of next inspection recommendation label (514.12.1)
- Presence of other required labelling (please specify) (Section 514)
- Examination of protective device(s) and base(s); correct type and rating (no signs of unacceptable thermal damage, arcing or overheating) (421.1.3)
- Single-pole protective devices in line conductor only (132.14.1; 530.3.2)
- Protection against mechanical damage where cables enter equipment (522.8.1; 522.8.11)
- Protection against electromagnetic effects where cables enter ferromagnetic enclosures (521.5.1)

DISTRIBUTION CIRCUITS

- Identification of conductors (514.3.1)
- Cables correctly supported throughout their run (522.8.5)
- Condition of insulation of live parts (416.1)
- Non-sheathed cables protected by enclosure in conduit, ducting or trunking (521.10.1)
- Suitability of containment systems for continued use (including flexible conduit) (Section 522)
- Cables correctly terminated in enclosures (Section 526)
- Examination of cables for signs of unacceptable thermal or mechanical damage / deterioration (421.1; 522.6)
- Adequacy of cables for current-carrying capacity with regard for the type and nature of installation (Section 523)
- Adequacy of protective devices: type and rated current for fault protection (411.3)
- Presence and adequacy of circuit protective conductors (411.3.1.1; 543.1)
- Coordination between conductors and overload protective devices (433.1; 533.2.1)
- Cable installation methods / practices with regard to the type and nature of installation and external influences (Section 522)
- Where exposed to direct sunlight, cable of a suitable type (522.11.1)
- Cables concealed under floors, above ceilings, in walls / partitions less than 50 mm from a surface, and in partitions containing metal parts
  1. installed in prescribed zones (see Section D. *Extent and limitations*) (522.6.101) or
  2. incorporating earthed armour or sheath, or run within earthed wiring system, or otherwise protected against mechanical damage by nails, screws and the like (see Section D. *Extent and limitations*) (522.6.101; 522.6.103)
- Provision of fire barriers, sealing arrangements and protection against thermal effects (Section 527)
- Band II cables segregated / separated from Band I cables (528.1)
- Cables segregated / separated from non-electrical services (528.3)
- Condition of circuit accessories (621.2(iii))
- Suitability of circuit accessories for external influences (512.2)
- Single-pole devices for switching in line conductor only (132.14.1; 530.3.2)
- Adequacy of connections, including cpc's, within accessories and to fixed and stationary equipment – identify / record numbers and locations of items inspected (Section 526)
- Presence, operation and correct location of appropriate devices for isolation and switching (537.2)
- General condition of wiring systems (621.2(ii))
- Temperature rating of cable insulation (522.1.1; Table 52.1)

FINAL CIRCUITS

- Identification of conductors (514.3.1)
- Cables correctly supported throughout their run (522.8.5)
- Condition of insulation of live parts (416.1)
- Non-sheathed cables protected by enclosure in conduit, ducting or trunking (521.10.1)
- Suitability of containment systems for continued use (including flexible conduit) (Section 522)
- Adequacy of cables for current-carrying capacity with regard for the type and nature of installation (Section 523)
- Adequacy of protective devices: type and rated current for fault protection (411.3)
- Presence and adequacy of circuit protective conductors (411.3.1.1; 543.1)
- Co-ordination between conductors and overload protective devices (433.1; 533.2.1)
- Wiring system(s) appropriate for the type and nature of the installation and external influences (Section 522)
- Cables concealed under floors, above ceilings, in walls / partitions less than 50 mm from a surface, and in partitions containing metal parts
    1. installed in prescribed zones (see Section D. *Extent and limitations*) (522.6.101)
    2. incorporating earthed armour or sheath, or run within earthed wiring system, or otherwise protected against mechanical damage from by nails, screws and the like (see Section D. *Extent and limitations*) (522.6.101; 522.6.103) or
    3. *for an installation not under the supervision of skilled or instructed persons, provided with additional protection by a 30 mA RCD (522.6.102; 522.6.103)
- Provision of additional protection by 30 mA RCD
    1. *for circuits used to supply mobile equipment not exceeding 32 A rating for use outdoors in all cases (411.3.3)
    2. *for all socket-outlets of rating 20 A or less provided for use by ordinary persons unless exempt (411.3.3)
- Provision of fire barriers, sealing arrangements and protection against thermal effects (Section 527)
- Band II cables segregated / separated from Band I cables (528.1)
- Cables segregated / separated from non-electrical services (528.3)
- Termination of cables at enclosures – identify / record numbers and locations of items inspected (Section 526)
    1. Connections under no undue strain (526.6)
    2. No basic insulation of a conductor visible outside enclosure (526.8)
    3. Connections of live conductors adequately enclosed (526.5)
    4. Adequately connected at point of entry to enclosure (glands, bushes etc.) (522.8.5)
- Condition of accessories including socket-outlets, switches and joint boxes (621.2 (iii))
- Suitability of accessories for external influences (512.2)

*Note: Older installations designed prior to BS 7671:2008 may not have been provided with RCDs for additional protection

ISOLATION AND SWITCHING

- Isolators (537.2)
    1. Presence and condition of appropriate devices (537.2.2)
    2. Acceptable location – state if local or remote from equipment in question (537.2.1.5)
    3. Capable of being secured in the OFF position (537.2.1.2)
    4. Correct operation verified (612.13.2)
    5. Clearly identified by position and /or durable marking (537.2.2.6)
    6. Warning label posted in situations where live parts cannot be isolated by the operation of a single device (514.11.1; 537.2.1.3)
- Switching off for mechanical maintenance (537.3)
    1. Presence and condition of appropriate devices (537.3.1.1)
    2. Acceptable location – state if local or remote from equipment in question (537.3.2.4)
    3. Capable of being secured in the OFF position (537.3.2.3)
    4. Correct operation verified (612.13.2)
    5. Clearly identified by position and /or durable marking (537.3.2.4)

- Emergency switching / stopping (537.4)
    1. Presence and condition of appropriate devices (537.4.1.1)
    2. Readily accessible for operation where danger might occur (537.4.2.5)
    3. Correct operation verified (537.4.2.6)
    4. Clearly identified by position and /or durable marking (537.4.2.7)
- Functional switching (537.5)
    1. Presence and condition of appropriate devices (537.5.1.1)
    2. Correct operation verified (537.5.1.3; 537.5.2.2)

CURRENT–USING EQUIPMENT (PERMANENTLY CONNECTED)

- Condition of equipment in terms of IP rating etc (416.2)
- Equipment does not constitute a fire hazard (Section 421)
- Enclosure not damaged/deteriorated so as to impair safety (621.2(iii))
- Suitability for the environment and external influences (512.2)
- Security of fixing (134.1.1)
- Cable entry holes in ceiling above luminaires, sized or sealed so as to restrict the spread of fire: List number and location of luminaires inspected (separate page)
- Recessed luminaires (downlighters)
    1. Correct type of lamps fitted
    2. Installed to minimise build-up of heat by use of "fire rated" fittings, insulation displacement box or similar (421.1.1)
    3. No signs of overheating to surrounding building fabric (559.5.1)
    4. No signs of overheating to conductors / terminations (526.1)

PART 7 SPECIAL INSTALLATIONS OR LOCATIONS

- If any special installations or locations are present, list the particular inspections applied.

APPENDIX 7 (Informative)

HARMONIZED CABLE CORE COLOURS

1 Introduction

The requirements of BS 7671 were harmonized with the technical intent of CENELEC Standard HD 384.5.514: *Identification*, including 514.3: *Identification of conductors*, now withdrawn.

Amendment No 2: 2004 (AMD 14905) to BS 7671:2001 implemented the following:

- the harmonized cable core colours and the alphanumeric marking of the following standards:
  HD 308 S2: 2001 *Identification of cores in cables and flexible cords*
  BS EN 60445:2000 *Basic and safety principles for man-machine interface, marking and identification of equipment terminals and of terminations*
  BS EN 60446:2000 *Basic and safety principles for the man-machine interface, marking and identification. Identification of conductors by colours or numerals.*

This appendix provides guidance on marking at the interface between old and harmonized colours and marking and general guidance on the colours to be used for conductors.

In the British Standards for fixed and flexible cables the colours have been harmonized. BS 7671 has been modified to align with these cables, but also allows other suitable methods of marking connections by colour (tapes, sleeves or discs), or by alphanumerics (letters and/or numbers). Methods may be mixed within an installation.

2 Addition or alteration to an existing installation

2.1 *Single-phase installation*

An addition or an alteration made to a single-phase installation need not be marked at the interface provided that:

   i) the old cables are correctly identified by the colour red for line and black for neutral, and

   ii) the new cables are correctly identified by the colour brown for line and blue for neutral.

2.2 *Two- or three-phase installation*

Where an addition or an alteration is made to a two- or a three-phase installation wired in the old core colours with cable to the new core colours, unambiguous identification is required at the interface. Cores should be marked as follows:

        Neutral conductors
        Old and new conductors: N

        Line conductors
        Old and new conductors: L1, L2, L3.

TABLE 7A – Example of conductor marking at the interface for additions and alterations to an a.c. installation identified with the old cable colours

Function	Old conductor		New conductor	
	Colour	Marking	Marking	Colour
Line 1 of a.c.	Red	L1	L1	Brown[1]
Line 2 of a.c.	Yellow	L2	L2	Black[1]
Line 3 of a.c.	Blue	L3	L3	Grey[1]
Neutral of a.c.	Black	N	N	Blue
Protective conductor	Green-and-yellow			Green-and-yellow

[1] Three single-core cables with insulation of the same colour may be used if identified at the terminations.

3   Switch wires in a new installation or an addition or alteration to an existing installation

Where a two-core cable with cores coloured brown and blue is used as switch wires, both conductors being line conductors, the blue conductor should be marked brown or L at its terminations.

4   Intermediate and two-way switch wires in a new installation or an addition or alteration to an existing installation

Where a three-core cable with cores coloured brown, black and grey is used as switch wires, all three conductors being line conductors, the black and grey conductors should be marked brown or L at their terminations.

5   Line conductors in a new installation or an addition or alteration to an existing installation

Power circuit line conductors should be coloured as in Table 51. Other line conductors may be brown, black, red, orange, yellow, violet, grey, white, pink or turquoise.

In a two- or three-phase power circuit the line conductors may all be of one of the permitted colours, either identified L1, L2, L3 or marked brown, black, grey at their terminations to show the phases.

6   Changes to cable core colour identification

TABLE 7B – Cable to BS 6004
(flat cable with bare cpc)

Cable type	Old core colours	New core colours
Single-core + bare cpc	Red or Black	Brown or Blue
Two-core + bare cpc	Red, Black	Brown, Blue
Alt. two-core + bare cpc	Red, Red	Brown, Brown
Three-core + bare cpc	Red, Yellow, Blue	Brown, Black, Grey

TABLE 7C – Standard 600/1000V armoured cable
BS 6346, BS 5467 or BS 6724

Cable type	Old core colours	New core colours
Single-core	Red or Black	Brown or Blue
Two-core	Red, Black	Brown, Blue
Three-core	Red, Yellow, Blue	Brown, Black, Grey
Four-core	Red, Yellow, Blue, Black	Brown, Black, Grey, Blue
Five-core	Red, Yellow, Blue, Black, Green-and-yellow	Brown, Black, Grey, Blue, Green-and-yellow

TABLE 7D – Flexible cable
to BS 6500

Cable type	Old core colours	New core colours
Two-core	Brown, Blue	No change
Three-core	Brown, Blue, Green-and-yellow	No change
Four-core	Black, Blue, Brown, Green-and-yellow	Brown, Black, Grey, Green-and-yellow
Five-core	Black, Blue, Brown, Black, Green-and-yellow	Brown, Black, Grey, Blue, Green-and-yellow

7 Addition or alteration to a d.c. installation

Where an addition or an alteration is made to a d.c. installation wired in the old core colours with cable to the new core colours, unambiguous identification is required at the interface. Cores should be marked as follows:

Neutral and midpoint conductors
Old and new conductors: M

Line conductors
Old and new conductors: Brown or Grey, or
old and new conductors: L, L+ or L-.

TABLE 7E – Example of conductor marking at the interface for additions and alterations to a d.c. installation identified with the old cable colours

Function	Old conductor		New conductor	
	Colour	Marking	Marking	Colour
Two-wire unearthed d.c. power circuit				
Positive of two-wire circuit	Red	L+	L+	Brown
Negative of two-wire circuit	Black	L-	L-	Grey
Two-wire earthed d.c. power circuit				
Positive (of negative earthed) circuit	Red	L+	L+	Brown
Negative (of negative earthed) circuit	Black	M	M	Blue
Positive (of positive earthed) circuit	Black	M	M	Blue
Negative (of positive earthed) circuit	Blue	L-	L-	Grey
Three-wire d.c. power circuit				
Outer positive of two-wire circuit derived from three-wire system	Red	L+	L+	Brown
Outer negative of two-wire circuit derived from three-wire system	Red	L-	L-	Grey
Positive of three-wire circuit	Red	L+	L+	Brown
Mid-wire of three-wire circuit	Black	M	M	Blue
Negative of three-wire circuit	Blue	L-	L-	Grey

APPENDIX 8 (Informative)
CURRENT-CARRYING CAPACITY AND VOLTAGE DROP
FOR BUSBAR TRUNKING AND POWERTRACK SYSTEMS

1   Basis of current-carrying capacity

The current-carrying capacity (I_n) of a busbar trunking or powertrack system relates to continuous loading and is declared by the manufacturer based on tests to BS EN 60439-2 (busbar trunking) or BS EN 61534-1 (Powertrack). The current-carrying capacity is designed to provide for satisfactory life of the system, subject to the thermal effects of carrying current for sustained periods in normal service.

Considerations affecting the choice of size of a busbar trunking or powertrack system include the requirements for protection against electric shock (see Chapter 41), protection against thermal effects (see Chapter 42), overcurrent protection (see Chapter 43 and sec 4 below) and voltage drop (see sec 5 below).

2   Rating factors for current-carrying capacity for busbar trunking systems

The current-carrying capacity of a busbar trunking system (I_n) can be affected by the ambient temperature and the mounting conditions (for example the orientation of the conductors).

Installation ambient temperature

If the ambient temperature exceeds 35 °C the rating factor k_α to be applied is obtained from the manufacturer of the busbar trunking system ($k_\alpha = 1$ for 35 °C).

The effective current-carrying capacity (I_z) at the new temperature is $k_\alpha \times I_n$.

Mounting attitude

The mounting factor k_β to be applied is obtained from the manufacturer of the busbar trunking system.

The effective current-carrying capacity (I_z) under the new mounting conditions is $k_\beta \times I_n$.

In a typical installation, both factors may have to be taken into account and the effective current-carrying capacity (I_z) then becomes $k_\alpha \times k_\beta \times I_n$.

NOTE 1: Factors k_α and k_β are the same as k_1 and k_2 respectively taken from BS EN 60439-2. The symbols k_α and k_β have been used to avoid confusion with the symbols k_1 and k_2 in Table 54.7.

NOTE 2: The effective current-carrying capacity for powertrack systems is the nominal rating declared by the manufacturer in accordance with BS EN 61534-1, under all normal conditions.

3   Effective current-carrying capacity

I_z must be not less than I_b, such that: $I_z \geq I_b$

where:

$\quad I_z \quad$ is the effective current-carrying capacity of the busbar trunking or powertrack system for continuous service under the particular installation conditions and

$\quad I_b \quad$ is the design current of the circuit.

4   Protection against overload current

The minimum operating current of the protective device should not exceed $1.45 I_z$. Where the protective device is a fuse to BS 88 series or a circuit-breaker to either BS EN 60947-2 or BS EN 60898, this requirement is satisfied by selecting a value of I_z not less than I_n, where I_n is the rated current or current setting of the device protecting the circuit against overcurrent.

5   Voltage drop

The voltage drop (V_d) for the busbar trunking or powertrack system is obtained from the manufacturer. It is usually expressed as mV/ampere/metre based on the line voltage of a 3-phase system, tabulated according to the value of the load-circuit power factor. The voltage drop given is calculated on the basis of a single load at the end of the run and, in this case, the total voltage drop = $V_d \times I_b \times L / 1000$ volts, where L is the length of run in metres.

In the case of an evenly distributed load (tapped off at intervals along the busbar trunking or powertrack system) then the voltage drop at the farthest tap-off point may be based on $0.5 V_d$, and is calculated by the above method.

In the case of an unevenly distributed load it will be necessary to calculate the voltage drop for each section between tap-off points and add them together to find the voltage drop at the furthest tap-off point.

APPENDIX 9 (Informative)
DEFINITIONS – MULTIPLE SOURCE, D.C. AND OTHER SYSTEMS

Fig 9 – Explanation of symbols used within Appendix 9

	Neutral conductor (N); midpoint conductor (M)
	Protective conductor (PE)
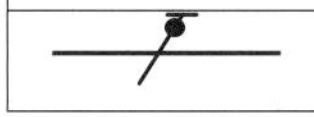	Combined protective and neutral conductor (PEN)

NOTE 1: The dotted lines indicate the parts of the system that are not covered by the scope of the Standard, whereas the solid lines indicate the part that is covered by the Standard.

NOTE 2: For private systems, the source and/or the distribution system may be considered as part of the installation within the meaning of this Standard. For this case, the figures may be completely shown in solid lines.

Fig 9A – TN-C-S multiple source system with separate protective conductor and neutral conductor to current-using equipment

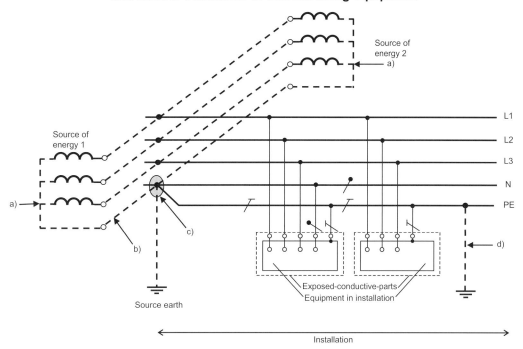

Fig 9B – TN multiple source system with protective conductor
and no neutral conductor throughout the system for 2- or 3-phase load

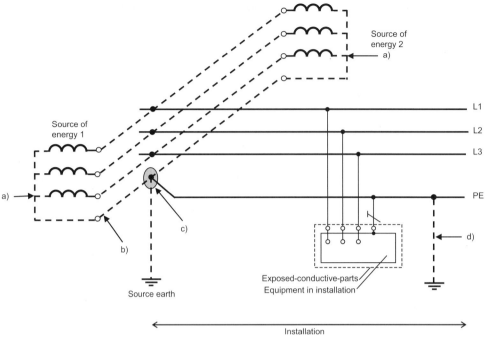

NOTES to Figures 9A and 9B

a) No direct connection from either the transformer neutral point or the generator star point to Earth is permitted.

b) The interconnection conductor between either the neutral points of the transformers or the generator star points is to be insulated. The function of this conductor is similar to a PEN; however, it must not be connected to current-using equipment.

c) Only one connection between the interconnected neutral points of the sources and the PE is to be provided. This connection is to be located inside the main switchgear assembly.

d) Additional earthing of the PE in the installation may be provided.

Fig 9C – IT system with exposed-conductive-parts earthed in groups or individually

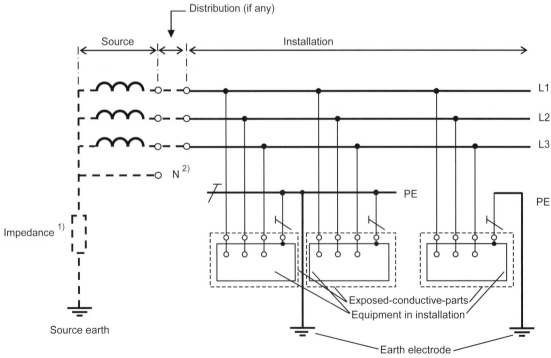

NOTES

Additional earthing of the PE in the installation may be provided.

1) The system may be connected to Earth via a sufficiently high impedance.

2) The neutral conductor may or may not be distributed.

Fig 9D – TN-S d.c. system
earthed line conductor L-
separated from the protective conductor throughout the installation

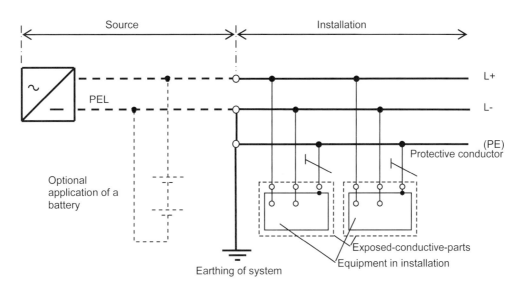

Fig 9E – TN-S d.c. system
earthed midpoint conductor M
separated from the protective conductor throughout the installation

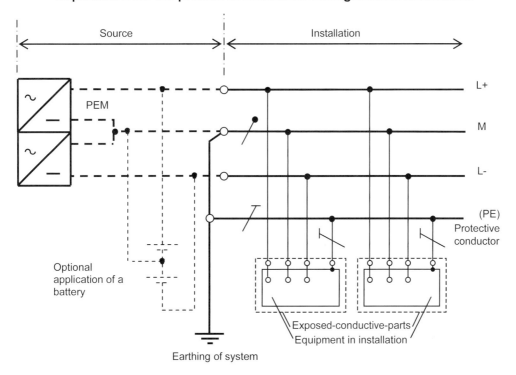

NOTE to Figures 9D and 9E

Additional earthing of the PE in the installation may be provided.

Fig 9F – TN-C d.c. system
earthed line conductor L– and protective conductor
combined in one single conductor PEL throughout the installation

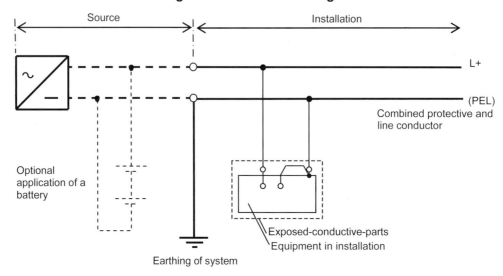

Source

Installation

L+

(PEL)
Combined protective and
line conductor

Optional
application of a
battery

Exposed-conductive-parts
Equipment in installation

Earthing of system

Fig 9G – TN-C d.c. system
earthed midpoint conductor M and protective conductor
combined in one single conductor PEM throughout the installation

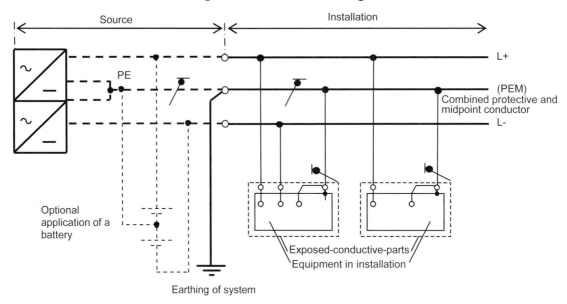

Source

Installation

L+

PE

(PEM)
Combined protective and
midpoint conductor

L-

Optional
application of a
battery

Exposed-conductive-parts
Equipment in installation

Earthing of system

NOTE to Figures 9F and 9G

Additional earthing of the PEL or PEM in the installation may be provided.

Fig 9H – TN-C-S d.c. system
earthed line conductor L– and protective conductor
combined in one single conductor PEL in a part of the installation

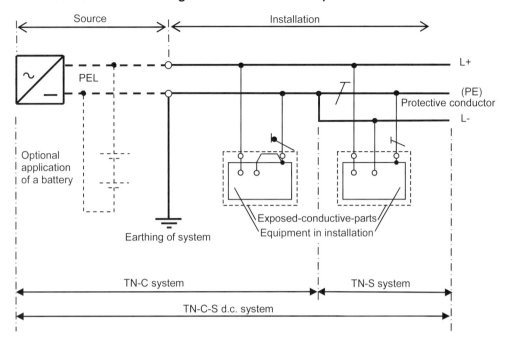

Fig 9I – TN-C-S d.c. system,
earthed midpoint conductor M and protective conductor
combined in one single conductor PEM in a part of the installation

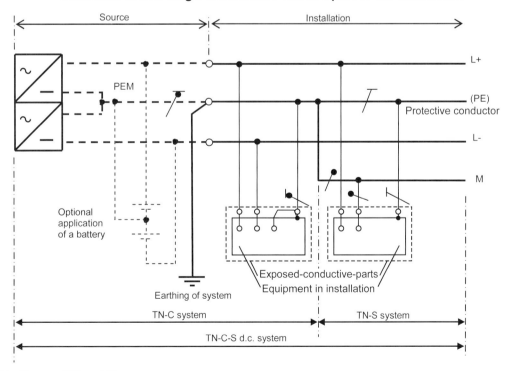

NOTES to Figures 9H and 9I

Additional earthing of the PE in the installation may be provided.

Regulation 8(4) of the Electricity Safety, Quality and Continuity Regulations 2002 states that a consumer shall not combine the neutral and protective functions in a single conductor in his consumer's installation.

Fig 9J – TT d.c. system

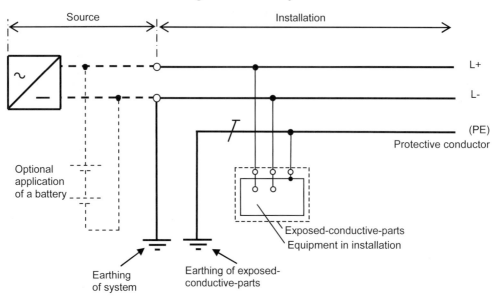

Fig 9K – TT d.c. system

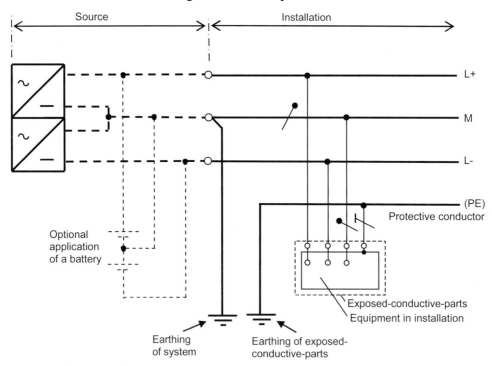

NOTE to Figures 9J and 9K

Additional earthing of the PE in the installation may be provided.

Fig 9L – IT d.c. system
earthed line conductor L- and protective conductor

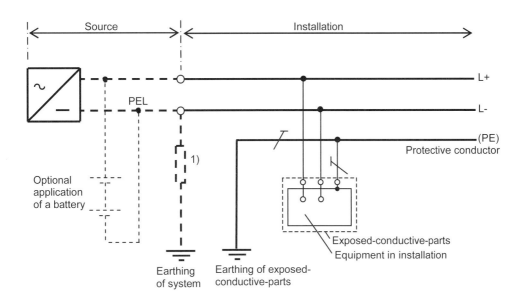

Fig 9M – IT d.c. system
earthed midpoint conductor M and protective conductor

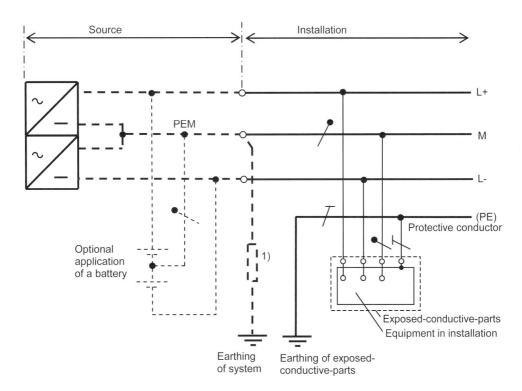

NOTES to Figures 9L and 9M

Additional earthing of the PE in the installation may be provided.

1) The system may be connected to Earth via a sufficiently high impedance.

APPENDIX 10 (Informative)
PROTECTION OF CONDUCTORS IN PARALLEL AGAINST OVERCURRENT

1 INTRODUCTION

Overcurrent protection provided for conductors connected in parallel should provide adequate protection for all of the parallel conductors. For two conductors of the same cross-sectional area, conductor material, length and disposition arranged to carry substantially equal currents the requirements for overcurrent protection are straightforward. For more complex conductor arrangements, detailed consideration should be given to unequal current sharing between conductors and multiple fault current paths. This appendix gives guidance on the necessary considerations.

NOTE: A more detailed method for calculating the current between parallel conductors is given in BS 7769 (BS IEC 60287).

2 OVERLOAD PROTECTION OF CONDUCTORS IN PARALLEL

When an overload occurs in a circuit containing parallel conductors of multicore cables, the current in each conductor will increase by the same proportion. Provided that the current is shared equally between the parallel conductors, a single protective device can be used to protect all the conductors. The current-carrying capacity (I_z) of the parallel conductors is the sum of the current-carrying capacity of each conductor, with the appropriate grouping and other factors applied.

The current sharing between parallel cables is a function of the impedance of the cables. For large single-core cables the reactive component of the impedance is greater than the resistive component and will have a significant effect on the current sharing. The reactive component is influenced by the relative physical position of each cable. If, for example, a circuit consists of two large cables per phase, having the same length, construction and cross-sectional area and arranged in parallel with unfavourable relative position (i.e. cables of the same phase bunched together) the current sharing may be more like 70/30 rather than 50/50.

Where the difference in impedance between parallel conductors causes unequal current sharing, for example greater than 10 % difference, the design current and requirements for overload protection for each conductor should be considered individually.

The design current for each conductor can be calculated from the total load and the impedance of each conductor.

For a total of m conductors in parallel, the design current I_{bk} for conductor k is given by:

$$I_{bk} = \frac{I_b}{\left(\dfrac{Z_k}{Z_1} + \dfrac{Z_k}{Z_2} + \cdots + \dfrac{Z_k}{Z_{k-1}} + \dfrac{Z_k}{Z_k} + \dfrac{Z_k}{Z_{k+1}} + \cdots + \dfrac{Z_k}{Z_m}\right)}$$

where:

I_b is the current for which the circuit is designed

I_{bk} is the design current for conductor k

Z_k is the impedance of conductor k

Z_1, Z_2 and Z_m are the impedances of conductors 1, 2 and m respectively.

For parallel conductors up to and including 120 mm² cross-sectional area (csa) the design current I_{bk} for conductor k is given by:

$$I_{bk} = I_b \frac{S_k}{S_1 + S_2 + \cdots + S_m}$$

where:

S_1, $\cdots$ S_m is the csa of the conductors and

S_k is the csa of conductor k.

In the case of single-core cables, the impedance is a function of the relative positions of the cables as well as the design of the cable, for example, armoured or unarmoured. Methods for calculating the impedance are given in BS 7769 (BS IEC 60287). It is recommended that current sharing between parallel cables is verified by measurement.

The design current I_{bk} replaces I_b in Regulation 433.1.1 as follows:

$$I_{bk} \leq I_n \leq I_{zk}$$

The value used for I_z in Regulation 433.1.1 is either:

(i) the continuous current-carrying capacity of each conductor, I_{zk}, if an overload protective device is provided for each conductor (see Figure 10A), hence:

$$I_{bk} \leq I_{nk} \leq I_{zk}$$

or

(ii) the sum of the current-carrying capacities of all the conductors, $\sum I_{zk}$, if a single overload protective device is provided for the conductors in parallel (see Figure 10B), hence:

$$I_b \leq I_n \leq \sum I_{zk}$$

where

I_{nk} is the rated current of the protective device for conductor k

I_{zk} is the continuous current-carrying capacity of conductor k

I_n is the rated current of the protective device

$\sum I_{zk}$ is the sum of the continuous current-carrying capacities of the m conductors in parallel.

NOTE: For busbar systems, information should be obtained either from the manufacturer or from BS EN 60439-2.
For powertrack systems, information should be obtained either from the manufacturer or from BS EN 61534.

**Fig 10A – Circuit in which an overload protective device is provided
for each of the m conductors in parallel**

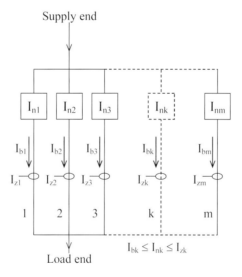

**Fig 10B – Circuit in which a single overload protective device is provided
for the m conductors in parallel**

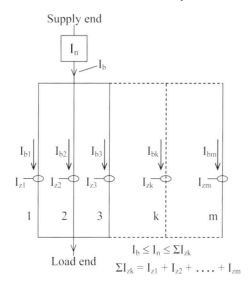

3   SHORT-CIRCUIT PROTECTION OF CONDUCTORS IN PARALLEL

Where conductors are connected in parallel, the possibility of a short-circuit within the parallel section should be considered.

If two conductors are connected in parallel and the operation of a single protective device may not be effective, then each conductor should have individual protection.

Where three or more conductors are connected in parallel then multiple fault current paths can occur and it may be necessary to provide short-circuit protection at both the supply and load ends of each parallel conductor. This situation is illustrated in Figures 10C and 10D.

Fig 10C – Current flow at the beginning of the fault   **Fig 10D – Current flow after operation of the protective device c_s**

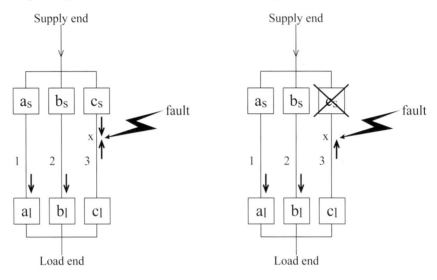

Figure 10C shows that, if a fault occurs in parallel conductor c at point x, the fault current will flow in conductors 1, 2 and 3. The magnitude of the fault current and the proportion of the fault current which flows through protective devices c_s and c_l will depend on the location of the fault. In this example it has been assumed that the highest proportion of the fault current will flow through protective device c_s. Figure 10D shows that, once c_s has operated, current will still flow to the fault at x via conductors 1 and 2. Because conductors 1 and 2 are in parallel, the current through protective devices a_s and b_s may not be sufficient for them to operate in the required time. If this is the case, the protective device c_l is necessary. It should be noted that the current flowing through c_l will be less than the current which caused c_s to operate. If the fault was close enough to c_l then c_l would operate first. The same situation would exist if a fault occurred in conductors 1 or 2, hence the protective devices al and bl will be required.

The method of providing protective devices <u>at both end</u>s has two disadvantages. Firstly, if a fault at x is cleared by the operation of c_s and c_l then the circuit will continue to operate with the load being carried by conductors 1 and 2. Hence the fault and subsequent overloading of 1 and 2 may not be detected. Secondly, the fault at x may burn open-circuit at the c_l side leaving one side of the fault live and undetected.

An alternative method to providing protective devices at both ends would be to provide linked protective devices at the supply end (Figure 10E). This would prevent the continued operation of the circuit under fault conditions.

Fig 10E – Linked protective devices installed at the supply end of the parallel conductors

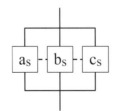

APPENDIX 11 *(Informative)*
EFFECT OF HARMONIC CURRENTS ON BALANCED THREE-PHASE SYSTEMS

Deleted by BS 7671:2008 Amendment No 1, content moved to Appendix 4 secs 5.5 and 5.6

APPENDIX 12 *(Informative)*
VOLTAGE DROP IN CONSUMERS' INSTALLATIONS

Deleted by BS 7671:2008 Amendment No 1, content moved to Appendix 4 sec 6.4.

APPENDIX 13 (Informative)
METHODS FOR MEASURING THE INSULATION RESISTANCE/IMPEDANCE OF FLOORS AND WALLS TO EARTH OR TO THE PROTECTIVE CONDUCTOR SYSTEM

1   GENERAL

Measurement of impedance or resistance of insulating floors and walls should be carried out with the system voltage to Earth and nominal frequency, or with a lower voltage of the same nominal frequency combined with a measurement of insulation resistance. This may be done, for example, in accordance with the following methods of measurement:

1)   a.c. system

 – by measurement with the nominal a.c. voltage, or

 – by measurement with lower a.c. voltages (minimum 25 V) and, additionally, by an insulation resistance test using a minimum test voltage of 500 V d.c. for nominal system voltages not exceeding 500 V and a minimum test voltage of 1000 V d.c. for nominal system voltages above 500 V.

 The following optional voltage sources may be used:
 a) The earthed system voltage (voltage to Earth) that exists at the measuring point
 b) The secondary voltage of a double-wound transformer
 c) An independent voltage source at the nominal frequency of the system.

 In options b) and c), the measuring voltage source is to be earthed for the measurement.

 For safety reasons, when measuring voltages above 50 V, the maximum output current should be limited to 3.5 mA.

2)   d.c. system

 – insulation resistance test by using a minimum test voltage of 500 V d.c. for nominal system voltages not exceeding 500 V

 – insulation resistance test by using a minimum test voltage of 1000 V d.c. for nominal system voltages above 500 V.

 The insulation resistance test should be made using measuring equipment in accordance with BS EN 61557-2.

2   TEST METHOD FOR MEASURING THE IMPEDANCE OF FLOORS AND WALLS WITH A.C. VOLTAGE

Current, I, is fed through an ammeter to the test electrode from the output of the voltage source or from the line conductor L. The voltage (U_x) at the electrode to Earth or to the protective conductor is measured by means of a voltmeter with an internal resistance of at least 1 MΩ.

The impedance of the floor insulation will then be: $Z_x = \dfrac{U_x}{I}$

The measurement for ascertaining the impedance is to be carried out at as many points as deemed necessary, selected at random, with a minimum of three. The test electrodes may be either of the following types. In case of dispute, the use of test electrode 1 is the reference method.

3   TEST ELECTRODE 1

The electrode comprises a metallic tripod of which the parts resting on the floor form the points of an equilateral triangle. Each supporting point is provided with a flexible base ensuring, when loaded, close contact with the surface being tested over an area of approximately 900 mm^2 and presenting a resistance of less than 5000 Ω.

Before measurements are made, the surface being tested is cleaned with a cleaning fluid. While measurements are being made, a force of approximately 750 N for floors or 250 N for walls is applied to the tripod.

Fig 13A – Test electrode 1

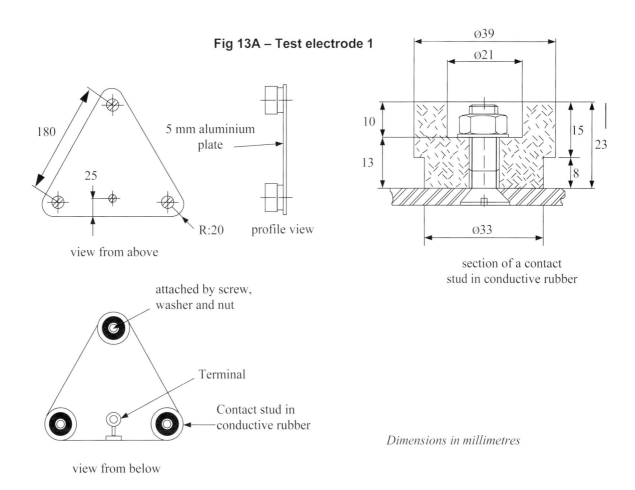

section of a contact
stud in conductive rubber

Dimensions in millimetres

view from above

view from below

4   TEST ELECTRODE 2

The electrode comprises a square metallic plate with sides that measure 250 mm, and a square of damped, water-absorbent paper, or cloth, from which surplus water has been removed, with sides that measure approximately 270 mm. The paper is placed between the metal plate and the surface being tested.

During measurement a force of approximately 750 N for floors or 250 N for walls is applied on the plate.

Fig 13B – Test electrode 2

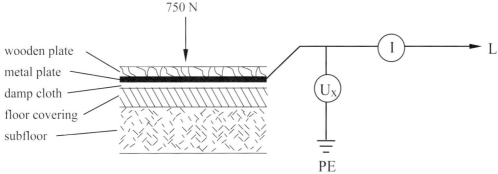

APPENDIX 14 (Informative)
MEASUREMENT OF EARTH FAULT LOOP IMPEDANCE: CONSIDERATION OF THE INCREASE OF THE RESISTANCE OF CONDUCTORS WITH INCREASE OF TEMPERATURE

When impedance measurements are made at ambient temperature the procedure hereinafter described may be followed to take into account the increase of resistance of the conductors with the increase of temperature due to load current, to verify, for TN and TT systems in which protection by automatic disconnection is provided by overcurrent devices, compliance of the measured values of earth fault loop impedance with the appropriate requirements of Regulation 411.4 or 411.5.

The requirements of Regulation 411.4.5 or 411.5.4, as appropriate, are considered to be met when the measured value of earth fault loop impedance satisfies the following equation:

$$Z_s(m) \leq 0.8 \times \frac{U_0}{I_a}$$

where:

$Z_s(m)$ is the measured impedance of the earth fault current loop up to the most distant point of the relevant circuit from the origin of the installation (Ω)

U_0 is the nominal a.c. rms line voltage to Earth (V)

I_a is the current causing the automatic operation of the protective device within the time stated in Table 41.1 or within 5 s according to the conditions stated in Regulation 411.3.2.3 (A).

Where the measured value of the earth fault loop impedance exceeds $0.8U_0/I_a$, a more precise assessment of compliance with Regulation 411.4.5 or 411.5.4, as appropriate, may be made, evaluating the value of the earth fault loop impedance according to the following procedure:

(i)   The line conductor to protective conductor loop impedance of the supply is first measured at the origin of the installation

(ii)   The resistances of the line conductor and protective conductor of the distribution circuit(s) are then measured

(iii)   The resistances of the line conductor and protective conductor of the final circuit are then measured

(iv)   The values of resistance measured in accordance with (ii) and (iii) are increased on the basis of the increase of conductor temperature, taking into consideration the design current, (I_b)

(v)   The values of resistance increased in accordance with (iv) are finally added to the value measured at (i) to obtain a realistic value of Z_s under earth fault conditions.

NOTE: Other methods are not precluded.

APPENDIX 15 (Informative)
RING AND RADIAL FINAL CIRCUIT ARRANGEMENTS, REGULATION 433.1

This appendix sets out options for the design of ring and radial final circuits for household and similar premises in accordance with Regulation 433.1, using socket-outlets and fused connection units. It does not cover other aspects of the design of a circuit such as:

– Protection against electric shock, Chapter 41

– Protection against thermal effects, Chapter 42

– Protection against overcurrent, Chapter 43

– Selection and erection of equipment, Part 5.

Fig 15A – Ring final circuit arrangements, Regulation 433.1.103

The load current in any part of the circuit should be unlikely to exceed for long periods the current-carrying capacity of the cable (Regulation 433.1.103 refers). This can generally be achieved by:

(i) locating socket-outlets to provide reasonable sharing of the load around the ring

(ii) not supplying immersion heaters, comprehensive electric space heating or loads of a similar profile from the ring circuit

(iii) connecting cookers, ovens and hobs with a rated power exceeding 2 kW on their own dedicated radial circuit

(iv) taking account of the total floor area being served. (Historically, a limit of 100 m^2 has been adopted.)

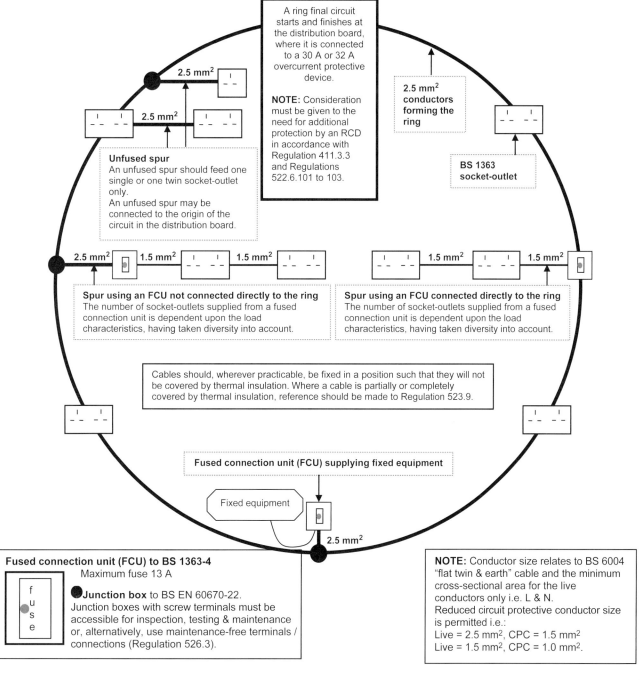

A ring final circuit starts and finishes at the distribution board, where it is connected to a 30 A or 32 A overcurrent protective device.

NOTE: Consideration must be given to the need for additional protection by an RCD in accordance with Regulation 411.3.3 and Regulations 522.6.101 to 103.

2.5 mm^2

2.5 mm^2

Unfused spur
An unfused spur should feed one single or one twin socket-outlet only.
An unfused spur may be connected to the origin of the circuit in the distribution board.

2.5 mm^2 conductors forming the ring

BS 1363 socket-outlet

2.5 mm^2 1.5 mm^2 1.5 mm^2

1.5 mm^2 1.5 mm^2

Spur using an FCU not connected directly to the ring
The number of socket-outlets supplied from a fused connection unit is dependent upon the load characteristics, having taken diversity into account.

Spur using an FCU connected directly to the ring
The number of socket-outlets supplied from a fused connection unit is dependent upon the load characteristics, having taken diversity into account.

Cables should, wherever practicable, be fixed in a position such that they will not be covered by thermal insulation. Where a cable is partially or completely covered by thermal insulation, reference should be made to Regulation 523.9.

Fused connection unit (FCU) supplying fixed equipment

Fixed equipment

2.5 mm^2

Fused connection unit (FCU) to BS 1363-4
Maximum fuse 13 A

fuse

● **Junction box** to BS EN 60670-22.
Junction boxes with screw terminals must be accessible for inspection, testing & maintenance or, alternatively, use maintenance-free terminals / connections (Regulation 526.3).

NOTE: Conductor size relates to BS 6004 "flat twin & earth" cable and the minimum cross-sectional area for the live conductors only i.e. L & N.
Reduced circuit protective conductor size is permitted i.e.:
Live = 2.5 mm^2, CPC = 1.5 mm^2
Live = 1.5 mm^2, CPC = 1.0 mm^2.

Fig 15B – Radial final circuit arrangements, Regulation 433.1

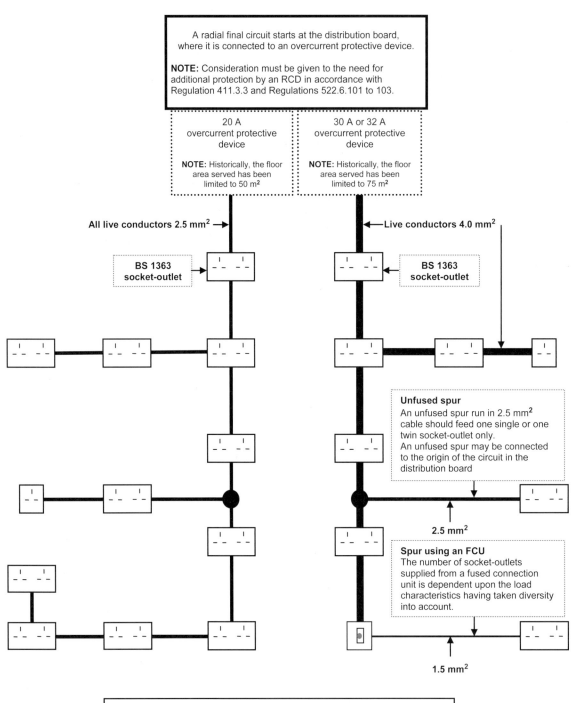

A radial final circuit starts at the distribution board, where it is connected to an overcurrent protective device.

NOTE: Consideration must be given to the need for additional protection by an RCD in accordance with Regulation 411.3.3 and Regulations 522.6.101 to 103.

20 A overcurrent protective device

NOTE: Historically, the floor area served has been limited to 50 m²

30 A or 32 A overcurrent protective device

NOTE: Historically, the floor area served has been limited to 75 m²

All live conductors 2.5 mm² →

← Live conductors 4.0 mm²

BS 1363 socket-outlet →

← BS 1363 socket-outlet

Unfused spur
An unfused spur run in 2.5 mm² cable should feed one single or one twin socket-outlet only.
An unfused spur may be connected to the origin of the circuit in the distribution board

2.5 mm²

Spur using an FCU
The number of socket-outlets supplied from a fused connection unit is dependent upon the load characteristics having taken diversity into account.

1.5 mm²

Cables should, wherever practicable, be fixed in a position such that they will not be covered by thermal insulation. Where a cable is partially or completely covered by thermal insulation, reference should be made to Regulation 523.9.

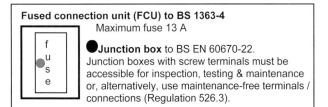

Fused connection unit (FCU) to BS 1363-4
Maximum fuse 13 A

⬤**Junction box** to BS EN 60670-22.
Junction boxes with screw terminals must be accessible for inspection, testing & maintenance or, alternatively, use maintenance-free terminals / connections (Regulation 526.3).

NOTE: Conductor size relates to BS 6004 "flat twin & earth" cable and the minimum cross-sectional area for the live conductors only i.e. L & N.
Reduced circuit protective conductor size is permitted i.e.:
Live = 4.0 mm², CPC = 1.5 mm²
Live = 2.5 mm², CPC = 1.5 mm²
Live = 1.5 mm², CPC = 1.0 mm².

APPENDIX 16 (Informative)
Devices for protection against overvoltage

Typical installation of a surge protective device (SPD) in a power distribution board for a TN-S system.

Fig 16A1 – SPD connected to the first overcurrent protective device (OCPD) to the incoming supply

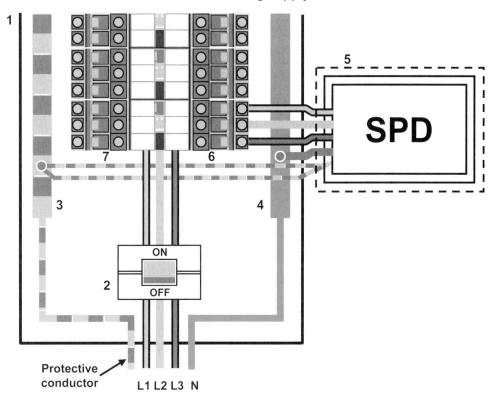

Key :

1 distribution board
2 main switch
3 earthing bar
4 neutral bar
5 enclosure for SPD
6 first OCPD
7 alternative first OCPD

NOTE: The OCPD provides a convenient means to protect the SPD and a means of isolation. As there is insufficient room within the distribution board the SPD is mounted in a separate enclosure for electrical safety. This enclosure is mounted directly alongside the distribution board to ensure the connecting leads are kept short. An additional local bonding connection is made to further minimize voltage drop on the connecting leads.

Installation of surge protective devices in TT systems, Connection Type 1 (CT 1)

Fig 16A2 – SPDs on the load side of an RCD [according to Regulation 534.2.5(i)]

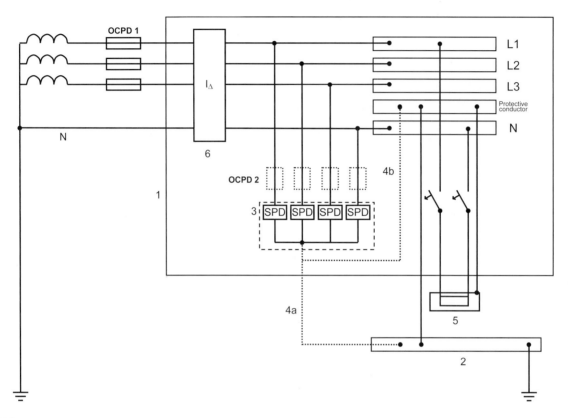

Key

1	Distribution board
2	Main earthing terminal or bar
3	Surge protective devices ensuring a protection level in accordance with overvoltage Category II
4	Earthing connection of surge protective devices, either 4a or 4b, whichever is the shorter route
5	Current-using equipment
6	Residual current protective device (RCD)

OCPD 1	Overcurrent protective device at the origin of the installation
OCPD 2	Overcurrent protective device

Installation of surge protective devices in TT systems, Connection Type 2 (CT 2)

Fig 16A3 – SPDs on the supply side of an RCD [according to Regulation 534.2.5(ii)]

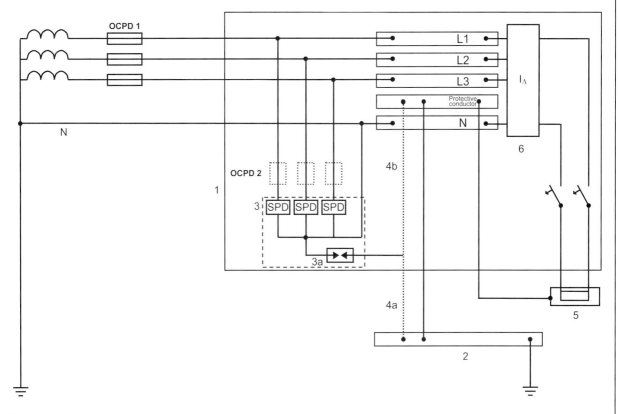

Key

1	Distribution board	OCPD 1	Overcurrent protective device at the origin of the installation
2	Main earthing terminal or bar		
3	Surge protective devices	OCPD 2	Overcurrent protective device
3a	Surge protective device (SPDs 3 and 3a in series ensuring a protection level in accordance with overvoltage Category II)		
4	Earthing connection of surge protective devices, either 4a or 4b, whichever is the shorter route		
5	Current-using equipment		
6	Residual current protective device (RCD) installed downstream of the surge protective devices		

Installation of surge protective devices in IT systems

Fig 16A4 – SPDs on the load side of an RCD

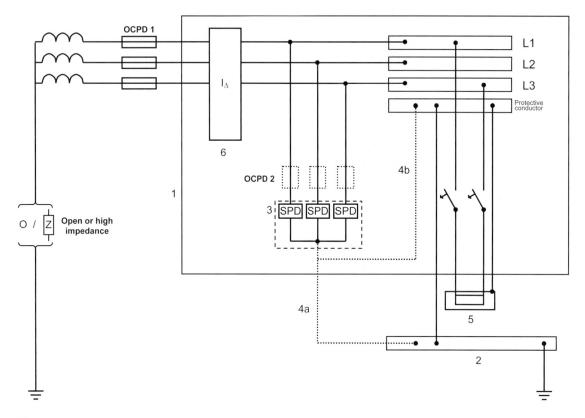

Key

1	Distribution board
2	Main earthing terminal or bar
3	Surge protective devices ensuring a protection level in accordance with overvoltage Category II
4	Earthing connection of surge protective devices, either 4a or 4b, whichever is the shorter route
5	Current-using equipment
6	Residual current protective device (RCD) installed upstream of the surge protective devices

OCPD 1	Overcurrent protective device at the origin of the installation
OCPD 2	Overcurrent protective device

Installation of Types 1, 2 and 3 SPDs, for example in TN-C-S systems

Fig 16A5 – Installation example of Types 1, 2 and 3 coordinated SPDs

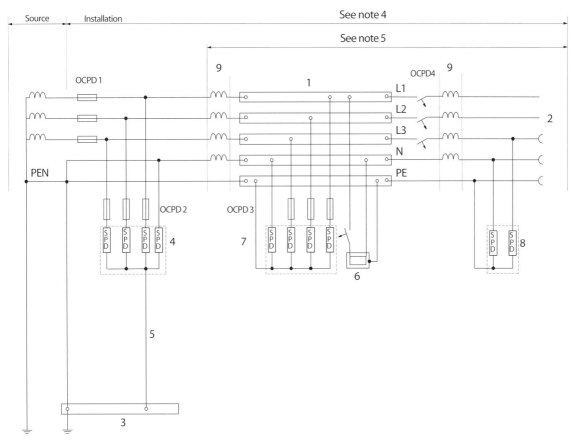

Key

1	Distribution board	7	Surge protective device, Type 2
2	Distribution outlet	8	Surge protective device, Type 2 or Type 3
3	Main earthing terminal or bar	9	Decoupling element or line length
4	Surge protective device, Type 1	**NOTE**:	If the cable length between the SPD types is short (refer to manufacturers' data), a decoupling element is employed to provide inductance for correct SPD co-ordination
5	Earthing connection (earthing conductor) of surge protective device		
6	Fixed equipment to be protected	OCPD 1, 2, 3,4	Overcurrent protective devices

NOTE 1: For further information reference should be made to CLC/TS 61643-12.

NOTE 2: SPDs 4 and 7 (or 7 and 8) can be combined in a single SPD.

NOTE 3: SPDs may require additional modes of protection for sensitive equipment.

NOTE 4: BS EN 62305-4 covers the protection of electrical and electronic systems within structures against lightning.

NOTE 5: Section 443 of BS 7671 deals with the protection of electrical installations against transient overvoltages of atmospheric origin, transmitted by the supply distribution system and against switching overvoltages generated by the equipment within the installation.

Typically, Type 1 SPDs are used at the origin of the installation, Type 2 SPDs are used at distribution boards and Type 3 SPDs are used near terminal equipment. Combined Type SPDs are classified with more than one Type, e.g. Type 1+2, Type 2+3. Type 1 SPDs are only used where there is a risk of direct lightning current.

The most important aspect in selecting an SPD is its limiting voltage performance (protection level U_p) during the expected surge event, and not the energy withstand (e.g. I_{imp}) which it can handle. An SPD with a low protection level will ensure adequate protection of the equipment, while an SPD with a high energy withstand may only result in a longer operating life.

TABLE 16A – Information on SPD classification

SPD according to BS EN 62305	SPD according to BS EN 61643-11
SPD tested with I_{imp}	Type 1
SPD tested with I_n	Type 2
SPD tested with a combination wave	Type 3

SPD tested with I_{imp} (BS EN 62305-4)

SPDs which withstand the partial lightning current (with a typical waveform 10/350 µs) require a corresponding impulse test current I_{imp}.

NOTE 1: For power lines, a suitable test current I_{imp} is defined in the Class I test procedure of BS EN 61643-11.

SPD tested with I_n (BS EN 62305-4)

SPDs which withstand induced surge currents with a typical waveform 8/20 µs require a corresponding impulse test current I_{nspd}.

NOTE 2: For power lines a suitable test current I_{nspd} is defined in the Class II test procedure of BS EN 61643-11.

SPD tested with a combination wave (BS EN 62305-4)

SPDs that withstand induced surge currents with a typical waveform 8/20 µs and require a corresponding impulse test current I_{sc}.

NOTE 3: For power lines a suitable combination wave test is defined in the Class III test procedure of BS EN 61643-11 defining the open-circuit voltage U_{oc} 1.2/50 µs and the short-circuit current I_{sc} 8/20 µs of a 2 Ω combination wave generator.

Table of figures

Table of tables

INDEX

439

E

Earth-
- connections to-   Sec 542
  - necessity for additional   331.1(xi)
- currents, high   543.7
- definition   Part 2

Earth electrode-
- caravan parks   708.553.1.14
- definition   Part 2
- mobile or transportable units   717.411.6
- network, definition   Part 2
- resistance-
  - allowance for corrosion   542.2.1
  - allowance for soil drying
    and freezing   542.2.4
  - area, definition   Part 2
  - definition   Part 2
  - TT and IT systems in general   411.5, 411.6
  - test of   612.7

Earth electrodes, selection of   542.2

Earth fault current-
- adequacy of earthing
  arrangements for   542.1.3.1 to 3
- definition   Part 2

Earth fault loop impedance-
- definition   Part 2
- external to installation,
  to be determined   313.1(iv)
- for automatic disconnection
  for fault protection, general   411.4, 411.5
- measurement   612.9, Appx 14
- reduced low voltage systems   411.8.3
- testing   612.9

Earth-free local equipotential
  bonding, protection by   418.2

Earth-free location *see Non-*
  conducting location

Earth leakage current *see*
  Protective conductor current;
  Leakage current

Earth loop impedance
  see Earth fault loop impedance   Part 2

Earth monitoring   543.3.4
    543.7.1.102 & 103

Earthed concentric wiring, definition
  (see also PEN conductor)   Part 2

Earthed equipotential bonding
  and automatic disconnection *see*
  Automatic disconnection of supply

Earthing-
- arrangements-
  - assessment of type of   Sec 312
  - selection and erection   Chap 54
  - suitability of   542.1.3.1
- combined protective and
  functional purposes   542.1.1, 543.5.1
- conductor, definition   Part 2
- conductors, electrode boilers   554.1
- conductors, selection and erection   542.3
- connections, warning notice at   514.13.1
- definition   Part 2
- exposed-conductive-parts-
  - FELV systems   411.7.3
  - general requirements   411.3.1.1
  - IT systems   411.6
  - reduced low voltage system   411.8.3
  - TN systems   411.4.2
  - TT systems   411.5
- high protective conductor currents   543.7
- impedance-
  - IT system   Figs 9C, 9L, 9M
      411.6.1
  - value of   531.3.1
      542.1.3.1

Earthing - *(cont'd)*
- neutral point or midpoint of
  reduced low voltage source   411.8.4.2
- neutral of supply
  (see also PEN conductors)   554.1.5 to 7
- prohibited, in earth-free
  local bonded location   418.2.3
- prohibited, in non-
  conducting location   418.1.3
- resistance-
  - provision for measurement of   542.4.2
  - variations in   542.2.1, 542.2.4
- system, types of   312.2
- terminal, main-
  - connection to Earth   542.1.100
  - definition   Part 2
  - provision of   542.4

Economic design of installation   311.1

Effects of the Regulations   Chap 12

Electric braking, with
  emergency stopping   537.4.2.2 note 2

Electric fence installations, the
  Regulations not applicable to   110.2(xii), 705.1 note

Electric shock-
- current
  see Shock current, definition   Part 2
- definition   Part 2
- emergency switching where risk of   537.4.1.2
- in case of fault *see Fault protection*
- in normal service
  see Basic protection
- protection against   Chap 41
- safety services   560.6.8.1, 560.7.6

Electric surface heating systems   Sec 753

Electric traction equipment
  see Railway traction equipment

Electrical connections
  see Connections, electrical

Electrical equipment, definition   Part 2

Electrical equipment *see Equipment*

Electrical Equipment (Safety)
  Regulations 1994   Appx 2 sec 7

Electrical installation, definition   Part 2

Electrical Installation Certificate-
- departures from Regulations
  to be noted   120.3, 133.5
- form of   Appx 6
- provision of   632.3

Electrical Installation Condition Report-
- authenticity of   631.4 & 5
- competent person to sign   631.4
- form of   Appx 6
- inspection schedule   Appx 6
- recording of defects, etc   634.2
- requirement for   631.2, 634.1
- schedules to accompany   634.1

Electrical interference   528.2 note

Electrical separation,
  protection by-   Sec 413, 418.3
- application of   413.1
- basic protection   413.2
- bonding of equipment   418.3.4
- disconnection where two faults occur   418.3.7
- exposed metalwork of circuit   413.3.6, 418.3.4
- flexible cables for   413.3.4, 418.3.6
- floor and ceiling heating systems,
  prohibited for   753.410.3.6
- for one item of equipment   413.1.2
- for several items of equipment   413.1.3, 418.3
- protective conductors for   418.3.6
- separation of circuit   413.3.2, 413.3.3
- socket-outlets for   418.3.5
- supplies for   413.3.2
- testing of   612.4.3
- voltage limitation   413.3.2
- wiring systems for   413.3.5

447

M

O

457

NOTES